Climate Change and Water Resources in Africa

Illustration: Photo taken by Khaled AbuZeid
Date: 16 June 2011—Location: Marchison Falls, Nile River, Uganda

Salif Diop · Peter Scheren · Awa Niang
Editors

Climate Change and Water Resources in Africa

Perspectives and Solutions Towards an Imminent Water Crisis

futurearth International Science Council Regional Office for Africa The African Academy of Sciences Springer

Editors
Salif Diop
Cheikh Anta Diop University
National Academy of Sciences
and Techniques of Senegal,
Dakar, Senegal

Awa Niang
Cheikh Anta Diop University
Dakar, Senegal

Peter Scheren
Lavardens
France

ISBN 978-3-030-61224-5 ISBN 978-3-030-61225-2 (eBook)
https://doi.org/10.1007/978-3-030-61225-2

This Springer imprint is published by the registered company Springer Nature Switzerland AG
The registered company address is: Gewerbestrasse 11, 6330 Cham, Switzerland

This book is dedicated to Nelson Rolihlahla Mandela "Madiba" (1918–2013), the first post-apartheid President of South Africa. He said, "put your thoughts in order by reflection, pen in hand. Appropriate yourself the power of the pen. Read, read, read every day, pen in hand."

Foreword by Felix D. Dakora

The warming rate of Africa over the last 27 years is the most alarming signal of the need for urgent action to mitigate its effects and implement strategies for climate resilience. Based on the fourth and fifth reports of the International Panel on Climate Change (IPCC), it is predicted that average temperatures in Africa will have increased between 2 and 4°C before the end of this century. At the same time, a decrease in rainfall of up to 20% as well as extreme events such as floods and droughts may eliminate considerable portions of low-lying lands, particularly affecting small island states.

There are many implications of climatic variability and change in Africa; one of the most key ones is the impact on water resources and hydrological systems, which directly affect water availability and, in turn, water security as a whole. These effects will negatively impact water provisioning, as well as rain-fed agricultural productivity in many parts of Africa, with variation in impacts occurring in specific places and localities. Various other changes and modifications are likely to occur in response to the significant variability in the global climate system, including extreme aridity in some cases and flooding in others. It is clear, therefore, that integrating climate change risks and opportunities into development decision-making is both a challenge and a necessity facing the African continent, particularly for the most vulnerable countries.

This book, *Climate Change and Water Resources in Africa*, sets out the many challenges and implications of climate change for freshwater resources in Africa, including its rivers, lakes, and aquifers. Under the influence of a range of human factors, the status of water resources in Africa has been changing for decades, playing out in disruption of water flow and variability, falling groundwater levels, changes in rainfall levels, and timing, and an overall decrease in water quality. Indeed, change is not new in this context. Climate change, however, will strongly accelerate the rate of change, affecting the ability of people and societies to respond in a timely manner to address their own needs.

This ability to respond to change is compounded by uncertainty relating to the impacts of climate change. While there are a number of models that attempt to predict these impacts, many are on a generalized scale and do not project localized impacts. Also, the models themselves operate with a high level of uncertainty and often predict contradictory outcomes, depending upon the model. Thus, African governments must manage these changes in the context of significant uncertainty. This requires adaptive management, encompassing continuous improvement underpinned by rigorous science, in order to understand the drivers of change over time and to be able to address them effectively.

With this in mind, the goal of this book is to offer a deeper analysis of the effects of climate on water resources in some of the most vulnerable areas in Africa, including approaches that may help reduce or mitigate the impacts of climate change. In this regard, while there is no quick fix to the pressures imposed on water resources by climate change, it is clear that increasing the resilience of ecosystems and communities to extreme events such as flooding and drought and integrating climate change risks and opportunities into development decision-making, will be key. It is also important that the wealthier countries assume responsibility for their historic greenhouse gas (GHG) emissions and support those countries that are most impacted by those emissions by contributing to their adaptations to climate change while reducing their own carbon footprints. Wealthy countries and those countries that are now major contributors to GHG emissions must make major and long-term contributions to development, technology transfer, training, and capacity building, notably in Africa. As a whole, this book intends to contribute to the debate around climate change in relation to water resource management on the African continent, and in particular, inform policy decisions and actions that will improve the ability of governments and communities to manage the challenges of climate change and variability in relation to the aquatic ecosystems upon which they depend.

Close collaboration among the editors, Prof. Salif Diop, Dr. Peter Scheren, and Dr. Awa Niang, has been the driving force behind, "Climate Change and Water Resource Management in Africa." The editors would also like to thank the authors who contributed their time and expertise to this important resource. It is our hope and belief that this book will provide important, actionable information to the public at large, and specifically to the many dedicated scientists, researchers, managers, decision-makers, and policymakers that are working together to safeguard, protect, and sustainably manage water resources in a changing climate in Africa.

Felix D. Dakora, Ph.D.
President of the African Academy of Sciences,
Professor, South African Chair in Agrochemurgy
and Plant Symbioses
Chemistry Department
Tshwane University of Technology
Pretoria, South Africa

Foreword by Daniel Nyanganyura

Recognizing the critical role of water in supporting human livelihood, economic growth and the very survival of the population, African governments developed the Africa Water Vision, describing an Africa that exhibits equitable and sustainable use and management of water resources for poverty alleviation, socioeconomic development, regional cooperation, and the environment. However, it is clear that this vision will become reality only when a range of challenges is properly addressed[1]. Moreover, the increasingly visible effects of climate change on water resources in Africa are compounding the challenge.

Consistent with global warming patterns, Africa is already experiencing significant spatial and temporal rainfall variability, with resulting challenges in terms of sustainable management of freshwater resources (Africa Water Vision for 2025, UNECA).

Large rural and semi-arid regions of the Sahel, central and eastern Southern Africa, and the Horn of Africa are at particular risk because they depend largely on rainfed agriculture. Increased climate variability, including more intense droughts and floods, increases temperatures, and reduces rainfall in these regions, putting these communities at significant risk. In particular, there is a need to protect the critical 'water towers' of central, west, and eastern Africa, (e.g., Congo forests; west Africa Fouta Djallon mountain areas; Eastern Africa and Ethiopian highlands), as deforestation and poor land-use management in these places have significantly decreased their effectiveness as critical catchment areas. Their protection and restoration are therefore critically important in order to secure continuous provisioning functions.

Recognizing the many weaknesses across the continent that complicate effective responses to climate change (e.g., limited institutional capacity, the scale of poverty, data scarcity, poor reliability of climate change models at local scale), immediate action is needed to improve community and societal resilience to climate change. Key approaches for good water management include alignment of water management plans with national development and poverty reduction strategies as well as building the required capacity to monitor and manage climate variability and its impacts.

---

[1]Reference to African Union-United Nations Framework on Implementation of Agenda 2063 and Agenda 2030 and Africa Water Vision for 2025, UNECA.

Continued water infrastructure and technology investments and the introduction of relevant incentives by African governments are also needed, as are investments by the private sector and international financing agencies, all of which have a role to play. Climate change is likely to produce increased numbers and intensities of floods in many areas, with a variety of management actions required to improve flood management, including early warning systems and rehabilitation of degraded catchments and flood-proofing of water supply and sanitation infrastructure, to reduce the risk of subjecting vulnerable communities to a lack of drinking water and functioning sanitation facilities.

In other parts of the continent, investing in increased water storage infrastructure and other water provisioning infrastructure may be a higher priority to avoid and/or overcome water shortages and periods of drought. Increased investment and access to information about appropriate irrigation technology (e.g., drip irrigation, rainwater harvesting) may be required to improve effective water use and productivity at the farm level in the face of climate change. Shifting from rainfed to irrigated agriculture will protect rural livelihoods and food security in many areas. Artificial groundwater recharge in areas experiencing over-exploitation and degradation is also needed. Overall, a systematic scaling up of locally appropriate solutions is critical to ensure area- and region-wide poverty alleviation, climate change adaptation, and economic development.

Flexible planning to promote long-term adaptation to climate change is also necessary. This requires access to accurate climate change information for key water-related development planning departments (agriculture, mining, power generation, municipalities, etc.) and strong alignment and cooperation between water departments and those responsible for development planning. For basins and aquifers, the primary allocations are among riparian states. At the sub-basin or local level, multiple water allocation systems operate in Africa, with many parallel systems, both formal and customary. All must be sufficiently flexible and cooperative to enable water allocation adjustments that address climate variability in the face of national development objectives, with simultaneous consideration of capacity constraints.

Water stakeholders also must be involved in the water resource management process to ensure full support for any approach to be implemented; adequate and timely information exchange between stakeholders and responsible authorities is particularly crucial. Climate change adaptation ranges from creating major storage infrastructure down to the household level, particularly in areas that governments fail to reach consistently or effectively. The simple and timely provision of information can assist communities and households to prepare for anticipated changes, including the strategic selection of new crops to grow, improvement of cropping and livestock management techniques, local water resource use, and protection and flood warnings. Improving and sharing knowledge and information about climate change vulnerability and adaptation involves complex interactions among local land and water resource characteristics, economic conditions, and often diverse livelihood capitals and strategies of individual households.

These approaches depend on improved science and information-sharing, particularly across vulnerable transboundary basins and aquifers. A major challenge that impacts Africa in particular is the weakness of models that predict climate change at the local level. Better modeling capacity is critical to ensure that management options and investment decisions are based on scientifically sound information. Accurately describing the current state, identifying emerging trends, and anticipating possible outcomes and associated vulnerabilities and risks also require appropriate monitoring systems to provide necessary data at an appropriate scale.

Against the backdrop of existing development challenges, substantial increases in financing to African countries are needed to improve land and water management systems and their capacity to adapt to climate change and to enhance resiliency. A range of new, innovative financing options is therefore required, including from government and private sources in developed countries.

A recurring theme in considering climate change is that no one is immune to its impacts, and that tackling it will require global collaboration, with emphasis on contributions from the industrial countries that have been responsible for most GHG emissions since 1850, and also from countries that are the current major contributors to GHG emissions. Wealthier developed countries, and major current GHG-emitting countries, must take the lead by

- Reducing their own GHG emissions, and meeting and/or exceeding the agreed emission targets;
- Meeting the financial commitments declared at global meetings and summits (i.e., ensuring that pronouncements made for political gain are matched by action and investment);
- Developing carbon pricing that provides incentives for transfer to sustainable sources of energy and phasing out perverse subsidies of all kinds;
- Implementing relevant measures to improve access by the poor to water, food, and energy; and
- Reviewing existing, and developing new, trade policies that support both development and technology transfer, in addition to training and capacity building, in Africa and developing countries elsewhere.

Overall, cutting emissions at their source aggressively, while addressing the impacts of global warming in the most vulnerable zones, is required. The obligation to act is therefore unequivocal: no matter what policy is in place; good intentions alone are not sufficient to address our grave imperative to take responsibility for emissions. There is an urgent need to act now to avoid severe climate change risks and impacts on water resources. At risk is nothing less than human livelihoods, as well as an existential threat to animal species that would likely be lost on an African continent that will soon be 2 °C warmer than pre-industrial levels.

Let's Act Now!

Dr. Daniel Nyanganyura
Regional Director, International Science,
Council/Former ICSU-Regional Office for Africa,
Pretoria, South Africa

Preface

The goal of *Climate Change and Water Resources in Africa* is to provide an overview of the many water resource management challenges, constraints, and opportunities for sustainable development in Africa. Water is very high on the agenda of the continent, particularly on issues of water supply and sanitation, and the water–energy–food nexus. Water is both an ecosystem 'good,' providing drinking water, irrigation and hydropower, and an ecosystem 'service,' enabling nutrient recycling and supporting habitats for fish and other aquatic organisms, as well as providing 'cultural services' such as scenic, recreational, and spiritual opportunities. Indeed, water is directly or indirectly important to almost every economic sector in Africa, including agriculture, manufacturing, trade, mining, tourism, and transport. The prospect of improving well-being in Africa is therefore critically dependent on its capacity to respond to water-related environmental changes—including climate change.

The IPCC recognizes Africa as one of the most vulnerable continents when it comes to the impacts of climate change—partly because of the projected nature of the climate change itself on the physical continent, and partly because of the significant levels of poverty and weak institutional capacity across the continent. Regarding the latter, the challenges of climate change overlay Africa's already fragile human and environmental conditions, with significant levels of poverty and service delivery that must be improved. Not only does climate change compound existing development pressures on limited water resources, but as climate change pressures intensify, they do so in the face of growing populations and economies, both of which place greater stress on existing water resources. Therefore, the challenge of managing water resources in Africa over the coming decades is both a climate change and a developmental challenge.

A consistent theme throughout the book is that the already-vulnerable poor population of Africa is most at risk from the impacts of climate change. The response must therefore focus in particular on increasing resilience at the household and community levels. Increased resilience will enable people, particularly those living in poverty, to respond effectively in order to recover faster from water-related disasters. The key elements of resilience are poverty eradication and access to appropriate information in a timely manner to support adaptation strategies. This approach will make climate change adaptation primarily a development challenge.

This book seeks to provide a basis for evidence-based policymaking, built on reliable data and sound information, to address the water-related challenges faced by Africa in the context of climate change and other stressors. It provides an overview of key water resources and climate change-related issues faced by Africa, including socioeconomic consequences, governance and understanding change in the context of institutional settings. The assessment is disaggregated first through a series of regional assessments, and subsequently in a series of detailed case studies, providing current scientific insight into various aspects of climate change-related freshwater challenges across the continent. The book furthermore discusses a number of tools and approaches for risk assessment and management. Finally, a number of critical research topics are proposed, including technology and innovation to address current and future challenges of climate change-related impacts on water resource management in Africa.

This book provides a platform for leading scientists from across the continent and beyond to contribute their expertise to address the challenges at hand. In this regard, we wish to thank the many contributors to this book for their relentless efforts and valuable contributions.

The Editors are extremely grateful to the African Academy of Sciences for supporting this project. The Editors are especially indebted to Elizabeth Marincola for her detailed editing of this book.

Dakar, Senegal — Prof. Salif Diop
Lavardens, France — Dr. Peter Scheren
Dakar, Senegal — Dr. Awa Niang

Acknowledgements to Peer Reviewers

Jean Albergel (Dr.), University of Montpellier, UMR LISAH, IRD, INRA, SupAgro-Montpellier France
Kassaye Asfawossen Asrat (Prof.), Member, African Academy of Sciences; School of Earth Sciences, Addis Ababa University, Ethiopia
Samuel Ndonwi Ayonghe (Prof.), University of Buea, Buea Cameroon
Charles Biney (Dr.), Director Ecosystem Environmental Solutions; No. 2 Yehans Crescent; East Legon, Accra
Peter Koefed Bjornsen (Dr.) , Former Director, UNEP-DHI Center water and Environment; Denmark
Adoté Blim Blivi (Prof.), CGILE, University of Lome, Togo
Michel Boko (Prof.), Abomey Calavi University, Benin
Francis Daniel Bougaire, African Development Bank, Ivory Coast
Todd Capson (Ph.D.), Consultant, Jussieu University, LOCEAN/IPSL, France
Thomas Chiramba (Dr.), UN-Habitat, Nairobi, Kenya
Honoré Dacosta (Dr.), Cheikh Anta Diop University, Department of Geography, Senegal
Bachir Diouf (Prof.), Cheikh Anta Diop University, Department of Geology, Senegal
Jacob O. Ehiorobo (Prof.), University of Benin, Benin City, Nigeria
Cheikh Becaye Gaye (Prof.), Cheikh Anta Diop University, Department of Geology, Senegal & National Academy of Sciences and Technology
Serge Janicot (Prof.), IRD, Sorbonne University, LOCEAN/IPSL, France
Anne Marie Lezine (Dr.), Jussieu University, LOCEAN/IPSL, France
Ibrahima Ly (Prof.), Cheikh Anta Diop University, Faculty of Law and Economical Sciences, Senegal
Walter Rast (Prof. Em), Texas State University
Desta Mebratu (Prof.), Member, African Academy of Sciences, Addis Ababa, Ethiopia
Clever Mafuta (Dr.), UNEP/GRID—Arendal, Norway
Dam Aimé Mogbante (Dr.), Global Water Parnership, Burkina Faso
Fred Mwango, Intergovernmental Authority on Development (IGAD), Republic of Djibouti

Ousmane Ndiaye (Dr.), ANACIM, Dakar, Senegal
Jacques André Ndione (Dr.), Centre de Suivi Ecologique, Senegal
Madiodio Niasse (Dr.), Consultant, Dakar, Senegal
Mbayang Thiam (MsC), Cheikh Anta Diop University—Senegal and Kwame Nkrumah University of Sciences and Technology, Kumasi - Ghana
Johnson Nkem Ndi (Dr.), African Climate Policy Centre Technology (ECA) Addis Ababa, Ethiopia
Imasiku Anayawa Nyambe (Prof.), Geology Department, Integrated Water Resources Management Centre, School of Mines, University of Lusaka, Zambia
Daniel Nyanganyura (Dr.), ROA/ICSU/Africa; Pretoria, South Africa
Allassane Ouattara (Prof.), University of Nangui Abrogoua, Côte d'Ivoire
Samuel Pare (Prof.), University of Ouaga 2, Ouagadougou, Burkina Faso; Member, African Academy of Sciences
Ajit Kumar Pattnai (Dr.), Vice President Wetlands International, South Asia, Odisha, India
Harrison Pienaar (Dr.), Hebei University of Engineering and CSIR; Pretoria 0001, South Africa
Richard Robarts (Dr.), World Water and Climate Change Foundation 1348 Crown Isle Blvd Courtenay, BC Canada.
Arona Soumare (Dr.), African Development Bank, Côte d'Ivoire
Pieter Van der Zaag (Prof.), IHE Delft Institute for Water Education, Delft, The Netherlands
Shem Wandiga (Prof.), Professor of university; Member, African Academy of Sciences, Nairobi, Kenya
Yongxin Xu (Prof.), Department of Earth Sciences, University of the Cape, Cap Town, South Africa
Pius Yanda (Prof.), START/IGBP, Dar-Es Salaam, Tanzania

Citations

"Action without thought is empty. Thought without action is blind"

Kwame Nkrumah (1909–1972), Prime Minister of Ghana from 1957 to 1960 and President of the Republic of Ghana from 1960 to 1966, and a Pan-African leader.

"there is unequivocal scientific evidence that our planet has been warming since the mid-20th century, primarily as a result of human activity. At the global level, we are still arguing about the sources and effects of climate change on the planet and on people, often driven more by self-interest than scientific evidence, while we in the developing world continue to suffer, even when we are not the culprits."

Amarakoon Bandara, Senior Economic Advisor, UNDP Zimbabwe in The Zimbabwe Human Development Report 2017 with a special focus on issues of climate change.

"rapid urbanization in Africa has placed enormous pressure on the continent's urban water supplies, and obviously this will likely be accentuated by the negative effects of climate change and its impact on freshwater resources. Effective solutions to urban Africa's water challenges can only succeed by improving water and sanitation infrastructure and by promoting integrated water and sanitation management… Viable solutions must be devised at the local level through a participatory approach involving all stakeholders. Indeed, community-based strategies including basic engineering to apply ecological principles and practices (for example, wastewater treatment and basic recycling of solid waste) is essential for addressing urban Africa's most acute water-related problems, especially in cities' most impoverished areas."

Salif Diop, from an interview provided to TWAS (The world academy of sciences for the advancement of science in developing countries) at the occasion of the World Water Day dedicated to Water and Cities—28 March 2011.

Contents

Editors and Contributors

About the Editors

Prof. Salif Diop has worked at the United Nations, in UNEP's Division of Early Warning and Assessment (DEWA), for the past 16 years as Senior Officer. His water expertise is in coastal oceanography, freshwater assessment, aquatic and marine issues, sustainable management and development. He earned a Third-Cycle Doctorate in 1978 from the University Louis Pasteur in Strasbourg, France, and a State Doctorate in 1986. He served a year-long sabbatical as a Senior Fulbright Scholar in the Division of Biological and Living Resources at The Rosenstiel School of Marine and Atmospheric Sciences of the University of Miami, Division of Biological and Living Resources, in 1986/87.

He is a member of multiple expert and working groups and attached to numerous scientific and research institutions. He has published more than 40 peer-reviewed articles and 7 books as main author or co-author, and was awarded a Nobel Peace Prize Certificate in 2007 for his contributions to the IPCC. In addition to his research articles, he has also contributed 140 technical documents, research reports, monographs, theses, abstracts, and book reviews. He is currently a University Professor and has served as one of the Vice-Chairs of the International Lakes Environment Committee Foundation (ILEC) Scientific Committee since November 2016 and member of the High-Level Panel Expert Group for a Sustainable Ocean Economy related to the Sustainable Ocean Initiative of the World Resources Institute since September 2019. He was named a Member of the National Academy of Sciences and Techniques of Senegal in 2006, a Member of the African Academy of Sciences (AAS) in 2009, and a Member of The World Academy of Sciences for the Advancement of Sciences in the Developing Countries (TWAS) in October 2010.

http://www.esalifdiop.org

Dr. Peter Scheren has been engaged in research, assessment, and management of both freshwater and marine ecosystems across Africa for over 25 years, in various parts of the continent. His passion is to bring the science, knowledge, and capacity for understanding of such systems together to guide policy, strategy, and governance systems that will help safeguard the ecosystem functions and values that strengthen the nature and the people that depend on them. He is the author and editor of many publications in this field.

He holds a Ph.D. from the University of Eindhoven and the University of Wageningen in the Netherlands, with a focus on Integrated Environmental Assessment of Water Systems. He has worked for various research institutions, consultancy offices and international organizations, and has managed several large-scale freshwater and marine programs in Africa for WWF, UN Environment, UNIDO, and UNDP. He is currently active as a freelance consultant, based in France, supporting the development of a range of strategic, transformational projects, and programs for WWF in their various fields of work across Africa.

Dr. Awa Niang is an Associate Professor of Continental Hydrology at Cheikh Anta Diop University, in Dakar, Senegal. She holds a first doctoral thesis from Cheikh Anta Diop University, with a focus on water quality management of Guiers Lake (Senegal River Delta). In 2014, she also earned a Ph.D. on recent hydrological and morphological changes in the Senegal River Estuary.

Her recent work focuses on the vulnerability and resilience of the coastal social ecosystems of the Senegal River Estuary. She participates in various projects and programs dealing with Integrated Water Resource Management, remote sensing applied to water resources and environmental management, and coastal and estuarine studies. She contributes to the design, implementation, and management of the Doctorate School on Water Quality and Uses. She leads the scientific Secretariat and communications of the doctoral school since its creation in 2008 and is now coordinating doctoral studies on continental hydrology and integrated water resource management. Since 2010, she was involved in the ACEWATER (African Networks of Centres of Excellence on Water Sciences) Project. Currently, in its second phase (ACEWATER2), the project is promoted by the African Ministers of Water, financed by DG DEVCO, implemented by DG JRC (project management and scientific cluster) with UNESCO (human capacity development cluster). The project aims at supporting the establishment of Human Capacity Development Programme of the AMCOW (African Ministers' Council on Water) in the Water Sector, strengthens institutional networking and improving research support to policymaking addressing the WEFE (Water–Energy–Food–Ecosystem) nexus. She is actually acting as coordinator of the Western Africa network (WANWATCE) of the Centers of Excellence on Water Sciences and Technology.

In the academic year 2018–2019, she was appointed as a technical adviser to the Ministry of Research, Higher Education and Innovation of Senegal, in charge of Academic Affairs.

awa10.fall@ucad.edu.sn

Contributors

Tamiru A. Abiye School of Geosciences, University of the Witwatersrand, Johannesburg, South Africa

Khaled M. AbuZeid Regional Water Resources Director, Centre for Environment & Development for the Arab Region (CEDARE), Cairo, Egypt

Abdou Ali AGRHYMET Regional Centre, Niamey, Niger

Abou Amani Water Sciences Division, UNESCO, Paris, France

J. Badenhorst Jones and Wagener Engineering and Environmental Consultants, Johannesburg, Rivonia, South Africa

Hannah Baleta WWF Consultant, Stanford, South Africa

Hylke Beck Princeton Climate Analytics, Princeton, USA

S. A. Bellal EGEAT Research Laboratory, University of Oran 2 Mohamed Ben Ahmed, Bir El Djir, Algeria

Mohammed Bila Lake Chad Basin Commission, N'Djamena, Chad

Aleix Capdevilla World Bank, Washington, DC, USA

P. A. B. Chabi Department of Geography, Abomey-Calavi University, Godomey, Benin

M. T. Chaibi National Research Institute for Rural Engineering Water and Forestry, Ariana, Tunisia

Ashok K. Chapagain Agricultural Economics, University of Free State, Bloemfontein, South Africa

V.G.P. Chimonyo Centre for Transformative Agricultural and Food Systems, School of Agricultural, Earth & Environmental Sciences, University of KwaZulu-Natal, Pietermaritzburg, Scottsville, South Africa

Jean Louis Chopart AgerConsult, Montpellier, France

Honoré Dacosta Laboratory of Morphology and Hydrology, Cheikh Anta Diop University, Dakar, Senegal

O. Dari EGEAT Research Laboratory, University of Oran 2 Mohamed Ben Ahmed, Bir El Djir, Algeria

Y. M. Diédhiou Department of Geography, Cheikh Anta Diop University, Dakar-Fann, Senegal

Sidy Dieye Département de Géographie, Faculté Des Lettres et Sciences Humaines, UCAD, Dakar-Fann, Senegal

B. I. Diomandé Department of Geography, Alassane Ouattara University, Bouaké, Côte d'Ivoire

C. Diop Department of Geography, Cheikh Anta Diop University, Dakar-Fann, Senegal

Penda Diop Anta Diop University, Dakar, Senegal

Salif Diop Cheikh Anta Diop University, National Academy of Sciences and Techniques of Senegal, Dakar, Senegal

J. M. Dipama Department of Geography, Joseph Ki-Zerbo University, Ouagadougou, Burkina Faso

Martin O. Eduvie National Water Resources Institute, Kaduna Nigeria, Nigeria

Bakary Faty Laboratory of Morphology and Hydrology, Cheikh Anta Diop University, Dakar, Senegal

Cheikh Faye Département de Géographie, UFR Sciences et Technologies, UASZ, Laboratoire de Géomatique et D'Environnement, Ziguinchor, Senegal

Luca Ferrini GIZ, Niamey, Niger

Colby Fisher Princeton Climate Analytics, Princeton, USA

T. Ghodbani EGEAT Research Laboratory, University of Oran 2 Mohamed Ben Ahmed, Bir El Djir, Algeria

M. Hadeid EGEAT Research Laboratory, University of Oran 2 Mohamed Ben Ahmed, Bir El Djir, Algeria

Mohamed Hamatan AGRHYMET Regional Centre, Niamey, Niger

G.P.W. Jewitt Centre for Water Resources Research, School of Agricultural, Earth & Environmental Sciences, University of KwaZulu-Natal, Pietermaritzburg, Scottsville, South Africa

Blanca Jimenez-Cisneros Water Sciences Division, UNESCO, Paris, France

B. L. C. N. Karambiri Department of Geography, Joseph Ki-Zerbo University, Ouagadougou, Burkina Faso

Harouna Karambiri Laboratory of Water, HydroSystems and Agriculture (LEHSA), International Institute for Water and Environmental Engineering (2iE), Ouagadougou, Burkina Faso

J. U. Kitheka School of Environment, Water and Natural Resources, Department of Hydrology and Aquatic Sciences, South Eastern Kenya University, Kitui County, Kenya

O. Koudamiloro Department of Geography, Abomey-Calavi University, Godomey, Benin

Francesca Kunedzimwe Department of Geography and Environmental Science, University of Zimbabwe, Harare, Zimbabwe

Samuel Kusangaya Department of Geography and Environmental Science, University of Zimbabwe, Harare, Zimbabwe

Khahliso C. Leketa School of Geosciences, University of the Witwatersrand, Johannesburg, South Africa

Babacar Lèye Laboratory of Water, HydroSystems and Agriculture (LEHSA), International Institute for Water and Environmental Engineering (2iE), Ouagadougou, Burkina Faso

T. Mabhaudhi Centre for Transformative Agricultural and Food Systems, School of Agricultural, Earth & Environmental Sciences, University of KwaZulu-Natal,

Pietermaritzburg, Scottsville, South Africa;
Centre for Water Resources Research, School of Agricultural, Earth & Environmental Sciences, University of KwaZulu-Natal, Pietermaritzburg, Scottsville, South Africa

Euan Mackway-Jones UNESCO, Paris, France

O. Mahjoub National Research Institute for Rural Engineering Water and Forestry, Ariana, Tunisia

Desmond Manatsa Geography and Environmental Science Department, Free State University, Bloemfontein, South Africa;
Geography Department, Bindura University of Science, Bindura, Zimbabwe

George A. Manful Climate Change Consultant, Cantonments, Accra, Ghana

Abel Vincent Manga Laboratory of Morphology and Hydrology, Cheikh Anta Diop University, Dakar, Senegal

Barbra Masunga Department of Geography and Environmental Science, University of Zimbabwe, Harare, Zimbabwe

Dominic Mazvimavi Institute for Water Studies, Dept. of Earth Sciences, University of the Western Cape, Bellville, South Africa

N. Mboya School of Environment, Water and Natural Resources, Department of Hydrology and Aquatic Sciences, South Eastern Kenya University, Kitui County, Kenya

Willis Memo Rift Valley Catchment Region, Water Resources Authority, Nakuru, Kenya

Anastasie Mendy Laboratory of Morphology and Hydrology, Cheikh Anta Diop University, Dakar, Senegal

Bernard Minoungou AGRHYMET Regional Centre, Niamey, Niger

Anil Mishra Water Sciences Division, UNESCO, Paris, France

Motlole Christopher Moseki Department of Water and Sanitation, Pretoria, South Africa

S. Mpandeli Water Research Commission of South Africa, Lynnwood Manor, Pretoria, South Africa

Jones Muli Kenya Marine and Fisheries Research Institute (KMFRI), Kampi Ya Samaki, Kenya

L. Nhamo Centre for Transformative Agricultural and Food Systems, School of Agricultural, Earth & Environmental Sciences, University of KwaZulu-Natal, Pietermaritzburg, Scottsville, South Africa;
Centre for Water Resources Research, School of Agricultural, Earth & Environmental Sciences, University of KwaZulu-Natal, Pietermaritzburg, Scottsville, South Africa;

Water Research Commission of South Africa, Lynnwood Manor, Pretoria, South Africa;
International Water Management Institute, Southern Africa (IWMI-SA), Pretoria, Silverton, South Africa

Awa Niang Cheikh Anta Diop University, Dakar, Senegal

Godfrey Ogonda OSIENALA (Friends of Lake Victoria), Kisumu, Kenya

Cornelius Okello Department of Environmental Sciences, Machakos University, Machakos, Kenya

S. Okoyo School of Environment, Water and Natural Resources, Department of Hydrology and Aquatic Sciences, South Eastern Kenya University, Kitui County, Kenya

Daniel O. Olago Institute for Climate Change and Adaptation & Department of Geology, University of Nairobi, Nairobi, Kenya

Christine Omuombo Institute for Climate Change and Adaptation & Department of Geology, University of Nairobi, Nairobi, Kenya

Obiero Ong'ang'a OSIENALA (Friends of Lake Victoria), Kisumu, Kenya

Yaw Opoku-Ankomah Climate Change Consultant, Cantonments, Accra, Ghana

Stuart Orr WWF Freshwater Practise Leader, Gland, Switzerland

I. Francis Oseke National Water Resources Institute, Kaduna Nigeria, Nigeria

Ming Pan Princeton Climate Analytics, Princeton, USA

Jackson Raini FlamingoNet, Nakuru, Kenya

P. Sagna Department of Geography, Cheikh Anta Diop University, Dakar-Fann, Senegal

Mor Talla Sall Senegalese Sugar Company, Richard-Toll, Senegal;
University of Liege, Liege, Belgium;
Anta Diop University, Dakar, Senegal

P. C. Sambou Department of Geography, Cheikh Anta Diop University, Dakar-Fann, Senegal

T. Sané Department of Geography, Assane Seck University, Ziguinchor, Senegal

Peter Scheren Lavardens, France

Mamoune Seck Senegalese Sugar Company, Richard-Toll, Senegal

A. Senzanje School of Engineering, University of KwaZulu-Natal, Pietermaritzburg, Scottsville, South Africa

M. A. Shamseddin Water Management and Irrigation Institute, University of Gezira, Wad Medani, Sudan

Justin Sheffield University of Southampton, Southampton, UK;
Princeton Climate Analytics, Princeton, USA

Munyaradzi D. Shekede Department of Geography and Environmental Science, University of Zimbabwe, Harare, Zimbabwe

G. Simpson Centre for Water Resources Research, School of Agricultural, Earth & Environmental Sciences, University of KwaZulu-Natal, Pietermaritzburg, Scottsville, South Africa;
Jones and Wagener Engineering and Environmental Consultants, Johannesburg, Rivonia, South Africa

Amadou Abdoul Sow Département de Géographie, Faculté Des Lettres et Sciences Humaines, UCAD, Dakar-Fann, Senegal;
Laboratory of Morphology and Hydrology, Cheikh Anta Diop University, Dakar, Senegal

Ibrahima Thiaw Laboratory of Morphology and Hydrology, Cheikh Anta Diop University, Dakar, Senegal

Nora Van Cauwenbergh IHE Delft, Delft, The Netherlands

Koen Verbist Water Sciences Division, UNESCO, Paris, France

E. W. Vissin Department of Geography, Abomey-Calavi University, Godomey, Benin

Andrea Vushe Namibia University of Science and Technology, Windhoek, Namibia

Joost Wellens University of Liege, Liege, Belgium

Eric Wood Princeton Climate Analytics, Princeton, USA

M. Yade Department of Geography, Cheikh Anta Diop University, Dakar-Fann, Senegal

Roland Yonaba Laboratory of Water, HydroSystems and Agriculture (LEHSA), International Institute for Water and Environmental Engineering (2iE), Ouagadougou, Burkina Faso

Cheick Oumar Zouré Laboratory of Water, HydroSystems and Agriculture (LEHSA), International Institute for Water and Environmental Engineering (2iE), Ouagadougou, Burkina Faso

Chapter 1
Introduction

Water Resource Management Within the Climate Change Context in Africa: Synthesis, Key Findings and Future Challenges

Salif Diop, Peter Scheren, and Awa Niang

Abstract This introductory chapter is an attempt to synthetize the rich and varied findings, insights and perspectives presented in the 20 peer-reviewed chapters of this book. The book as a whole provides a rich compilation of observations, views, case studies and recommendations from scientists across the African continent and beyond. It presents a representative overview of the current knowledge of climate change/water resource interactions in Africa, illustrated by a variety of regional, national and local level case studies, discussions of tools and methodologies, as well as considerations for governance and institutional settings and specific management approaches. Based on the varied analyses and findings presented in this book, this introductory chapter also presents a number of options and recommended actions for improved management of water resources within the context of climate change in Africa. Of particular importance, in this regard, is the emerging need to address the economic and financial aspects of climate change adaptation, recognizing that adaptation is a development challenge that is adding to the existing poverty and development-related challenges on the continent. Additional and substantial increases in financing are therefore required in order to improve the adaptive capacity and resilience of rural households as well as land and water management systems.

Keywords Climate change · Freshwater resources management · Impacts of extreme events · Disasters and risks · Vulnerability · Adaptation · Resilience · Tools and methodologies · Response strategies · Transboundary · Regional integration

---

S. Diop (✉)
Cheikh Anta Diop University, National Academy of Sciences and Techniques of Senegal, Dakar, Senegal
e-mail: salif.diop@ansts.sn

P. Scheren
Lavardens, France

A. Niang
Cheikh Anta Diop University, Dakar, Senegal

S. Diop et al. (eds.), *Climate Change and Water Resources in Africa*,
https://doi.org/10.1007/978-3-030-61225-2_1

1 Introduction

The IPCC has recognized that Africa is among the continent's most vulnerable to climate change impacts, partly because of projected climatic change itself, and partly because of the complicating poverty levels and the paucity of institutional capacity across the continent (Christensen et al. 2007). According to the ND-Gain index for climate change vulnerability, eight of the ten most vulnerable countries in the world are African, with the top five all in Africa (Notre Dame Global Adaptation Initiative 2019).

The challenges of properly addressing climate change impacts in Africa is not only because of the compounding pressures of development on limited water resources, but also because climate change pressures are intensified by growing populations and economies, both placing yet-greater stresses on existing water resources. Thus, the requirements of managing water resources in Africa over the coming decades are both a climate change and a development challenge (UNEP 2009). As the already-vulnerable rural areas are populated by the people most at risk from climate change, any action that increases the resilience of these communities will help them respond more effectively to the impacts of climate change, enabling them to integrate climate change adaptation into development planning and decision-making frameworks. Within this context, this book, provides an overview of the current challenges related to water resource management in the context of climate change faced by some of the most vulnerable areas in Africa, and approaches to help alleviate them.

This introductory chapter provides cross-cutting insights, observations and findings of the individual contributions presented in this book. The book, in turn, provides a rich compilation of observations, views, case studies and recommendations from scientists across the African continent and beyond, including an overview of current knowledge of climate change/water resource interaction, illustrated by a variety of regional, national and local level case studies, discussion of tools and methodologies, governance and institutional settings, and specific management approaches. Based on these analyses and findings, this book also presents options and recommended actions for improved management of water resources within the context of climate change in Africa.

2 The Challenges of Water Resource Management in the Context of Climate Change in Africa

Climate change in Africa is projected to result in significant changes in the demand for, and availability of, water. The case studies presented in this book show that large parts of Africa are already subject to seasonally variable hydrology, and geographically uneven distribution of water resources. While pressures on Africa's unique ecosystems, population and economy continue to grow, this situation is further compounded by climate change, to which Africa is particularly vulnerable (Chap. 2). Indeed, Africa carries a particular burden of climate change: while the continent is the least polluting and smallest emitter of greenhouse gases (less than 5% of total worldwide emissions), she is projected to be the most impacted among all continents, with drastic consequences for her water resources.

Although climate change projections for Africa, like all climate projections, have large margins of uncertainty, it is clear that climate change will bring more frequent and more intense water-related disasters to many parts of Africa, a continent already prone to floods and droughts, with dramatic consequences for critical ecosystem goods and services, and, therefore, its population and development (Chaps. 3, 4, 5 and 17). While countries like Mozambique, South Africa, Malawi and Zimbabwe are already experiencing recurring floods and droughts, projected increases in the frequencies and magnitudes of such events through 2050, coupled with rising populations in these regions, suggest a particular vulnerability of weather sensitive sectors, especially water resources and agricultural production (Chap. 5). Because of the difficulty of predicting climate change impacts on water systems with accuracy, flexibility must be built into planning in order to manage changing climate conditions over the long term.

Within the context of water resource management, it must also be recognized that the rapidly increasing populations in Africa, and their transitional economic developmental needs, will impose growing demands and pressures on already stressed freshwater resources. Rivers and groundwater resources, as well as lakes and reservoirs, are readily usable sources of large volumes of freshwater which, if properly managed, will continue to serve as an important basis for the health, wellbeing and economic growth of these populations (Chap. 2). A related factor is the impact of aging infrastructure for water storage and delivery of these lentic water systems, an issue of obvious importance to both developed and developing countries (Chaps. 6, 8, 12 and 15).

3 Toward Improved Management of Water Resources in the Context of Climate Change in Africa

While it is important to acknowledge the many weaknesses across the continent that may complicate effective response to climate change, including limited institutional capacity, poverty, the paucity of data and limited modeling of climate change impacts on a local scale, it is equally important to take immediate action to improve the resilience of communities and societies to climate change impacts. An overview and synthesis of key approaches in this regard is presented in the following sections.

3.1 Create Enabling Policy and Institutional Conditions

Climate change-related freshwater management strategies must be based on water management plans that are aligned with national development and poverty reduction strategies. Moreover, enabling conditions to manage the challenges of climate variability must be created. Such measures include the creation and implementation

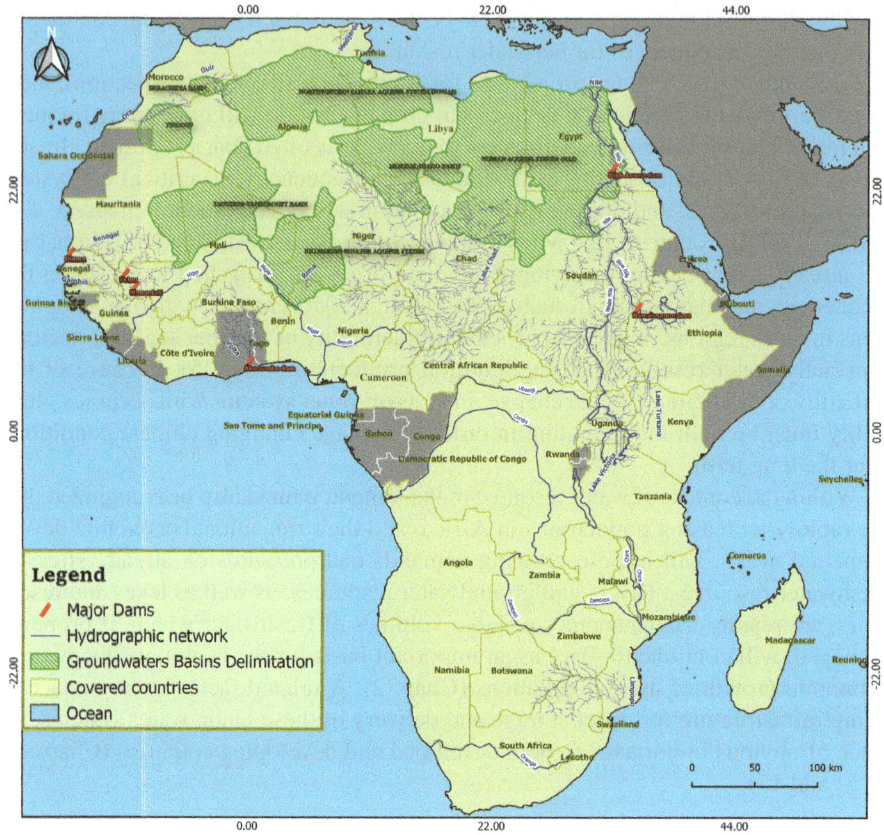

Countries covered in this book: Algeria; Angola; Benin; Botswana; Burkina Faso; Cameroon; Comoros; Côte d'Ivoire; Egypt; Ethiopia; Gambia; Ghana; Guinea; Guinea Bissau; Kenya; Lesotho; Libya; Madagascar; Malawi; Mali; Morocco; Mauritius Mauritania; Mozambique; Namibia; Niger; Nigeria; Uganda; Central African Republic; Democratic Republic of Congo; Senegal; Seychelles; Soudan; South Africa; Swaziland; Tanzania; Chad; Tunisia; Zambia; Zimbabwe

Fig. 1 Geographical location of African Regions considered in the book

of appropriate legislation, including transboundary agreements, to ensure that the institutional capacity to manage water resources and services exists, including the training and development of sufficient skilled and experienced staff to manage water effectively, and is supported by appropriate financial resources to develop, operate and maintain the necessary water infrastructure to respond to climate change. Provision of accurate, accessible and timely public information is a critical part of any water management plan. Given the limited human and financial resources in many parts of Africa, leaders must prioritize management of the most vulnerable areas and the most critical issues, avoiding the risk of spreading limited resources over too large an area or addressing too many issues at once. It is also critical that transboundary water resource management addresses multiple transboundary basins that are vulnerable to the impacts of climate change simultaneously (Chaps. 6, 15, 17 and 19).

Management of the water-related impacts of climate change requires alignment between national development objectives and water availability. Because of the difficulty of accurately predicting climate change impacts on water management (IPCC 2014, 2018), plans must be structured with sufficient flexibility to facilitate adaptation to a changing climate over the years (Chaps. 18, 19, 20 and 21). This requires access to relevant and up-to-date climate change information for key water-related development planning departments. It is also critical to integrate local resource development and management plans into broader-scale planning. Given the particular vulnerability of women to climate change impacts, development and climate change response plans must proactively address gender differences, and incorporate strategies to protect and support women, girls and children in particular (Chaps. 15 and 20).

Substantial increases in financing are also needed to improve adaptive rural household and land and water management system capacities and resilience. As public resources in African countries are limited, financing options in addition to government sources are needed, including from private sources and public sector funding from developed countries.

3.2 *Invest in Ecological Infrastructure*

As a first line of defense, investments in natural infrastructure, for example, the protection and rehabilitation of aquifers, lakes, reservoirs and wetlands, can contribute significantly to increasing resilience to climate change. Vital ecosystem functions and services are under great pressure from population growth, energy demands, exploitive land-use practices and other factors, resulting in deforestation and land degradation. Climate change compounds these challenges. Thus, increasing land and water management resilience calls for integrated ecosystem-based approaches, including sustainable land-use management, the designation of water protected areas and effective management of natural water storage systems (Chaps. 5, 6 and 21).

It is also crucial to secure vital freshwater ecosystem functions through appropriate environmental flow and reserve regulation (Chaps. 2 and 19). The objective of such regulation is to secure an appropriate flow of rivers and the reserves of other aquatic ecosystems (in particular groundwater reservoirs and lakes) through appropriate water allocation, regulation of dam operations, etc. Although stakeholders sometimes interpret such approaches as being in direct competition with human development needs, they in fact present an opportunity to maintain important water-related ecosystem services for the overall benefit of society.

3.3 Invest in Climate-Smart Infrastructure and Technologies

A key feature of the case studies presented is to invest in water management infrastructure and technology. In many areas, climate change is likely to bring increased flooding and/or drought. Considered decisions must guide flood management, including development of early warning systems and rehabilitation of degraded catchments (Chaps. 4 and 5). Also, under flood conditions, water services infrastructures (for water and sanitation) may be damaged, leaving communities vulnerable to poor quality or lack of drinking water, and compromised or eliminated sanitation facilities. Flood-proofing of the water supply and sanitation infrastructure in vulnerable areas must therefore be part of any management plan.

At the same time, some areas will experience decreased rainfall and increased drought. In many parts of Africa, the storage of rainfall is insufficient to grow the local economy. Even without accounting for climate change, Africa requires more water storage capacity—both large dams and small storage facilities—in order to overcome the impacts of frequent droughts (Chaps. 12 and 15). With the possibility of climate change extending the duration and increasing the intensity of droughts, water storage needs become even more acute. For these reasons, procurement of financial resources to develop infrastructure remains a critical challenge, with important roles to play by African governments, the private sector and international financing agencies. At the farm level, increased investment and access to information about appropriate irrigation technology, including drip irrigation and rainwater harvesting, is required to improve water use and productivity in the face of climate change. In many areas, a shift from rainfed to irrigated agriculture may be necessary to protect rural livelihoods and food security (Chaps. 3 and 5). In the case of groundwater, targeted policies and cooperation in research and development are required to overcome obstacles such as the high costs of well construction and limited understanding of groundwater resources that currently restrict development of groundwater irrigation in many parts of Africa (Chaps. 6, 10 and 18).

Artificial recharge of groundwater is an important tool for the sustainable management of resources in danger of over-exploitation and degradation. This is one of the many potential methods for using groundwater as a buffer to ensure the broader function of retaining and intercepting rainfall and runoff, conserving it in the soil or storing it underground as a means of supplementing aquifers during dry periods (Chaps. 3, 6 and 21). But the increasing impacts of climate change on water resources complicates groundwater recharge in many parts of Africa[1]

Finally, with more frequent and more intense water-related disasters likely, water systems in many parts of Africa that are already prone to floods and droughts attributable to climate change must redouble their efforts to improve disaster preparedness, including well-developed early warning systems and post-disaster intervention plans, to ensure the resilience of society in the face of climate change (Chap. 18).

---

[1] Working Group II to the Fourth Assessment Report of IPCC 2008.

3.4 Improve Science and Information

Science and information communications are essential to the ability of communities and nations in vulnerable water basins and aquifers, in particular, to adequately abate and adapt to the effects of climate change on water resources (Chaps. 19 and 21). In order for public information to be effective, it must define and describe the current state, identify emerging trends, and anticipate potential vulnerabilities and risks. This requires an appropriate monitoring system, which can deliver necessary information at the appropriate scale (Chaps. 9, 12 and 17). This system must extend beyond simply monitoring climate trends, to also monitoring the status of water resource, to the early detection of trends, and to monitoring relevant environmental variables and processes related to water and to ecosystem-based climate change adaptation. Commensurate with the critical role of groundwater in climate change adaptation, better understanding and coherent, accessible region-wide information sharing on groundwater resources is necessary (Chap. 6).

Any effective information system incorporates early warning systems, particularly for floods (Chap. 17). These systems, built upon better forecasting, are a prerequisite for adaptation, particularly for predicting and preventing the effects of floods, droughts, tropical cyclones and other extreme events, as well as for practical planning purposes such as determining planting dates to coincide with the beginning of the rainy season, and foreseeing potential disease outbreaks in areas prone to epidemics. Improved early warning systems and their application will, therefore, reduce vulnerability to risks associated with climate variability and change.

The development of state-of-the-art monitoring systems is obviously limited by the financial and human capacity constraints of institutions in Africa. Protocols must be developed that enable the exchange of information across the continent, enhancing the understanding of climate change, and facilitating adaptation to climate change between and within countries.

A particular challenge remains the management of transboundary waters under increasing stress, complicated by high levels of political instability and conflict (Chaps. 16 and 19). There are important lessons to be learned across the continent from indigenous approaches for improved water management and adaptation to climate change (Chap. 20); this illustrates that the ability to respond to changes will depend on the ability to learn from one another, to share information, and to develop a body of African experience and knowledge about managing the impacts of climate change (Chap. 21).

As mentioned previously, a central challenge is the lack of models to predict climate change at the local level. It is critical that the capacity to model climate change is enhanced so that management options can be based on scientifically sound information. Achieving this goal will require increased investments in the science of climate change, including understanding its impacts, its trends and methods to adapt and remediate them. One approach may be to create African climate change centres of excellence that would be charged with developing African capacity and expertise to tackle challenges related to climate change, regionally as well as locally, to monitor water quality, collect data, identify good water management approaches, reinforce traditional adaptation mechanisms and provide early warning systems to assist local communities that experience frequent climatic hazards and adverse environmental changes.

3.5 Work at Multiple Scales

There exists a range of water allocation systems across Africa, with many parallel formal and informal systems working at different scales (Chaps. 6, 7, 8, 9, 10, 11, 12, 13, 14, 15, 16 and 17). All of these must be sufficiently flexible to enable water allocation adjustments to address climate variability while accounting for national development objectives, while also being sufficiently simple to be implemented effectively and managed within capacity constraints.

In transboundary lakes, basins and aquifers, the first level of allocation is to riparian states (Chaps. 6, 15 and 17). In basins facing water stress as a result of climate change, it is particularly important that effective transboundary water allocation systems are put in place, supported by accurate, shared data on the status of the basin. While there are a number of agreements in place in transboundary basins in Africa (Chaps. 3 and 19), there are also a large number of basins for which no such agreements currently exist. Even where such agreements are in place, some lack effective dispute resolution mechanisms, and many lack effective institutional capacity at the national or transboundary level for effective implementation and optimal sharing of water resources.

Local groundwater resources that have been generally neglected also require urgent strengthening of institutional structures at continental and regional scales (Chaps. 2, 3, 4 and 5). Solutions to climate change and related development challenges in many areas will come from both transboundary cooperation and from greater exploitation of regional comparative advantages. This will enable development opportunities within the context of a region, rather than a country. Climate change is a key driver to expand regional integration across the continent.

Further, regional integration should be seen in a broader context than just water. There is an opportunity for the development of regional public goods, such as transport infrastructure, markets, regional power pools, trade arrangements and food security that can provide substantial value to building regional and local resilience to climate change (FAO 2018).

A range of water allocation systems operates in Africa at the sub-basin or local level, with parallel formal and customary systems in many countries (Chaps. 6, 7, 8, 9, 10, 11, 12, 13, 14, 15, 16, 17, 18 and 19). It is important that these systems be sufficiently flexible to enable adjustments in allocation to manage climate variability and climate change in support of national development objectives. It is also critical that stakeholders be fully involved in the water resource management process so that there is support for actions, and so that information exchange between stakeholders and authorities can be quickly responsive to emergent situations and provide optimal adaptive action. However, the key challenge in developing responsive institutions lies in building the adaptive capacity of such institutions. In fact, effective response to climate change relies on both the ability to accurately predict the changing climate and its water-related impacts, and on the ability to respond to change. This combined approach enables innovation at all levels, and creates flexible and responsive water management systems. Flexible allocation systems that adjust to short-term climate variability and longer term climate change are of critical importance (Claussen et al. 2003).

3.6 Decentralize Adaptation

Adaptation takes place on a number of levels, ranging, for example, from the creation of major storage and flood prevention infrastructure to the household level, particularly in rural areas (Niang et al. 2014). While governments may not be able to consistently extend the necessary services to vulnerable populations to protect them from climate change, information itself can assist communities and households to prepare themselves for coming changes. Information and training for rural communities is therefore particularly important because of their vulnerability, and because they are often left out of the information loop. Information can include alternative crops and climate-smart production methods, improved livestock management techniques, local water resource use and protection, and flood protection and warning systems (Chap. 15).

Gender mainstreaming and understanding the particular vulnerabilities of women should also form a key part of climate change adaptation strategies (Chaps. 15 and 20). Women are particularly at risk from climate change and natural disasters, being physically less able to escape from floods, for example, because of their need to carry babies or small children. Women in countries with the greatest cultural differences between the roles and expectations of women and men are most vulnerable in this regard. Thus, it is essential that disaster plans are gender-sensitive and address the particular cultural expectations and norms of women. Also, it is critical that investments in infrastructure and technology reflect the priorities and needs of women and men, and that women and men are actively involved in decisions relating to infrastructure development.

Access to groundwater is perhaps the most critical factor enabling both rural and urban populations to maintain sustainable livelihoods (Chap. 6). Groundwater is strategically important for adaptation because it is the key resource for local coping strategies. Because it has remained a poorly understood resource, however, groundwater is still insufficiently considered in the Integrated Water Resource Management (IWRM) and Integrated Lake Basin Management (ILBM) approaches, as well as generally in development planning. This deficiency needs to be addressed strategically through appropriate regional, sub-regional and national policies, proper integration into IWRM/ILBM processes, structures and institutions, and appropriate prioritization in adaptation initiatives.

With focus on adaptation on the local level, attention to local institutions is critically important in designing adaptation projects and policies. Households and social groups rely on such institutions to deploy specific adaptation practices. Also, a systematic scaling-up of locally appropriate solutions is key to ensuring area- and region-wide impacts on poverty alleviation, climate change adaptation and economic development (Chap. 20).

3.7 Consider the Water/Energy/Food/Health Nexus

A number of African countries depend on hydropower, even though the hydropower potential of Africa is still hugely underdeveloped (Chap. 18). As indicated in the chapters that discuss regional issues, however, hydropower is under threat in some areas from diminishing stream flows and/or increased flow variability. As a result, the "climate change-proofing" of current infrastructure is an important measure to protect the energy supply of many countries and, therefore, to protect economic and social development. Such measures might include, for example, amended operating rules to consider changing rainfall patterns, adding height to dam walls, and/or changes to environmental flow releases. At the same time, new hydropower development must be implemented with a clear understanding of the potential impacts of climate change accounting for these impacts.

For these reasons, it is important to integrate water and energy security plans that take into account the likely impacts of climate change. It is particularly critical to coordinate effectively at the basin and national level in the large number of transboundary basins in Africa across the water-energy-food nexus. Furthermore, and at the core of the 2030 Sustainable Development Goals/SDG's and 2063 African Agenda "The Africa We Want", the water/energy/food/health and ecosystems nexus is increasingly recognized as core to addressing the issue of climate change and securing the wellbeing of the many millions of people in Africa without access to basic services such as water and sanitation, energy, food and health (Chaps. 18 and 20).

4 Conclusions

A number of general conclusions may be drawn. First, it is evident that vulnerability to climate change risks and impacts, as well as opportunities for adaptation and mitigation in Africa, are shaped by complex land and water resource characteristics, economic conditions and the often highly diverse livelihoods of different regions and corresponding strategies of individual households. Thus, effective adaptation planning for and implementation of land and water management systems is highly context-specific and therefore demands an appropriate level of knowledge and understanding of local conditions.

Adaptation should be knowledge-based, integrating scientific and local factors. Local knowledge systems that guide people to adapt to climate conditions in many places have evolved from the experience of generations. However, historical understanding of local conditions is of limited use when there exist significant short- and long-term uncertainties about precipitation and water availability.

Moreover, effective and equitable adaptation to climate change in land and water management is dependent on good communication. Such communication must be in a form that users can understand, is suitably packaged and presented, and is culturally sensitive to all decision-makers and those impacted at every level.

In order to address the economic and financial aspects of climate change, it is also important to recognize that adaptation is a development challenge that is adding to already substantial poverty and development challenges on the continent. Additional and substantial increases in financing are therefore needed to improve the adaptive capacity and resilience of rural households and land/water management systems. To achieve this, a full range of financing options must be utilized, including innovative financing mechanisms, from both private sources and government. Moreover, public funds must also come from wealthier countries and from those countries that are now major contributors to GHG emissions.

The United Nations Framework Convention on Climate Change (UNFCCC) negotiations address mitigation and adaptation, and may present important financing opportunities. For example, the scope of the Clean Development Mechanism and/or emerging market mechanisms for sustainable development should be expanded to address issues relevant to Africa, such as reforestation, agro-forestry and soil carbon sequestration and requires substantial financing. The principle of shared but differentiated responsibility as defined under the UNFCCC suggests the necessity to take into account equity based on historical and current consumption of resources to fight climate change. To put this into perspective, the African Development Bank (AfDB 2012), predicts adaptation costs in Africa to be on the order of US$20–$30 billion per annum over the next ten to twenty years (2020–2030). But as of 2012, only about $350 m of adaptation funding had been approved for spending in Africa, of which just $130 m had been disbursed (United Nations ECA 2019). Considerable additional investment will therefore be required to bridge the gap.

"Can Africa really fund its own responses to climate change impacts as unconditional contributions using its domestic resources, irrespective of how marginal this could be? As Africa is the region bearing the greatest burden of climate change, did Africa over-pledge, and how feasible is it for African countries to internalize their Nationally Determined Contributions in their current form? Is it ethically and morally acceptable to put at risk the impoverishment of many more people by deploying scarce national resources for the implementation of NDCs?"

Nkem (2016) in "Unpacking Africa's enigma with NDCs", September 2016, Science*Policy*Africa

References

AfDB (2012) The cost of adaptation to climate change in Africa. UNECA, Addis Ababa, 41 p

Christensen JH, Hewitson B, Busuioc A, Chen A, Gao X, Held I, Jones R, Kolli RK, Kwon WT, Laprise R, Magaña Rueda V, Mearns L, Menéndez CG, Räisänen J, Rinke A, Sarr A, Whetton (2007) Regional Climate Projections. In: Solomon S, Qin D, Manning M, Chen Z, Marquis M, Averyt KB, Tignor M, Miller HL (eds) Climate change, 2007: The physical science basis contribution of working group I to the fourth assessment report of the intergovernmental panel on climate Change. Cambridge University Press, Cambridge, United Kingdom and New York, NY, USA

Claussen M, Brovkin V, Ganopolski A (2003) Climate change in Northern Africa: the past is not the future. Clim Change 57(1–2):99–118

FAO (2018) The state of world fisheries and aquaculture 2018—meeting the sustainable development goals. Rome. Licence: CC BY-NC-SA 3.0 IGO. http://www.fao.org/3/i9540en/I9540EN.pdf

IPCC (2014) Summary for policymakers. In: Field CB, Barros VR, Dokken DJ, Mach KJ, Mastrandrea MD, Bilir TE, Chatterjee M, Ebi KL, Estrada YO, Genova RC, Girma B, Kissel ES, Levy AN, MacCracken S, Mastrandrea PR, White LL (eds) Climate change 2014: impacts, adaptation, and vulnerability. part a: global and sectoral aspects. contribution of working group II to the fifth assessment report of the intergovernmental panel on climate change. Cambridge University Press, Cambridge, United Kingdom and New York, NY, USA, pp 1–32

IPCC (2018) Summary for policymakers. In: Masson-Delmotte V, Zhai P, Pörtner HO, Roberts D, Skea J, Shukla PR, Pirani A, Moufouma-Okia W, Péan C, Pidcock R, Connors S, Matthews JBR, Chen Y, Zhou X, Gomis MI, Lonnoy E, Maycock T, Tignor M, Waterfield T (eds). Global warming of 1.5 °C. An IPCC special report on the impacts of global warming of 1.5 °C above pre-industrial levels and related global greenhouse gas emission pathways, in the context of strengthening the global response to the threat of climate change, sustainable development, and efforts to eradicate poverty. World Meteorological Organization, Geneva, Switzerland, 32 pp http://www.ipcc.ch/pdf/special-reports/sr15/sr15_spm_final.pdf)

Niang I, Ruppel OC, Abdrabo MA, Essel A, Lennard C, Padgham J, Urquhart P (2014) Africa. In: Barros VR, Field CB, Dokken DJ, Mastrandrea MD, Mach KJ, Bilir TE, Chatterjee M, Ebi KL, Estrada YO, Genova RC, Girma B, Kissel ES, Levy AN, MacCracken S, Mastrandrea PR, White LL (eds) Climate change 2014: impacts, adaptation, and vulnerability part b: regional aspects. contribution of working group II to the fifth assessment report of the intergovernmental panel on climate change. Cambridge University Press, Cambridge, United Kingdom and New York, NY, USA, pp 1199–1265

Nkem J (2016) In: Unpacking Africa's enigma with INDCs; Science*Policy*Africa; Quarterly Newsletter of the African Academy of Sciences; pp 14 and 15; Sept 2016. Nairobi, Kenya

Notre Dame Global Adaptation Initiative (2019). https://gain.nd.edu/our-work/country-index/

United Nations ECA (2019) Economic report in Africa 2019; fiscal policy for financing sustainable development in Africa; 169 p (published in March 2019 by United Nations Economic Commission for Africa P.O. Box 3001 Addis Ababa, Ethiopia)

UNEP (2009) Kenya: Atlas of our changing environment. United Nations Environmental Programme, Nairobi, Kenya, pp 1–39

Chapter 2
Africa-Wide Trends in Development and Water Resources Through a Climate Change Lens

Hannah Baleta, Stuart Orr, and Ashok K. Chapagain

Abstract African economies are growing, with current economic growth on the continent driven by sectors that are highly dependent upon reliable water quantity and quality. However, water availability across Africa varies climatologically, geographically and in relation to infrastructure development. The distinction between blue and green water footprints is particularly relevant to understanding and managing variability, resilience, institutional governance and infrastructure development in the future. As the continent develops, demands for and impacts on water resources are likely to increase rapidly. As Africa's economy grows, people become more affluent, cities expand and the climate changes. These drivers are likely to impact negatively on the quality of water available across the continent, putting freshwater habitats that are currently largely intact, and biodiversity in general, at increasing risk. These pressures are especially applicable to "water towers" on which water quantity and quality in Africa is dependent. In addition to the water towers, each freshwater ecosystem must support the diversity of biota that exists in Africa as well as the economy that supports the people who live there. In order to grow and develop sustainably, Africa needs resilient infrastructure and adaptable institutions. Depending on the "development trajectory", the risks and opportunities associated with water and climate change may vary among countries. Water-related investment choices made over the next decade will have a substantial influence on the long-term development trajectory of any African country. Better allocation of resources through considered system-scale decisions will lead to a greener future where infrastructure development is not a trade-off against freshwater ecosystems. To reach the ideal balance, it is essential to develop governance mechanisms and related institutions that support responsive and coherent local resource-use planning, while making the most of opportunities

H. Baleta (✉)
WWF Consultant, Stanford, South Africa
e-mail: Hannah.baleta@gmail.com

S. Orr
WWF Freshwater Practise Leader, Gland, Switzerland
e-mail: Sorr@wwfint.org

A. K. Chapagain
Agricultural Economics, University of Free State, Bloemfontein, South Africa
e-mail: a.chapagain@gmail.com

S. Diop et al. (eds.), *Climate Change and Water Resources in Africa*,
https://doi.org/10.1007/978-3-030-61225-2_2

to manage trade-offs at a national level. Trade-offs for water use among different sectors, such as agriculture, energy, urban growth and manufacturing, will become increasingly urgent. It is imperative that decisions regarding water resources across Africa that are made now and in the near future take into account the impact of the water, energy and food nexus within each country and across the continent. In particular, healthy, free-flowing rivers and freshwater habitats are critical to ensure the sustainability of Africa's water, energy and food in the face of the growing demands that are aggravated by climate risk.

Keywords Water-energy-food (WEF) nexus · Green water · Blue water · Grey water · Climate change · Development · Water infrastructure · Water institutions

1 Africa's Economy and Water Use

Africa's economy is growing. The average economic growth is estimated to have risen from 2.3% in 2017 to 2.7% in 2018, barely keeping up with population growth, partly due to economic softening in Nigeria, South Africa and Angola—Africa's three largest economies (World Bank 2018). However, the outlook for GDP growth and per capita GDP growth on the continent is projected to improve, albeit unevenly across regions, as seen in Fig. 1 below.

Current economic growth in Africa is driven by sectors that are highly dependent on reliable water quantity and quality. Water is critical not only for drinking, but also for the production of energy through hydropower and coal generation (cooling) and the production of food, whether rain-fed or irrigated. An existential threat to Africa is that water is central to all aspects of water, energy and food (WEF)

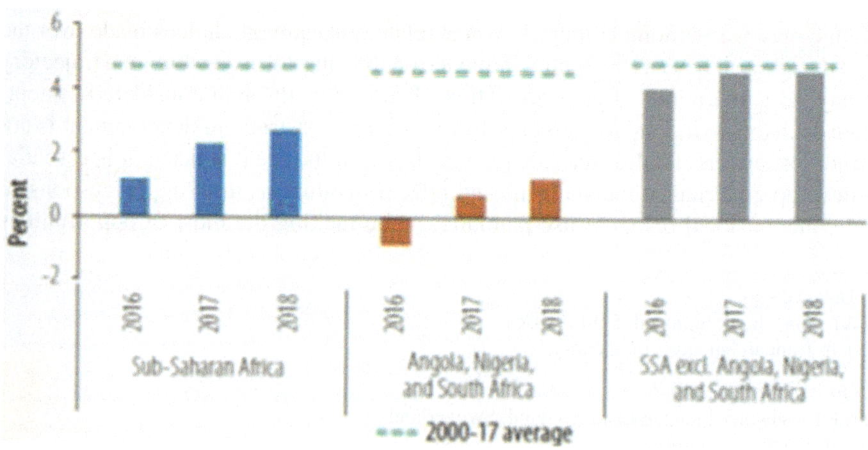

Fig. 1 The current and projected average GDP growth in Sub-Saharan Africa. Period 2016–2018 (*Source* World Bank 2018)

needed for development and growth. As economic development and the expected growth in urban population takes place, the demand on water for drinking, energy and food will continue to increase. In parallel, impacts on the freshwater habitats of Africa (WWF and SAB 2018) will grow. According to the FAO (2010), overall water withdrawals, in particular the blue water footprint, for agriculture represents 81% of all water withdrawals in Africa, with the industrial sector at 4% and domestic water use at 15%. With growth projections in Africa, these withdrawals are likely to shift towards increasing industrial and domestic water use.

Water availability across Africa varies enormously. The North and South ends of the continent and the Horn of Africa are arid, while the middle third of the continent is relatively wet. Furthermore, temperatures vary significantly year-to-year and intra-seasonally. Critical to the quality and availability in Africa are "water towers" (Fig. 2).

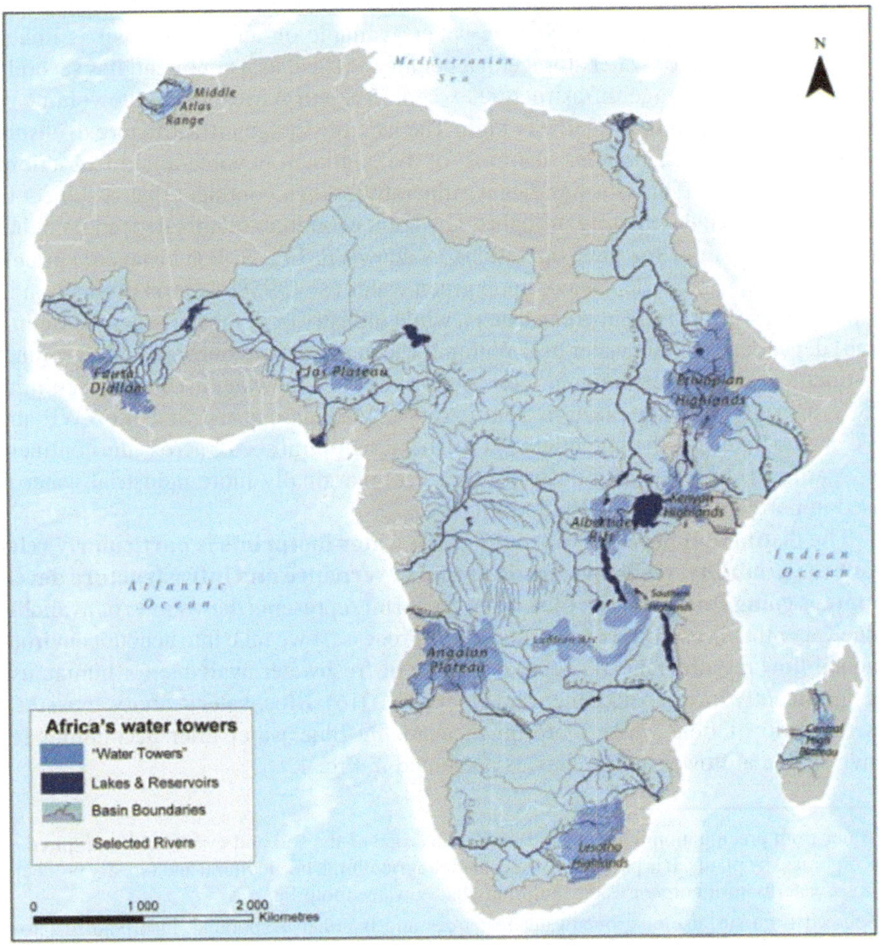

Fig. 2 Selected water towers in Africa (UNEP 2010)

These forested upland areas store and contribute a very high proportion of the water in some of Africa's major rivers. Some of these areas are under threat from, for example, deforestation, which will have major impacts on the quality and quantity of available water in these areas. Many of these towers provide water in transboundary basins so that their fate—protection or destruction—has impacts across national borders. The 400,000 hectare (ha) Mau Forest Complex in Kenya, for example, is the largest of Kenya's five water towers. It is the single most important water catchment in the Rift Valley and western Kenya. Water from the Mau Forest Complex supports agriculture, hydropower, urban water supply, tourism, rural livelihoods and wildlife habitats all through western Kenya. The water flows into Lake Victoria and ultimately the White Nile (MEWNR 2015). The total economic value of the Mau Forest Complex is estimated to be in excess of 1.3 billion USD a year (Cameron 2010).

In Africa, water use differs by sector, and also by geography and infrastructure development. Water use across Africa is categorized by green[1] water use and blue[2] water use. The amount of blue water available on an annual basis is linked closely to the level of water storage in a country. Africa, as a continent, has severely underdeveloped storage infrastructure, except in South Africa where the volume of water stored per capita is relatively high. The lack of storage infrastructure to ensure water during dry years means that most of the continent is vulnerable to variations in annual rainfall. This is a significant vulnerability on a continent that is subject to regular and often devastating droughts. Different economic sectors depend on either green of blue water for their operations as shown in Fig. 3. For instance, agriculture in Africa is more dependent upon green water use (>95% of food production in sub-Saharan Africa is from green water), while industry or domestic use is (by definition) dependent on blue water use. Within these generalities, there are some regional distinctions across the continent. For example, of the blue water used for agriculture, 75% of irrigation takes place in South Africa, Madagascar and Sudan (WWF and IFC n.d). Therefore, the potential to grow the agricultural sector across the continent is significant. In terms of industrial water use, marginally more industrial water is used in northern Africa (WWF and IFC n.d).

The distinction between blue and green water footprints is particularly relevant to variability, resilience, institutional governance and infrastructure development going forward. The blue water footprint represents the volume of available blue water that is used and evaporated in the process. If we take into account environmental flow requirements, the actual volume of freshwater available for human use is significantly lower (Mekonnen and Hoekstra 2016). Blue water scarcity, measured as the ratio of blue water footprint to available blue water after accounting for environmental flow requirements, is presented in Fig. 4.

---

[1] Water from precipitation that is stored in the root zone of the soil and evaporated, transpired or incorporated by plants. It is particularly relevant for agricultural, horticultural and forestry products. https://waterfootprint.org/en/water-footprint/what-is-water-footprint/.

[2] Sourced from surfac.e or groundwater resources and is either evaporated, incorporated into a product or taken from one body of water and returned to another, or returned at a different time. Irrigated agriculture, industry and domestic water use can each have a blue water footprint. https://waterfootprint.org/en/water-footprint/what-is-water-footprint/.

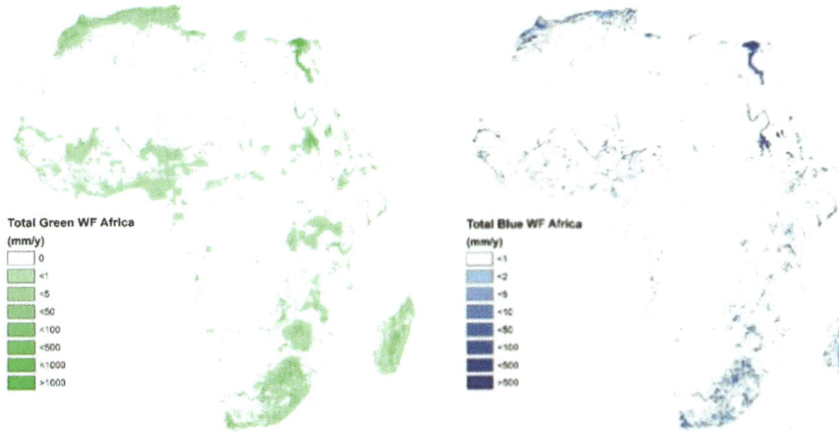

Fig. 3 Green (left) and blue (right) annual water footprint across Africa for all sectors (Water Footprint Network n.d)

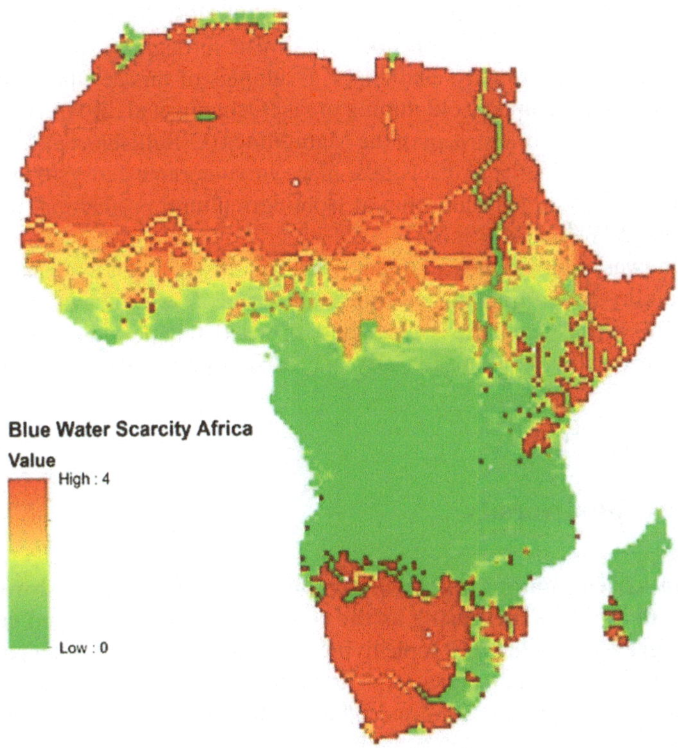

Fig. 4 Blue Water Scarcity of Africa (*Source* Mekonnen and Hoekstra 2016)

Demands for and impacts on water resources are likely to increase rapidly as Africa's economy grows, people become more affluent, cities expand and the climate changes. These drivers are likely to have a negative impact on the quality of water available across the continent. Population growth alone will continue to drive increased water demand. Africa's population growth rate is relatively high, particularly in west, central and eastern Africa, where predicted annual population growth rates are over 2.5%. This population growth will place increasing pressure on water resources, particularly in countries that are already facing water stress in West and East Africa (WWF and IFC n.d). As Africa's economy grows, people become more affluent, triggering changes in lifestyle and consumption patterns that are more resource-intensive, similar to that in developed nations. As a result of both population and economic growth, surface and groundwater are vulnerable to a range of water quality impacts across the continent. For instance, in 2001 nearly 500 million hectares of land were moderately to severely degraded due to population pressure, destructive inappropriate land use, poor land management and agricultural practices and drought (UNECA 2006). Land degradation causes turbidity, poor agricultural practices contribute to eutrophication, and poor sanitation and solid waste management result in pollution of water resources located close to dense settlements or industrial zones. Mining, although not widespread throughout the continent, has driven significant water quality challenges. Examples of this are the Vaal River in South Africa, downstream of gold mining areas (Ochieng et al. 2010), and the Mara Wetlands in Tanzania, downstream of the Mara Mines (O'Sullivan et al. 2016). Often these negative water quality impacts take place in vulnerable ecosystems including wetlands that are under particular threat from agricultural production, mining and urban sprawl (WWF and IFC n.d) (Fig. 5).

Due to the variations in measurement, water quality is much harder to map than water availability. In addition, there is little monitoring of raw water quality across the continent. This means that there is a lack of data on existing pollution and pollutants, and very poor tracking of more recently appreciated pollutants, such as endocrine disruptors.

2 Africa's Biodiversity

Pressure on Africa's unique and, in many places, largely intact freshwater habitats and biodiversity is increasing as the continent's population and economies continue to grow and develop. People living across Africa are fundamentally dependent upon the health of their ecosystems, in particular rivers and healthy freshwater systems, for subsistence agriculture and fisheries. The livelihoods of 70% of the continent's population is in agriculture as smallholder farmers working on parcels of land that are, on average, less than two hectares (AGRA 2017). These ecosystems require continued protection and sustainable management. As a result of developments taking place across the continent, Africa's aquatic ecosystems face a range of threats, from urbanization, deforestation, introduction of exotic plants and animals into rivers,

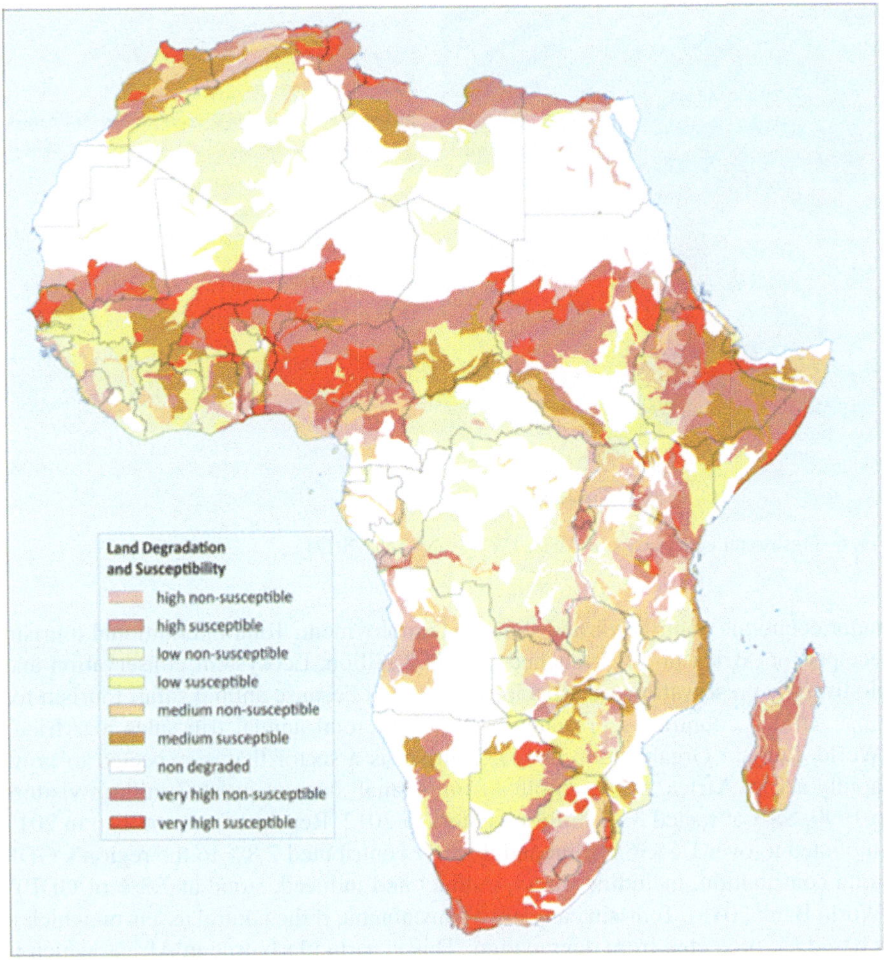

Fig. 5 Land degradation and susceptibility (UNEP 2010)

dams and lakes, over-fishing and increasing pollution. For instance, "twenty one percent (21%) of all freshwater species are threatened within Africa. Ninety one percent (91%) of these species (4,539) are endemic to the continent and are, therefore, also globally threatened" (Darwall et al. 2011).

The plethora of diverse freshwater ecosystems shown in Fig. 6, in addition to biodiversity, supports agricultural, industrial and domestic growth across the continent. These competing water uses must be managed alongside support of ecosystems on the continent.

Each freshwater ecosystem in Africa is responsible for supporting the diversity of biota as well as their respective economies and societies. For example, rich biodiversity supports the majority of the tourism sector on the continent, a

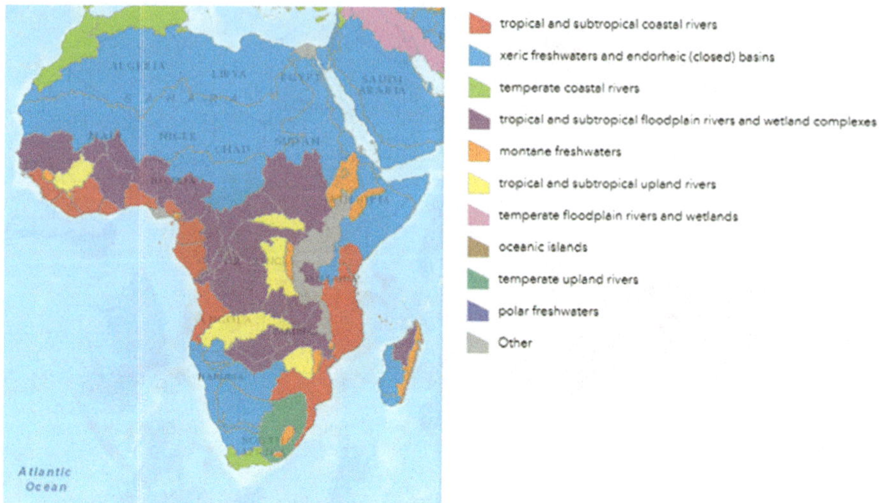

Fig. 6 Freshwater ecosystems of Africa (WWF and TNC 2013)

major economic contributor and source of employment. Total international tourism receipts for Africa in 2013 reached US$34.2 billion. Ecosystem conservation and biodiversity preservation are especially important because animal safari tourism for most African countries represents "80% of the total annual trip sales to Africa" (World Tourism Organization 2014). Tourism is a sector that is expected to grow rapidly across Africa. For example, "from a small base of just 6.7 million visitors in 1990, SSA attracted 33.8 million visitors in 2012. Receipts from tourism in 2012 amounted to over US$36 billion and directly contributed 2.8% to the region's GDP (total contribution, including direct, indirect and induced, stood at 7.3% of GDP)" (World Bank 2013). Tourism can only be sustainable if the natural assets on which it is based are protected from degradation. This is particularly true in Africa, which is marketed as a nature, wildlife, resort and cultural heritage destination (World Bank 2013).

Freshwater fisheries are another example of how livelihoods in Africa are dependent upon healthy rivers and water systems. Inland catches, which account for 25% of global catches, are especially important for food security in several countries in Africa. On average, 2.6 million tons (2006–2014) of freshwater fish are harvested across Africa's inland waters, according to a recent report by the FAO (2018). Only 17% of freshwater fish production in Africa is through aquaculture: the majority of fish production is still wholly dependent upon healthy catchments and ecosystems. In 2016, 5.6 million people in Africa were employed in the fisheries sector. Moreover, healthy inland freshwater fisheries have additional benefits, for instance, inland fisheries may represent a form of biological disease vector control. "Mosquitofish, carp and tilapia have been used in many areas to control vectors of diseases such as malaria, Zika and bilharzia through predation on the hosts of the parasites" (FAO

2018). A recent study on the Mekong has shown that should fisheries fail there (as a result of hydropower development along the river), the replacement protein and lysine from wild caught fisheries would require a significant expansion of agricultural land, which may not be available. Cambodia, for example, would require as much as 2,409,000 ha (a 43% increase) of agricultural land to replace lost lysine, and an additional 1,556,000 ha to replace protein lost by the expansion of the current crop and livestock composition of the country (Orr et al. 2012). A similar assumption can be made in Africa in terms of the additional land needed to replace wild caught fisheries. Furthermore, this represents an increase in the blue water footprint (to irrigate crops and feed livestock) rather than use of green water, which is related to wild caught fisheries.

3 Africa's Climate

Large parts of Africa are subject to seasonally variable hydrology and geographically uneven distribution of water resources. This is compounded by climate change, to which Africa is particularly vulnerable. These risks are furthermore compounded by the diversity of hydro-climatic zones across the continent; different economic sectors, such as agriculture, manufacturing and mining, are subject to different climate risks. A recent study (2011), which models the impact of climate change on economic growth, finds that water remains a significant obstacle to growth in both developed and developing countries irrespective of current water scarcity (Distefano and Kelly 2017). It concludes that efficiency gains from greater investment in demand management approaches, such as reducing the use of water-intensive goods and services, less waste and shifting water footprints to regions that are more resilient to the impact of climate change, are required. This underscores that water security risk mitigation in Africa must account for both infrastructure and management.

Climate change projections for Africa, like all climate projections, have large margins of uncertainty, but clearly trend to warming. It is therefore important to recognize that macro-scale estimates may vary significantly at regional or sub-regional scale. Nonetheless, historical data shows a warming trend across the continent since the 1960s. According to a high Representative Concentration Pathway (RCP) estimate, warming projections using moderate projections indicate that extensive areas of Africa will exceed warming by 2 °C by the last two decades of this century. It is likely that land temperatures across Africa will rise faster than global land temperature averages, particularly in more arid regions, and that the rate of increase of minimum temperatures will exceed the rate of increase of maximum temperatures (Niang et al. 2014). This air temperature warming alone has significant implications for agriculture and the agri-business sector, particularly in relation to increased crop water requirements, changes in growing seasons and water availability in general (Figs. 7 and 8).

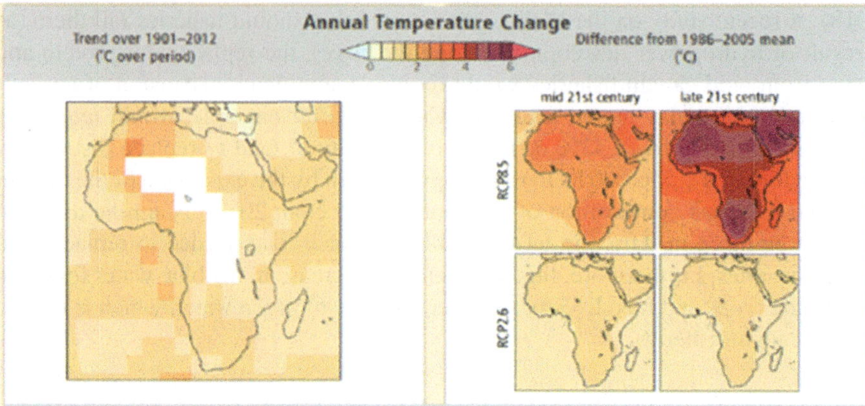

Fig. 7 Observed and projected changes in annual average temperature in Africa (Niang et al. 2014)

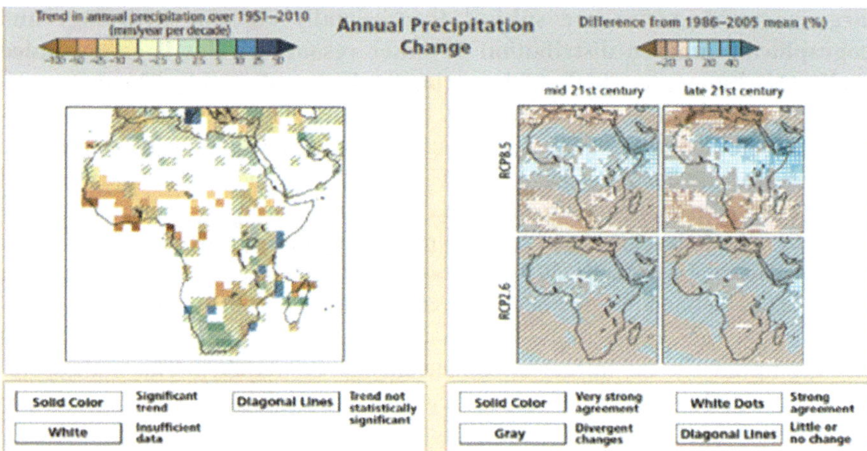

Fig. 8 Observed and projected changes in annual average precipitation in Africa (Niang et al. 2014)

Projecting rainfall changes associated with climate change in Africa is more difficult as there is no agreement between all the different climate models.[3] Key projections indicate

- very likely decreases in mean annual precipitation over southern Africa and the Mediterranean region of northern Africa in the mid- and late-twenty-first- century periods;

[3]Over two-thirds of models for the African continent do not indicate changes in mean annual precipitation that exceed the baseline variability of the models for mid- or late twenty-first-century periods for RCP2.6 (Niang et al. 2014).

- likely increases in mean annual precipitation over areas of central and eastern Africa;
- consistent annual and seasonal drying over the northern African region (including North of Morocco, Algeria, Libya, Egypt and Tunisia);
- a significant decrease over the northern basin of Tunisia in the median and 10th and 90th percentiles of precipitation in winter and spring seasons, and
- rainfall in East Africa that is at variance with the observed rainfall pattern.

Climate projections show the possibility of increased flooding and drought extremes. This has significant impacts on agricultural production in countries the weather patterns of which are impacted by these systems. An increase in the frequency of extreme events has implications for decisions about infrastructure, often requiring larger, more engineered solutions that are more expensive and can have negative environmental impact. Extreme flood, drought and disease events will increase migration pressures on refugee settlements and urban informal areas as well as fauna and flora survival.

The impact of climate change on streamflow must be considered in addition to its impact on rainfall. The amount of rain in a river that is converted into actual streamflow depends on a number of factors, including the vegetation in the catchment, the level of land degradation, the nature of the soil, slope and temperature. In South Africa, for example, only around 9% of rainfall ends up as river streamflow. The IPCC predicts that across the continent, streamflow will change from −15% to + 5% by 2050. In southern Africa in particular, almost all countries will experience a decrease in streamflow. Other areas are less reliably predictable: for example, there is no clear understanding of how streamflow in the Nile will be affected by climate change. Changes in streamflow are especially pertinent for sectors such as agriculture, manufacturing and mining, because streamflow is often translated into blue water that is dammed, piped and then irrigated or treated for domestic consumption or industrial use. The extent of streamflow change is also impacted by development and land use change in parallel with the catchment. For these reasons, management of the nature and extent of agriculture, manufacturing, mining and other land use will fundamentally change how much water is available for use.

Climate change and green water use in agriculture poses risks as well as opportunities. Erosion is a particular risk to rain-fed agriculture in the presence of climate change. In Egypt, climate change could reduce crop production by up to 28% for soybeans and 11% for rice by 2050. A reduction of 20% in the annual crop-growing period in semi-arid areas is projected by 2050. At the same time, there may also be positive impacts on agriculture, such as the projected increase in rainfall for some parts of tropical or eastern Africa (Niang et al. 2014). But coastal zones and estuaries are at particular risk from sea level rise, changes in run off and changing temperatures. A decrease in rainfall, for example, may significantly change the distance to which saltwater penetrates upstream in a river. As a result, coastal agriculture, including palm oil and coconut plantations in Benin and Cote d'Ivoire, may be affected by inundation and soil salinization. In Kenya, a 1 m sea level rise could result in a US$500 million loss of income from mangoes, cashew nuts and coconuts (Pegasys 2011).

Climate change also threatens blue water availability as a result of growing domestic consumption through urbanization. A growing body of literature generated since the AR4 suggests that climate change in Africa will have a relatively modest impact on water scarcity relative to other drivers, such as population growth, urbanization, agricultural growth and land use change (Niang et al. 2014). In Africa, increased urbanization over the next century will take the form of unplanned settlements. This will place increasing strains on already weak infrastructure and management systems. Climate change will exacerbate the effects of poor management of stormwater in unplanned settlements, which will result in flash flooding and increased ponding, which may act as a vector for water-related diseases.

4 Africa's Water Infrastructure and Institutions

In order to grow and develop sustainably going forward, Africa needs resilient infrastructure and adaptable institutions that are able to manage uncertainty. Infrastructure development in the water sector offers significant opportunities for growth and employment. For instance, "investment in small-scale projects providing access to safe water and basic sanitation in Africa could offer an estimated economic return of about US$28.4 billion a year, or nearly 5% of gross domestic product (GDP) of the continent" (UN Water 2016). However, infrastructure without capable institutions to manage investments is not acceptable, and could possibly do more harm than good. Moreover, the potential to increase productivity of resources may be limited: there are 1082 dams in South Africa with a total storage capacity of 65% of the annual runoff, and the country has just a handful of suitable sites left for development. Even these are only questionably suited to dam construction due to the high environmental costs and low projected benefit (Van Vuuren 2012). Natural climate variability and anthropogenic climate change both compound the management of water resources and require significant investment in water information, planning, governance and hard and soft infrastructure. Depending on the stage of economic development of the country, risks will differ in relation to the state of infrastructure development and institutional adaptability.

The risks and opportunities associated with water and climate change also depend on a country's "development trajectory." The stages of economic development of specific countries are characterized as **developing, emerging and developed**.

- **Developing economies**, with low levels of industrial and infrastructural development, are generally more focused on resource cultivation or extraction, such as agriculture or mining. The country's development goals typically focus on improving livelihoods, ensuring food security and developing commodity-driven exports. Agriculture is generally the biggest consumer of green water, making these countries particularly vulnerable to climate change; they are also generally in need of investment in infrastructure.

- **Emerging economies** that are growing rapidly are generally in the process of a shift from cultivation or extraction towards manufacturing. Although rural poverty challenges remain, these countries are generally undergoing increased urbanization with economic and social development leading to improved human welfare indices. Development tends to focus on ensuring sustained investment, diversification and adding value to economic activity and trade, while providing livelihoods in rural and peri-urban areas. Water and air quality may be rising on the political agenda. In this phase, blue water consumption begins to grow in industrial and urban areas; in some cases, this dynamic may increase the challenge of balancing infrastructure and institutional governance and management issues.
- **Developed economies** have generally shifted to a consumption-based, service-dominated economy supported by trade and by specialization in manufacturing and commodities, with high rates of urbanization. The focus here is on promoting growth through trade, sometimes seeking improved environmental quality, while maintaining employment, welfare and aging infrastructure. The emphasis is on use of blue water for irrigation and industrial or domestic consumption. In this phase, countries are trending to institutional mechanisms and governance from infrastructure investment to manage water use.

Water, energy and food-related challenges are different depending on a country's stage of development. This is especially true for the sectors on which countries are most economically dependent. Economic resilience is dependent upon a country's focus on the particular challenge associated with its development trajectory, and the climate- and water-related issues associated with them. For instance, developing countries may be more dependent upon green water because of their emphasis on agricultural productivity. In the face of a changing climate, the consideration of appropriate infrastructure to ensure productivity amid uncertainty is necessary. However, for more developed economies, infrastructure may already be established, in which case their focus may be on trade-offs among sectors, and managing water resources through improved governance and management frameworks.

Water-related investment choices over the next decade will have a substantial influence on the long-term development trajectory of African countries. The water-energy-food (WEF) nexus, manifested through trade-offs for water use among agriculture, energy, cities and manufacturing, will increasingly come to the fore. The concept of thresholds for development is especially important on a continent where the potential for further growth exists, but needs to be balanced alongside ecosystem health and sustainability. Currently, across Africa, there are developing, emerging and more developed economies that have reached different thresholds. The continent still requires significant development, to create a healthy social foundation. However, this can conflict with ecological ceilings, which are already beyond sustainability in some cases. Economic development decisions will fundamentally impact upon our ability to create a healthy social foundation without overstressing ecological conditions.

Development trade-offs for Africa will require management of the WEF nexus and the need to optimize natural resources for agriculture, manufacturing, urbanization and mining, together with other economic sectors including tourism, ecosystem services for livelihoods and the environment itself. Infrastructure and institutional decisions are not easily modified once made. This is especially true in Africa, where large-scale infrastructure and institutional decisions are being made in the early stages of development that may lead to weak resilience at later stages, sometimes as a result of climate change (Conway et al. 2018). Countries that have historically developed extensive water, energy and food-related infrastructure without addressing coherence or resilience questions often create significant financial sustainability challenges, especially because the infrastructure is frequently ill-suited to current conditions of population pressure and/or climate variability. In some instances, countries subject to such conditions have removed inappropriate or stranded infrastructure assets to improve natural ecosystem resilience and reduce the associated financial burden. Strengthening institutions and ensuring flexibility in the management of water, energy and agricultural land resources (through development planning) enables countries to adapt to evolving development needs and uncertain climate futures. This is closely related to demands for robust water, energy, agricultural and transportation infrastructure (grey and green) to support economic production and household access, within the context of limited financial and human resources.

5 Africa's Future Is Now

Africa faces a particular conundrum due to the vulnerability of rural populations across the continent that have limited access to infrastructure. Access to water for people living in rural environments influences the livelihoods and household security of half the population of Africa. However, as noted previously, the continent is growing and developing. Sectors including agriculture, manufacturing and mining present water-, energy- and food-related trade-offs on the continent. The development challenge is further compounded by the risks of climate change and water insecurity. Available water is becoming increasingly unusable for human consumption due to pollution and accelerated evaporation.

Decision-makers must consider the multi-faceted implications of development and growth on the continent. Priorities must account for the implications of development on the natural ecosystem functioning of the continent, because many countries in Africa are in the developing or emerging stage of their economies. Large proportions of these populations depend on healthy catchments and river systems for their food and livelihoods. One particular manifestation of the development-ecosystem trade-off are the impacts of dams, bridges and roads on the status of free-flowing rivers in Africa, an important indicator of river ecosystem health. Cambodia would require up to a 155% increase in pasture lands to replace lost fish with lysine from grazing livestock, while Laos could replace lost fish lysine by doubling pork or bovine production. For Laos, this would require an increase in pasture land of up to 47% for grazing livestock. If we were to lose the free flowing and healthy rivers and catchments of Africa, would we be able to replace the lost protein and micro-nutrients?

Better allocation of resources at system scale will lead to a greener development future. Decisions regarding where, how and when to carry out agriculture, manufacturing, mining or urbanization across Africa must be considered through the nexus lens. In addition, green, blue or grey water footprint implications must be taken into account. Africa needs to meet minimum development thresholds without exceeding ecological thresholds to feed and power the continent. Africa should benefit from being behind more developed continents on the development curve, by avoiding the mistakes developed economies have made in their development-environment trade-offs.

It is essential to develop governance mechanisms and institutions that support responsive and coherent local resource-use planning, while optimizing nexus trade-offs at the national level. The greatest tension between a country's early development decisions and its later resilience lies in the stewardship of each nation's natural capital. Some depletion of natural assets may be a necessary early strategy to create greater security of livelihoods. But soil quality, freshwater availability and quality, and forest resources all make fundamental contributions to a nation's productivity. New sources of insecurity, and enduring, weakened resilience emerge from continued economic dependence on the destruction of natural capital. Resilient economic systems are those that benefit from, reinforce and replenish the natural systems on which they ultimately depend.

Water-related investment choices made over the next decade will have a substantial influence on the long-term development trajectory of African countries. Trade-offs among water for agriculture, energy, cities and manufacturing will emerge increasingly. It is imperative that decisions regarding water resources across Africa take into account the impact of the water, energy and food nexus within countries as well as within the continent. Healthy, free-flowing rivers and freshwater habitats must support the sustainability Africa's water, energy and food in the face of growing resource demands that are compounded by climate risks.

References

Alliance for a Green Revolution in Africa (AGRA) (2017) Africa agriculture status report: the business of smallholder agriculture in sub-saharan Africa (5)

Cameron K (2010) Putting a value on Mau Forest. https://www.cabi.org/environmentalimpact/news/21066

Conway D, Curran P, Gannon K (2018) Policy brief climate risks to hydropower supply in Eastern and Southern Africa. http://www.lse.ac.uk/GranthamInstitute/wp-content/uploads/2018/08/Climate-risks-to-hydropower-supply-in-eastern-and-southern-Africa_Conway-et-al.pdf

Darwall W, Smith K, Allen D, Holland R, Harrison I, Brooks E (eds) (2011) The diversity of life in African freshwaters: under water, under threat. an analysis of the status and distribution of freshwater species throughout mainland Africa. IUCN, Cambridge, United Kingdom and Gland, Switzerland

Distefano T, Kelly S (2017) Are we in deep water? Water scarcity and its limits to economic growth. Ecol Econ 142:130–147. https://doi.org/10.1016/J.ECOLECON.2017.06.019

FAO (2010) Aquastat Database. http://www.fao.org/nr/water/aquastat/data/query/results.html

FAO (2018) The state of world fisheries and aquaculture 2018—meeting the sustainable development goals. Rome. http://www.fao.org/3/i9540en/I9540EN.pdf

Mekonnen M, Hoekstra A (2016) Four billion people facing severe water scarcity. Sci Adv

Niang I, Ruppel OC, Abdrabo MA, Essel A, Lennard C, Padgham J, Urquhart P (2014). Africa. In: Barros VR, Field CB, Dokken DJ, Mastrandrea MD, Mach KJ, Bilir TE, Chatterjee M, Ebi KL, Estrada YO, Genova RC, Girma B, Kissel ES, Levy AN, MacCracken S, Mastrandrea PR, White LL (eds) Climate change 2014: impacts, adaptation, and vulnerability. part b: regional aspects. contribution of working group ii to the fifth assessment report of the intergovernmental panel on climate change. Cambridge University Press, Cambridge, United Kingdom and New York, NY, USA, 2014

O'Sullivan JJ, Lupakisyo Mwalwiba G, Purcell PJ, Turner JN, Mtalo F (2016) Assessing sediment and water quality issues in expanding African wetlands: the case of the Mara River, Tanzania. Int J Environ Stud 73(1): 95 107. https://doi.org/10.1080/00207233.2015.1116226

Ochieng G, Seanego E, Nkwonta O (2010) Impacts of mining on water resources in South Africa: A review. Sci Res Essays 5:3351–3357

Orr S, Pittock J, Chapagain A, Dumaresq D (2012) Dams on the Mekong River: Lost fish protein and the implications for land and water resources. Global Environ. Change. https://doi.org/10.1016/j.gloenvcha.2012.06.002

Pegasys (2011) Water footprint and competitive advantage in the Nile Basin. https://studylib.net/doc/8684417/water-footprint-and-competitive-advantage-and-trade-in-th

UNECA African Water Development Report (2006) Economic Commission for Africa, United Nations. http://www.uneca.org/awich/AWDR_2006.htm

UNEP (2010) Africa Water Atlas, division of early warning and assessment (DEWA). United Nations Environment Programme (UNEP), Nairobi, Kenya. http://www.unep.org/pdf/africa_water_atlas.pdf

UN-Water (2016) World Water Development Report. http://www.unwater.org/water-drives-job-creation-economic-growth-says-new-world-water-development-report/

Van Vuuren L (2012) In the footsteps of giants–exploring the history of South Africa's large dams. Water Research Commission, South Africa. ISBN 978-1-4312-0212-6 http://www.wrc.org.za/wp-content/uploads/mdocs/Footsteps%20of%20giants_web.pdf

World Bank (2013) Africa tourism report. https://www.worldbank.org/content/dam/Worldbank/document/Africa/Report/africa-tourism-report-2013-overview.pdf

World Bank (2018) Africa's Pulse (English). Washington, D.C., World Bank Group. http://documents.worldbank.org/curated/en/881211538485130572/Africas-Pulse

World Tourism Organization (2014) Towards measuring the economic value of wildlife watching tourism in Africa, briefing paper. UNWTO, Madrid. http://sdt.unwto.org/content/unwto-briefing-wildlife-watching-tourism-africa

WWF and SAB (2018) Africa's watershed moment: how better management can underpin Africa's development. https://wwf.panda.org/?309071/Africas%2DWatershed%2DMoment

WWF and TNC (2013) Freshwater ecoregions of Africa. https://panda.maps.arcgis.com/apps/webappviewer/index.html?id=e145fa8ba32d4f76a4ab28fdc8d414f3, http://www.feow.org/

Water Footprint Network (n.d). https://waterfootprint.org/en/resources/interactive-tools/water-footprint-assessment-tool/

WWF and IFC (n.d) Water risks in Sub-Saharan Africa

Chapter 3
Water Resource Availability and Quality in the North Africa Region Under Climate Change

O. Mahjoub, M. T. Chaibi, and M. A. Shamseddin

Abstract Countries lie within the north of Sahara are the most African driest sub-region, with an average annual rainfall ranging between 50 mm/year in the southern part and 1000 mm/year in a few parts of the extreme Northwest. These conditions result in limited and unevenly distributed water resources. In Northern African countries, people and economies are increasingly threatened by the progressive change in climatic conditions over the past decades that has resulted in growing pressure on freshwater resources. Climate change is regarded as an additional burden that affects water resources of all related environments, like soil and ecological goods and services. This chapter is meant to provide a clear picture of the status of water resources and an assessment of progress made so far by Northern African countries in addressing the challenges related to the sustainable management of water resources under climate change. It also reports on current and future trends of non-conventional water resource (re)use to preserve the environment and public health from waterborne disease and to guarantee food security under variable climate conditions. Current and anticipated climate change policies and strategies are examined together, accounting for national and regional strategies for mitigating impacts of extreme events on water quality and pollution sources. Selected technical solutions and research challenges that impact water resource development and management in North Africa are discussed, accounting for current and future climate change and variability.

Keywords Climate change · Water resources · North africa · Non-conventional water · Quality

---

O. Mahjoub · M. T. Chaibi (✉)
National Research Institute for Rural Engineering Water and Forestry, BP10, 2080 Ariana, Tunisia
e-mail: thameur.chaibi@gmail.com

O. Mahjoub
e-mail: olfama@gmail.com

M. A. Shamseddin
Water Management and Irrigation Institute, University of Gezira, Wad Medani, Sudan
e-mail: shams_id@yahoo.com

S. Diop et al. (eds.), *Climate Change and Water Resources in Africa*,
https://doi.org/10.1007/978-3-030-61225-2_3

1 Introduction

Northern Africa is the Africa's driest sub-region and characterized by its limited water resources with an average annual rainfall ranging between 50 mm/year in the Southern parts and 1000 mm/year in a few parts of the extreme Northwestern area. Available conventional renewable water resources from surface water and groundwater are insufficient to sustain growing water demand. Existing challenges to water management under scarcity are compounded by the availability and seasonality of freshwater resources and overall environmental protection presented by climate change.

This chapter provides an evaluation of the progress made in the water sector by countries of the North Africa sub-region. It treats the principal aspects of water resources while providing an assessment of water resource development and utilization, including conventional (surface water and groundwater) and non-conventional (brackish water and treated wastewater) resources. It also addresses the main technical solutions that can be implemented to improve water supply and management. It highlights the research required to adequately respond to current and emerging concerns about the sustainability of water resources in the region.

2 Assessment of Water Resources

This section gives an overview of the status of conventional water resources, including renewable and non-renewable resources, in North African countries, with a quantitative assessment and analysis of critical problems related to water resource development. The focus is on updating data on water resources and on highlighting the risks that arise from climate change impacts.

2.1 Renewable Water Resources

Regional Assessment of water resources in the North Africa is based on information collected from the national reports completed from the AQUASTAT FAO database (FAO 2016) on water resources, and the World Bank data (2018). The results are presented in Table 1.

The North Africa region has a renewable water potential of about 115 km^3, out of which 47.3 km^3 are produced within the region. The total internal renewable water includes 35.66 km^3 of surface water and 18.2 km^3 of groundwater.

Water resources are unevenly distributed across the region: Egypt has slightly more than half of the total exploitable water resources available in the region, more than twice as much as the potential of the four countries that are least well-endowed with water resources (Mauritania, Algeria, Tunisia, and Libya).

Table 1 Water resources potential in North Africa region

Country	P (mm)	IRGW (km3/year)	IRSW (km3/year)	IRW (km3/year)	TRW (km3/year)	ARFW (m3/cap/year)
Algeria	89	4.49	9.76	11.25	11.67	324.3
Egypt	51	1.3	0.5	1.8	57.3	694.2
Libya	56	0.61	0,2	0.71	0.71	109
Mauritania	92	0.3	0.1	0.4	11.41	3219
Morocco	346	10	22	29	29.1	898.6
Tunisia	207	4.49	3.1	4.19	4.59	433.7
Total	–	18.19	35.66	47.35	114.78	–

P: Average rainfall, IRGW: Internal renewable groundwater, IRSW: Internal renewable surface water, IRW: Internal renewable water, TRW: Total renewable water, ARFW: Annual renewable freshwater available m3/capita. Source: FAO (2016); World Bank (2017)

Egypt and Morocco have a water potential of less than 1000 m^3/cap/year, the level generally agreed to represent chronic water scarcity, in which the lack of water begins to hamper economic development and human health and well-being. Three other countries (Algeria, Tunisia, and Libya) have already reached the threshold for absolute water scarcity, with freshwater availability falling below 500 m^3/cap/year (Fig. 1).

Figure 1 illustrates that the renewable freshwater availability is at about one-third of 1960 levels and is expected to further decrease by half of the current levels by the next decade (Intergovernmental Panel on Climate Change 2014). Only Morocco and Libya are independent of external surface water resources while Mauritania

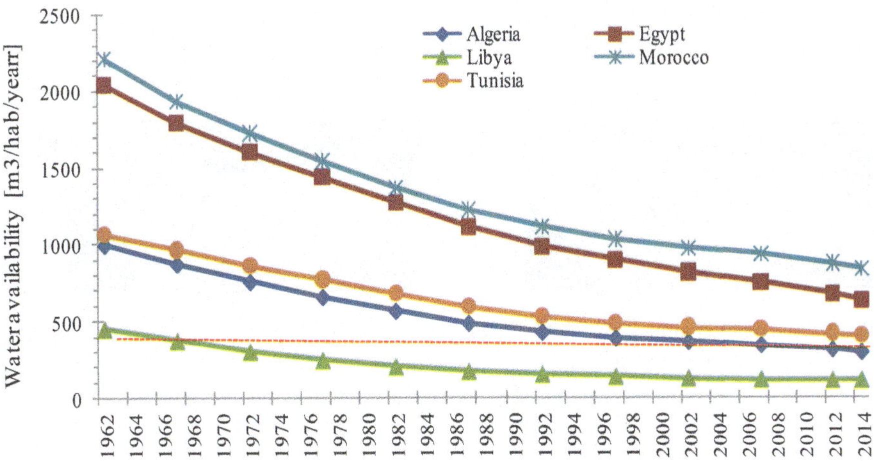

Fig. 1 Renewable freshwater resources per capita (m3/cap/year). *Source* FAO (2016); World Bank (2017)

and Egypt are highly dependent on external resources (Senegal and Nile Rivers) by 96% and 97%, respectively. Tunisia and Algeria, where 79% and 53% of domestic water resources have been already mobilized, respectively, are less dependent on transboundary water resources, to meet only 9% and 3% of their needs, respectively. North Africa is highly exposed to climate change through changing weather patterns coupled with frequent droughts. Based on projections of climate change in the region using different climate models and scenarios drawn from those used by the IPCC (2014), most countries in the region will experience a decrease in precipitation, resulting in a potential water deficit, which is already apparent in the current climate and is anticipated to become an even larger deficit in the future (Fig. 2). In the Maghreb countries, for instance, the increase in mean annual temperature is likely to exceed 2 °C, with maximum projected increases up to 6 °C, based on Representative Concentration Pathways (RCP) of 2.6 and 8.5, respectively, by the year 2100 (United Nations Development Programme, UNDP, and Global Environment Fund, GEF, 2018). Precipitation projections are highly variable: up to 40% of the reduction is forecasted to occur in the North African Region by the end of the twenty-first century. The largest decrease in precipitation is projected to occur in Morocco, with an average estimated decrease of 20 mm (9%) in 2020–2030 and about 45 mm (20%) in 2040–2050 (Arab Environment 2008). Algeria is also vulnerable to climate change, with an anticipated decrease in precipitation of about 11% for both periods. Egypt and Libya are projected to experience the smallest annual precipitation changes and can depend on precipitation within the country, and therefore will be minimally impacted by any change in precipitation. Egypt, Libya, Morocco, and Tunisia are the most exposed African countries to the Sea Level Rise (SLR). About 1 m of SLR would subject 12% of the agricultural land to risk in Egypt. If no mitigation and protection/adaptation measures were taken against the SLR in Egypt, billions of US dollars of annual damage is projected by 2100 (UNDP and GEF 2018). Also, a decrease in rainfall of 10% will lead generally to an increase in virtual water imports of about 15% in the region (Besbes et al. 2010).

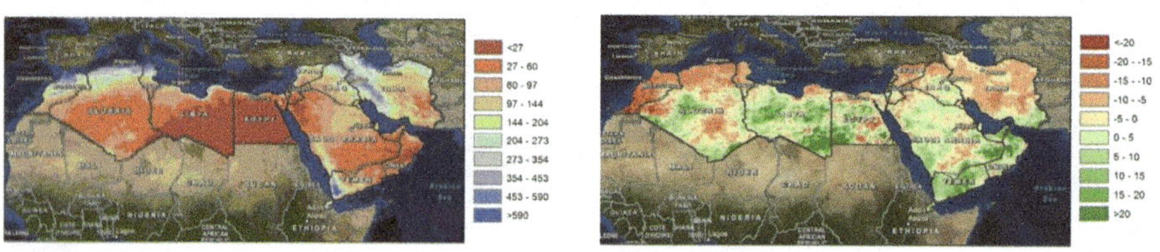

Fig. 2 Precipitation projections under climate change. Left: Precipitation under current climate and right: Precipitation anomaly 2040–2050 [%]. *Source* After (Terink et al. 2013)

2.2 Non-renewable Water Resources

A non-renewable water source is defined as groundwater reservoirs with a negligible rate of recharge for human uses. Although groundwater reservoirs may allow for the storage of huge quantities of water accumulated during the pluvial periods of Quaternary, they are not sustainable in the long term, because the current rate of recharge will slowly deplete aquifers compared to discharge ones. Moreover, the increase in the cost of pumping, as well as the deterioration of the water quality in some areas, may also make the abstraction of fossil water less attractive with time.

Except for Morocco, all sub-region countries have limited renewable water resources lie on top of important non-renewable groundwater basins which overlap with neighboring countries. The major shared aquifers in the sub-region are the Nubian Sandstone Aquifer System and The North-West Sahara Aquifer System (Fig. 3).

The Nubian Sandstone Aquifer System is one of the largest connected aquifer systems in the world and is a major North African transboundary groundwater basin, shared by Libya, Egypt, Chad, and Sudan. It covers an area of around 2 million km^2 and contains approximately 150,000 km^3 of economically exploitable groundwater. Over recent decades, these aquifers have been increasingly exploited by Egypt and intensive projects in Libya, e.g., "The Great Man-made River Project." The North-West Sahara Aquifer System (NWSAS) is approximately half the size of the Nubian Sandstone Aquifer System; the NWSAS is shared by Algeria, Tunisia, and Libya, composing of two main aquifers, the Complex Terminal (CT) and the deeper Continental Intercalaire (CI), with a total estimated volume of 60,000 km^3 (Bakhbakhi 2006). The groundwater depletion in both aquifer systems is estimated to be about

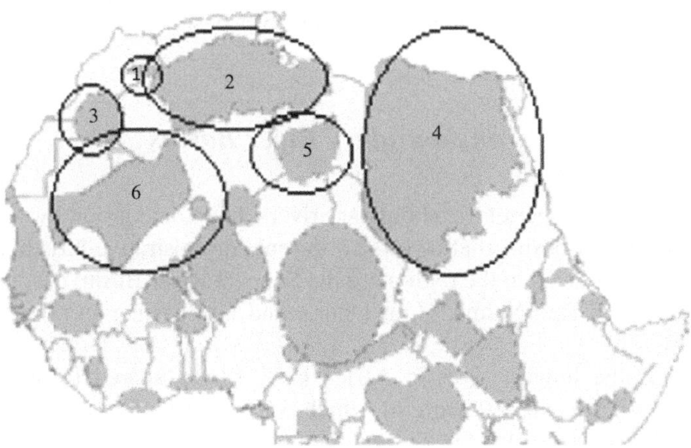

Fig. 3 Main international shared aquifers in North Africa. 1: Errachidia Basin, 2: North-Western Sahara Aquifer System, 3: Tindouf Aquifer, 4: Nubian Sandstone Aquifer System, 5: Mourzouk-Djado Basin, 6: Taoudeeni Basin. After [International Groundwater Resources Assessment Centre (IGRAC), 2012]

1 mm depth depletion of an equivalent layer of water over the total horizontal extent of these aquifer systems.

North Africa also has a small aquifer system called Murzuq Basin. This aquifer is located mainly in the southwest part of Libya, with small parts in Algeria, Chad, and Niger. It covers an area of 450,000 km^2 and contains an estimated 4,800 km^3 of exploitable groundwater.

Data on non-renewable water sources in the region are often limited and, in some cases, ignored, due to the incomplete information provided by national water authorities. This results in unreliable estimates, especially where these resources are treated as an unrestricted resource and users behave in an individually competitive manner. Libya is a remarkable example: the mining extraction consumes 90% of its groundwater supply.

Mazzoni et al. (2018) have developed a model to quantify and forecast water deficits and depletion in the main exploitable fossil aquifers in North Africa under different climatic and socioeconomic scenarios from 2016 to 2050. Their results show that Egypt and Libya will experience, in the upcoming few decades, severe water deficits of about 45% and 90% of their current water budget in 2050, respectively, and may reach a total depletion level in 200–350 years under the currently extraction rates.

3 International Shared River Basins

This section covers transboundary water resources, including international river and aquifer basins shared by the countries of the North African region. It explores the impacts of climate variability through changes in precipitation and temperature of the Nile, which is considered the main African river basin.

3.1 Assessment of Transboundary River Basins

North Africa has a number of transboundary rivers, lakes, and groundwater systems, as well as in-country basins that traverse different administrative jurisdictions. The major shared river basins in the region are the Senegal River Basin; Medjerda River Basin; Tafna, Oued Ben Naima, Guir, Doura, and Dra basins; and the Nile River Basin (Fig. 4).

The Nile is the longest river in Africa (6,700 km), shared by eleven countries covering an area of 8.9 km^2, of which 35% contributes to the Nile Basin, sustaining life for 20% of Africa's population (The Nile Basin Initiative (NBI) 2017). The NBI is the coordination platform shared by the Nile riparian countries since 1999. The basin is divided generally into two systems: the Blue Nile (87% of the basin area), which includes four sub-basins (Blue/Abbay, Atbra/Tekeze, Sobat/Baro/Akobo, and Main Nile) whose total seasonal runoffs contribute almost

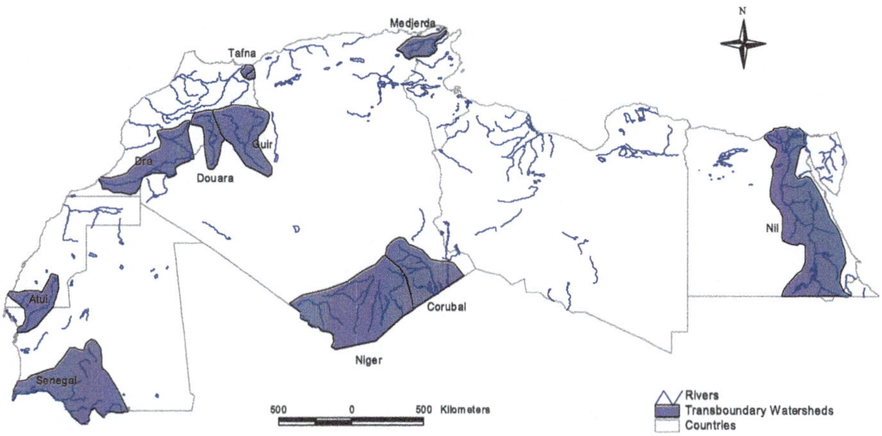

Fig. 4 International shared river basins crossing the region countries borders. *Source* PNUD (2016)

85% of Nile River flows. The Equatorial System, which originates in the Equatorial Lake Plateau, has six sub-basins (Lake Victoria, Victoria-Nile, Lake Albert, Bahar Eljebel, Bahr Alghazal, and White Nile). The Equatorial System contains the world's second-largest wetland/Sudd, where the White Nile River loses 57% of its flows by evaporation before joining the Blue Nile in Khartoum (Sudan), forming the main Nile River that flows North to meet the Mediterranean Sea in Alexandria (Egypt) (NBI 2017).

3.2 *Climate Change Impacts on the Nile River*

The 2008 independent report of Blackmore and Whittington (2008) stresses that any decision pertinent to Nile water has profound effects on poverty, development, and security in the region. This demands a cooperative framework to manage prevailing water scarcity conditions. Passive decisions and pathways that occur as a result of climate change would likely have profoundly detrimental impacts on the region and will drive water management. Water supply and demand in a changing climate will be a critical concern for assuring long-term sustainability in the region.

The Nile River traverses substantially different climatic zones and ecosystems, ranging from super-humid to desert. Annual rainfall shows a gradual increase in the North-South direction with the mean for the basin of 571 mm, peaking in July–August with a variation coefficient of 86% during the period 1901–2015. Spatially, the equatorial basins are receiving almost 67% of the annual rainfall of the whole basin. Furthermore, downstream countries (Sudan and Egypt) are the driest parts in the region and rely heavily on the transboundary-generated runoff.

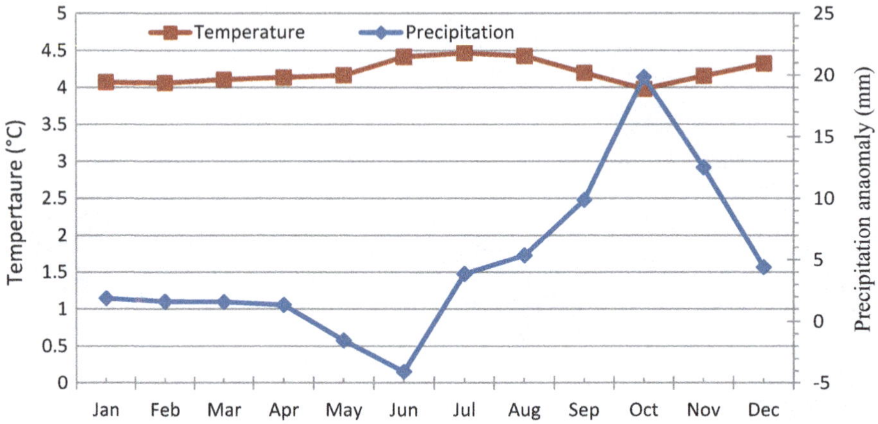

Fig. 5 Monthly anomalies in median temperature and rainfall by the year 2099, relative to 1986–2005, the Nile Basin

The Nile Basin has been subjected to frequent wet-dry cycles for thousands of years. Projected climate change impacts on the basin's rainfall vary in spatio-temporal patterns. Figure 5 presents temperature and rainfall projections from 35 Global Climate Models (GCMs), downloaded from the World Bank Climate Portal under unmitigated scenario conditions, the RCP 8.5. Wetter conditions are being projected by the year 2099 (an increase of 57 mm in the basin's annual median rainfall), accounting for an increase in monthly rainfalls, with the exception of May and June. This is coupled, however, with a substantial increase of 4.2 °C in temperature, suggesting an increase of 13.6 mm in rainfall per 1 °C increase in temperature. Consequently, total rainwater volume of 1,413 km^3/year for a basin area of 3,173,324 km^2 is projected. This is 9% lower than the current volume documented in Nile-based studies (1600–2000 km^3/year), attributed to differences in time windows considered and in delineated areas of sub-basins.

The climatic changes in rainfall and temperature have impacted Evapotranspiration (ET), the second-largest hydrological flow (Bastiaanssen et al. 2014), only marginally. Current basin-wide actual ET is estimated at 1800–2000 km^3/year (Bastiaanssen et al. 2014; FAO 2011; Kirby et al. 2010). The study of Alemu et al. (2015) states a significant positive relationship ($P < 0.05$) between ET and rainfall for 65% of the basin, showing contradicted trends in ET. Analysis of simple linear temperature-ET and rainfall-ET relationships based on monthly datasets results in a total ET volume of 2,337 and 2,215 km^3/year, respectively, for the basin area, compared to a baseline of 2,013 km^3/year, an increase of 16 and 10%, assuming no change in current land uses. As a drought adaptation measure, however, the irrigated area in the eastern Nile Basin is projected to increase by 3.8 million ha with crop water requirements of 49 km^3/year, which is almost equal to the total Blue Nile annual flow (Beyene et al. 2009). Thus, better water use efficiency is urgently required. Generally, the runoff corresponds to changes in ET and rainfall. Hagemann et al. (2011) indicated that a bias-corrected projection of increased rainfall is associated with increased ET in the Nile Basin, and estimate the current climatology runoff (rainfall-ET) during the period 1961–1990 at 155 km^3/year for a basin area of 3,173,324 km^2 compared to the current Nile flows of 84 km^3/year measured at the High Aswan Dam (Egypt). Using the same approach with unbiased corrected data sets, the projected wide-basin runoff is estimated to be 83 km^3/year by the year 2099 (a 24% decrease relative to the baseline) with nil contribution from the Main Nile, White Nile, and the Bahr Eljebel and Bahr Alghazal sub-basins. In contrast, the Blue Nile sub-basin provides the highest contribution, up by nearly 67%, followed by Lake Victoria (15%). Thus, the increased ET would outweigh the increased rainfall. Regionalization of climate change impacts on the hydrology of the Nile Basin requires sophisticated bias-corrected and observational datasets, which may change the projected runoff signal. The availability of observational datasets is a real challenge considering the limited number of gauge stations and the lack of regional data sharing measurement protocols so far. Nevertheless, the statistical bias correction approach could be successful (Nawaz et al. 2010).

4 Non-conventional Water Sources

This section provides an overview of water resource quality and its vulnerability to climate change. It highlights current and future trends of the (re)use of unconventional water resources, including desalinated water and wastewater.

4.1 Desalinated Water

Desalination of seawater and brackish water can be used to meet the increasing demand for freshwater. Prospects for desalination vary greatly among the North African countries, depending on their incomes and water scarcity. Algeria and Libya, with high availability and low costs of energy, are likely to be the main drivers for future seawater desalination. Current information indicates that desalination capacity in the region is about 6.8 10^6 m^3/day, largely contributed by Algeria at a rate of about 2.8 10^6 m^3/day. For the time being, Morocco and Egypt have sufficient conventional water resources to satisfy most municipal needs. However, desalination tends to be necessary for coping with serious shortages in some coastal areas, such as the Egyptian Red Sea coast and Southern Morocco, with production capacities of 1.995 and 0.568 million^3/day, respectively. Tunisia, with its limited energy and water resources, has considered desalination to be a tool to bridge the gap between water supply and demand within an integrated water resource management framework, rather than a solution to solve water scarcity in the country. The capacity of desalination is estimated to be 200,000 m^3/day. Figure 6 shows the distribution of the overall desalination capacity among countries of the region.

Fig. 6 Desalination capacities per country. *Source* GWI (2017)

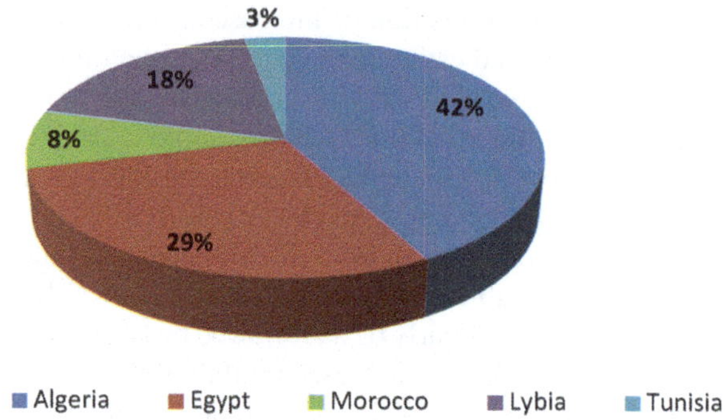

Desalination plants present particular vulnerability to climate impacts. Over the expected lifetime of desalination facilities, generally located on the coast, sea levels could rise significantly, which affects desalination plant design and operation (Cooley et al. 2006). Notwithstanding this reality, consideration of the innovative design necessary to adapt to climate change is rarely discussed before permission is granted for plant construction. Major studies have reported that the energy requirements of desalination that is powered by fossil fuels contribute to the increase of greenhouse gas emissions and climate change. Moreover, using water desalination plants introduces serious environmental risks, including an increased threat to sea life and disruption of the ecosystem.

Desalination plants could be made more energy-efficient and powered by renewable energies. There are several examples of desalination plants using renewable systems to provide heat or electrical energy. Most of them are demonstration plants with a capacity of less than 50 m^3/day (Chaibi 2000). One of the largest renewable energy-driven desalination plants is currently under construction in Morocco. This plant is designed for drinking and irrigation with a capacity of 275,000 m^3/day, with the potential for capacity expansion of up to 450,000 m^3/day.

In many parts of the region, alternative water sources could be provided to meet future water needs at lower economic and environmental cost than desalination. These alternatives include optimization of existing water supplies by adopting water conservation and efficiency, and implementation of inter-basin water transfer.

4.2 Wastewater and Reuse

Wastewater management is one of the pillars of the "Integrated Water Resources Management" (IWRM) concept promoted by Target 6.5.1 of SDG 6 (Sustainable Water and Sanitation); in order to succeed, wastewater reuse must be integrated into water strategies and be included in each country's water budget (League of the Arab States 2019). Within this framework, wastewater treatment, water recycling, and demand management measures have to be introduced to overcome the challenges of inadequate supply (UN Water 2008). By 2020, 50% of countries in the region are expected to be applying Best Available Technologies (BAT) and Best Environmental Practices (BEP) in non-conventional water projects and have national systems to integrate water reuse in their national IWRM plans. A regional program will be implemented by 2020 to increase the technological awareness on water reuse in the North Africa Region as a whole. However, the feasibility of wastewater reuse depends on the development of local capacity for making the required technology affordable and environmentally friendly (League of the Arab States 2019). There is the danger that climate change will introduce adverse impacts on water reclamation, either directly (by compromising technological performance) or indirectly (by interfering with management and operation activities), posing a threat to the wastewater reclamation industry. The impact of temperature, rainfall and SLR, greenhouse gases (carbon dioxide and methane formed from the anaerobic decomposition of organic matter and nitrous oxide from the nitrification and denitrification processes) are not new, having been discussed in numerous studies worldwide. Wastewater generally has a buffer capacity to tolerate slight fluctuations of temperature, but temperatures outside of the optimal range can affect biological processes, especially nitrifying bacteria (Vo et al. 2014). Warm temperatures in particular may create conditions conducive to corrosion of raw wastewater pipelines, by forming hydrogen sulfide, which may increase fermentation of solids in sludge thickeners, which causes odor issues. Regarding rainfall, heavy rain can affect the performance of wastewater treatment processes by increasing the pollutant concentrations, suspended materials, and sediments at the beginning of the storm event (Vo et al. 2014).

In general, Northern African countries have evolved very differently and at a different pace in developing their water sector. Even though some countries have established or are establishing water plans (policies, master plans, and strategies), they are still struggling to implement a water policy that integrates reuse in the IWRM strategy. Apart from Egypt and Tunisia that are the most advanced in this process, few countries in the region have been able to implement substantial wastewater treatment and reuse programs (Jagannathan et al. 2009). Tunisia has made steps toward integrating wastewater reuse in its water management strategy since the early 1990s, and Egypt adopted a national plan that incorporates integrated water resources management. Algeria has started to integrate reuse in its water policy, while for Libya, there is no indication that wastewater reuse is fully integrated into its water policy. The least advanced country is Mauritania (AWC/UNDP/CEDARE 2005) because of a lack of data on water policy as a whole. In general, the reasons for poor strategy development and implementation are political, institutional, financial, and technical.

Tunisia has shifted from water resource mobilization to a water demand management-based strategy through the introduction of incentives, such as pricing and financing, in addition to technical, legal, and institutional measures (Ben Abdallah 2003). In Tunisia, the reuse of reclaimed water has been an integral part of the National Water Resources Strategy since the 1990s, after the severe drought there in 1989. In the early 2000s, a National Strategy for Wastewater Reuse was developed to promote reuse for agricultural irrigation and other outlets. In 2018, the country had 119 treatment plants producing mainly secondary biologically treated effluents (Gharbi 2012). It is estimated that for the sanitation sector, greenhouse gas emissions from the production and management of wastewater and sludge contribute about 84% of the methane, while 16% is due to energy consumption related to sanitation services. Moreover, it is estimated that 0.79 million tCO_2 will be emitted by the sanitation sector in 2020 and 0.94 million tCO_2 are expected for the year 2030. Tunisia is planning to stabilize greenhouse gas by 2050. These measures are included under the program, "Nationally Appropriate Mitigation Action" (NAMA), supported by Germany (Fadhel 2017). Of the existing 119 wastewater treatment plants (WWTPs), only 30 WWTPs are supplying 31 irrigated areas with water for agricultural reuse. Taking into account the rate of intensification, the percentage of reuse for agricultural irrigation and municipal purposes (golf and garden irrigation) is around 9%, while the five-year National Development Plan (2016–2020) targeted a rate of reuse of 50%. Today, about 8100 ha of fruit trees and fodder crops, and 1490 ha of landscape (1040 ha of golf course and 450 ha of green areas) are irrigated by secondarily treated effluent (Gharbi 2012). The regulatory framework and the Tunisian National Quality Standards (NT 106.03) were set in 1989, and may also have to be revised to account for actual use. In the meantime, a decree was passed in 2018 with more relaxed standards related to the discharge of effluents in water bodies.

In addition to agricultural reuse, aquifer recharge using reclaimed water is being adopted to replenish depleted aquifers and to counterbalance the decline in groundwater level caused by over-pumping, especially in the coastal areas. Over-pumping is expected to increase in the coming years due to climate change. Because of the uneven distribution of rainfall over time and the increasing trend of extreme events, there are seasonal excesses of surface water. The availability of treated effluents on the supply side, including Managed Aquifer Recharge (MAR), can protect, prolong, sustain, or augment groundwater supplies. As one of a suite of IWRM strategies, this expands local water resources, reduces evaporation loss, and helps to replenish depleted aquifers (Dillon et al. 2014).

In 2018, a National Action Plan on reuse was launched within the Water Partnership Program, supported by the World Bank, to promote reuse as an integral part of water resource management in Tunisia (SCP 2017). The plan was developed because the country has undergone several extreme events,, i.e., drought and flash floods, between 2016 and 2018, resulting in considerable damage to the agricultural sector and infrastructure. Indeed, the 30 main dams in the country were filled to only 28% of capacity in November 2017, and 32% in November 2018. This water shortage has forced the government to restrict use of irrigation water and growing certain crops that require high water consumption (ONAGRI 2018).

Egypt has progressed in integrating wastewater reuse in its water policy. Since the Nile provides Egypt with 97% of its agricultural irrigation requirements, a huge amount of drainage water coming from agriculture is considered to be wastewater and is reused indirectly by mixing it with treated and untreated domestic wastewater, as well as with industrial effluents disposed in agricultural drainage canals (Abdel Wahaab and Omar 2011). However, the Nile Basin is already forecast to undergo wetter conditions and an increase of up to 4.2 °C in temperature by 2099. As of 2011, wastewater produced was about 7 billion m^3, of which 3.7 billion m^3 was untreated, 2.4 billion m^3 was secondarily treated, 0.9 billion m^3 was primarily treated and only 0.068 billion m^3 was tertiarily treated. Of 3.368 billion m^3 of treated wastewater, only 0.271 billion m^3 was reused directly for agriculture, while the remaining amount was disposed into the drainage network. 2011 is considered to be the baseline for developing a strategic future vision. Under this strategy, 5.82 billion m^3 will be used directly in agricultural expansion areas, while 5.53 billion m^3 will be disposed into drains. About 0.6 million ha will be reclaimed for cultivation, according to the 2030 Sustainable Agriculture Strategy, with a total average annual water requirement of about 5.42 billion m^3. Based on this vision, the secondarily treated wastewater produced in 2030 will satisfy water requirements. The Strategic Vision estimates an additional 0.61 million ha available to reclaim based on the remaining potential of secondarily treated wastewater of 0.4 billion m^3 from desert front governorates and 5.53 billion m^3 from delta governorates at an estimated water requirement of about 9800 m^3/ha/year (Abu Zeid and Elrawady 2012).

Currently, Egypt has 400 WWTPs, producing more than 10.67 billion m^3 of effluents (WWHC 2018). About 40% of sewage water produced is not treated, because its reuse is of concern (Qadir et al. 2009). Unofficial wastewater reuse is significant and uncontrolled by the government and poses threats to human health and the environment. Indeed, only 56% of the urban population is connected to sewage systems (WWHC 2018), and a significant volume of wastewater enters directly into water bodies without any treatment. Reuse of treated wastewater is limited to government schemes to irrigate forests and establish a green belt around the capital (ATH Group 2009). The Egyptian Code for the Use of Treated Wastewater in Agriculture (501/2005) regulates the direct use of wastewater, but not its discharge into drains (Abdel Wahaab and Omar 2011). The Ministry of Agriculture and Land Reclamation has implemented reform that encourages reuse for woodlands. About 1,000 acres have been irrigated in Luxor, Qena, New Valley, Edfu, Ismailia, Sadat City, and South Sinai (Zimmo and Imseih 2010). In addition to forest areas, several crops are irrigated using reused water, including sunflower, jatropha, and casuarinas (Abdel Wahaab and Omar 2011).

In Morocco, the baseline scenario shows that most basins will experience a water deficit by 2030. In the past and through the present day, the lack of water resources and the increasing demand of food production has constrained a large part of the population to use raw reclaimed water for irrigation, sometimes mixed with water from wadis, but rarely properly treated. Hence, the use of raw wastewater for agricultural irrigation is a common practice (Zimmo and Imseih 2010). However, the practice started regressing after the establishment of the National Programme for Sanitation and Wastewater Treatment (PNAL) in 2006. PNAL has substantially raised the number of wastewater treatment plants from 18 in 2005 to 117 in 2016. The majority (74%) of these WWTPs are operating an extensive process based on natural stabilization ponds. Today, 50% and 37% of the effluents are secondarily and tertiarily treated, respectively. The waste sector as a whole contributes around 0.05% of greenhouse gas emissions and output has increased by 2% between 1990 and 2012.

Most of the wastewater produced by inland towns is used to irrigate about 7,235 ha of crops after insufficient or even no treatment. There exist about 70 irrigated areas, located mainly in Marrakech (2000 ha), Meknes (1400 ha), and Oujda (1175 ha). Some pilot projects were launched in Fez, Al Attaouia, and Drarga, which include construction of innovative wastewater treatment plants (UN Water 2008). In 2017, 45% of the produced effluents were collected and treated (compared to 8% in 2005) and 60% of effluents were discharged into the sea, while the rest went to other water bodies (DRPE 2017).

About 70 million m^3 of raw wastewater has been used every year to irrigate fodders, cereals, and fruit trees. Irrigated lands are usually located at downstream effluent discharge points, around treatment plants. Irrigation of vegetables is forbidden in principle, but the ban is generally not respected (Chouk-Aallah and Hamdy 2005). Recent research studies have contributed to improving the reuse for various purposes. A total of 47.6 million m^3 of treated effluents are planned for golf courses (52%), agricultural irrigation (25%), and industrial use (15%) (DRPE 2017). In 2016, about 550 million m^3 of effluents were produced. This figure is expected to reach more than 600 million m^3 in 2020 and about 740 million m^3 in 2030. The potential volume available for reuse by 2030 is estimated at 325 million m^3, with a majority in the Souss Massa and Sebou watersheds and a target rate of reuse of 90% (DRPE 2017).

In Algeria, treated wastewater represents a promising alternative water resource that can be used to irrigate water shortage areas and replenish aquifers in regions where groundwater is overexploited. In 1985, 660 million m^3 were discharged because of deficient sanitary services and the low rate of connection to the sewer system. In 2007, only 48% of the population was connected to urban wastewater treatment plants. Consistent with these deficiencies, the four priority population types for the construction of WWTPs are (i) communities of more than 100,000 inhabitants; (ii) those situated upstream of dams already existing or in construction (iii) cities along the coast; and (iv) agglomerations of fewer than 10,000 inhabitants. In 2009, about 86% of the urban population had access to sanitation. A program was implemented to rehabilitate the sewage network in 12 cities, construct 44 sewage treatment plants, and build 42 stabilization ponds to reach a treatment capacity of 1.2 billion m3/year by 2020 (Raouan Hacen et al. 2011). As of 2016, 136 WWTPs with a capacity of 1.4 million m^3/day were producing 207 million m^3 per year of treated effluents per year (Khachaba 2017), compared to 700 million m^3 per year of wastewater treated in 2011 (ONA 2012). Natural and aerated ponds are the main treatment processes in place with 70 WWTPs, in addition to activated sludge (63 WWTPs) (Khachaba 2017). The waste sector accounts for 10% of greenhouse emissions in Algeria.

The rate of reuse of treated wastewater in Algeria is still very low, with only 3.2% of the produced effluents reused (Guardiola-Claramonte et al. 2012). The situation in Algeria is a challenging one, although the National Plan for Water 2005 provides a good foundation for the implementation of Integrated Water Resources Management (AWC/UNDP/CEDARE 2005). The reuse of treated wastewater was abandoned for a long time in some regions, because of malfunctioning wastewater treatment plants. With the restoration of the old treatment plants and the construction of new plants, several irrigation projects using treated wastewater are being studied or already realized. Currently, the treated effluents of 18 WWTPs, corresponding to a volume of 21 million m^3, are used for the irrigation of 11,212 ha of agricultural land. The Sanitation Utility (ONA) has put in place strategies to monitor wastewater quality produced by treatment plants and is planning to upgrade some of them with tertiary treatment in order to allow non-restrictive reuse with minimum risk to human health (Khachaba 2017). Two projects are underway: (i) Hennaya in Tlemcen, with an irrigated area of 912 ha of fruit trees and fodders; and (ii) Boumerdès, with 125 ha of fruit trees. The current program projects that 150 to 216 plants will be functioning in 2020, with a capacity of 1200 million m^3/year of clear water. 100,000 ha are expected to be irrigated from these new sources.

In Libya, wastewater barely represents 1% of total water resources (Abu Zeid and Elrawady 2012). From the 1970s until the 1990s, the availability of treatment capacity trended up to 484,735 m^3/day in 1985. In the period 1965–1995, 25 treatment plants were constructed. However, only three were operating effectively, which resulted in a large gap between the designed and the actual capacity of treatment (Wheida and Verhoeven 2005). In general, wastewater is discharged partially treated or untreated into the environment. Data on the quality of the produced and discharged wastewater are not available due to a lack of trained people and facilities (Wheida and Verhoeven 2005) to produce this information. In 1999, a volume of 546 million m^3 of wastewater was produced and only 40 million m^3 was treated and reused (Femia and Werrell 2013). Tripoli and Benghazi used to be the main areas of reuse of reclaimed water with 6000 ha, where its application was limited to irrigated crops, fruit trees, and animal fodder (Wheida and Verhoeven 2005).

In the 1990s, many of the wastewater treatment plants ceased functioning because of lack of maintenance, especially in big cities like Benghazi and Tripoli. In other regions, the lack of a sewage network and the poor access of the population to good sanitation has resulted in the discharge of urban sewage water directly into the environment, including by large industries. Similarly, small industry with unregulated discharge has caused environmental problems (DG Environment European Commission 2006). In some big cities like Tripoli, runoff is not collected, finding its way to the sea with a heavy load of pollutants, according to the National Water Resources Management Strategy, a five-year national program launched in 2004 to enable all Libyan cities to connect to the sewer network. However, in North African countries, political unrest has limited the availability of updated data on the sector.

For Mauritania, one of the drought effects is the lack of clean water for tens of thousands of rural people, with associated health problems. Very few data are available on wastewater collection, treatment, and reuse for irrigation. In 1999, 1 million m^3 of wastewater was collected and treated, but not reused. A single treatment plant was operating with a capacity of 2000 m^3/day (ECA Office for North Africa 2005). About 65% of the population had no access to sanitation (Abu Zeid and Elrawady 2012) with no clean water in rural areas, unsurprisingly the source of significant health problems (African Development Bank 2012). In 2007, only 38% of the population had access to sanitation, with the greatest severity in rural areas. With assistance by international programs, Mauritania planned to rehabilitate its treatment plants. In 2010, 21.8% of the population had access to improved sanitation (African Development Bank 2015). Recently, Mauritania was encouraged to implement an IWRM, including projects to improve access to water and sanitation.

5 Water Resource Quality

In North African countries, irrigated agriculture was developed to counterbalance the complete reliance of food and non-food crop production on rainfall, which has become increasingly variable across time (distribution across seasons), space (geographic distribution), and amount (quantity and intensity). Water quality is also a crucial factor in irrigated agricultural production and food security. A decline in water quality may even threaten the political stability of the region, especially where numerous countries share the resource, which is the case in the southern part of North Africa.

Within a country, there are conflicting interests between economic development and climate change agendas. Hence, the effects of climate change are aggravating existing quantitative and qualitative pressures on water resources. Consequently, the availability of surface and groundwater are expected to decline with a direct effect on water allocation and the environment. A drop in available water quantity can be driven by both the degradation of the water quality, making insufficient the available amounts of water resources suitable for a given use, and the environmental and health risks to humans and the ecosystem.

Water resource quality in North Africa is largely affected by multiple sources of pollution that can be either natural or anthropogenic; the latter is mainly caused by domestic and economic activity like industry, urbanization, tourism, and agriculture. Since anthropogenic causes are predominant, they result in a wide range of environmental and health impacts across the region. Point and non-point sources of pollution releasing physical (organic matter, suspended matter), chemical (nutrients, heavy metals, emerging pollutants), and biological (microorganisms) contaminants are widespread.

One of the most important water quality parameters regularly monitored is salinity. It reflects on the load of mineral species in the water. Any increase in salinity and/or change in the composition of salts may have negative impacts on the sustainability of a water resource and activities that depend on them, including domestic uses.

The economies of North African countries rely on the agricultural sector, which consumes about 80% of the countries' water resources. The expansion of irrigated agricultural land and the use of traditional irrigation techniques, e.g., surface irrigation, is inevitably leading to the use of large amounts of irrigation water and the loss of large amounts by both runoff and drainage.

Climate change can substantially affect the regenerative capacity of the environment, which reduces water quality. The lack of surface water due to severe drought and high evaporation increases reliance on groundwater resources. However, groundwater resources will be diminished to satisfy agricultural water demand. Salinization of irrigation water, estuaries, and freshwater systems, and seawater intrusion in coastal aquifers, could threaten the sustainability of urban and agricultural activity due to the substantial reduction of freshwater availability. Shallow aquifers are substantially affected by salinity, leading to agricultural soil salinization and loss of fertility, which will reduce the amount of land that is suitable for cultivation, and loss of land in the long term. In Algeria, the salinity of surface waters can reach up to 1.5 g/L, but in general is around 1 g/L. Groundwater is generally non-saline worldwide, but it can reach up to 8 g/L. In Tunisia, around 60% of water resources are located in the North, of which about 70% has salinity below 1.5 g/L. More than 80% of groundwater has salinity exceeding 1.5 g/L, with 30% of the shallow aquifers with salinity higher than 4 g/L. Groundwater, especially the aquifers located in the Southern part of the country, has shown an excess of sulfates and chlorides due to rock weathering (MARHP 2017). Water resources in the South of the country are composed mainly of three deep transboundary aquifers. The terminal complex aquifer is depleting from intensive exploitation of groundwater; salinity varies from 1.5 g/L in Nefzaoua to 4 g/L on the peninsula of Kebili. The intercalary continental aquifer has salinity that varies from 2.5 to 3 g/L at the Chott Fedjej to 5 g/L at El Borma, while the Djeffara aquifer has salinity between 3 and 8 g/L (CEDARE 2014). Therefore, salinity represents the main hurdle to a number of economic activities and even to the supply of potable water. Climate change is forecast to affect the quality of these resources, especially in the South. This has led the government to build desalination treatment plants to assure sufficient water supply (CEDARE 2014; MARHP 2017).

Apart from salinity, there are multiple water quality parameters that are monitored on a regular basis. The largest surface water area in the North African region is the Nile River in Egypt. It used to be a freshwater river, with less than 0.5 mg/L of salinity (Elnazer et al. 2018). Now, it is under great anthropogenic pressure. The major sources of water pollution to the Nile are the discharge of untreated or partially treated wastewater, including industrial and domestic effluents, and the release of drainage water from agriculture, leaching pesticides and nutrients. Water quality degradation has become of serious concern. Many factors contribute to this pollutant load, including flow, population density associated with the low rate of connection to sanitation systems (illegal construction with leaking pipes), industrial activity, and social and economic conditions. Microbiological pollutants are frequently detected in some drains and discharge into the Nile as raw effluents that usually find their way into agricultural drains (Dahshan et al. 2016; Abdel-Satar et al. 2017). Several studies highlight the variability of water quality of the Nile River. Relatively good water quality is found in the upstream part at Dakahlia Governorate, while in the Damietta Governorate, various sources of pollution are observed, including domestic wastewater, industrial discharge, agricultural runoff, and drainage water (Badr et al. 2016). In Alexandria and Behiera governorates, Mahmoudia canal receives water from the Rosetta branch, which in turn receives domestic and agricultural waste from the Zarcon Drain and other non-point sources. In that area, the situation has become problematic since potable water is supplied by the Mahmoudia canal. Water quality analysis shows that ammonia concentrations exceed permissible limits. Moreover, the Chemical Oxygen Demand (COD) and the Biochemical Oxygen Demand (BOD) exceed national standards, making them unsuitable for drinking water and also for irrigation and fisheries (Abdullah and Hussona 2014). According to WHO, Egypt has one of the highest rates of death related to water pollution, because 95.5% of Egypt's population drinks improperly treated water due to poor sanitation and widespread pollution. Contamination with cyanides represents a threat to human health in Assiut (Hassan and Farghali 2018). Water quality has also deteriorated downstream of El-Salam Canal (one of the largest agricultural canals) due to the polluted water discharged from canal feeders (e.g., the El-Serw and Bahr Hadous drains) (Assar et al. 2019). Studies on water quality from Qena to Sohag districts show high concentrations of Pb and to some extent Cd, Cr, As, Cd, Cr, and Pb, making the water unsuitable for drinking. However, it is likely still suitable for irrigation, despite high concentrations of As and Cr (Alnazer et al. 2018). Agricultural drainage water represents the largest amount of wastewater in Egypt flowing into the Nile River: 4.9% of total water resources. Drainage water and wastewater of domestic, industrial, and commercial activities mix in the agricultural drains and are reused for irrigation (Abdel Wahaab and Omar 2011), which represent a threat to consumers of the resulting crops.

In Morocco, the overall quality of the river basin water shows that almost 50% of surface and groundwater resources are of fairly good to excellent quality. Therefore, protection of water quality is a government priority. Morocco started experiencing quality degradation of its surface and groundwater resources in the early 1990s. The discharge of domestic and industrial wastewater and the widespread use and release of fertilizers and pesticides in agriculture contributes tremendously to the pollution of available water sources in the country. At the same time, the population of Morocco increased three-fold in 50 years, with noticeable rural flight, giving rise to slums on the borders of urban areas. These populations, with no access to clean water and sanitation, are exposed to waterborne a disease which affects about 30% of the population with gastrointestinal infections, malaria, and typhoid. A recent report of the World Bank and the Institute of Health (2018) highlights that Algeria is one of the world's most polluted countries, at the same level as Mexico, China, and India, resulting in a high rate of mortality. A cholera outbreak in 2018 was caused by the consumption of polluted water from Sidi Kabir Wadi in Tipaza. The Kebir River was adversely affected by pollutants and discharge of effluents without any treatment. Chemical and physical degradation were caused by agricultural and industrial activities; the lack of sanitation was a major polluter (Amamra and Khanchoul 2019). The evaluation of surface water quality along the Mencha River shows contamination by domestic sewage water from drains of local suburbs. Leakage from septic tanks is evident by the presence of fecal contamination registered by several monitoring stations, with an increasing pollution trend along the river downstream (Krika and Krika 2017). In Southern Algeria, the quality of water resources is poor, with high salinity and excessive fluoride concentrations. Meanwhile, groundwater resources in the Sahara and in particular in the Northern part of the country are highly mineralized and often of poor quality, thus requiring membrane filtration (nano and reverse osmosis) to supply the large majority of the population of the Sahara (Ramdani et al. 2012). Moreover, several aquifers located in the nearby urban centers are shown to be contaminated by the infiltration of contaminated surface water. This is the case in the El-Hadjar (Annaba) aquifer in the Northern part of the country, which is vulnerable to pollution from two wadis: Seybouse and Meboudja (Chaoui et al. 2015).

Owing to its location between Algeria and Libya, Tunisia shares surface and groundwater resources with its neighboring countries. Among these are Medjerda Stream in the North and deep aquifers in the South. Serious drought observed in recent years has threatened to damage the sensors of the monitoring network installed in the Medjerda stream and its watershed. The latter is undergoing great pollution pressure by the urban and agro-food industries that may impact water resources, soil, and biodiversity. Yearly, around 37 million m^3 of urban and industrial effluents are discharged in the whole watershed, corresponding to 60,000 tons of COD, 21,600 tons of DBO5 and 17,400 tons of suspended matter. In addition, significant quantities of chemicals comprising fertilizers, pesticides, and heavy metals are released through runoff and drainage. Significant contamination of shallow groundwater with nitrates was observed in agricultural areas, more specifically in the Northern part of the country. This non-point pollution of water resources is attributable to an absence of outreach to farmers to assist them to better manage fertilizers. The situation is aggravated by the release and application of manure and intensive livestock production (MARHP 2017).

In Tunisia, recent analysis by the National Agency of Environmental Protection (2017) provides evidence of the degradation of the quality of a number of waterways, not only from liquid discharge emanating from industrial units, but also from dumping of solid waste. In recent years, studies have been launched to inform better management of solid waste in rural areas in order to limit the impact on the environment. Despite an increase in awareness of the pollution threat to the environment, lack of funding is a barrier to implementation of action plans. Water resource quality monitoring is performed according to a program that includes surface water (dams, streams, canals, lakes) and groundwater (shallow and deep) as well as treated wastewater (MARHP 2017). The revision of national standards for discharge, initially developed in 1989, was a major achievement in 2018.

Adaptation to climate change means ensuring adequate supply and quality of water in the face of intensifying risks. One key element of adaptation is diversification of water management strategies, which is reliant on multiple water resources of different qualities to serve various uses. Alternative water resources like the reuse of TWW for irrigation or aquifer recharge and desalination of brackish water and seawater have been adopted. Nevertheless, these solutions require a strong institutional, regulatory, and technical foundation to be environmentally friendly and sustainable. In fact, water policies in several countries of the region have advocated IWRM as the preferred governance approach for existing resources (Sowers et al. 2011).

6 Technical Solutions and Research Challenges

Adaptive technical measures and the resulting capacity of water resources is likely to differ according to economic, social, and environmental conditions, as well as availability and access to technology in all sectors and levels, from local to national. In many cases, technology adoption is limited by inadequate financial resources and knowledge.

Given these constraints, a wide range of ambitious programs and technical measures have been initiated by regional governments to improve water supply and management, including the improvement of water efficiency, and the control or reallocation of water consumption. These initiatives are the central ambition of many development programs. They include mainly

- Maximizing available surface and renewable groundwater resources by building large and hillside dams, and constructing and rehabilitating tube wells, particularly in Tunisia, Algeria, and Morocco;
- Improving the water efficiency of rain-fed and irrigated agriculture to enhance production and preserve water in all areas. This is an integral part of the approach of almost all countries where the overall irrigation efficiency is usually less than 50%;
- Promoting and adopting modern irrigation projects, including creating small and medium facilities, and extending, renovating, and modernizing existing ones;
- Developing and extending interregional water transfer networks among countries such as Algeria, Morocco, and Tunisia, where water resources are not uniformly distributed;
- Protecting water quality and access to safe water and adequate sanitation. This is mainly a concern of the Mauritania government, which is still lagging far behind other regional countries: only one-third of the population has access to safe, drinkable water, and adequate sanitation;
- Increasing structures for water harvesting and groundwater recharge at large-, medium-, and small-scale facilities across all countries;
- Improving irrigation efficiencies and use of efficient irrigation systems;
- Saving water by reinforcing and improving water economy techniques for efficient irrigation;
- Encouraging the adequate reuse of wastewater as an alternative water resource. This practice has not been sufficiently adopted in almost any country due to technical and socioeconomic constraints;
- Developing brackish and seawater desalination plants, particularly in Tunisia, Egypt and Libya, where the total amount of domestic water is scarce and almost fully exploited, and where the transfer from other regions is not economically feasible. Algeria and Morocco, where resources are most vulnerable to climate change, could face a serious problem of water quality; and
- Strengthening mechanisms to improve data and information availability as a tool to guide planning, monitoring, and assessment of water resources, in particular for joint management of international shared aquifers and river basins.

Research and technology development in the water sector is considered key to the sustainable development of the region. Even though there is considerable variation in reporting of water research from country to country, the northern part of Africa is generally considered to be in a better position than countries in Sub-Saharan Africa to invest in research, technology, and innovation. These include infrastructure, budgeting, and training. Surveys of water management research in the region indicate the status of individual countries as follows:

- **Egypt.** Because renewable water resources are supplied mainly by the Nile River, and that source is almost fully exploited with the quality of available resources progressively degrading, *Egyptian research institutes have focused their interest on research programs on i*ntegrated water management of river basins from upstream to the coast for the sustainability of water resources along the river. This includes drainage management and reuse of wastewater for irrigation and water quality;
- **Morocco** has focused its research strategy mainly on the characterization and conservation of water resources, economic water management, and geographical information systems. However, momentum is growing for the need for conservation on the demand side, improving efficiency of water use, rather than the increase of supply. Particular attention is concentrated on the integration of water quality and quantity, as well as collaborative problem solving;
- In **Tunisia**, a particular effort has been directed to the development of research on water resource assessment and use of available water, management of water resources mostly for the agriculture sector and the development of hydro-agricultural models and cropping systems to optimize water resources. Research on development of marginal water use is also an important topic being investigated by Tunisian research institutes;
- In **Algeria,** the main problems facing the water resource sector concern inadequate water management and poor waste management practices leading to water pollution and desertification. The research strategy to which the Algerian government has committed takes into account these aspects, and encourages the establishment of numerous ambitious research programs, including on water supply management, recovery and production of water, and distribution;
- In **Libya**, research and higher education generally suffer from a lack of appropriate planning mechanisms and procedures. This has led to increasing dependence on skilled foreign scientists in many fields, including water research and management. Today, such planning is just starting to emerge, with very limited water resource management and research, which is not sufficiently oriented to resolving crucial problems facing the country. Nevertheless, there is focus on some fundamental issues, including delivery of a sustainable supply of water to the growing population, management of complex problems associated with sharing the water supply with riparian countries and sustainable use and management of groundwater and seawater intrusion problems, and

- In **Mauritania**, research is hindered by the lack of human resource management and the absence of clear planning programs, as well as by insufficient material resources which have produced very limited impact on water management. Despite these constraints, the government has established that understanding the economical use of water resources, and monitoring both the quality and quantity of water, is a priority.

Water research institution networks and technology transfer have been successfully developed in the region. These networks involve mainly countries from both sides of the Mediterranean, rather than within Africa.

Indeed, the need for sustainable water resource management constitutes a potentially key opportunity for cooperation across Africa to jointly lower the risk of conflict within the region as a result of a more integrated and prosperous North Africa.

7 Conclusions and Recommendations

The northern part of the African continent as a whole suffers from serious water scarcity. The six countries of North Africa are below the African average at 260 m^3/cap/year. Five of the six countries are subject to a serious water deficit, with water shares of less than 500 m^3/cap/year. In the region, Egypt and Mauritania are 95% dependent on transboundary water resources. Groundwater withdrawal very often exceeds natural replenishment, resulting in a progressive decline of the aquifers and in a deterioration of water quality.

The water situation in North Africa is expected to worsen due to the growing needs generated by population growth, urbanization, and the effects of climate change on the availability and variability of water resources. Because of the increasing demands of water use, in particular for agriculture estimated at an average of 80% of water consumption, non-conventional water supplies have been developed progressively to offer additional fit-for-purpose water.

Particular attention has been paid by regional governments to improve access to water by focusing on water governance, and institutional and technical solutions. In the last decade, all North African countries have taken up the challenge to improve water resource management by adopting water policies through a master plan for sustainable water resource management. These policies are different among countries, and the effectiveness of their implementation is dependent upon significant commitment and cooperation among water sector partners.

High-level recommendations for water resource management and increasing effectiveness of water use in the region are

- Making water-saving a key element of water policy, adopting modern irrigation technologies and controlling and reducing unaccounted-for water;
- Introducing water conservation strategies and encouraging water harvesting by returning to traditional, small-scale infrastructures and robust water and soil conservation systems, especially in arid areas;

- Sharing water resources in an equitable manner through transfers of water among hydraulic basins and dams developed increasingly in Algeria, Tunisia, and Morocco;
- Prioritizing the management of water demand rather than supply for optimization of already mobilized water resources, and avoiding environmental pollution;
- Decentralizing water management responsibility through the creation of structures for local participative management of water resources. This approach has demonstrated considerable advantages in, for example, Morocco and Tunisia;
- Engaging the private sector, currently adopted in Morocco and partly in Tunisia, in building, operating, and maintaining water facilities. This implies strengthening the role of government regulation;
- Applying incentives and sanctions to achieve and encourage conservation of resources, including water pricing for the various sectors to include targeted subsidies to low-income households;
- Reusing wastewater combined with artificial recharge as a framework to offer alternate solutions in many countries. Desalination using efficient and environmentally friendly renewable energies, coupled with inexpensive desalination technologies, could be conducive to producing freshwater on both medium and small scales. This solution is recommended for supplying more than 50% of the rural population living in the region, and
- Promoting cooperation for sustainable management of transboundary water resources and creating a framework of cooperation specific to the region. The ongoing experience of Algeria, Libya, and Tunisia in coordinating management of the North-Western Sahara Aquifer System—supported program implemented by the Sahara and Sahelian Observatory (OSS) is considered a potential model to improve and enhance communication among key stakeholders.

The countries of the North Africa region should play to their strengths, benefitting from coordination and taking examples from successful, applicable experiences from fields including building dams, and supplying drinking water supply and sanitation. A framework must be created to create suitable cooperation in the region, taking into account their common problems of water availability and quality in the region. Strengthening education and training at all levels and advancing data collection and sharing is also critical.

References

Abdel Wahaab R, Omar M (2011) Wastewater reuse in Egypt: opportunities and challenges (Part II). Expert consultation on waste water management in the Arab word, Dubai-UAE, 22–24 May 2011

Abdel-Satar A, Ali M, Gohera M (2017) Indices of water quality and metal pollution of Nile River Egypt. Egyp J Aqua Res 43:21–29

Abdullah A, Hussona S (2014) Water quality assessment of Mahmoudia canal in Northern West of Egypt. J Poll Effects Control 2:121

Abu Zeid K, Elrawady M (2012) 2nd Arab state of water Report 2012, Water resources management Program, CEDARE and Arab Water council, ISBN: 978 977 90 3806 3, ISSN: 2357 0318

African Development Bank (2012) Mauritania: results-based country strategy paper (RBCSP) 2011–2015. http://www.afdb.org/filea.-RBCSP%25202011-2015x.pdf. Accessed 10 Nov 2012

Alemu H et al (2015) Evapotranspiration in the Nile Basin: identifying dynamics and drivers, 2002–2011. Water. https://doi.org/10.3390/w7094914

Amamra A, Khanchoul K (2019) Water quality of the Kebir watershed, Northeast of Algeria. J Clean Was 3:28–32

Arab Environment (2008) Future challenges, report of the Arab forum for environment and development. In Mostapha K, Najib T, Saab W (eds.)

Assar W, Ibrahim MG, Mahmod W, Fujii M (2019) Assessing the agricultural drainage water with water quality indices in the El-Salam Canal mega project Egypt. Water 11:1013

ATH Group (2009) MEDA countries (Egypt, Lebanon, Morocco, Syria and Tunis). Identification and removal of bottlenecks for extended use of wastewater for irrigation or for other purposes, RG/2008-01/FTF

AWC/UNDP/CEDARE (2005) Status of integrated water resources management (IWRM) plans in the Arab region

Badr E, El-Sonbati M, Nassef H (2016) Water quality assessment in the Nile River, Damietta Branch, Egypt. Catrina 8:41–50

Bakhbakhi M (2006) Nubian sandstone aquifer system. In: Foster S, Loucks D (eds) Non-renewable groundwater resources: a guidebook on socially sustainable management for water-policy makers. IHP-VI, Series on Groundwater No. 10. Paris/Reading (UNESCO/IAH, 2006), 75 p

Bastiaanssen WG, Karimi P et al (2014) Earth observation based assessment of the water production and water consumption of Nile basin agroecosystems. Remote Sens. https://doi.org/10.3390/rs6 1110306

Ben Abdallah S (2003) La réutilisation des eaux usées traitées en Tunisie. 1ère Partie. Point de départ, conditions-cadres et stratégie politique d'eau (The reuse of treated watewater in Tunisia (1st part). Starting point

Besbes M, Chahed J et al (2010) Book water and sustainability in Arid Regions: bridging the gap between physical and social sciences, vol 7. In: Schneier-Madanes G, Courel MF (eds) Springer Science+Business Media

Beyene T, Lettenmaier D, Kabat P (2009) Hydrologic impacts of climate change on the Nile River basin: implications of the 2007 IPCC scenarios. Clim Change. https://doi.org/10.1007/s10584-009-9693-0

Blackmore D, Whittington D (2008) Opportunities for cooperative water resources development on the Eastern Nile: risks and rewards. An independent report of the scoping study team to the Eastern Nile Council of Ministers, http://dalewhittington.web.unc.edu/files/2015/12/Blackmore-Whittington-EN-Scoping-Study-Final-2009.pdf. Accessed 23 Nov 2017

CEDARE (2014). Tunisia water sector. Monitoring and evaluation rapid assessment report, Monitoring and evaluation for water in North Africa (MEWINA) project, water resources management program, CEDARE. http://web.cedare.org/wp-content/uploads/2005/05/Tunisia-Water-Sector-Monitoring-and-Evaluation-Rapid-Assessment-Report.pdf. Accessed 26 June 2019

Chaibi MT (2000) An overview of solar desalination for domestic and agriculture water needs in remote arid areas. Desalination. https://doi.org/10.1016/S0011-9164(99)00197-6

Chaoui W, Attoui B, Benhamza M, Bouchami T, Alimi L (2015) Water quality of the plain of El-Hadjar Wilaya of Aannaba (Northern Algeria). Energy Proc 74:1174–1181

Choukr-Allah R, Hamdy A (2005) Wastewater treatment and reuse as a potential water resource for irrigation, Alger, Algeria, pp 101–124

Cooley H, Gleick P, Wolff G (2006) Desalination, with a grain of salt A California perspective. In: Ian H (ed) Pacific Institute for Studies in Development, Environment and Security, pp 1–83. www.pacinst.org

Dahshan H, Megahed A et al (2016) Monitoring of pesticides water pollution-The Egyptian River Nile. J Environ Health Sci Eng 14:15

DG Environment European Commission (2006) Support to DG environment for development of the Mediterranean depollution initiative "Horizon 2020." Review of ongoing and completed activities
Dillon P, Escalante E, Tuinhof A (2014) Management of aquifer recharge and discharge processes and aquifer storage equilibrium Ground-water Governance. Thematic Paper 4: a global framework for action. Gef, World Bank, Unesco-IHP, FAO, IAH. http://www.groundwatergovernance.org/fileadmin/user_upload/groundwatergovernance/docs/Thematic_papers/GWG_Thematic_Paper_4.pdf Accessed 20 June 2015
DRPE (2017) Plan national de réutilisation des eaux usées. Secrétariat auprès du ministre de l'équipement, du transport, de la logistique et de l'eau chargé de l'eau. Rapport final (National plan for wastewater reuse. Secretariat of the ministry of equipment, transport and logistics in charge of water. Final report
ECA Office for North Africa (2005) Water resources development in North Africa. Twentieth meeting of the intergovemental committee of experts. Tangier, Morocco, pp 13–15
Elnazer A, Mostafa A et al (2018) Temporal and spatial evaluation of the River Nile water quality between Qena and Sohag Cities Egypt. Bull Nat Res Centre 42:1–7
Fadhel I (2017) Tunisia's strategy on climate change development in the energy sector. http://www.environnement.gov.tn/PICC/wp-content/uploads/Imed-Fadhel-Tunisia%E2%80%99s-Strategy-on-Climate-Change.pdf. Accessed Nov 2018
FAO (2011) Information products for Nile Basin Water Resources Management. http://www.fao.org/3/a-i2615e.pdf. Accessed 12 Dec 2018
FAO (2016) AQUASTAT. http://www.fao.org/nr/water/aquastat/main/index.stm. Accessed 26 Nov 2018
Femia F, Werrell C (2013) Climate change before and after the arab awakening: The cases of Syria and Libya. In: Werrell CE, Femia F (eds) The Arab spring and climate change. A climate and security correlations series
Gharbi N (2012) Wastewater and reuse for agricultural purposes in Tunisia. In: Kick-off meeting of the emerging pollutants in water and wastewater in Tunisia (EMPOWER-Tunisia) Project
Guardiola-Claramonte M, Sato T, Choukr-Allah R, Qadir M (2012) In integrated water resources management in the mediterranean region. In: Choukr-Allah R, Ragab R, Rodriguez-Clemente R (eds) Dialogue towards new strategy. Springer, Dordrecht
GWI (2017) Desalination market, global water intelligence, Media Analytics Ltd, Oxford, UK17, Media Analytics Ltd., Kingsmead House, Oxford, OX1 1XX, UK
Hagemann S, Chen C, Haerter J (2011) Impact of a statistical bias correction on the projected hydrological changes obtained from three GCMs and two hydrology models. J Hydrometeorol. https://doi.org/10.1175/2011JHM1336.1
Hassan D, Farghali R (2018) Cyanide pollution in different water sources in Assiut, Egypt: levels, distributions and health risk assessment. Res J Environ Sci 12:213–219
International Groundwater Resources Assessment Centre (IGRAC) (2012) Transboundary aquifers of the world. http://www.un-igrac.org/publications/456/. Accessed 10 Dec 2018
IPCC (2014) Contribution of working group II to the fifth assessment report. Cambridge University Press, Cambridge, United Kingdom and New York
Jagannathan N, Mohamed A, Perry C (2009) Introduction: beyond WRM-unbundling water management in MNA countries. In: The International bank of reconstruction and development/The World Bank/ Middle East and North Africa (MNA) Region, Washington DC
Khachaba R (2017) Le potentiel méconnu de l'assainissement pour la préservation de la ressource hydrique. Journées techniques sur la réutilisation des eaux usées épurées. 21– 22 Mai 2017, Algiers, Algeria
Kirby M, Eastham J, Mainuddin M (2010) Water-use accounts in CPWF basins: simple water-use accounting of the Nile Basin. In: CPWF Working Paper: Basin Focal Project series, BFP03 (The CGIAR Challenge Program on Water and Food
Krika A, Krika F (2017) Assessment of surface water quality of Mencha River, Northeastern Algeria by index method. Caspian J Environ Sci 15:157–164

League of the Arab States (2019) The Arab strategy for water security in the Arab Region-meeting the future challenges and needs of sustainable development (2010–2030). The Action Plan. https://www.susana.org/en/knowledge-hub/resources-and-publications/library/details/3431. Accessed 3 March 2019

MARHP (2017) Rapport National du Secteur de l'Eau. (National Report of the Water Sector). République Tunisienne

Mazzoni A, Heggy E, Scabbia G (2018) Forecasting water budget deficits and groundwater depletion in the main fossil aquifer systems in North Africa and the Arabian Peninsula. Glob Environ Change. https://doi.org/10.1016/j.gloenvcha.2018.09.009

Nawaz N, Bellerby T, Sayed M, Elshamy M (2010) Blue Nile runoff sensitivity to climate change. Open hyd. https://doi.org/10.2174/1874378101004010137

Nile Basin Initiative (2017) Nile Basin Resources Atlas, http://nilebasin.org/index.php/information-hub/technical-documents/44-nile-basin-water-resources-atlas. Accessed 21 Nov 2017

ONA (2012) http://www.ona-dz.org. Accessed 06 January 2012

ONAGRI (2018) Observatoire National de l'Agriculture. www.onagri.nat.tn/ Accessed July 2018

PNUD (2016). Le développement humain pour tous : Rapport sur le développement humain 2016/PNUD. New York, États-Unis

Qadir M, Wichelns D et al (2009) The challenges of wastewater irrigation in developing countries. Agr Wat Manag 97:561–568

Ramdani A, Djellouli H, Aït Yala N, Taleb S, Benghalem C, Mahi A. Khadraoui (2012) Physico-chemical water quality in some regions of southern Algeria and pretreatment prediction. Proc Eng 33:335–339

Raouan Hacen O, Belhouari B et al (2011) Contribution à la mise en évidence de l'expérience de l'Algérie dans la gestion intégrée des ressources en eau. In: Euro-Mediterranean Scienc Congress on Engineering (EMSCE 2011), pp 125–130

SCP (2017) Etude préalable à un plan national « réutilisation des eaux usées traitées » pour la Tunisie. Diagnostic de l'existant (Prestudy to a national plan on wastewater reuse for Tunisia. Diagnosis of the current status

Sowers J, Vengosh A, Weinthal E (2011) Climate change, water resources, and the politics of adaptation in the Middle East and North Africa. Clim Change 104:599–627

Terink W, Immerzeel W, Droogerset P (2013) Climate change projections of precipitation and reference evapotranspiration for the Middle East and Northern Africa until 2050. Int J Clim. https://doi.org/10.1002/joc.3650

UN (2008) Status report on integrated water resources management and water efficiency plans, Prepared for the 16th session of the commission on sustainable development May 2008, United Nations. https://www.unwater.org/publications/status-report-integrated-water-resource-management-water-efficiency-plans-csd-16/ Accessed 3 March 2019

UNDP, GEF (2018) http://www.undp.org/content/dam/undp/library/Climate%20and%20Disaster%20Resilience/Climate%20Change/Arab-States-CCA.pdf. Accessed 10 Nov 2018

Vo P, Ngo H et al (2014) A mini-review on the impacts of climate change on wastewater reclamation and reuse. Sci Tot Environ 494:9–17

Wheida E, Verhoeven R (2005) Wastewater treatment and its application as a water supply in Libya. In: Paper presented at the WSTA 7th gulf water conference, Kuwait

WHO (2018) Cholera-Algeria. In: Disease outbreak news. https://www.who.int/csr/don/14-september-2018-cholera-algeria/en/. Accessed 20 June 2019

World Bank (2018) http://databank.worldbank.org/data/home.aspx. Accessed 27 Nov 2018

WWHC (2018) Holding company for water and wastewater https://www.hcww.com.eg/. Accessed 12 Nov 2018

Zimmo O, Imseih N (2010) Waste water treatment and reuse in the mediterranean region. Springer, Berlin, p 155

Chapter 4
Climate Change and Water Resources in West Africa: A Case Study of Ivory Coast, Benin, Burkina Faso, and Senegal

P. Sagna, J. M. Dipama, E. W. Vissin, B. I. Diomandé, C. Diop, P. A. B. Chabi, P. C. Sambou, T. Sané, B. L. C. N. Karambiri, O. Koudamiloro, Y. M. Diédhiou, and M. Yade

Abstract Climate change is a major challenge for humanity due to its numerous impacts on people and the environment, which requires multiple strategies to address it. West Africa, particularly Ivory Coast, Benin, Burkina Faso, and Senegal are confronted with the full impact of that challenge. This study seeks to analyze climate

---

P. Sagna (✉) · C. Diop · P. C. Sambou · Y. M. Diédhiou · M. Yade
Department of Geography, Cheikh Anta Diop University, POB 5005, Dakar-Fann, Senegal
e-mail: pascalsagna@hotmail.com

C. Diop
e-mail: cheikh83.diop@ucad.edu.sn

P. C. Sambou
e-mail: piero0036@yahoo.fr

Y. M. Diédhiou
e-mail: dyayamansour@yahoo.fr

M. Yade
e-mail: yademadiop@hotmail.com

J. M. Dipama · B. L. C. N. Karambiri
Department of Geography, Joseph Ki-Zerbo University, 03 POB 7021, Ouagadougou, Burkina Faso
e-mail: jmdipama@yahoo.fr

B. L. C. N. Karambiri
e-mail: bienvenuechantal@gmail.com

E. W. Vissin · P. A. B. Chabi · O. Koudamiloro
Department of Geography, Abomey-Calavi University, 01 POB 526, Godomey, Benin
e-mail: exlaure@gmail.com

P. A. B. Chabi
e-mail: philippe_chabi@yahoo.fr

O. Koudamiloro
e-mail: olivierkoudamiloro@gmail.com

B. I. Diomandé
Department of Geography, Alassane Ouattara University, 01 POB v18, Bouaké, Côte d'Ivoire
e-mail: beh.ibrahimdiomande@gmail.com

S. Diop et al. (eds.), *Climate Change and Water Resources in Africa*,
https://doi.org/10.1007/978-3-030-61225-2_4

change by looking at the climate characteristics and rainfall duration recorded at forty selected stations. Senegal and Burkina Faso stretch over three climatic zones that include the Sahelian zone in the North, the North-Sudanian zone in the center, and the South-Sudanian zone in the South. For Ivory Coast and Benin, the northern and central areas are, respectively, located in the North-Sudanian and South-Sudanian climate zones, whereas the South is covered by the Guinean zone. In all four countries, break tests and water balance are characterized by the Standardized Precipitation Index, which measures changing water resources. Our research shows significant rainfall variability, with important impacts on water resources. For data from Ivory Coast from 1921 to 2016, there occurred a sharp decrease in the early 1980s; a similar decrease occurred in the late 1960s in Benin, Burkina Faso, and Senegal. This led to a noticeable decrease in surface water, while underground water, especially deep water, was less affected. However, a slight increase in rainfall is noted in Burkina Faso and Senegal in the 1990s. As local populations have grown aware of the negative effects of rainfall changes, they have initiated action to protect and manage water resources in a sensible and efficient way to preserve incomes from farming, grazing, and fishing. For this context, implementation of appropriate strategies likely to alleviate the harmful consequences of climate change must be tackled head-on. This can only be achieved through a wider knowledge of climate trends, coordination of stakeholders, and a broader dialogue to find the best communal solutions.

Keywords Climate change · Water resources · Impacts · Adaptation strategies · West africa

1 Introduction

Climate change is a global challenge for humanity because of its devastating impacts and the multiple adaptation and mitigation strategies required to address it. West Africa in general, and more acutely Ivory Coast, Benin, Burkina Faso, and Senegal, suffers particular effects of climate change with a rise in temperatures, irregular rainfall, variable rainy seasons, sea-level rise with significant coastal erosion, decrease in water resources, etc. Given the importance of farming and its role in the economic and social development of these countries, the major climate characteristics of this area remain the spatio-temporal and interannual changes in precipitation, which have become a permanent concern for local populations.

Indeed, since the severe drought (Leroux 1995) of the Sahelian zone, which began in 1968 and gradually spread to the South, including the Sudanian and Guinean zones (Sagna et al. 2015), the four countries selected for this study are experiencing a climatic and environmental crisis that jeopardizes their future in many sectors.

T. Sané
Department of Geography, Assane Seck University, POB 523, Ziguinchor, Senegal
e-mail: tsane@univ-zig.sn

Important studies have been conducted that lend a better understanding of precipitation and climate dynamics of circulation and rainstorms (Leroux 1995; Sagna 2005; Leroux 2010; Sagna et al. 2012, 2018). The spatio-temporal evolution of rainfall, including effects of drought in the 1970s and 1980s, is also well documented (L'Hôte et al. 2002; Nicholson et al. 2012; Kaboré et al. 2017). Research on runoff has shown a decrease in flows, directly correlated with a decrease in rainfall (Oyebande and Odunuga 2010). Groundwater is also affected by rainfall variability (Barrat 2012; Hammond Murray-Rust and Fakhruddin 2014). Other studies emphasize the importance of rainwater for human activity in West Africa (Diop 1999; Sène 2007; Sané et al. 2008; Sultan 2011; Diallo et al. 2012; Camara et al. 2013; Sagna et al. 2015; Sultan et al. 2015; Sané 2017), demonstrating the dependence of rural livelihoods on rainwater, especially over the last fifty years, as well as the negative impacts of irregular rainfall on rural development and the living conditions of farmers. The impacts of climatic hazards have also been a major political and scientific concern (Sultan et al. 2015; Sané 2017), insofar as the socioeconomic foundation of the region rests on factors directly or indirectly dependent on climatic conditions: farming, grazing, and fishing. Strategies to address climate hazards have been analysed (Houssou-Goé 2008; Kpadonou et al. 2012; Vissin 2013; Sambou 2015; Dipama 2016; Gautier and Denis 2016; Callo-Concha 2018). Gautier and Denis 2016 reviewed literature on adaptation in West Africa based on 37 publications on Senegal, 66 on Burkina Faso and 18 on Benin—three of the four countries included in our study. These studies reveal that the most studied sector is farming. They entail changes in crops and varieties, as well as changes in practices such as sowing period, water conservation techniques, and relocation of fields, to conserve and optimize water.

Despite the significant amount of research on climate change and its impacts and adaptation strategies for West Africa, some issues remain unclear, especially on the current trend of rainfall and its manifestations that affect the livelihoods of local communities who still depend on rainwater. The consequences of interannual variations in rainfall significantly influence water resources with all the attendant consequences for the livelihoods of broad populations, especially subsistence farmers. Given this critical situation, the current state of climate change must be studied in West Africa, particularly in Ivory Coast, Burkina Faso, Benin, and Senegal, because of the poor quality of available information.

The aim of this study is to analyze the trend of annual rainfall variations and their tangible impacts on the environment. This will be achieved by the analysis of time-series data from the four selected countries, as well as by observations of water flows and groundwater. The impacts of water deficit and water excess on socioeconomic activities will also be studied, and potential strategies to improve response to changes in water resources in the context of climate change will be discussed.

2 Data and Methodology

2.1 The Four Countries Included in the Study

The four countries chosen for this study, Ivory Coast, Benin, Burkina Faso, and Senegal (Fig. 1), experience climatic similarities and differences. The similarities are in circulation mechanisms and disturbances; the differences are mainly related to rainfall and temperature. As for surface atmospheric circulation, trade wind and monsoon circulations have been observed to alternate, contributing different proportions of wind frequencies. Average flows for the year 2014 illustrate this alternation:

- 13.8% for trade wind and 86.2% for monsoon in Ivory Coast;
- 18.4% for trade wind and 81.6% for monsoon in Benin;
- 40.5% for trade wind and 59.5% for monsoon in Burkina Faso, and
- 58% for trade wind and 42% for monsoon in Senegal.

Precipitation is mainly from squall lines, especially in Senegal and Burkina Faso, while the proportion from the Intertropical Convergence Zone (ITCZ) is very high in the south of Ivory Coast and Benin. There are also orographic and thermal factors, as well as cyclonic disturbances, all of which have impacts on precipitation. Overall, given the dynamics of precipitation, the southern countries (Ivory Coast and Benin) have higher levels of rainfall than the northern countries (Burkina Faso and Senegal).

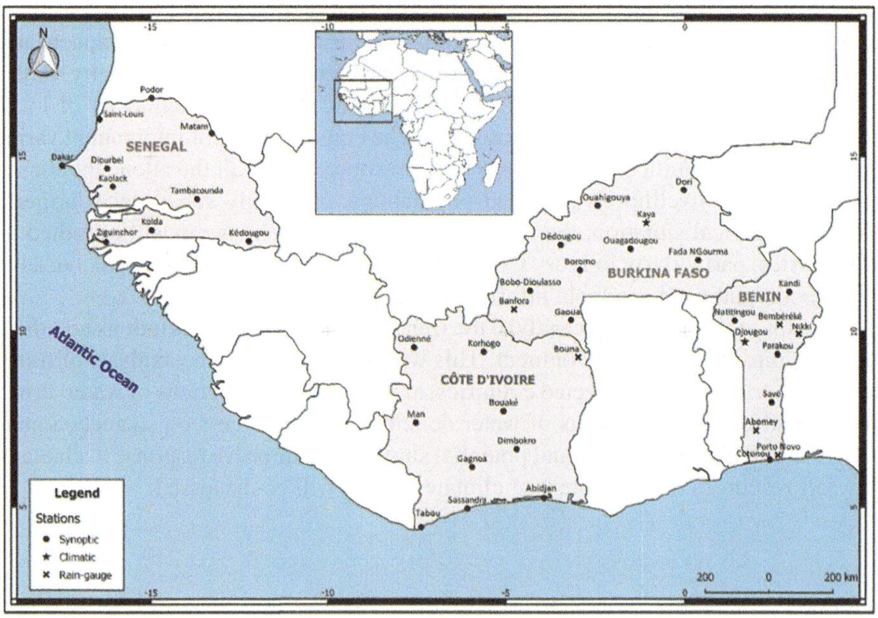

Fig. 1 Countries studied

Therefore, in terms of recorded rainfall and rainy seasons, there are three climatic zones in Senegal and Burkina Faso: the Sahelian zone in the north, the North-Sudanian zone in the center and the South-Sudanian zone in the south. For Ivory Coast and Benin, the South-Sudanian zone spans the northern and central areas while the southern area is located in the Guinean zone. In those countries, interannual variability of rainfall was used to determine the evolution of water resources and especially their scarcity because of drought that recurs in the area.

2.2 Data

In each of the four countries, ten stations were selected based on their geographic distribution, time-series lengths, meteorological significance, and data reliability. There were a total of forty stations (Fig. 1 and Table 1).

At most stations, observations began in the 1920s. For consistency, the study applied a starting year of 1921 and ending year as 2016 for all four countries, a time series of 96 years. The data came from various national meteorological services in the countries and the Inter-African Committee of Hydraulic Studies through the Overseas Office of Scientific and Technical Research (ORSTOM). For each country, the Standardized Precipitation Index (SPI) was calculated to determine if a given year is in excess (SPI > 0) or in deficit (SPI < 0) of the baseline standard. Although the SPI has been designed to detect droughts and is still widely used to characterize dry phases (Agnew 2000; WMO 2016), it allows an estimation of both dry and wet periods (McKee et al. 1993; WMO 2012), thus enabling the determination of breaks and phases in rainfall. The SPI is calculated as follows:

$$\mathbf{SPI} = \frac{P - \bar{P}}{\sigma} \tag{1}$$

- where **P** is the average rainfall of the year in the country;
- $\bar{P}$ is the average rainfall in the 1921–2016 time series in the country; and
- σ is the standard deviation of the annual rainfall series in the country.

Rainfall-break detection remains important in any analysis of climate change. Rainfall series were subjected to Pettitt's (1979) and Lee and Heghinian's (Lee and Heghinian 1977) break-detecting tests. Pettitt's approach is nonparametric because it makes no assumptions about the underlying distribution of data derived from the Mann–Whitney test (Pettitt, 1979). Pettitt's procedure is applied here to certify the research findings of our study. As a result, the absence of a break in the series (xi) of size T constitutes the null hypothesis (no break in the time series). When conducting this test, we assumed that for any moment t between 1 and T, the time series (xi) i = 1 at t and t + 1 at T belong to the same population. The variable to be tested is the maximum in absolute value of the variable Ut, with T defined by

Table 1 Stations in the four countries of the study area

Country	Station	Latitude	Longitude	Altitude (m)	Type of station
Ivory Coast	Odienné	09° 30' N	07° 34' W	432	Synoptic
	Korhogo	09° 27' N	05° 38' W	300	Synoptic
	Bouna	09° 16' N	03° 00' W	316	Rain-gauge
	Bouaké	07° 41' N	05° 02' W	369	Synoptic
	Man	07° 24' N	07° 31' W	340	Synoptic
	Dimbokro	06° 39' N	04° 42' W	110	Synoptic
	Gagnoa	06° 08' N	05° 57' W	205	Synoptic
	Tabou	04° 25' N	07° 22' W	4	Synoptic
	Sassandra	04° 57' N	06° 05' W	50	Synoptic
	Abidjan	05° 15' N	03° 56' W	7	Synoptic
Benin	Kandi	11° 08' N	02° 56' E	290	Synoptic
	Bembéréké	10° 12' N	02° 40' E	491	Rain-gauge
	Natitingou	10° 19' N	01° 23' E	460	Synoptic
	Nikki	09° 56' N	03° 12' E	402	Rain-gauge
	Djougou	09° 42' N	01° 40' E	439	Climatic
	Parakou	09° 21' N	02° 36' E	392	Synoptic
	Savé	08° 02' N	02° 29' E	198	Synoptic
	Abomey	07° 11' N	01° 59' E	260	Rain-gauge
	Porto-Novo	06° 29' N	02° 37' E	20	Rain-gauge
	Cotonou Ville	06° 21' N	02° 26' E	5	Rain-gauge
Burkina Faso	Dori	14° 02' N	00° 02' W	288	Synoptic
	Ouahigouya	13° 35' N	02° 26' W	329	Synoptic
	Kaya	13° 06' N	01° 05' W	313	Climatic
	Dédougou	12° 28' N	03° 28' W	308	Synoptic
	Fada Ngourma	12° 04' N	00° 21' W	292	Synoptic
	Ouagadougou	12° 22' N	01° 31' W	296	Synoptic
	Boromo	11° 44' N	02° 55' W	264	Synoptic
	Bobo-Dioulasso	11° 10' N	04° 18' W	432	Synoptic
	Gaoua	10° 20' N	03° 11' W	333	Synoptic
	Banfora	10° 37' N	04° 46' W	270	Rain-gauge
Senegal	Podor	16° 38' N	14° 56' W	7	Synoptic
	Saint-Louis	16° 03' N	16° 27' W	4	Synoptic
	Matam	15° 38' N	13° 15' W	15	Synoptic
	Dakar	14° 44' N	17° 30' W	27	Synoptic
	Diourbel	14° 39' N	16° 14' W	7	Synoptic
	Kaolack	14° 08' N	16° 04' W	6	Synoptic

(continued)

Table 1 (continued)

Country	Station	Latitude	Longitude	Altitude (m)	Type of station
	Tambacounda	13° 46' N	13° 41' W	49	Synoptic
	Ziguinchor	12° 33' N	16° 16' W	26	Synoptic
	Kolda	12° 53' N	14° 58' W	35	Synoptic
	Kédougou	12° 34' N	12° 13' W	165	Synoptic

(ORSTOM 1973)

$$U_{t,T} = \sum_{i=1}^{t} \sum_{j=t+1}^{T} D_{ij} \quad (2)$$

where

- D_{ij} = sgn(Xi-Xj) with sgn(X) = 1 if d > 0; 0 if d = 0, and -1 if d < 0.
- With d = Xi - Xj.

The rejection of the null hypothesis supposes an estimation of the date of break which is given by the moment t which defines the maximum in absolute value of the variable $U_{t,T}$. This tool is used regularly because of its precision and robustness, as well as assurance of high levels of reliability.

The hypothesis of a relatively wet period in the countries selected for this study is also verified with Hubert's segmentation procedure (Hubert et al. 1989). Tests for detecting breaks and the segmentation procedure were applied with KhronoStat software developed by the *Maison des Sciences de l'Eau* of the Research Institute for Development (Boyer 2002).

Several climate and hydrologic models have been used to analyze rainfall and water resources, particularly in West Africa. These models include, for example, the Regional Climate Models (RCMs) and the Global Climate Models (GCMs), the hydrologic model GR2M and the Water Flow and Balance Simulation Model (WaSiM). Based on relatively well-calibrated reference situations, projections of rainfall and water resources improve management of water resources.

Surveys were carried out in several areas and on many aspects of the lives of both rural and urban populations, so as to better understand the impacts of current climate change on socioeconomic activities. Endogenous and state strategies are also deployed by several stakeholders to address climate change. Findings vary in their relevance and effectiveness from one country to another, and from one local area to another.

3 Research Findings

3.1 *Interannual Rainfall Variability*

Application of the different statistical tools on rainfall records in the 96-year study period resulted in findings that differed among the four countries.

Rainfall in Ivory Coast showed two break years in the study period: 1981 for the Standardized Precipitation Index and Pettitt's test, and 1982 for Lee and Heghinian's test (Figs. 2, 3, and 4). Compared to the former, the latter two have periods of unequal duration with different characteristics. The wet period, which extends from 1921 to 1981, shows an average rainfall of 1538 mm, while the dry period, from 1982 to 2016,

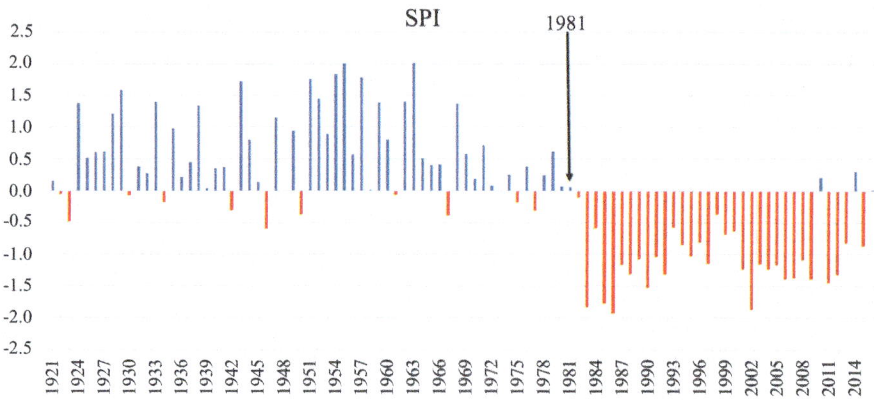

Fig. 2 Evolution of the Standardized Precipitation Index in Ivory Coast from 1921 to 2016

Fig. 3 Evolution of Pettitt's test results in Ivory Coast from 1921 to 2016

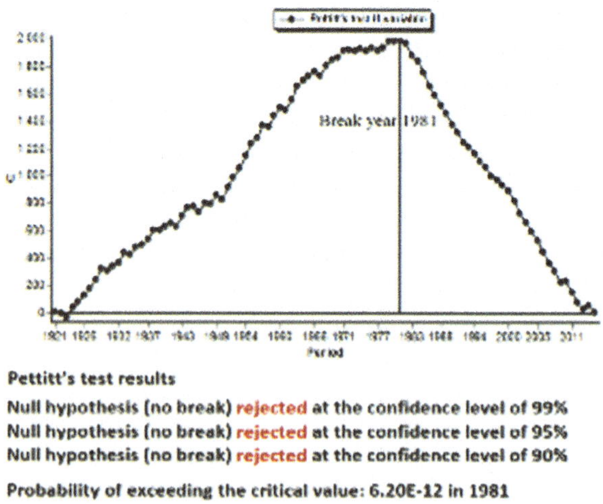

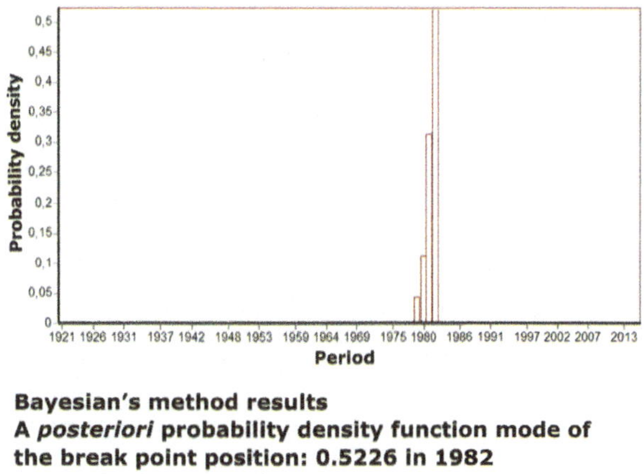

Fig. 4 Evolution of Lee and Heghinian's test results in Ivory Coast from 1921 to 2016

shows an average of 1139.8 mm. The difference between the two periods is 398.2 mm, which corresponds to a drop of 25.9% in rainfall. The rainiest years are concentrated between 1951 and 1963: 1951 with 1833.5 mm, 1954 with 1851.3 mm, 1955 with 1893.7 mm, 1957 with 1838.7 mm and 1963 with 1895 mm. The driest years, apart from 2002 (928.2 mm), are also close to one another: 1983 with 937.3 mm, 1985 with 952.5 mm, 1986 with 911.3 mm and 1990 with 1011.4 mm. Between the wettest year (1963) and the driest year (1986), the difference is 983.7 mm. In the entire series, 51 years are in excess and 45 years in deficit. Rainfall seems to continue to diminish.

In Benin, the application of the Standardized Precipitation Index and the tests of Pettitt and Lee and Heghinian reveal a break in 1969 within the series from 1921 to 2016 (Figs. 5, 6, and 7). Two periods stand out in this period: the first, which extends for 49 years from 1921 to 1969, with an average rainfall of 1276.5 mm, is wet, while the second, which extends for 47 years, between 1970 and 2016 with an average rainfall of 1122.1 mm, is dry. The difference between the two periods is

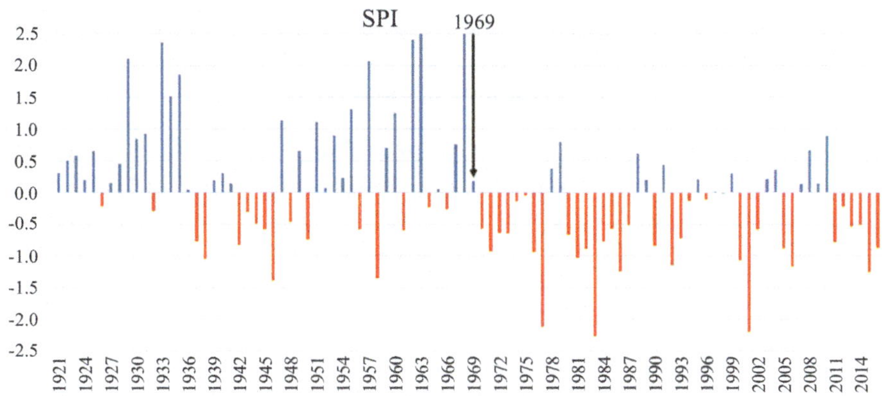

Fig. 5 Evolution of the Standardized Precipitation Index in Benin from 1921 to 2016

Fig. 6 Evolution of Pettitt's test results in Benin from 1921 to 2016

Pettitt's test results

Null hypothesis (no break) rejected at the confidence level of 99%
Null hypothesis (no break) rejected at the confidence level of 95%
Null hypothesis (no break) rejected at the confidence level of 90%

Probability of exceeding the critical value: 2,67E-04 in 1969

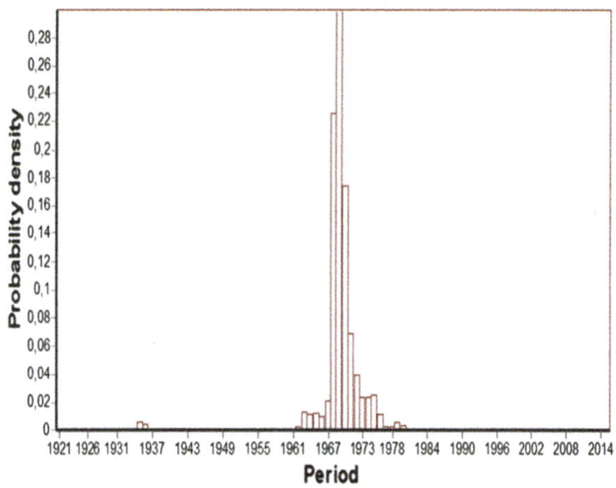

Fig. 7 Evolution of Lee and Heghinian's test results in Benin from 1921 to 2016

Bayesian's method results

A *posteriori* probability density function mode of the break point position: 0,2993 in 1969

154.4 mm: 12.1%. The six rainiest years, with an SPI higher than 2, are found in the first period, with 1567 mm in 1929, 1611.3 mm in 1933, 1561.1 mm in 1957, 1618.9 mm in 1962, 1647.5 mm in 1963, and 1683.7 mm in 1968. The driest years are more scattered, with 960.8 mm in 1946, 967.8 mm in 1958, 835 mm in 1977, 809.7 mm in 1983, and 821 mm in 2001. The difference between the wettest year (1968) and the driest year (1983) is 874 mm. Overall, analysis of the rainfall series reveals 47 years of excess, including 33 in the first period and 14 in the second, and 49 years of deficit, including 16 in the first period and 33 in the second. More recent years remain affected by drought.

In Burkina Faso, from the years 1921 to 2016, there was a rainfall break in 1969, confirmed by the Standardized Precipitation Index, the tests of Pettitt and Lee and Heghinian, and by the segmentation of Hubert (Figs. 8, 9, and 10). Thus, we found that the period from 1921 to 1969 was on the whole wet, with an average rainfall of 927.4 mm, marked by a dry period between 1970 and 2016, despite a significant increase from 2008 to 2016, with an average of 805.3 mm. The difference in rainfall between the two periods mentionned above is 122.1 mm and corresponds to a decrease in rainfall of 13.2%. The wettest years are scattered within the first period, with 1033.6 mm in 1922, 1064.6 mm in 1933, 1044.9 mm in 1951, 1048.7 mm in

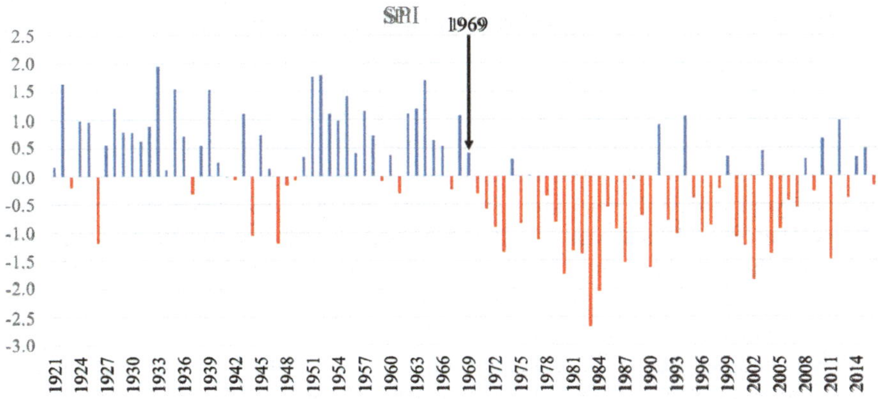

Fig. 8 Evolution of the Standardized Precipitation Index in Burkina Faso from 1921 to 2016

Fig. 9 Evolution of Pettitt's test results in Burkina Faso from 1921 to 2016

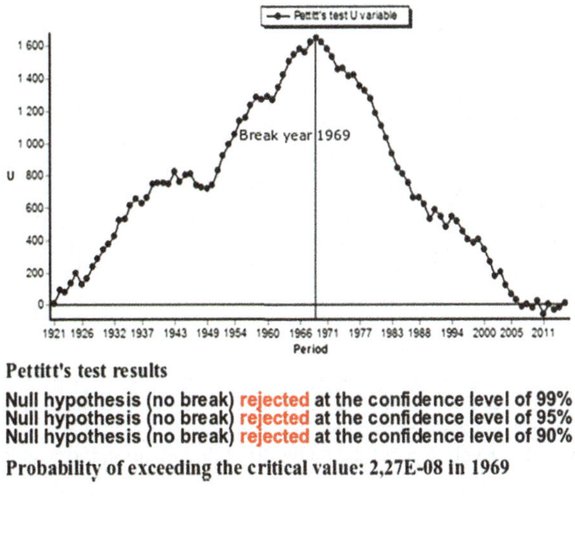

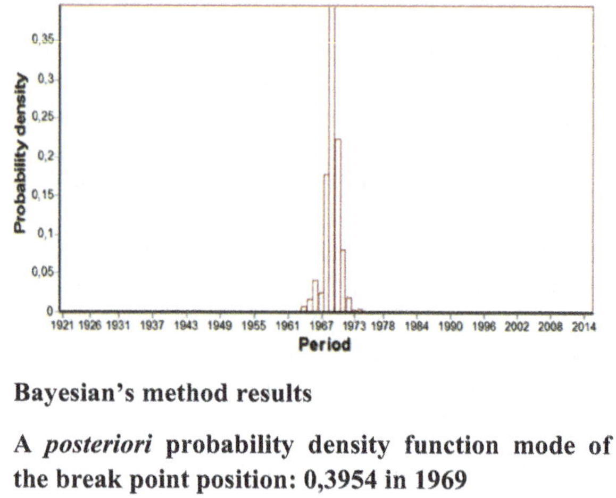

Fig. 10 Evolution of Lee and Heghinian's test results in Burkina Faso from 1921 to 2016

1952, and 1039.3 mm in 1964. The driest years in the second period are closer to each other, with 691.5 mm in 1980, 598.5 mm in 1983, 662.6 mm in 1984, 704.4 mm in 1990, and 681.6 mm in 2002. The difference between the rainiest year (1933) and the driest year (1983) is 466.1 mm. This series of rainfall measurements in Burkina Faso is on average 48 years greater than the average of the period 1921–2016, with 37 years in the wet period and 11 years in the dry period and 48 years lower, including 12 in the first period and 36 in the second. Despite this relative balance, recent years have marked an increase in rainfall.

In Senegal, the Standardized Precipitation Index and the tests of Pettitt and Lee and Heghinian indicate a rainfall break in 1967 (Figs. 11, 12, and 13). The break reveals two major periods: a generally wet period from 1921 to 1967, and a relatively dry period from 1968 to 2016. The average rainfall is 832 mm for the first period and

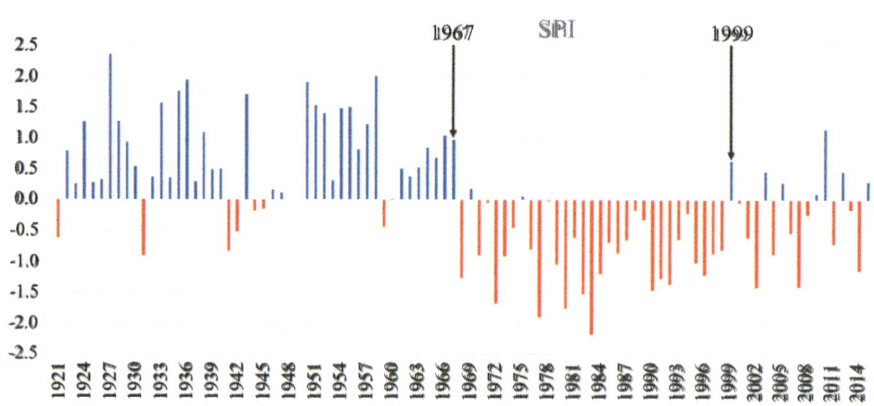

Fig. 11 Evolution of the Standardized Precipitation Index in Senegal from 1921 to 2016

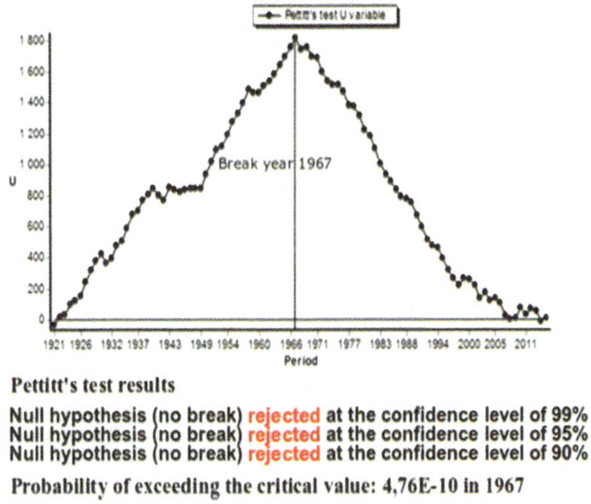

Fig. 12 Evolution of Pettitt's test results in Senegal from 1921 to 2016

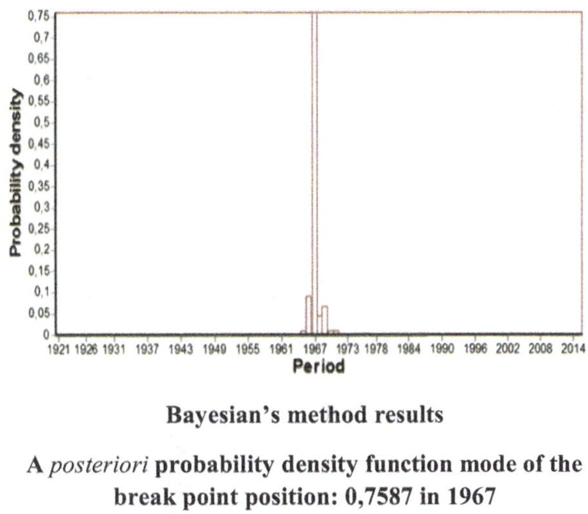

Fig. 13 Evolution of Lee and Heghinian's test results in Senegal from 1921 to 2016

639.4 mm for the second—a rainfall decrease of 192.6 mm (23.1%). The rainiest years were 1927, 1936, 1950, and 1958, with respective rainfalls of 1076.5 mm, 1016.6 mm, 1011.2, mm and 1026 mm. The driest years were 1972, 1977, 1980, and 1983, with respective rainfalls of 488.1 mm, 456.3 mm, 475.8 mm, and 414.9 mm. The difference between the wettest year and the driest year was 661.6 mm. The dry period, the beginning of which was confirmed in 1968 by the segmentation of Hubert, was subdivided into two sub-periods, the first extending from 1968 to 1998 with an average rainfall of 603.3 mm, and the second from 1999 to 2016 with an average of 701.6 mm. The 1999–2016 sub-period represents a slight increase in rainfall of 98.3 mm (14%) compared with the previous period. Over the entire series, 47 years are in excess, of which 38 are during the wet period, and 49 years are in deficit, of which 40 are during the dry period. It is notable that droughts still occur in Senegal.

3.2 Impacts of Rainfall Variation on Water Resources

The alignment of rainfall analyses with reality is visible by the impacts of changes in water resources on livelihoods. It is imperative to be able to predict the supply of and demand for water resources over a given period to be able to create and maintain sustainable and efficient water management.

3.2.1 Impacts of Rainfall Variation on Runoff and Lakes

In Ivory Coast, an overall downward trend in water resources correlates with a decrease in rainfall. For example, in the Sassandra River basin, Coulibaly et al. (2018) used the Representative Concentration Pathway (RCP) 4.5 and RCP 8.5 climate models and the Regional Climate Model 4 (RegCM4) of the Abdus Salam International Center for Theoretical Physics (ICTP) to obtain data to integrate into the GR2M hydrologic model. By 2030, 2050, 2070, and 2090, annual flows will decrease compared to the 1961–1980 reference period. The decline in flows follows the same progression as the climate projections of the RegCM4 model. The RCP models 4.5 and 8.5 also show a decline of 6.9–24.1% in flows from the Guinean source to the Ivorian parts of the Sassandra between 2030 and 2090. According to Koffi and Diomandé (2015), Lake Kossou (a reservoir on the Bandama River in Central Ivory Coast) declined by 37.4%, a loss of 20,394 hectares (Fig. 14). Its area has decreased from 54,583 hectares in 2002 to 34,189 hectares in 2014.

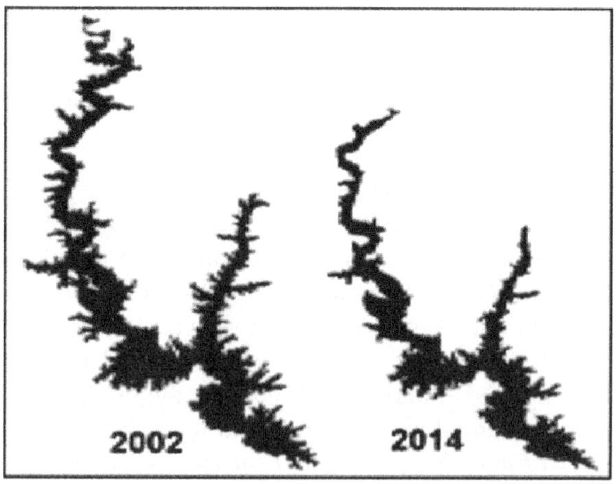

Fig. 14 Comparative area of Lake Kossou between 2002 and 2014

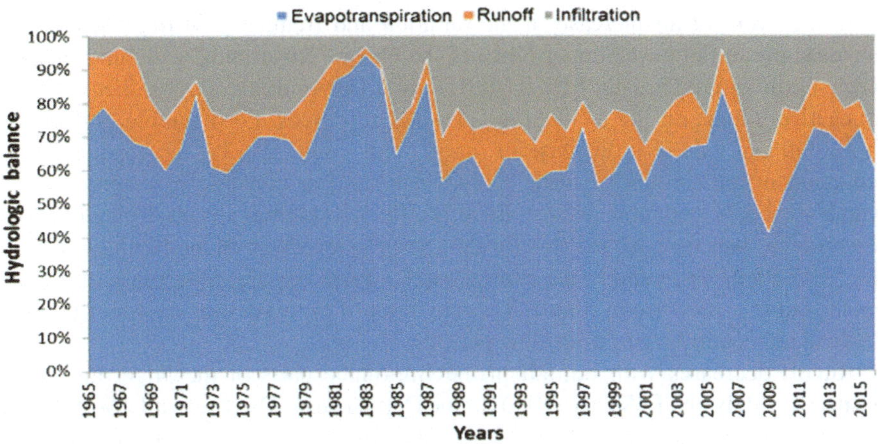

Fig. 15 Hydrologic balance of the Ouémé basin from 1965 to 2016

In Benin, the hydrologic balance of the Ouémé basin from 1965 to 2016 indicates that evapotranspiration accounts for ≥50% of the hydrologic balance annually, whereas storage is around 35–40% (Fig. 15). As a result, the majority of rainfall evaporated, while 11–13% became runoff. These findings align with the simulated findings (TopAMMA) from Le Lay (2006), which found that evapotranspiration was 50% and runoff was 12%, with underground stock change of 38%. There was a decrease in flow during dry years (early 1970s and 1980s). Water balance in the series showed an increase in evapotranspiration during those years, while runoff and infiltration decreased. The output of the BenHydro model (Speth et al. 2010) for the entire Ouémé River basin clearly shows the impact of the decrease in rainfall on surface runoff by 2050. From 1980 to 2050, the model predicts a steady decline in the flows of the Ouémé at Bonou, except for 2031–2040, that shows a slight increase (MEPN 2008) (Fig. 16). By combining 65 projections of 24 climate models, Essou and Brissette (2013) demonstrate that annual flows of the Ouémé will drop by 3–5% by 2099.

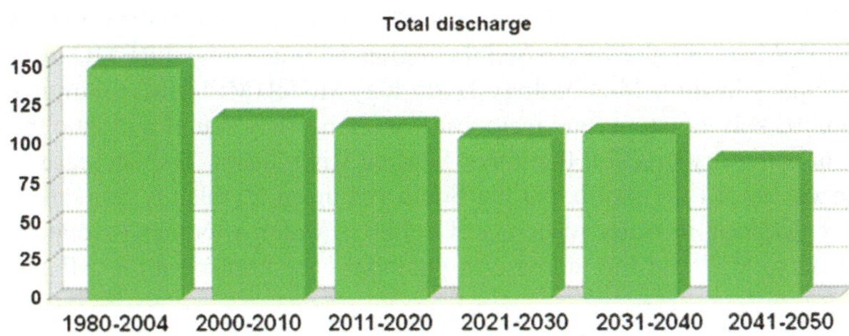

Fig. 16 Evolution of discharges of the Ouémé at Bonou at different time horizons (MEPN 2008)

In the event of an increase in the duration and frequency of dry seasons and decrease in rainfall in the upper basin of the Ouémé River and the Beninese Niger River basin as reported by MEHU (2011) and Lawin et al. (2013), there will be a negative impact on the availability of water resources. For example, a decrease in rainfall can lead to a decrease in runoff. This could also be accompanied by a reduction in the amount of water stored in natural or constructed reservoirs into which seasonal rivers flow. Hence, there is a likelihood of recurrent drying of water reservoirs as was the case in 2014 for the Djougou reservoir. In the Beninese part of the Niger River basin and in the central and southern regions, a decrease in rainfall could lead to a decrease in runoff. The late onset of rainy seasons could introduce a delay in the occurrence of high-flow periods.

Runoff decreases with rainfall in the Burkinabe parts of the Niger and Volta basins (El Vilaly and El Vilaly 2013). In the Dapola basin, there was a 1.5% decrease in decadal discharges after 1971. Yira et al. (2017) used a set of six regional and global climate models (RCMs and GCMs) to predict future rainfall and runoff in the Dano basin. The data are then integrated into the Water Flow and Balance Simulation Model (WaSiM). Two scenarios of greenhouse gas concentrations (the Representative Concentration Pathways RCP 4.5 and RCP 8.5) were selected. Findings show that there are uncertainties about the rainfall and runoff relationship in the basin over the period 2021–2050 compared to the period 1971–2000.

In the Senegal River basin, the relatively dry period from 1960 to 1996 resulted in a decrease of 30–40% in water availability following a decrease of 20% in rainfall (Oyebande and Odunuga 2010). In the upper basin, the average discharge during 1971–2010 (after the rainfall break) decreased by 34–54%, depending on the waterway, compared to 1904–1969 discharges (before the break). The HadRM3P and RCA climate models and the GR2M hydrologic model allow us to estimate climatic and hydrologic changes. They show a general downward and cyclical trend of flows for the 2030, 2060, and 2090 horizons compared to the 1961–1990 period (Diakité 2017). In fact, variations in flows are twice as high as changes in rainfall (Le Lay and Galle 2005). Magistro and Lo (2001) find a halving of the flows of the Senegal River between the 1950s and the 1990s. Bodian et al. (2013) used four 2007 models by the IPCC and a hydrologic model (GR2M) to study the impact of climate change on the upper Senegal River basin. According to three climate models (CSMK3, HadCM3 and MPEH5), there will be a gradual decrease in discharge from 2030 to 2090 (Bodian et al. 2013). Mbaye et al. (2015) used the regional climate model REMO as input to the Max Planck Institute Meteorology-Hydrology Model (MPI-HM) to simulate flow rates, runoff, soil moisture, and evapotranspiration. They found a drastic decrease of 50% in water resources in the upper Senegal River basin. The decrease is more pronounced toward the North. To the South (Guinean Highlands), models do not show any variation in water resources. Decline in water resources is related to the decrease in rainfall and the increase in potential evapotranspiration (Mbaye et al. 2015).

3.2.2 Impacts of Rainfall Variation on Groundwater Levels

In West Africa, data collected between 1992 and 2005 by the AMMA project (Multidisciplinary Analysis of the African Monsoon) show that there is no direct correlation between annual rainfall and groundwater recharge. This is due to the intensity of precipitation, which is more decisive. Moderate rains lead to more infiltration than intense and light rains (Hammond Murray-Rust and Fakhruddin 2014). Estimating the impacts of precipitation on groundwater is difficult to achieve in West Africa because locally collected data are scarce and the rates of change in rainfall, runoff, and infiltration can be highly variable, even in small areas. However, we note that anthropogenic factors, particularly irrigation, have a greater influence on the decrease in groundwater levels than climate change, according to a simulation in Mali. The increase in the groundwater level of the Continental Terminal in Niger averaging 0.2 m/year since the 1960s shows that the climate-groundwater relationship is not simple (Barrat, 2012).

In Benin, the comparative analysis of the two representations (Fig. 17) demonstrates the dynamics of Lake Nokoué bathymetry between 1980 and 2004.

The analysis of bathymetry maps made by Tessier (1980) (at left) and Dakpogan (2005) (at right) shows that Nokoué Lake experienced depth decreases between 1980 and 2004. In 1980, depths reached about 2 m in the West and in the North. But in 2004, depths dropped below 2 m. Only in the central depression and at the entrance of the channel did they reach or exceed 2 m (Dakpogan 2005).

The 1970s and 1980s were characterized by a decline in groundwater recharge. Sites at Ouagadougou (Burkina Faso) in 1987 and Fô-Bouré (Benin) in 2006 showed a decrease in groundwater. Deep groundwater is less affected because its recharge process is more complex. In the Volta basin, assuming no change in water demand, groundwater recharge is projected to drop by approximately 50% in the 2100 horizon, in relation to a temperature increase of 3.8 °C and a rainfall increase of 20% (Hammond Murray-Rust and Fakhruddin 2014).

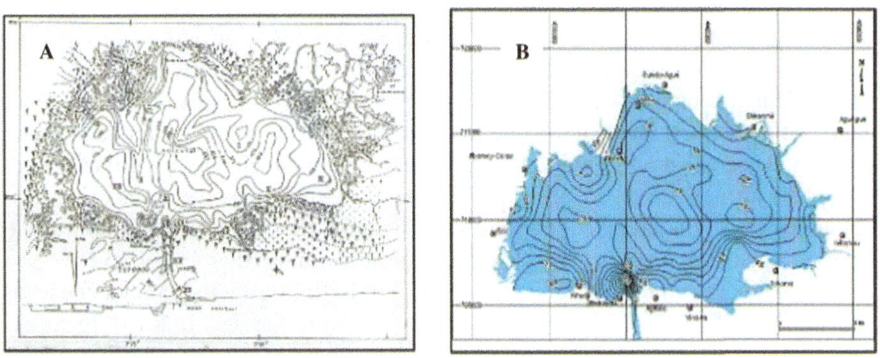

Fig. 17 Bathymetry of Lake Nokoué in 1980 (**a**) (Tessier 1980) and 2004 (**b**) (Dakpogan 2005)

In Senegal, the impact of the rainfall deficit on groundwater resources has been greatest on the northern coast (the Niayes) where almost all of the country's truck farms are located. In this location, decreases in groundwater levels occurred mainly in the 1970s and 1980s (Aguiar et al. 2010; Sambou 2015): the droughts of the 1970s and 1980s caused the groundwater level to decrease from 50 to 22 m between 1958 and 1984 (Aguiar et al. 2010). Malou (2004) noted that this was the most severe documented drying trend of groundwater reserve. During the droughts, groundwater recharge was very low (Dasylva and Cosandey 2005). Moreover, the drop was exacerbated by losses through evapotranspiration. On the horst of Ndiass (far south of the northern coast), the fall in groundwater levels, over the last 25 years, caused mainly by human activity, varied from 20 to 25 m (MEPN 2005).

The combination of declining rainfall, increased temperature (evaporation and evapotranspiration), the intensity of showers (which favors runoff to infiltration), and the multiple human uses of water result in scarce resources and low groundwater levels globally. The drying up of wells in many villages is evidence of this phenomenon.

3.3 Impacts of Rainfall Variation on Socioeconomic Activities

Rainfall variability has repercussions for farming, grazing, and fishing in West Africa. Impacts vary from one crop to another and by region.

3.3.1 Impacts of Rainfall Variations on Farming

In Ivory Coast, farmers mainly plant two types of crops: cocoa, a perennial crop grown in the southern forest area, and yams, a food crop grown in the central and northern savannah. Dependence on rainwater for these crops correlates to coefficients of 0.88 for cocoa production and 0.75 for yams. In general, the long rainy season is decreasing, with recorded drops from six to four months. During half of the years in deficit, the water requirement of cocoa trees was not met during the long rainy season. This prolonged water stress resulted in the death of both young and adult plants, which in turn led to a decline in the production of butter from cocoa beans. In recent decades, yam production was also negatively affected by irregular rainfall. Among the farmers surveyed, about 55% marketed part of their production. However, they unanimously acknowledged that their incomes from yam production had not met their needs for years because few farmers earned more than Euros 152.

In Benin, corn is the main food crop in the South. In Couffo, the rainwater excess, especially in 2007–2008, caused loss of part of the corn harvest. In low-lying areas, some farmers lost all of their crops (Houssou-Goé 2008). In the Ouémé basin, cotton, groundnut, and sorghum are predicted to grow better with the reduction in the length of the cropping season forecast for the 2030 horizon. On the other hand, crops vulnerable to drought, like cassava, yam, corn, and rice, will be negatively impacted.

A modest decline in yield was observed for legumes (soya and bean) and sweet potatoes. Incomes across all crops declined.

In most parts of Burkina Faso, changes in early and late season, dry spells, heavy rains and pest and disease attacks affect farm productivity through physical damage to plants (Ouoba 2013). Indeed, in the case of a late or false start of the rainy season, there is poor growth and/or death of seedlings, the appearance of worms and caterpillars, termite attacks, etc. Heavy rains are equally devastating. Excessive rainfalls generally cause flooding, the uprooting of stems, the rotting of seeds (especially at the beginning of the season), asphyxia, and sometimes non-ripening of crops. Losses in agricultural production due to flooding of cultivated fields were estimated at Euros 2,748,727 in 1992 and Euros 97,474,891 in 1994 (MECV 2007). These factors, which lead to a decline in agricultural production, raise the crucial issue of food security. The drop-in rainfall also induces a drop-in millet yields, particularly in Dori in the Sahel. In the southern areas, corn yields fell sharply due to the July, August, and September water deficit (MECV 2007). In fact, corn is very sensitive to drought: the slightest shortage of rainfall can jeopardize the whole harvest.

In Senegal, too, the interannual variability of rainfall affects crops. A delay in the start of the rainy season and the annual rainfall deficit in 2014, for example, compromised the harvest of millet and groundnuts in Louga, Thiès, and Mbour (in the northwest of the country). Yield declines of 8–31% were recorded. In the center (Kaolack) and the south (Ziguinchor) of the country, where rain deficits were milder, yield decreases were less severe (Diop et al. 2016). From 1961 to 2013, yields fell significantly in Louga and Kébémer in the north-central area of the country (Sambou 2015). Even in areas where fields are irrigated, such as the Senegal River Valley and the Anambé Basin in Upper Casamance, the development of farming depends on rainwater, since its abundance or scarcity determines the availability of water resources and, therefore, yields. In the south (communities of Bona and Diacounda), in 2007, the production of millet and corn decreased, respectively, by 38% and 40% because of the rainfall deficit, compared to 2006 (Diédhiou 2008). Rice production also dropped by 20% in the same year. From 1981 to 2015, the year 2007 was noteworthy, with one of the shortest rainy seasons on record.

3.3.2 Impacts of Rainfall Variations on Pastoral Activities

Grazing prevails mainly in the Sahelian and Sudanian zones of the four countries, where the amount of rainfall and its effects vary significantly. These factors include late rainfalls, heavy rains, pockets of drought, early cessation of rains, and short rainy seasons. Late rainfalls result in water shortages and, therefore, less pasture area, as well as skin diseases and diarrhea in some animals, which often result in death. These effects are also observed when the rainy season is too short. In Senegal, for example, a reduction in the number of ponds compared to the pre-break period of 1968 is observed by farmers in the areas of Sakal and Ndande (in the northwest of the country). In contrast, increased precipitation favors milk production. The recent increase in rainfall has allowed an increase in milk production, particularly between 2009 and 2010, with 2% more than the 2005–2010 average (Sambou 2015).

3.3.3 Impacts of Rainfall Variation on Inland Fisheries

With the reduction of volume and area of water bodies following unfavorable changes in climate parameters, production of inland fisheries has declined in the countries from which information is available. In Ivory Coast, fish production in Béoumi (on the left bank of Lake Kossou) decreased from 1,142 tons in 2009 to 320 tons in 2015, according to the Departmental Agency of the Ministry of Animal and Fish Production (MPAH 2016). This magnitude represents a threat to self-sufficiency and food security in that area. Burkina Faso, which has more than 200,000 hectares of surface water including rivers and ponds, and 1,450 dammed lakes and reservoirs that can be used for fishing and fish production in general (Bationo 2015), is similarly affected. Indeed, precipitation fluctuation, fill rate of water bodies, drying up of waterways, disappearance of certain species, and the opening of the flood evacuation valves that facilitates the migration of fish are factors that help explain decline in production. In Senegal, inland fish catches are estimated at 37,000 tons in 1999. This activity is practiced in both freshwater and brackish water (MPTM 2001). Fish farming has been integrated with rice production since the 1900s (Niane and Ndong 2015). The decline in inland fish catches is mainly due to droughts and changes in river flow rates. Dams, overexploitation, dilapidated pirogue fleets, fishing techniques, etc., are other disabling factors (MPTM 2001).

3.4 Adaptation Strategies

West African populations have been grappling with rainfall variability for a long time. As a consequence, people have developed strategies to find water in all seasons, regardless of precipitation volume and temporal distribution. These initiatives are now supported by states and international institutions.

In the Sahelian zone, water collection technologies are many. Farmers collect rainwater from roofs and channel spring water to reservoirs. In case of drought, stored water is used for about five months, depending on the capacity of the tank (Ajani et al. 2013). In vegetable gardens, water is stored in basins fed by PVC pipes connected to water pumps. This technique reduces water loss compared to irrigation canals (Nkwade et al. 2014). Water can also be stored in retention basins. In Bobo-Dioulasso, residents use pumps and increase the depth of wells to address a lack of water (Toe 2014). In Senegal, the deepening and cleaning out of wells ensure a continuous water supply (UICN 2011).

3.4.1 Changes in Crops and Varieties

Two main government departments provide technical guidance to farmers growing perennial and food crops in Ivory Coast. Firstly, the National Center for Agronomic Research (CNRA: Centre National de Recherche Agronomique) is remarkably effective in researching and implementing new plant varieties adapted to local climate requirements. This department, through its many research sites, innovates in crops and varieties. For example, in the cocoa sector, several varieties with improved resistance to rainfall and climate variations have emerged. Secondly, the National Rural Development Support Agency (ANADER: Agence Nationale d'Appui au Développment Rural), another government department, provides support and expertise to the farming community on the ground, by offering new varieties of cocoa from the National Center for Agronomic Research (CNRA). These efforts support production with high yields by using climate-resistant crops. These new, improved varieties of cocoa are gradually replacing traditional varieties that have been grown in orchards over thirty years. The ANADER also trains farmers in agroforestry techniques and codes of conduct, and in the proper use of plant protection products to control pests and destructive diseases. For instance, cocoa pod mirids have a negative effect on yield and production. In a food crop sector like yams, both agencies are involved and help introduce early varieties.

In Benin, the choice of varieties and crops is decisive for harvests. Farmers use improved seed varieties to adapt food or market-garden production to climate change—for example, cowpea with four vegetative phases that correspond to variable water requirements. For this reason, farmers are increasingly using early varieties (*TVX-1850-01-F*) that require 65 days from planting to harvest and semi-late (*IT81D-1137*) varieties of 70–80 days, while gradually reducing the use of late varieties (*TN98-63*) of 90–100 days. Early varieties such as 75-day corn are grown in some areas to cope with the rainfall deficit. The same is true for the 90-day groundnut (*TS-32-1*), which is adopted instead of the 120-day groundnut (*69–101*). The early 75-day corn crop is grown in association with beans at the expense of 90-day corn crops for the sole purpose of overcoming constraints related to irregular and inadequate rainfall. Early varieties thus have the advantage of reaching their full development before the onset of floods or droughts. However, a major limitation of this approach remains the supply of seeds that doesn't meet demand.

In Burkina Faso, farmers are abandoning local varieties of cereal such as sorghum, millet, and corn. The new early varieties do not exceed 60 days to ripen. Creeping plants of the *Cucurbitaceae* family are also grown, supported by agricultural research institutes. The lowlands, which are used for rice-growing during the rainy season, are exploited for corn production or market gardening in the off-season (GWP 2010).

In Senegal, too, early varieties are preferred. The adoption and diversification of crop varieties, especially those developed by research institutes such as the Senegalese Institute for Agricultural Research (ISRA: Institut Sénégalais de Recherches Agricoles), allow farmers to better adapt to the shortening of rainy seasons and the recurrence of dry spells. Growing cowpeas (*Vigna unguiculata*) has become widespread in Senegal because of their quick growth ($\leq$60 days). *Souna* and *IBV*

8004 millet varieties ripen in 75 and 95 days, respectively. The *55-437* and *fleur-11* varieties of groundnut can complete their cycles in 90 days (MA 2012).

3.4.2 Changes in Farming Practices

In Ivory Coast, several changes in farming practices have been implemented because of irregular rainfall. We will consider two major events that resulted in changes in the cropping calendar and the reorientation of farming lands in the region of Bouaké in the heart of the savannah. New farming techniques, in addition to schedule modifications, have included staking fruit trees and mulching mounds. The farming calendar had usually begun in March with the clearing and removal of grass in new plots. This practice has become obsolete due to evolution of the temporal distribution of rainfall in the region. A one-month delay at the beginning of the rainy season has shifted this process from March to April. New agro-climatic conditions compel farmers to reorient toward increasingly colonized lowlands and to rice-growing and market gardening.

Farming changes have also been made in Benin in order to maintain good harvests. Crop association is a system that consists of growing on small farms with several crops in the same plot, generally in very fertile backyard gardens. This system allows farmers to diversify production and avoid repeated weeding. The crop associations most frequently encountered in the study area are shown in Table 2.

Crop association also responds to erosion caused by rainwater on fragile ferruginous soils. Moreover, it has the enormous advantage of achieving a significantly higher yield than for crops grown separately. In rural communities, this technique is used because different plants draw different nutrients from the soil, and some plant residues produce nutrients for other plants. In the event of a break in the rainy season, plants with low water requirements can still produce acceptable yields (Houndénou 1999) and in case of excessive rainfall, plants that demand water are more resistant. Crop association also optimizes a short rainy season. Other forms of adaptation being explored are the exploitation of lowlands, ridging, crop rotation—notably with yams, cotton, and cassava—and the abandonment of the empirical farming calendar that is no longer suitable to the changing climate.

Table 2 Crop associations in the Ouémé Watershed

Associated crops						
Yam + Millet	Yam + Sorghum	Yam + Corn	Yam + Bean	Yam + Okra + Chili	Corn + Millet	Corn + Sorghum
Corn + Beans	Sorghum + Groundnut	Corn + Groundnut + Cassava	Cassava + Sorghum	Cassava + Millet	Cassava + Millet + Beans	Cassava + Yam + Sorghum

Source Koudamiloro (2017)

In Burkina Faso, plowing reduces runoff by holding water in the soil. When the plants reach an advanced vegetative stage, ridging, with well-known techniques such as *zaï*, *demi-lune,* and mulching, allows the concentration of water in furrows. These techniques use stony barriers, bunds, micro-dams, lowland development, etc. (GWP 2010).

In Senegal, because of irregular precipitation, especially pronounced from 1968 to 1998, farmers, on their own initiative or with the support of non-governmental organizations, have developed strategies through crop diversification and conversion to other economic sectors. Farmers are unable to accurately assess the effects of precipitation until August, when it is too late to sow on the plateau, so they generally take advantage of the opportunities offered by the topography of their respective environments, and develop practices that respond to the topo-sequence and the amount of time ponds remain in different segments of the slope. Thus, lowland areas are exploited with rice varieties adapted to the limited amount of water available in short rainy seasons, which clears the banks while exposing rice-growing valleys to silting. Moreover, the development of market gardening in rural areas, especially in the dry season, indicates the awareness of farming communities to climate hazards.

3.4.3 Crop Insurance as a Response to Changes in Rainfall

Climate insurance works by taking into account the level from which bad climatic conditions start to negatively impact plants, known as the **trigger threshold**, and

Table 3 Range of index insurance products in Ivory Coast, Burkina Faso, and Senegal

Countries	Crops	Covered risks	Input data	Precision
Ivory Coast	Corn	Drop-in production	Vegetation, NDVI*, Satellite Emodis	5 km × 5 km
Burkina Faso	Multi-cereal*	Water deficit (3 phases)	Climate, Rainfall-estimate Satellite NOAA ARC 2	10 km × 10 km
Senegal	– Groundnut – Corn – Millet – Rain-fed rice	– Sowing failure – Water deficit (3 phases) – Dry spells	Climate, Anacim's* weather stations, automatic and mechanical	5-to-7-km radius of the rain-gauge
	Multi-cereal	Water deficit (2 phases)	Climate, Rainfall-estimate Satellite NOAA ARC 2	10 km × 10 km

(Kara and Weber 2017)

* Multi-cereal: several cereals

* ANACIM: National Agency of Civil Aviation and Meteorology (Senegal)

* NDVI: Normalized Difference Vegetation Index

the level at which climatic conditions are intolerable for plants, when they enter a state of total water stress, known as the **exit threshold** (Shadreck et al. 2017). These two thresholds are generally observed during different growing phases of crops in the rainy season. When the trigger threshold is reached, minimum insurance compensation is paid. The amount increases progressively to a maximum of 100% when the meteorological status corresponds to the exit threshold.

Index climate insurance was introduced in 2011 in Ivory Coast, Burkina Faso, and Senegal. In Benin, it existed from 2013 to 2015. Covered risks are generally limited to sowing failure and lack of water during the different growing phases: vegetative, flowering, and ripening. Two other phases can also be covered, either the beginning or/and the end of the cropping season, as is the case with the satellite index (Table 3).

Between 2011 and 2017, payouts varied from one year to another depending on weather conditions. The year with the most payouts was 2014 (Table 4).

Table 4 Index insurance statistics between 2011 and 2017 (PlaNet Guarantee)

Ivory Coast								
Years	2011	2012	2013	2014	2015	2016	2017	Total
Insured farmers						242	850	1092
Premiums (Euros)						5 999	20 238	260236
Claims (Euros)						2 723	–	2 723
Claims ratio	45% (end 2016)							
Benin								
Years	2011	2012	2013	2014	2015	2016	2017	Total
Insured farmers			48	1 099	1 200			2 347
Premiums (Euros)			812	10 178	11 113			22 103
Claims (Euros)			811,8	6 879				7 691
Claims ratio	35% (end 2015)							
Burkina Faso								
Years	2011	2012	2013	2014	2015	2016	2017	Total
Insured farmers	194	1 340	2 332	8 298	5 565	7 476	4 080	29 091
Premiums (Euros)	3 615	16 859	55 615	95 684	48 714	36 104	25 282	278 256
Claims (Euros)		17 059	20 262	88 116	54 202	7 414	–	187 053
Claims ratio	74% (end 2016)							
Senegal								
Years	2011	2012	2013	2014	2015	2016	2017	Total
Insured farmers		60	2 326	5 660	8 962	12 957	14 247	44 212
Premiums (Euros)		1 615	43 170	92 031	189 496	294 581	337 445	958 339
Claims (Euros)		0	23 122	68 334	71 651	114 738	–	277 844
Claims ratio	45% (end 2016)							

"–" means no data available

In Benin, insurance appears to be the most effective tool for sustainable risk management in farming. Indeed, the purpose of managing farming-related risk is to ensure the sustainability of farms while reducing their exposure to risk and supporting farmers who are coping with exceptional disaster. It involves defining levels of risk exposure and corresponding degrees of coverage. The Mutual Crop Insurance of Benin (AMAB: Assurance Mutuelle Agricole du Bénin) is the multi-risk crop insurance syndicate established in that country. The main crops covered are corn, cashews, cotton, cassava, groundnuts, cowpeas, and yams. To participate, farmers must first subscribe to a multi-risk crop insurance policy after completing simple membership formalities and payment of a premium of Euros 12 for six months of coverage. However, farmers surveyed indicate that improvements are needed to optimize the operation of the AMAB.

3.4.4 Adaptation of Breeders

In Ivory Coast, to alleviate the effects of climate change, cattle farmers in the region of Boundiali, in the north of the country, changed the raising of livestock by strengthening the feed system with crop byproducts. This resulted in the evolution of a breeding system from purely nomadic to increasing adoption of transhumance (cycling livestock between highlands and lowlands seasonally). To boost the development of pastoral activity in Boundiali, for example, government departments (the former SODEPRA and ANADER) built agro-pastoral dams on both sides of the territory of Boundiali. But exposure to eutrophication and lack of maintenance have rendered most of these dams unusable today. The livestock manager at the regional branch of ANADER in Boundiali reports that some of them are drying up.

In Benin, strategies developed by herdsmen to adapt to climate crises and the effects on their livelihoods are transhumance, use of legume straw for cattle feed, sharing drinkable water with livestock, and unlawful invasion of protected areas.

In Burkina Faso, transhumance, previously unknown in some areas, has become a widespread practice with rainfall variability. Villages have been built in transhumance corridors—evidence of the colonization of these areas by herders. The lack of grazing land has compelled farmers to practice semi-intensive farming and to store fodder. Herds are being recomposed, with hardy species (goats, donkeys, and camels) supplanting cattle and sheep. Herders also tend to sell part of their herds during the dry years to buy food and increase the chance of survival of their remaining animals (GWP 2010).

In Senegal, an early warning system had been set up in partnership with the Ecological Monitoring Center (CSE: Centre de Suivi Ecologique), based on regional agro-meteorologic and hydrologic data. In the Ferlo (northern Senegal), climate information warns farmers of the lack of pasture or ponds to prevent herd mortality. Alerts are broadcast through community radios, weekly markets, and on the Internet. A bushfire alert system was tested using mobile phone messaging. To prevent conflict with farmers, information on the location of potentially relevant events along pastoral corridors is provided systematically (Martin 2015). Climate constraints, which impose challenges to pastoral activity and the social practices associated with them, are compelling pastoral communities and public authorities to increasingly develop better coping strategies. A focus on ponds, especially in this part of the Sahel, has been and remains an effective albeit partial response to the difficulties of pastoral livelihoods, which are to a large extent impacted by climate variability. In addition, the migration to cities, once regarded as the preserve of farmers, now appears to be a survival strategy for pastoral communities.

3.4.5 Adaptation of Inland Fisheries

In Ivory Coast, as fishermen are losing their livelihoods because of climate change impacting their sector, some have opted for alternation and others have left the sector for other livelihoods altogether. Surveys reveal that in 2016, 73% of fishermen have opted for alternative livehoods to fishing, including farming, such as market gardening. Of those who have abandoned fishing, most are Gôdè aboriginals; the majority of Bozo Malian fishermen have remained in the fishing sector. Our surveys found, however, that 27% of Bozo Malian fishermen have completely converted to another activity. Cashews remain the most popular crop; former fishermen indicate that cashew farming is attractive because the natural environment is conducive to successful cashew production, which generates high incomes.

In Benin, fishermen have developed different strategies to resist the effects of climate change, including deepening of ponds, use of appropriate fishing equipment, prohibition of the use of pesticides in waterways, and the protection of ponds and other water bodies by ceasing all fishing activity there to allow for restoration of the fish population. Fishermen report that deepening ponds increases the amount of water and reduces the rise in water temperature. The use of appropriate fishing equipment is also a strategy to address the effects of climate change on fish production. Fishermen report that prohibition of fine mesh nets promotes sustainable exploitation of ponds, but the ban has been only modestly effective, with 60% of fishermen continuing to use these nets despite their destruction of aquatic biodiversity.

In Senegal, to address declining fish catches, fishermen convert temporarily or permanently to farm work with the development of modern farming and rice growing in northern Senegal (MPTM 2001). The first trials of aquaculture, to mitigate the effects of lower catch volumes, were conducted in the 1980s in the Senegal River Valley (Niane and Ndong 2015). Aquaculture has proved to help mitigate climate vagaries, affecting rivers, lakes, and other water bodies, and is now well-established in Senegal. Among its objectives established in 2001, the Ministry of Fisheries and Maritime Transportation has created a Fisheries and Aquaculture Institute at Cheikh Anta Diop University in Dakar. The Ministry is making plans to share experience in aquaculture with countries including Mali, Ivory Coast, and Burkina Faso. It also hopes to develop a partnership with the FAO through regular participation in the activities of the Committee for Continental Fisheries and Aquaculture (CCFA).

4 Conclusion

Several methods have allowed analysis of trends in climatic evolution in Ivory Coast, Benin, Burkina Faso, and Senegal. The decrease in rainfall corresponds to that of total availability of water. Although the Sahel has gone through spells of drought, wet periods have also been documented. Similarly, parts of the Sudanian and Guinean climate zones have experienced rainfall breaks during different time series. The inclusion of recent changes in longer periods of climate change in the Sahelian, Sudanian, and Guinean zones of West Africa has mitigated their exceptional character. In contrast, climate and hydrologic models from the IPCC predict increases or decreases for the coming decades, until 2100. Thus, even if there were a likelihood of rainfall increase in the Sahel in the 1990s and 2000s, West Africa remains subject to elusive climate variability. The 2014 deficit, in a context of relatively improved precipitation, has shown that any reactive strategy based on models, or even findings based on global averages, may prove unreliable. We must act and think locally, and now.

Crops such as yams in Ivory Coast and Benin, as well as inland fisheries, show that countries with abundant rainfall not only remain vulnerable to rainfall deficits, but are also susceptible to floods. Cross-sectoral mobility and strategies independent of climate are, therefore, necessary for effective adaptation. Climate information must, therefore, be associated with seasonal migrations and the development of alternative activities (agroforestry, solar energy production, ecotourism, etc.). Variety maps must offer a plurality of varieties for a given location so that farmers can use the most appropriate crop or variety for any given growing season. Water sobriety is a relevant alternative to manage water resources. Since most of the loss of irrigation water is through evaporation and infiltration, increasing available water resources may allow for the exploitation of more land.

The management of water resources in West Africa faces the challenges of spatial and temporal distribution. Crop insurance can mitigate the effects of unequal distribution by financially offsetting difficult periods and deprived climatic regions.

Transboundary rivers are natural systems of water distribution in West Africa. Analyses indicate that in the region's water tower (the Fouta Djallon Mountains), only slight variations in rainfall and runoff are recorded, compared to the rest of the region. Despite the effort to exploit water resources of transboundary rivers, political and financial obstacles have interfered with exploiting the full potential of those rivers. People living in areas located more than one kilometer away from waterways prefer to use groundwater, which is now easier to access by drilling and pumping (solar energy is used in some areas in Senegal).

Climate change patterns in West Africa demonstrate the crucial importance of hydrological potential (both surface water and groundwater), despite multiple signals and warnings. The effective management of water resources improves many opportunities in farming, and also in inland fishing. In addition, it helps restore the population balance between deserted places and overpopulated capital cities. The climate context of West Africa reveals previously unknown facts about the relationship among climate, water, and society. Population growth, constant demand for water in cities and factories, and changes in lifestyle are all factors that increasingly affect water resources.

References

Agnew CT (2000) Using the SPI to identify drought. Drought Network News (1994–2001), National Drought Mitigation Center, University of Nebraska, Lincoln
Aguiar LAA, Garneau M, Lézine AM, Maugis P (2010) Evolution de la Nappe des Sables Quaternaires dans la région des Niayes du Sénégal (1958–1994): relation avec le climat et les impacts anthropiques. Sécheresse 21:97–104
Ajani EN, Mgbenka RN, Okeke MN (2013) Use of indigenous knowledge as a strategy for climate change adaptation among farmers in sub-Saharan Africa: implications for policy. Asian J Agricultural Extension, Economics & Sociology 3:23–40
Barrat JM (2012) Changements climatiques en Afrique de l'Ouest et conséquences sur les eaux souterraines. Banque africaine de Développement-Observatoire du Sahara et du Sahel, Abidjan, Côte d'Ivoire
Bationo AF (2015) Poisson au Burkina: La production nationale couvre seulement 20% des besoins. Ecodufaso du 5 août 2015. http://:www.ecodufaso.com. Accessed 11 nov 2018
Bodian A, Dezetter A, Dacosta H, Ardoin-Bardin S (2013) Impact du changement climatique sur les ressources en eau du haut bassin du fleuve Sénégal. Revue de Géographie du Laboratoire Leïdi 11:236–251
Boyer JF (2002) KhronoStat Statistical time series analysis software. Maison des Sciences de l'Eau, Institut de Recherche pour le Développement, Montpellier
Callo-Concha D (2018) Farmer perceptions and climate change adaptation in the West Africa Sudan Savannah: reality check in Dassari, Benin, and Dano, Burkina Faso. Climate. 10.3390/cli6020044
Camara M, Diédhiou A, Sow BA, Diallo MD, Mbaye I, Diallo I (2013) Analyse de la pluie simulée par les modèles climatiques régionaux de CORDEX en Afrique de l'Ouest. Sécheresse 24:14–28
Coulibaly N, Coulibaly TJH, Mpakama Z, Savané I (2018) The impact of climate change on water resource availability in a trans-boudary basin in West Africa: the case of Sassandra. Hydrology. https://doi.org/10.3390/hydrology5010012
Dakpogan N (2005) Contribution à la restauration écologique du lac Nokoué. Mémoire de maîtrise en géographie, Université Abomey-Calavi, FLASH, DGAT, Bénin
Dasylva S, Cosandey C (2005) L'exploitation de la Nappe des Sables Quaternaires pour l'alimentation en eau potable de Dakar: une offre compromise par l'insuffisance de la recharge pluviométrique. Géocarrefour 4:249–358. https://doi.org/10.4000/geocarrefour.1385
Diakité MM (2017) Caractérisation de la variabilité hydroclimatique dans le haut bassin du fleuve Sénégal. Thèse de doctorat, Université Cheikh Anta Diop, Dakar, Sénégal
Diallo I, Sylla MB, Camara M, Gaye AT (2012) Interannual variability of rainfall over the Sahel based on multiple regional climate model simulations. Theor Appl Climatol
Diédhiou YM (2008) Impacts de l'hivernage 2007 sur les cultures dans le sud de la communauté rurale de Bona. Mémoire de Maîtrise, Université Cheikh Anta Diop, Dakar, Sénégal
Diop C, Sambou PC, Sagna P, Yade M (2016) Précipitations de l'hivernage 2014 dans la partie occidentale du Sénégal et leurs impacts sur les rendements de mil et d'arachide. Actes du colloque en hommage au professeur Fulgence Afouda, Risques et catastrophes climatiques: Vulnérabilité et adaptation en Afrique de l'Ouest, Vol. 2: Agroclimatologie, Abomey-Calavi, Benin, pp 79–90
Diop M (1999) Caractérisation du facteur hydrique en agriculture pluviale au Sénégal : cas de l'arachide et du mil. Thèse de doctorat, Université de Bourgogne, Dijon

Dipama JM (2016) Changement climatique et agriculture durable au Burkina Faso: stratégies de résilience basées sur les savoirs locaux. Rapport d'étude. Innovation Environnement Développement en Afrique (IED Afrique) et Promouvoir la Résilience des Économies en zone Sémi-Arides (PRESA), Dakar, Sénégal

El Vilaly A, El Vilaly MA (2013) Climate change and water resources in West Africa: transboundary river basins. Report by Tetra Tech ARD, Burlington, Vermont, United States Agency for International Development (ARCC: African and Latin American Resilience to Climate Change Project), Washington DC

Essou GRC, Brissette F (2013) Climate change impacts on the Ouémé River, Benin, West Africa. Earth Sci Clim Change. 10.4172/2157-7617.1000161

Gautier D, Denis D (2016) Locatelli B (2016) Impacts of drought and responses of rural populations in West Africa: a systematic revue. WIREs Clim Chang 7:666–681. https://doi.org/10.1002/wcc.411

GWP: Global Water Partnership/West Africa (2010) Inventaires des stratégies d'adaptation au changement climatique des populations locales et échanges d'expériences de bonnes pratiques entre régions du Burkina Faso. GWP/WA, Ouagadougou, Burkina Faso

Hammond Murray-Rust D and Fakhruddin SHM (2014) Climate change and water resources in West Africa: an assessment of groundwater management. Report by Tetra Tech ARD, Burlington, Vermont, United States Agency for International Development (ARCC: African and Latin American Resilience to Climate Change Project), Washington DC

Houndénou C (1999) Variabilité climatique et maïsiculture en milieu tropical humide. L'exemple du Bénin, diagnostic et modélisation. Thèse de doctorat, Université de Bourgogne, Dijon

Houssou-Goé SSP (2008) Agriculture et changements climatiques au Bénin : risques climatiques, vulnérabilité et stratégies d'adaptation des populations rurales du département du Couffou. Thèse pour l'obtention du diplôme d'ingénieur agronome, Université Abomey-Calavi, Bénin

Hubert P, Carbonnel JP, Chaouche A (1989) Segmentation des séries hydrométéorologiques - Application à des séries de précipitations et de débits de l'Afrique de l'Ouest. J Hydrol 110:349–367

Kaboré PN, Ouédraogo A, Sanon M, Yaka P, Some L (2017) Caractérisation de la variabilité climatique dans la région du Centre-Nord du Burkina Faso entre 1961 et 2015. Climatologie 14:82–95

Kara A, Weber S (2017) L'assurance indicielle en Afrique de l'Ouest. Risques. Les cahiers de l'assurance 109:23–28

Koffi KE, Diomandé BI (2015) Stress hydrique sur le lac de Kossou dans un milieu physique en mutation, in ANYASA, Université de Lomé, Laboratoire de Recherche sur la Dynamique des Milieux et des Sociétés, 4:95-110

Koudamiloro O (2017) Vulnérabilité aux risques hydroclimatiques et stratégies d'adaptations des populations dans le bassin de Ouémé à Bétérou (Bénin). Thèse de doctorat, Université de Ouaga 1 Pr Joseph Ki-Zerbo

Kpadonou RAB, Adégbola PY, Tovignan SD (2012) Local knowledge and adaptation to climate change in Ouémé Valley, Benin. African Crop Sci J 20(s2):181–192

Lawin AE, Akponikpè PBI, Jalloh A, Nelson CG, Thomas ST (2013) Benin. In. Jalloh A, Gerald CN, Timothy ST, Zougmoré R, Roy-Macauley H (ed) West African agriculture and climate change: comprehensive analysis, pp 53–77

Lee FS, Heghinian M (1977) A shift of the mean level in a sequence of independent normal random variables: a Bayesian approach. Technometrics 19(4):503–506

L'Hôte Y, Mahé G, Somé B, Triboulet JP (2002) Analysis of a Sahelian annual rainfall index from 1896 to 2000: the drought continues. Hydrol Sci 47:563–572

Le Lay M (2006) Modélisation hydrologique dans un contexte de variabilité hydroclimatique : une approche comparative pour l'étude du cycle hydrologique à méso-échelle au Bénin. Thèse de Doctorat, Institut National Polytechnique, Grenoble

Le Lay M, Galle S (2005) How changing rainfall regimes may affect the water balance: a modeling approach in West Africa. Proceedings of the symposium held during the seventh International

Association of Hydrological Sciences' (IAHS) scientific assembly, Regional hydrological impacts of climate change, Hydroclimatic variability, Foz do Iguaçu, Brezil. IAHS, 296, pp 203–210

Leroux M (1995) Dynamique de la Grande Sécheresse Sahélienne. Revue de Géographie de Lyon 3–4:223–232

Leroux M (2010) Dynamic analysis of weather and climate. Springer-Praxis Publishing Chichester, UK, Atmospheric Circulation, Perturbations, Climatic Evolution

MA: Ministère de l'Agriculture du Sénégal (2012) Catalogue variétale. MA, Dakar

Magistro J, Lo M (2001) Historical and human dimensions of climate variability and water resource constraint in the Senegal River Valley. Clim Res 19:133–147

Malou R (2004) Impact du climat sur les ressources en eau souterraines en zone soudano-sahélienne. Thèse de Doctorat, Université Cheikh Anta Diop, Dakar, Sénégal

Martin A (2015) Les solutions face aux changements climatiques. Habbanae le journal des donateurs d'Agronomes & Vétérinaires Sans Frontières. No 17, 3e trimestre, septembre 2015

Mbaye ML, Hagemann S, Haensler A, Stacke T, Gaye AT, Afouda A (2015) Assessment of climate change impact on water resources in the Upper Senegal Basin (West Africa). Am J Clim Change 4:77–93

McKee TB, Doesken NJ, Kleist J (1993) The relationship of drought frequency and duration to time scales. In: Proceedings of the eighth conference on applied climatology. Anaheim, California, 17–22 Jan 1993, American Meteorological Society

MECV: Ministère de l'Environnement et du Cadre de Vie du Burkina Faso (2007) Programme d'Action National d'Adaptation à la variabilité et aux changements climatiques (PANA). Ouagadougou, Burkina Faso

MEHU: Ministère de l'Environnement, de l'Habitat et de l'Urbanisme du Bénin (2011) Deuxième communication nationale de la république du Bénin sur les changements climatiques. Cotonou, Bénin

MEPN: Ministère de l'Environnement et de la Protection de la Nature du Sénégal (2005) Rapport sur l'état de l'environnement au Sénégal, Dakar, Sénégal

MEPN: Ministère de l'Environnement et de la Protection de la Nature du Bénin (2008) Programme d'Action National d'Adaptation aux effets néfastes des changements climatiques, Cotonou, Bénin

MPAH: Ministère de la production animale et halieutique (2016) Développement de la pêche continentale in « Plan stratégique de développement de l'élevage, de la pêche et de l'aquaculture en Côte d'Ivoire (PSDEPA 2014–2020) », Tome I: Diagnostic–Stratégie de développement–Orientations

MPTM: Ministère de la Pêche et des Transports Maritimes du Sénégal (2001) Stratégie de développement durable de la pêche et de l'aquaculture. Tome 2, Dakar, Sénégal

Niane A, Ndong D (2015) National aquaculture sector overview. Senegal. National aquaculture sector overview Fact Sheets. In : Département des pêches et de l'aquaculture de la FAO. Rome http://www.fao.org/fishery/countrysector/naso_senegal/fr. Updated 17 July 2015. Accessed 19 Oct 2018

Nicholson SE, Dezfuly AK, Klotter D (2012) A two-century precipitation dataset for the continent of Africa. Bull Am Meteor Soc 93:1219–1231. https://doi.org/10.1175/BAMS-D-11-00212.1

Nkwade A, Magistro J, Horjus P (2014) Adaptation to the Sahel's changing climate: local Approaches: an economic and technical feasibility analysis of adaptation techniques in Niger, Burkina Faso and Mali. Catholic Relief Services, Baltimore, Maryland

ORSTOM (1973) Précipitations journalières de l'origine des stations à 1965, Comité Interafricain d'Etudes Hydrauliques (CIEH). République française, Paris

Ouoba AP (2013) Changements climatiques, dynamique de la végétation et perception paysanne dans le Sahel burkinabé. Thèse de doctorat, Université de Ouagadougou, Burkina Faso

Oyebande L, Odunuga S (2010) Climate change impact on water resources at the transboundary level in West Africa: the cases of the Senegal, Niger, and Volta Basins. Open Hydrol J 4:163–172

PAM Programme Alimentaire Mondiale (2018) Rapport de suivi de l'indemnisation assurance agricole indicielle dans les régions de Tambacounda et Kolda. Programme initiative en faveur de la résilience des communautés, PAM, Dakar, Sénégal

Pettitt AN (1979) A non-parametric approach to the change-point problem. App Statist 28(2):126–135
Sagna P (2005) Dynamique du climat et son évolution récente dans la partie Ouest de l'Afrique occidentale. Thèse de doctorat d'Etat ès Lettres, Université Cheikh Anta Diop, Dakar, Sénégal
Sagna P, Yade M, Sambou PC (2012) Migrations de l'Equateur Météorologique, fréquences de la mousson et importance des précipitations au Sénégal en 2008 et 2009. Annales de la Faculté des Lettres et Sciences Humaines 42/B:59–75
Sagna P, Ndiaye O, Diop C, Diongue Niang A, Sambou PC (2015) Les variations récentes du climat constatées au Sénégal sont-elles en phase avec les descriptions données par les scénarios du GIEC? Pollution atmosphérique, Climat, santé, société, A propos de la COP 21: la parole aux parties-prenantes (227), https://doi.org/10.4267/pollution-atmospherique.5320
Sagna P, Chabi PAB, Diop C, Yabi B, Afouda F, Sané T (2018) Migrations de l'Equateur Météorologique, présence de mousson et précipitations au Bénin en 2014. Mélanges en hommage aux Professeurs HOUSSOU CS, HOUNDAGBA CJ, THOMAS O, La géographie au service du développement durable, Volume 1, Risques hydro-climatiques, santé des populations et changements climatiques dans les pays de l'Afrique de l'Ouest, pp 404–417
Sambou PC (2015) Évolution climatique récente, impacts et stratégies d'adaptation des populations dans les arrondissements de Sakal et de Ndande, dans la région de Louga. Thèse de doctorat unique, Université Cheikh Anta Diop, Dakar, Sénégal
Sané T (2017) Vulnérabilité et adaptabilité des systèmes agraires à la variabilité climatique et aux changements sociaux en Basse-Casamance (Sud-Ouest du Sénégal). Thèse de Doctorat unique en cotutelle internationale, Université Paris-Diderot-Paris 7 – Université Cheikh Anta Diop de Dakar, France-Sénégal
Sané T, Diop M, Sagna P (2008) Étude de la qualité de la saison pluvieuse en Haute-Casamance (Sud Sénégal). Sécheresse 19(1):23–28
Sène IM (2007) Impacts des changements climatiques sur l'agriculture au Sénégal: dynamiques climatiques, économiques, adaptations, modélisation du bilan hydrique de l'arachide et du mil. Thèse de Doctorat de 3e cycle, Université Cheikh Anta Diop, Dakar, Sénégal
Shadreck M, Groenendaal H, Dugger C (2017) Risk modeling for appraising named peril index insurance products. A guide for practitioners. Directions in development. World Bank Group, Washington DC. 10.1596/978-1-4648-1048-0
Speth, P, Chritoph M, Dickrüger B (2010) Impacts of global change on the hydrological cycle in west and northwest Africa. Springer, Theidelberg, Dordrecht London, New York
Sultan B (2011) L'étude des variations et du changement climatique en Afrique de l'Ouest et ses retombées sociétales. Habilitation à Diriger des Recherches (HDR), Université Pierre et Marie Curie, Paris
Sultan B, Lalou R, Sanni MA, Oumarou A, Soumaré MA (eds) (2015) Les sociétés rurales face aux changements climatiques et environnementaux en Afrique de l'Ouest. IRD Éditions, Paris
Tessier H (1980) Étude d'environnement lagunaire du domaine margino-littoral sud-Bénin. Étude hydrologique préliminaire du lac Nokoué. Bull Inst Géol 25:149–166
Toe CL (2014) Perception des variations climatiques et stratégies d'adaptation des populations de Bobo-Dioulasso et de sa banlieue en vue d'une gestion durable des ressources édaphiques et hydriques. Mémoire de fin d'étude pour l'obtention du diplôme de master en gestion et aménagement des écosystèmes forestiers, Université Polytechnique de Bobo-Dioulasso, Burkina Faso
UICN (2011) Intégration de l'adaptation au changement climatique dans les stratégies de réduction de la pauvreté en Afrique de l'Ouest. Rapport synthèse des études de capitalisation des connaissances, pratiques, stratégies et technologies locales d'adaptation au changement climatique au Burkina Faso, Mali et Sénégal. Rapport final, Gland, Switzerland
Vissin EW (2013) Déterminants de la vulnérabilité et stratégies d'adaptation aux changements climatiques dans le département de l'Alibori au Bénin. Revue de géographie du Laboratoire Leïdi Dynamiques des territoires et développement 10:252–266

World Meteorological Organization (WMO) (2012) Standardized precipitation index. User Guide. WMO-No. 1090, Geneva

World Meteorological Organization (WMO) (2016) Handbook of drought indictors and indices. WMO-No. 1173, Geneva

Yira Y, Diekkrüger B, Steup G, Bossa AY (2017) Impact of climate change on hydrological conditions in a tropical West African catchment using an ensemble of climate simulations. Hydrol Earth Syst Sci 21:2143–2161

Chapter 5
Climate Change Impact on Hydrological Regimes and Extreme Events in Southern Africa

Samuel Kusangaya, Dominic Mazvimavi, Munyaradzi D. Shekede, Barbra Masunga, Francesca Kunedzimwe, and Desmond Manatsa

Abstract Climate change is expected to exacerbate current stresses on water resources resulting from population growth, economic factors and land use changes, impacting on the availability of water resources and the sustainability of local ecosystems. In southern Africa, changes in the intensity, frequency and duration of weather extremes have resulted in recurrent droughts, floods, tropical cyclones among other phenomena. For example, the El Niño-Southern Oscillation effect has intensified in the past four decades (1900–2010) resulting in increased frequency of floods and droughts in Southern Africa. Furthermore, the projected increase in the frequency of droughts is likely to reduce flows and water storage. On the other hand, some regions may experience an increase in the frequency and magnitude of extreme precipitation events associated with flooding (and extreme high flows). Countries like Mozambique, South Africa, Malawi and Zimbabwe are already experiencing recurrent floods and droughts. Projected increases in the frequencies and magnitudes of such events through to 2050, coupled with rising populations in the region, suggests that the vulnerability of weather-sensitive sectors, particularly water resources and agricultural production will increase. This will affect aquatic ecosystems, water supply, irrigation, leisure, and hydropower generation. The implications of these findings are as follows: (1) Water availability is likely to be compromised, (2) Water management measures are likely to be tightened. (3) Dependence on rain-fed agriculture is likely to increase food insecurity for most southern African countries as a result of reduced agricultural production. (4) Similarly, electricity generation is likely to

S. Kusangaya (✉) · M. D. Shekede · B. Masunga · F. Kunedzimwe
Department of Geography and Environmental Science, University of Zimbabwe, P.O Box MP 167, Mt Pleasant Harare, Zimbabwe
e-mail: kusangayas@yahoo.com

D. Mazvimavi
Institute for Water Studies, Dept. of Earth Sciences, University of the Western Cape, Bellville, South Africa

D. Manatsa
Geography and Environmental Science Department, Free State University, Bloemfontein, South Africa

Geography Department, Bindura University of Science, Bindura, Zimbabwe

S. Diop et al. (eds.), *Climate Change and Water Resources in Africa*,
https://doi.org/10.1007/978-3-030-61225-2_5

be reduced due to increased droughts and enhanced evapotranspiration as a result of the high temperatures. It is thus concluded that southern Africa is deemed to be a particularly vulnerable region on the African continent, as multiple biophysical, political and socioeconomic stresses interact to heighten the region's susceptibility to climate change and variability and constrain its adaptive capacity. Thus, there is need to increase resilience of ecosystems and communities in light of the anticipated changes in hydrological regimes and extreme events.

Keywords Climate change · Hydrological regimes · Extreme hydrological events · Resilience of ecosystems and communities · Southern Africa

1 Introduction

The southern Africa region is predominantly semi-arid and characterised by both high seasonality and inter-annual variability of rainfall, river flows and groundwater recharge. The growing human population including population in urban areas, expansion of irrigated areas to improve food security, and industrialization are increasing the water demand while water availability is progressively becoming constrained. Some of the southern African countries such as South Africa and Zimbabwe are already considered water-stressed countries. Future increases in water demand will result in other countries in this region following suit. However, the available surface and ground water resources depend on the balance between precipitation and evapotranspiration, the very aspects which are severely impacted on by climate change.

However, climate change will have spatially differentiated effects on precipitation in various parts of southern Africa that will impact on available water resources. Changes in the seasonality of rainfall will alter hydrological regimes thus affecting the availability of water resources during both rainy and dry seasons (Kusangaya et al. 2018). Several studies have demonstrated that climate change will cause intensification of the hydrological cycle (Schulze 2005; Huntington 2006; Syed et al. 2010; Warburton et al. 2010). This will alter the occurrence of rain days, duration of wet and dry spells (Kusangaya et al. 2018).

Apart from agriculture (Pielke 2007; Crane et al. 2011), climate change induced changes on the availability of water resources will affect other sectors such as health (Githeko et al. 2000; Huntingford et al. 2007; Vermeulen et al. 2012), ecosystems and biodiversity (Eriksen and Watson 2009), and energy generation (Magadza 1996; Yamba et al. 2011). The patterns of weather-related hazards will be affected by climate change and thus impacting on human settlements causing loss of life, social disruption and economic hardships (Hulme et al. 2001; Desanker 2002). Such impacts will be felt more by the resource poor in Southern Africa (Magadza 1996; Ngigi 2009).

Nevertheless, there are still uncertainties regarding the rate of climate change, and the spatial and temporal distribution of these impacts over southern Africa.

However, the persistence of the already observed changes in climate over the past century (Solomon et al. 2007; IPCC 2014) like the recent increase in the occurrence of extreme events that are accompanied by enhanced magnitude and severity entails amplified impacts on future hydrological regimes. Thus, there are still uncertainties about the specific effects of climate change in different river basins in southern Africa. This constrains development of river basin specific adaptation measures. Despite these uncertainties, there is a need to have an improved understanding of potential hydrological effects of climate change on water resources, e.g. seasonality of the available water resources, patterns of extreme events such as floods and droughts. This chapter aims to increase awareness about the potential hydrological effects of climate change in southern Africa.

2 Water Resources in Southern Africa

The Southern African Development Community (SADC) comprises the following countries: Angola, Botswana, the Democratic Republic of Congo, Lesotho, Madagascar, Malawi, Mauritius, Mozambique, Namibia, Seychelles, South Africa, Swaziland, United Republic of Tanzania, Zambia and Zimbabwe. Apart from Sychelles, Mauritius and Madagascar, the rest of the Southern African countries have 15 shared river catchments (Fig. 1).

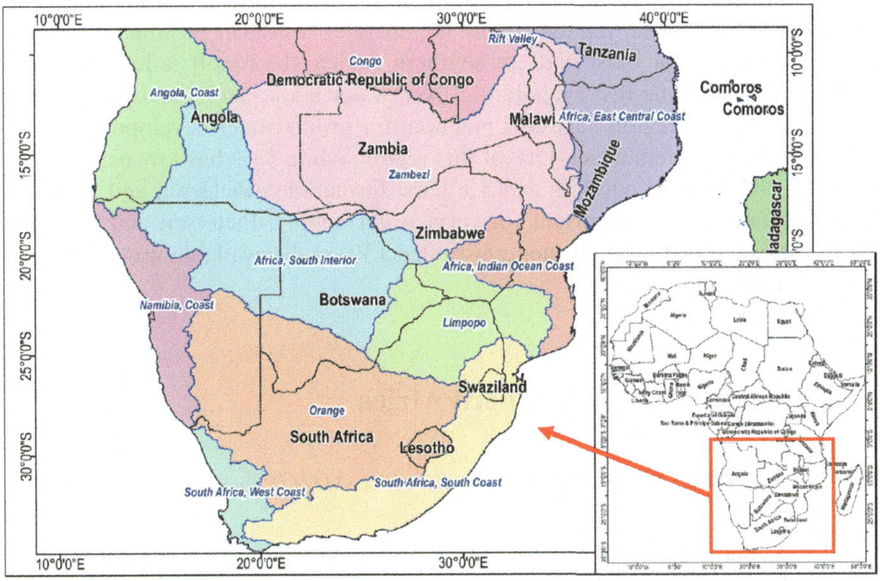

Fig. 1 Location of southern African countries and the major transboundary river catchments

According to Tshimanga (2012) river catchments with considerable total renewable water resources will play an important role in regional cooperation. About seventy percent (70%) of surface water resources are in 15 transboundary river catchments (Fig. 1 and Table 1). Five SADC countries have water resource dependency ratios of over 50%, that is, they rely on water generated outside their borders to supply more than half of their total water requirements (Kistin et al. 2009; Malzbender et al. 2009). Kusangaya et al. (2017) summarised the water resources characteristics for the Southern African region as follows:

1. About 70% of the available freshwater resources are from shared water resources.
2. Water is unevenly distributed spatially and temporally with abundant water resources in the northern and eastern areas while the south western areas are predominantly dry.
3. Most of the rivers have highly seasonal flows with surplus water during the rainy season, and water deficit during the dry season.
4. Surface water sources are sensitive to inter-annual variation of rainfall, and frequent drought periods, most of the arid and semi-arid parts experience dire shortages of water.
5. There is a very low conversion rate of rainfall to runoff with most parts having runoff coefficients less than 0.15. Most of the available runoff especially in South Africa is generated from the few mountainous catchments.

Most parts of southern Africa lying south of the 17°S latitude receive less than 650 mm/yr of rainfall with the exception of areas of high altitudes like hilly and mountainous areas. North of this latitude, the average rainfall varies mostly from 900–2000 mm/yr. Mean annual runoff is generally less than 50 mm/yr south of the 17°S latitude. For most parts of southern Africa, the runoff is less than 10% of the rainfall with the rest evaporating from water bodies, soils and plants. Most parts of the SADC region have low potential for groundwater development. Intergranual porous aquifers cover 45% of this region, while 33% has low permeability aquifers, 19% fissured aquifers and 3% karst formations (Pietersen and Beekman 2016. Groundwater is important for rural water supply. Pietersen and Beekman (2016) estimated that groundwater constituted 13% of the available water resources in SADC or 7199 m^3/yr/capita.

3 Climate Change in Southern Africa

3.1 Temperature Changes in Southern Africa

Analysis of both remote sensed and observed temperature data in southern Africa shows that the region has experienced a warming trend (Hughes and Balling 1996; Unganai 1996; Kruger and Shongwe 2004; Warburton and Schulze (2005a); New et al. 2006; Manatsa and Reason 2016) (Table 1). Kruger and Sekele (2013) found

Table 1 Summary of studies on observed temperature changes in Southern Africa

Author (s)	Temperature parameters analysed	Study Area	Conclusions
Monyela (2017)	Surface air temperature	South Africa	Summer drought in the complex South African rainfall regime. However, there was the inter-annual rainfall variability in South Africa
Archer van Garderen (2017)	Surface air temperature	Southern Africa	Southern Africa's seasonal rainfall and temperatures are strongly linked to ENSO (e.g. Climate driver of southern African seasonal-to-inter-annual variability
Lenton et al. (2017)	Surface air temperature	Southern Africa	Temperature swings to hit poor countries hardest During season with maximum insolation, temperature variability increases by ~15% per degree of global warming in Southern Africa
MacKellar et al. (2014)	Observed trends in maximum and minimum temperature	South Africa	There is a clear signal of increased temperatures since 1950 Maximum temperatures have increased significantly throughout the country for all seasons Increases in minimum temperatures exist for most of the country Except central interior, where minimum temperatures have decreased significantly

(continued)

Table 1 (continued)

Author (s)	Temperature parameters analysed	Study Area	Conclusions
Kruger and Sekele (2013)	Annual mean minimum Annual mean maximum Diurnal temperature range	South Africa	Warm extremes increased and cold extremes decreased for all of the weather stations The trends vary on a regional basis, both in magnitude and statistical significance Western SA, parts of the northeast and east of SA, show relatively stronger increases in warm extremes and decreases in cold extremes than elsewhere in the country 79% of the weather stations experienced increases in warm extremes which is statistically significant at the 5% level for maximum temperature 61% of weather stations experienced significantly negative trends, as for minimum temperature The strongest negative trends are found in the north and east (with a mean of −2.07% per decade), the southwestern Cape (with a mean of −1.67% per decade), as well as most of the Northern Cape province (with a mean of −1.28% per decade).

(continued)

Table 1 (continued)

Author (s)	Temperature parameters analysed	Study Area	Conclusions
Department of Environmental Affairs, (2013)	Maximum temperature Minimum temperature Mean annual temperature		Mean annual temperatures have increased by at least 1.5 times the observed global average of 0.65°C reported by IPCC AR4 for the past five decades Maximum and minimum temperatures show significant increases annually, and in almost all seasons. A notable exception is the central interior (zone 3, Vaal), where minimum temperatures have been increasing less strongly, and some decreases have been observed High temperature extremes have increased significantly in frequency, and low temperatures have decreased significantly in frequency annually and in most seasons across the country, but particularly in the western and northern interior of the country. The rate of temperature change has fluctuated, with the highest rates of increase from the middle 1970s to the early 1980 s, and again in the late 1990s to middle 2000s

(continued)

Table 1 (continued)

Author (s)	Temperature parameters analysed	Study Area	Conclusions
Tshiala et al. (2011)	Annual mean minimum Annual mean maximum Annual seasonal minimum and maximum Diurnal temperature range	Limpopo Province, South Africa	An increase of 0.12°C/decade in the mean annual temperature 13% of the catchments showed negative trends, while 87% showed positive trends in annual mean temperature 20% of catchments showed negative trends, while 80% of catchments showed positive trends in their diurnal temperature range Mean seasonal temperature trends of the thirty catchments were found to be 0.180C/decade and 0.090C/decade for winter and summer, respectively
Collins (2011)	Annual mean minimum Annual mean maximum Diurnal temperature range	Africa	Significant increasing temperature trends were found in Southern Hemisphere of Africa among other regions For the months June–August, South Africa saw significantly warmer temperatures in the period 1995–2010 than in the period 1979–94 When the two most recent decades are compared with the period 1979–90, warming is observed and is concentrated in the most recent decade, from 2001 to 2010 Over the period 1979–2010, there is a significant (to 1% level) upward linear trend noted with an estimate of 0.16 0C/decade.

(continued)

Table 1 (continued)

Author (s)	Temperature parameters analysed	Study Area	Conclusions
Harrison et al. (2011)	Daily minimum and maximum temperature	Mozambique	Increases in average mean temperature Increases in, both minimum and maximum temperature At Tete station, a decrease in average maximum and 90th percentile daytime temperatures
Collins (2011)	Annual mean minimum Annual mean maximum Diurnal temperature range	Africa	Significant increasing temperature trends were found in Southern Hemisphere of Africa among other regions For the months June–August, South Africa saw significantly warmer temperatures in the period 1995–2010 than in the period 1979–94 When the two most recent decades are compared with the period 1979–90, warming is observed and is concentrated in the most recent decade, from 2001 to 2010 Over the period 1979–2010, there is a significant (to 1% level) upward linear trend noted with an estimate of 0.16 0C/decade
Morishima and Akasaka (2010)	Annual mean minimum Annual mean maximum Diurnal temperature range	Southern Africa	Annual mean surface temperature has shown an increasing trend across the whole region with particularly large rates of increase in Namibia and Angola Surface temperature from Namibia to south-eastern South Africa tending to increase from July to October

(continued)

Table 1 (continued)

Author (s)	Temperature parameters analysed	Study Area	Conclusions
Aguilar et al. (2009)	Annual mean minimum and maximum Diurnal temperature range	Zimbabwe	A decrease in cold extremes and an increase in warm extremes
New et al. (2006)	Daily temperature (maximum and minimum)	SADC	Temperature extremes show patterns consistent with warming over most of the region A large proportion of stations showed a statistically significant warming trend for all temperature indices The regionally averaged occurrence of extreme cold (5th percentile) days and nights decreased by 3.7 and 6.0 days/decade, respectively The occurrence of extremely hot (95th percentile) days and nights increased by 8.2 and 8.6 days/decade, respectively The average duration of warm (cold) has increased (decreased) by 2.4 (0.5) days/decade and warm spells Diurnal temperature range (DTR) showed consistent increases in a zone across Namibia, Botswana, Zambia and Mozambique, coinciding with more rapid increases in maximum temperature than minimum temperature extremes

(continued)

Table 1 (continued)

Author (s)	Temperature parameters analysed	Study Area	Conclusions
Warburton et al. (2005b)	Annual mean minimum Annual mean maximum Summer mean maximum temperature Winter mean minimum temperature Extreme temperatures	southern Africa n (ZA, SW, LS)	Stations in the Western Cape, KwaZulu-Natal, Northern Cape and Northern Province indicate a warming of at least 0.2°C in 1980–2000 in comparison to 1950–1970 period A cooling of winter means suggests that, although the annual means of temperature may be increasing, the winter temperatures on average may be reducing There is an overall decrease in variability of, both daily minimum and maximum temperatures between 1950–1970 and 1980–2000
Kruger and Shongwe (2004)	Annual mean minimum Annual mean maximum Mean annual temperature Mean seasonal temperature	South Africa	23 (of 26) stations showed positive trends in annual mean maximum temperature, 13 statistically significant with trends higher for central stations than those closer to the coast 21 stations showed positive trends in annual mean minimum temperatures with 18 statistically significant The annual average temperature data series of 24 of the stations showed positive trends with 18 of them significant Trends of mean seasonal temperature showed that temperature trends are not consistent throughout the year

(continued)

Table 1 (continued)

Author (s)	Temperature parameters analysed	Study Area	Conclusions
Hulme et al. (2001)	Annual mean minimum Annual mean maximum Diurnal temperature range	Africa	Warming through the twentieth century has been at the rate of about 0.5°C per century with slightly larger warming in the June–August (JJA) and September–November (SON) seasons than in December–February (DJF) and March–May (MAM) The 6 warmest years in Africa have all occurred, since 1987 with 1998 being the warmest year The long-term increase in annual DTR in Zimbabwe is due almost entirely to increases during the November–February wet season; trends during the rest of the year have been close to zero Temperatures during the 1990s are higher than they were earlier in the century and are currently between 0.2 and 0.3°C warmer than the 1961–1990 average

(continued)

Table 1 (continued)

Author (s)	Temperature parameters analysed	Study Area	Conclusions
King'uyu et al. (2000)	Mean surface minimum Mean surface maximum, air temperatures	Eastern Africa (ZW, BW, MD, ML, MZ, ZA)	There were large geographical and temporal variations in the observed trends with some neighbouring locations at times indicating opposite trends The Mozambique channel region showed cooling during, both night time and daytime A significant feature in the temperature variability patterns was the recurrence of extreme values

out that warm extremes increased while cold extremes decreased for South Africa. Hulme et al. (2001) further established that temperatures were higher in the 1990s than previously in the century and are currently between 0.2 and 0.3 °C warmer than the 1961–1990 average. An increasing trend across the whole region, with particularly large rates of increase in Namibia and Angola has been identified. Overall, for Southern Africa, minimum temperatures are rising faster than maximum temperatures and there is a notable decrease in cold extremes while warm event extremes are on the increase.

An increase in temperature typically causes the intensification of the hydrological cycle (Trenberth 1999; Allen and Ingram 2002; Schulze 2005; Huntington 2006; Syed et al. 2010), as a result of the increase in turbulence and evaporation, as well as rainfall (Warburton et al. 2005b). This potentially increases the frequency of occurrence of droughts and floods (Schulze 2011) which may lead to changing rainfall patterns and hence the spatial and temporal distribution of runoff, soil moisture and ground water reserves. As such, temperature changes have a direct bearing on water resources availability within Southern Africa. A summary of the analysis of temperature changes over Southern Africa is given in Table 1.

3.2 *Rainfall Changes in Southern Africa*

Climate change detection studies on rainfall have been carried out using observed, interpolated and remote sensing derived rainfall. Nicholson et al. (2017) found a shift from relatively wet conditions of the 1920s–1950s to dry conditions from the 1970s onward. For the northernmost part of southern Africa, including Zimbabwe, an increase in rainfall in the 1970s of 6% followed by a reduction of 5% in the eighties is shown by Nicholson (2001). Morishima and Akasaka (2010) analysed rainfall data for the 1979–2007 period for southern Africa and concluded that annual rainfall has decreased over the African continent from the equator to 20°S, as well as in Madagascar resulting in a shorter and weaker rainfall season in southern Africa with rainfall in Angola, Zambia and Namibia tending to decrease from December to March. New et al. (2006) reported that spatially coherent increases in consecutive dry days were found over much of southern Africa in the past decades of the twentieth century. However, analysis of historical rainfall trends for Zimbabwe (1933–2000) by Mazvimavi (2010b) showed that there was no statistically significant rainfall reduction in Zimbabwe. This, however, contrast sharply with earlier results from Unganai (1996) who concluded that annual rainfall in Zimbabwe had declined by 10% between 1900 and 1994. Elsewhere, in South Africa, Hewitson and Crane (1996) reported increase in rainfall in regions of South Africa where orographic influences dominate based on 50 years of data (1950–1999). Increases in late summer dry-spell duration for much of the summer rainfall region were also detected in the same study.

Combined results from rainfall trend analysis suggest that changes in rainfall are contrasting and subject to considerable uncertainty, regarding the extent (spatial and temporal) and magnitude of change. This is largely because rainfall is characterised by high inter-annual variability over Southern Africa. The insufficient data records for southern Africa also make analysis challenging (Mazvimavi 2010a). These limitations render detection of rainfall changes due to climate change difficult. However, as pointed out by Warburton and Schulze (2005a) changes in rainfall have important implications for the hydrological cycle and for water resources since rainfall is the main driver of variability in the water balance, upon which humans and the environment depend. A summary of the results of rainfall analysis is given in Table 2.

4 Occurrence of Water-Related Extreme Events in Southern Africa

Several studies from southern Africa have shown that extreme hydrological events are increasing in frequency and magnitude (e.g. Kabat et al. 2002; Reason and Keibel 2004; Kadomura 2005; Patt and Schröter 2008a; Yanda 2010; Goodess 2013). Several studies show that the El Niño-Southern Oscillations are becoming more intense as a result of climate change (e.g. McMichael et al. 2006; Collins et al. 2010; Manatsa and Behera 2013). For example, in Southern Africa, the frequency of droughts is projected to increase and will most likely increase the frequency of extreme low flows and low storage episodes (Desanker and Magadza 2001). These will inevitably affect aquatic ecosystems, water supply, irrigation, leisure, and hydro power generation. On the other hand, Kusangaya et al. (2014) reported that some areas in southern Africa have, and will, experience an increase in the frequency and magnitude of heavy precipitation events which in most cases will result in flooding (and extreme high flows) (Figs. 2 and 3).

Thus, from other reports (Field et al. 2012) it is observed that (a) Africa has the second-highest number of extreme events after Asia, and, (b) Africa has the highest number of hydrological extreme events. A significant number of these hydrological (e.g. floods) and climatological (e.g. droughts) are occurring in Southern Africa. For southern Africa countries, it can be seen that (a) the occurrence of natural disasters in the last two decades (1991–2000 and 2001–2010) is by far higher than for the previous decades, with the 2001–2010 decade having the highest occurrences (Fig. 2) and (b) the occurrence of these water-related disasters has been increasing at an exponential rate. The number of natural disasters in the last decade is in most cases equivalent to the total number of disasters in the previous 3 decades (Fig. 3). It is therefore concluded that there is an increasing trend of extreme events in southern Africa (Kabat et al. 2002; Patt and Schröter 2008b). The classification of extreme water-related events used in this chapter was adopted from Centre for Research on the Epidemiology of Disasters (CRED) (Table 3)

Table 2 A summary of studies on rainfall changes in Southern Africa

Author (s)	Period of analysis	Study area	Degree of change (Conclusion)
(Bathiany et al. 2018)	Until 2100	Southern Africa	The study reveals that most of the increased temperature fluctuations in the tropics are associated with droughts—an extra threat to food and water supplies
Monyela (2017)	2014/2015 and 2015/2016	South Africa	Rainfall variability in southern Africa is affected most by El Nino
Banze et al. (2017)	1979 and 2010	Southern Africa	The study concluded that the details of climate change trends and forecasts for southern Africa can be difficult to discern from the high level of natural variability in temperature, rainfall and runoff
David Nash (2017)	late nineteenth to early twentieth century	Southern Africa	Fluctuations in both summer rainfall zone (SRZ) and winter rainfall zone (WRZ) rainfall are linked to the variability of sea-surface temperatures in the oceans surrounding southern Africa and are modulated by the interplay of large-scale modes of climate variability, including the El Niño-Southern Oscillation (ENSO), Southern Indian Ocean Dipole and Southern Annular Mode

(continued)

Table 2 (continued)

Author (s)	Period of analysis	Study area	Degree of change (Conclusion)
MacKellar et al. (2014)	1960–2010	South Africa	Observed trends in seasonal and annual total rainfall, number of rain days and daily maximum and minimum temperature were calculated for a number of stations in South Africa for the period 1960–2010. Maximum temperatures have increased significantly throughout the country for all seasons and increases in minimum temperatures are shown for most of the country Regionally aggregated trends for six water management zones covering the entire country are not evident for total rainfall, but there are some significant trends for the number of rain days
Jury (2013)	1900–2100	Southern Africa	Although rainfall trends in the satellite era were minimal in many data sets because of drought in the early 1980s, there was a significant downtrend in the Institut Pierre Simon Laplace (IPSL) simulation of −0.013 mm/day per year. When averaging the longer data sets together over the twentieth century, the southern African rainfall trend was −0.003 mm/day per year. Other key features of the analyses include a poleward drift of the sub-tropical anticyclones and a +1.5 mm/year rise in sea-surface height along the coast.

(continued)

Table 2 (continued)

Author (s)	Period of analysis	Study area	Degree of change (Conclusion)
Love et al. (2010)	1930–2005	Limpopo Basin, Zimbabwe	The annual average of maximum daily temperatures showed a significantly increasing trend High inter-annual variability of rainfall was noted Total annual rainfall, days of heavier rainfall and number of longer dry spells showed a significant drying trend for the whole time series 80% of stations showed a change point to a drier regime from around 1980 characterised by fewer wet days and increased number of dry spells
Mazvimavi (2010b)	1892–1940 1992–2000	Zimbabwe	Rainfall in Zimbabwe has high inter-annual variability and currently any change due to global warming is not yet statistically detectable The Mann-Kendal Test did not identify a significant trend at all the 40 stations and, therefore, there is no proof that the average rainfall at each of these stations has changed. All the other stations revealed no changes over time in, both the extreme low and high rainfall at the annual interval. Therefore, there is no evidence that the frequency and severity of droughts has changed during the 1892 to 2000 period

(continued)

Table 2 (continued)

Author (s)	Period of analysis	Study area	Degree of change (Conclusion)
Morishima and Akasaka (2010)	1979–2007	southern Africa	Annual rainfall has decreased over the African continent from the equator to 20ºS, resulting in shorter and weaker rainy season in southern Africa The spatial and temporal structures of trends in rainfall have apparent seasonality with rainfall in Angola, Zambia and Namibia tending to decrease from December to March There are no apparent trends over the continental region south of 20ºS, except in the north-eastern part of South Africa
Aguilar et al. (2009)	1955–2006	Zimbabwe	Zimbabwe showed no significant increases in rainfall over the last half-century
Shongwe et al. (2009)	1941–1997 1901–2000	southern Africa	Study investigates likely changes in mean and extreme rainfall over southern Africa More severe droughts in the southwest of southern Africa and enhanced rainfall farther north in Zambia, Malawi and northern Mozambique This study shows that changes in the mean rainfall vary on relatively small spatial scales in southern Africa and differ between seasons

(continued)

Table 2 (continued)

Author (s)	Period of analysis	Study area	Degree of change (Conclusion)
Kane (2009)	1900–1998	South Africa	Annual values showed considerable year-to-year fluctuations (50–200% of the mean), while five-year running means showed long-term fluctuations (75–150% of the mean) Running means over 21 years did not indicate linear trends, upwards or downwards. Instead, considerable oscillations were seen with magnitudes different in different regions (5–25%) In the largest part of South Africa, there has been no real evidence of changes in rainfall over the last century, there are, however, some identifiable areas, where significant changes in certain characteristics of rainfall have occurred over the period 1910 to 2004
Conway et al. (2009)	1901–2001	Africa + southern Africa	High inter-annual variability in southern Africa Modest inter-annual rainfall and river flow variability and quite low runoff coefficients Mean rainfall between 1931–1960 and 1961–1990 is fairly stable, as is river flow in the Zambezi, whilst in the Okavango it recorded a 14% increase

(continued)

Table 2 (continued)

Author (s)	Period of analysis	Study area	Degree of change (Conclusion)
Chamaillé-Jammes et al. (2007)	1928–2005	Hwange, Zimbabwe	No significant long-term trend in mean annual rainfall But droughts have actually worsened over the study period Dry years (0.1th quantile) became between 20 and 50% drier
Kruger (2006)	1910–2004	South Africa	In the largest part of South Africa there has been no real evidence of changes in rainfall over the last century Areas with significant decreases in annual rainfall are the northern part of Limpopo Areas with significant increases in annual rainfall include the northern part of North-West and an area covering parts of the Western Cape, Northern Cape and Eastern Cape
New et al. (2006)	1961–1999	southern Africa	No consistent or statistically significant trends across the region Regionally averaged total rainfall has decreased, but is not statistically significant Decrease in average rainfall intensity and an increase in dry-spell length

(continued)

Table 2 (continued)

Author (s)	Period of analysis	Study area	Degree of change (Conclusion)
Fauchereau et al. (2003)	1920–2000	Southern Africa	Inter-annual variability has increased, since the late 1960s and droughts have become more intense and widespread At seasonal time-scale, in the observation, southern Africa as a whole does not experience a definitive trend toward desiccation or more moist conditions
Richard et al. (2001)	1950–1988	Southern Africa	The 1970–1988 period had more variable rainfall and more widespread and intense droughts than the 1950–1969 period The January–March rainfall during the twentieth century does not show any long-term trend of desiccation nor abrupt shift over the southern African region Since the beginning of the 1980s, the annual rainfall amount has slightly decreased The strongest rainfall anomalies for the January–March inter-annual rainfall (more than 2 standard deviations) are found at the beginning of the century (e.g. severe drought of 1922) and during recent decades Anomalies are weaker (not more than 1 standard deviation) between 1935 and the middle of the 1960s. Between 1970 and 1988, droughts are more extensive, affecting northern Zambia and Mozambique more significantly Droughts are also more intense in Namibia and South Africa

(continued)

Table 2 (continued)

Author (s)	Period of analysis	Study area	Degree of change (Conclusion)
Hulme et al. (2001)	1900–2000	Southern Africa	More moderate drying of 5 to 15% per century noted along the Mediterranean coast and over large parts of Botswana and Zimbabwe and the Transvaal in southeast Africa
Nicholson (2001)	1970–1989	Africa	Rainfall percent departure from the long-term mean Southern Africa - North (1970–79) +6% and (1980–1985) −5% Southern Africa–South (1970–79) +10% and (1980–1985)−7% Southern Africa–West (1970–79) +26% and (1980–1985) −12%
Richard et al. (2001)	1950–1988	South Africa	Examination of many parameters (number of rain days, mean and median rainfall by rain day and values of the 90th and 95th percentile) revealed shifts in daily rainfall distribution from approximately the 1970s. The south-central part of the country (former Transkeï) seems to have experienced increased intense rainfall, accompanied by decreased number of rain days, while the north-eastern part shows decreased values of the 95th percentile

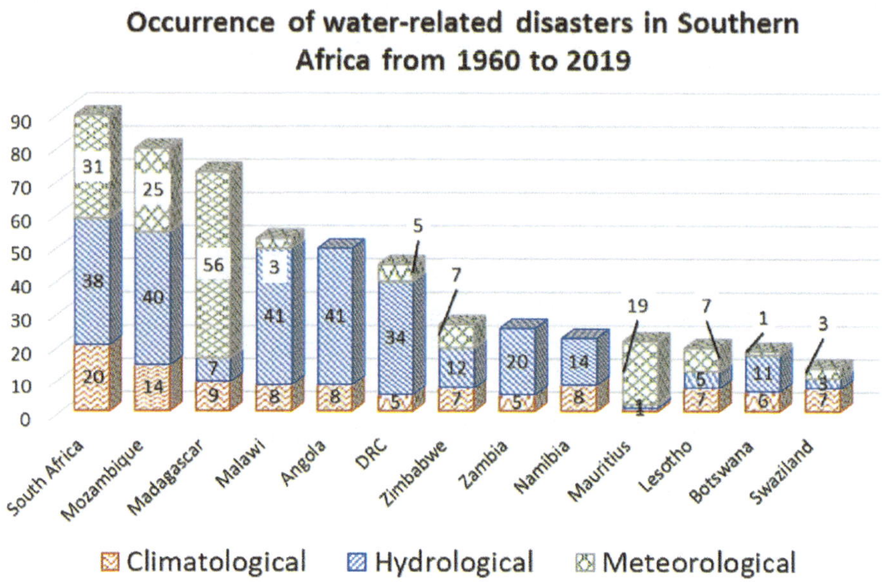

Fig. 2 Total number of recorded water-related disasters between 1960 and 2019 for Southern Africa [Source Data: Source: EM-DAT: The Emergency Events Database—Universite catholique de Louvain (UCL)—CRED, D. Guha-Sapir—www.emdat.be, Brussels, Belgium, (accessed July 4 2019)]

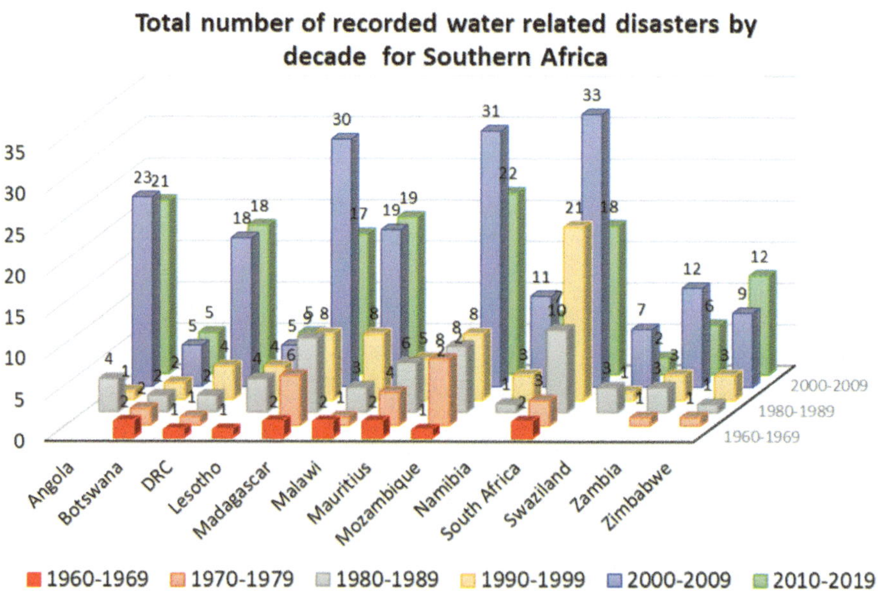

Fig. 3 Trends in the occurrence of water-related disasters for southern Africa by decade. [Source Data: Source: EM-DAT: The Emergency Events Database—Universite catholique de Louvain (UCL)—CRED, D. Guha-Sapir—www.emdat.be, Brussels, Belgium, (accessed July 4 2019)]

Table 3 Definition of Water-related disasters

Disaster subgroup	Definition	Disaster main type
Meteorological	A hazard caused by short-lived, micro- to meso-scale extreme weather and atmospheric conditions that last from minutes to days	Extreme temperature storm
Hydrological	A hazard caused by the occurrence, movement, and distribution of surface and subsurface freshwater and saltwater	Flood landslide
Climatological	A hazard caused by long-lived, meso- to macro-scale atmospheric processes ranging from intra-seasonal to multi-decadal climate variability	Drought wildfire

(*Source* The international Disaster Database (https://www.emdat.be/classification) of the Centre for Research on the Epidemiology of Disasters (CRED)

4.1 Floods and Droughts

The frequency and magnitudes of both floods and droughts will be affected by climate change as a result of changes in the intensity and frequency of rainfall. These events cause considerable damages to life and infrastructure. Floods and droughts also cause indirect adverse effects such as epidemics (Kundzewicz et al. 2008; Knox et al. 2010). Floods due to tropical cyclones frequently affect Malawi, Mozambique and Zimbabwe. Recent floods due to Cyclone Idai caused over a thousand deaths and extensive damage to infrastructure particularly for rural communities and poor urban households. Tropical cyclone activity in southern Africa has been recorded with increased frequency in recent years causing massive damage and loss of life in Mozambique, Zimbabwe, Zambia, Swaziland, South Africa, Madagascar and Mauritius (Reason and Keibel 2004; Klinman and Reason 2008) (Fig. 4 and Table 4). Tadross et al. (2005) and New et al. (2006) showed evidence of changing rainfall seasonality and extreme events. These observations ultimately, signify a significant shift of climate variables from the recent past.

Droughts cause food insecurity and considerable damage to economic production (Zinyowera and Unganai 1992; Kabat et al. 2002; Patt et al. 2008b; Tirado et al. 2010). For example, six warmest years on record for Zimbabwe occurred since 1987, with corresponding increased frequency of droughts in southern Africa since 1990 causing massive drops in crop yields in Zimbabwe, Malawi and Zambia (Rouault and Richard 2005; Manatsa et al. 2008; Gizaw and Gan 2017).

Figure 5 shows the trends in the occurrence of water-related disasters for selected Southern African countries. What is evident is that the occurrence of droughts has been increasing exponentially for the following countries: Lesotho, Malawi, Mozambique, South Africa and Madagascar (Fig. 5). Floods occurrence also exhibits the same pattern as droughts as can be seen for South Africa, Zambia, Mozambique, Malawi, Botswana and Angola. Recently, the region has been experiencing heat waves and extensive wildfires, phenomena that had never been recorded in southern

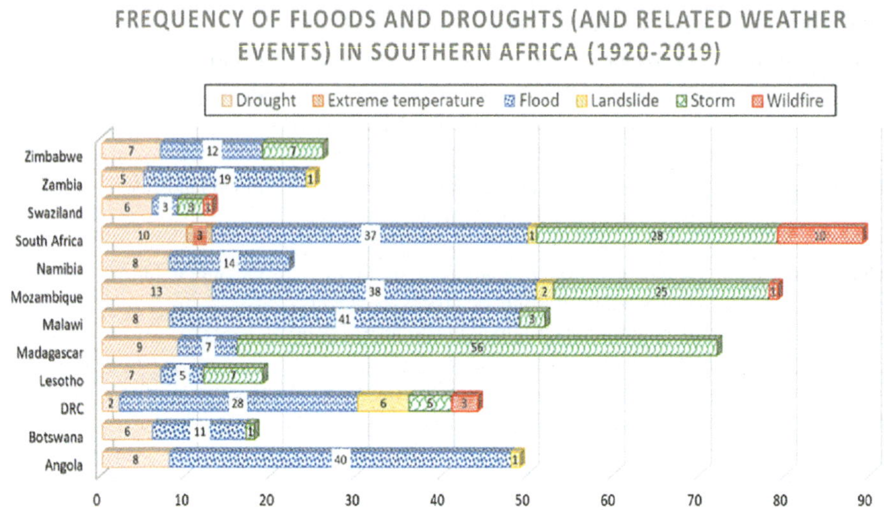

Fig. 4 Frequency of floods and droughts (and related weather events) in southern Africa (1920–2019)

Africa (Table 4 and Fig. 4). The upsurge in heat waves and wildfires are attributable to increasing temperatures. The increasing cases of recorded epidemics are also closely related to the increasing occurrence of floods due to increased cyclone activity as in the case of Mozambique, Zimbabwe, Lesotho and Malawi (Fig. 5).

5 Hydrological Regimes

River discharge is affected by several drivers such as landuse changes, water withdrawals and climate variations (Richter et al. 1996; Mathews and Richter 2007). The increase of the global temperature affects precipitation (P) and the evaporative demand (E) both of which determine the availability of water as shown in Eq. (1)

$$P = Q - E \pm \Delta S, \tag{1}$$

where Q = runoff, ΔS = change of both surface and subsurface water storages.

Precipitation and the evaporative demand set the broad limits to the amount of water available as runoff or river flows, and the replenishment of surface and subsurface storages. Climate change will affect rainfall and temperature, resulting in changes in the available water resources (Q) (Wuebbles et al. 2010). An increase in temperature will cause an increase in the evaporative demand (E) and thus affecting both runoff and subsurface water. In combination with the increasing demographic pressure and low adaptive capacity, these changes could, therefore, have significant impacts on people and sectors that depend on the availability of water. Future climate

Table 4 Occurrence of water-related disasters in southern Africa by decade

	1950–1959	1960–1969	1970–1979	1980–1989	1990–1999	2000–2009	2010–2019	Total
Angola	**3**	**18**	**28**					**49**
Drought	3	3	2					8
Flood		14	26					40
Landslide		1						1
Botswana			**8**	**10**				**18**
Drought			5	1				6
Flood			3	8				11
Storm				1				1
DRC				**44**				**44**
Drought				2				2
Flood				28				28
Landslide				6				6
Storm				5				5
Wildfire				3				3
Lesotho				**6**	**13**			**19**
Drought				2	5			7
Flood				3	2			5
Storm				1	6			7
Madagascar					**70**	**2**		**72**
Drought					9			9
Flood					7			7

(continued)

Table 4 (continued)

	1950–1959	1960–1969	1970–1979	1980–1989	1990–1999	2000–2009	2010–2019	Total
Storm					54	2		56
Malawi						**52**		**52**
Drought						8		8
Flood						41		41
Storm						3		3
Mozambique						**79**		**79**
Drought						13		13
Flood						38		38
Landslide						2		2
Storm						25		25
Wildfire						1		1
Namibia						**22**		**22**
Drought						8		8
Flood						14		14
Seychelles						**4**		**4**
Flood						2		2
Storm						2		2
South Africa						**15**	**74**	**89**
Drought						4	6	10
Extreme temperature							3	3
Flood						7	30	37

(continued)

Table 4 (continued)

	1950–1959	1960–1969	1970–1979	1980–1989	1990–1999	2000–2009	2010–2019	Total
Landslide							1	1
Storm						4	24	28
Wildfire							10	10
Swaziland							**13**	**13**
Drought							6	6
Flood							3	3
Storm							3	3
Wildfire							1	1
Zambia							**25**	**25**
Drought							5	5
Flood							19	19
Landslide							1	1
Zimbabwe							**26**	**26**
Drought							7	7
Flood							12	12
Storm							7	7
Total	**3**	**18**	**36**	**60**	**83**	**174**	**138**	**512**

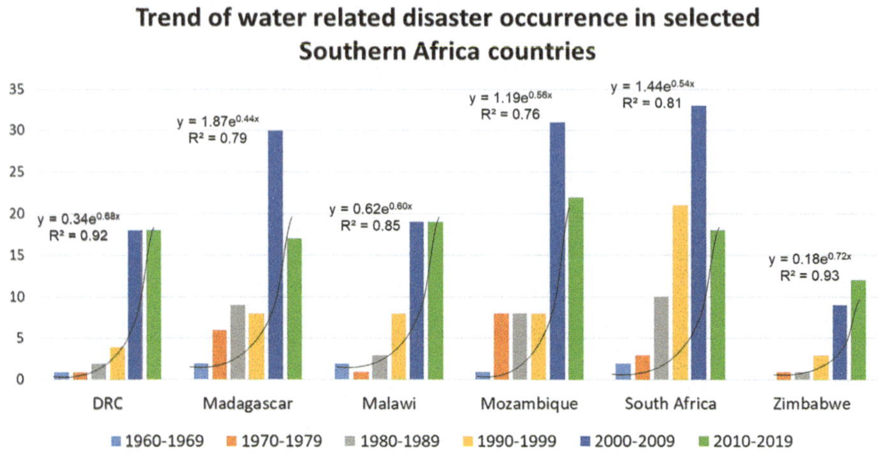

Fig. 5 Trend of water-related disaster occurrence in selected southern Africa countries

change is projected to have a significant impact on the hydrological regime, water resources and their quality in many parts of the world. The water-dependent ecosystems are exposed to the risk of climate change through altered precipitation and evaporation patterns. Table 5 provides an overview of the potential effects of climate change on hydrological regimes in southern Africa.

6 Discussion

6.1 Climate Change Impact on Water Availability

Globally, approximately four billion people experience severe water scarcity, during at least some extended periods of the year (Mekonnen and Hoekstra 2016). With the expected increase in supply and demand pressures such as land degradation, pollution, population growth and urbanisation, the availability of fresh water is likely to be compromised. Climate change has the potential to impose additional pressures in some regions such as southern Africa (Gardner-Outlaw and Engelman 1997). Kusangaya et al. (2017) showed that fourteen African countries are already experiencing water stress while 11 others will be water stressed by the year 2025, and climate change is likely to exacerbate this. By 2020, an additional 75–250 million people in Africa are projected to be exposed to increased water stress (Solomon et al. 2007). According to Vörösmarty et al. (2005) water stress is high for 25% of Africa's population with a further 13% experiencing water stress due to recurring droughts. For SADC countries, using the Falkenmark indicator of water stress (Falkenmark 1989), Lesotho, Malawi, South Africa and Zimbabwe are already regarded as water stressed (Fig. 6). Based on projected population for the 2010 to 2050 period by the

Table 5 Summary of hydrological regimes in Southern Africa

Author	Area of study	Period of analysis	Conclusion
Gyamfi et al. (2016)	South Africa	2000–2013	Runoff is increased by urbanisation and Agriculture with the former as a higher determinant of runoff.
Conway et al. (2009)	Sub-Saharan Africa	1931–1990	Many basins are marked with spatial and temporal variations in runoff strength – There is weak almost random behaviour of runoff
Schulze (2000)	Southern Africa	1975–1992	There is high spatial temporal variability in runoff
Taylor et al. (2003)	South Africa	1998–2002	– Low flow season is extended to include May and November–December flows. For all winter low flow months there are substantial reductions of occurrences of streamflows within the RVA target range, yet only slight reductions in summer high flow months
UNESCO (1997)	Southern Africa	1942–1990	– Annual river flow series for 1942 to 1990 identify late 1940s as being characterised by below normal conditions and the 1970s has a period above normal conditions. Runoff conditions during preceding winters is weak
Saraiva et al. (2015)	South Africa	1950–2011	There is high variability of streamflow because of competing water demands from irrigated agriculture, energy, forestry and industries. The impacts of these demands, relative to the natural flow regime, appear significant
Ashton et al. (2001)	South Africa	19195–1998	Hydrological cycle has significant changes in the runoff and discharge of water in rivers
Todd et al. (2011)	Africa (Sahel)	1900–2007	above-average precipitation during the 1950–1960s and persistently low precipitation during the 1970–1990s (Dai et al. 2004a), resulting in devastating droughts and low discharge in rivers

(continued)

Table 5 (continued)

Author	Area of study	Period of analysis	Conclusion
UNESCO (2004)	Southern Africa	2000–2003	Rainfall-runoff models were identified as potentially useful tools that could provide a vital component in the regional assessment of water resources availability enquired for setting up and running models
Fanta et al. (2001)	Southern Africa	1940–2000	The study has shown that runoff in the region ranges from over 320 mm year"1 in the Lower Zambezi and the highlands of Tanzania to less than 10 mm year"' in the deserts of Namibia and the Kalahari. There is evidence of declining runoff in parts of Zambia, Angola, Mozambique and the High Veld in South Africa. The recent decline seems to have started from around 1975
Sarron (2005)	South Africa	1996–2004	Farming activities in Wetlands reduce discharge of surface rivers since wetlands are river sources
Troy et al. (2007)	South Africa	2007	Landuse change accounts for small scale observed increase in runoff at catchment scale
Bullock (1992)	Zimbabwe, Zambia and Malawi	1992	Dambos reduce flows in certain regions of Zimbabwe, Malawi and Zambia

(continued)

Table 5 (continued)

Author	Area of study	Period of analysis	Conclusion
Rateb et al. (2017)	Africa	2017	In African, hydrological regimes climate conditions may contribute to Grace errors more than size of regime does Measurement and leakage errors are a result of region's aridity Relying on scale factor in water storage, estimation is insufficient for ground water storage
Sarron (2005)	South Africa	1970–2005	Continued destruction will result in decrease in water quality and less reliable supplies (especially during dry season), increased severe flooding and lower agricultural productivity
Schulze and Pike (2004)	South Africa	As of 2002	The distribution of median values in the 41-year 129 record indicates that there is greater constancy of the Mkomazi streamflows for the low flow season(e.g. 4.0 $m^3.s^{-1}$) There is a greater degree of inter-annual variation than for the high flow season (e.g. 34.6, 49.3 and 32.2 $m^3.s^{-1}$ for January, February and March). There is also a greater degree of inter-annual variation for high flow months (CDs of 1.02, 1.21 and 1.31 for December, January, February) than for low flow months (CDs of 0.65, 0.60 and 0.62 for May, June and July). Extreme rainfall events in late September 1987 have influenced the inter-annual variation of streamflows at the end of the low flow season, resulting in October

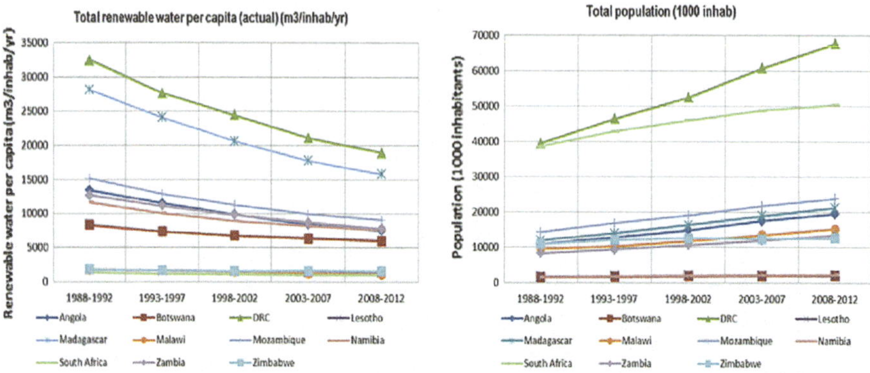

Fig. 6 Total renewable water and respective population changes for southern Africa [Data Source: FAO, 2012. AQUASTAT database—[accessed 19/11/2012, http://www.fao.org/nr/water/aquastat/data/]

United Nations, Department of Economic and Social Affairs, Population Division, Mazvimavi (2010a) showed that Tanzania will fall within the moderate water stress category by 2020 without climate change. By the year 2050, Malawi, South Africa, Tanzania and Zimbabwe will be in the high-stress category even without climate change (Mazvimavi 2010a). Thus, climate change will increase pressure on water resources which are already stressed. The population at risk of increased water stress in Africa is estimated to increase to 350–600 million people by the 2050s (Solomon et al. 2007) and up to 10% of SADC population. The water sector and related sectors will have to adapt to scarce water resources. Adaptation in the water sector could be done through conservation, improved efficiency, technological change and adoption of integrated water resources management (Matondo 2010; Schulze 2011).

6.2 Extreme Events

Several factors operating at different spatial and temporal scales have led to various extreme events in Southern Africa. Some of these extreme events are attributed to or exacerbated by climate change. While there is unequivocal evidence that the climate is changing, there are divergent views on the causes and attributed consequences of the obtaining climate change (Arnell 1999). For example, whilst some areas are experiencing an increasing number of frequent dry spells, others are experiencing totally the opposite, i.e. they are experiencing an increase inconsecutive wet spells that span a relatively long period of time. So far a number of studies from southern Africa have shown that extreme hydrological events are increasing in frequency and magnitude (e.g. Kabat et al. 2002; Reason and Keibel 2004; Kadomura 2005; Patt and Schröter 2008a; Yanda 2010; Goodess 2013). For example, the El Niño-Southern Oscillation effect has continued to strengthen in recent decades: e.g. 2000–2010

(Timmermann et al. 1999; Manatsa et al. 2011) resulting in more floods and droughts. Climate change is being attributed to the intensification of the El Niño-Southern Oscillations (e.g. McMichael et al. 2006; Collins et al. 2010; Manatsa and Behera 2013). The frequency of droughts is projected to increase and will most likely lead to an increase in the frequency of extreme low flows and low storage episodes (Desanker and Magadza 2001). Extreme low flows and low storage episodes inevitably affect aquatic ecosystems, water supply, irrigation, leisure and hydro power generation.

Given that our understanding of climate change, which points towards an increasing frequency, timing and intensity of extreme hydrological events, whose impacts are also expected to be widespread (Easterling et al. 2000; Groisman et al. 2005; Osbahr et al. 2008; Yanda 2010; Yamba et al. 2011; Goodess 2013) as exemplified by the recent (2019) impacts of cyclones Idai and Kenneth. Knowledge about the characteristics of extreme hydrological events is valuable for planning and designing infrastructure. Both urban and rural areas contain many types of hydraulic engineering structures (such as dams, levees, water distribution networks, water collection networks, sewage collection networks, storm water management, etc.) which need to be designed to accommodate peak flows of a certain magnitude in order to function safely at a given level of risk (Schulze et al. 2014). Should the structures fail, especially where human settlement is dense, there are potential societal, economic and environmental consequences. Table 6 summaries the estimated costs associated with cyclonic activity in Southern Africa.

6.3 Vulnerability to Climate Change

The combination of high climate variability and limited livelihoods options due to high levels of poverty affecting a large proportion of the population contributes to vulnerability to climate change. High climate variability that results in frequent regional exposure to floods and droughts contribute significantly to the vulnerability to current and future climate shocks to a population with limited resilience to these shocks. Large proportions in rural and urban areas do not have access to adequate access to water and sanitation. When extreme weather events such as floods and drought occur, they exacerbate an existing problem. Climate change will worsen these problems. Although efforts are being made to improve water resources management in southern Africa, the existing systems such dams developed for satisfying domestic, industrial and municipal uses are still not able to satisfy the demand in most parts of southern Africa. This is due to inadequate development of infrastructure and human capacity for managing these systems. Climate change will exacerbate these problems.

A large proportion of the southern African population has very limited livelihood options due to poverty and limited resource base. When floods and droughts occur they are not able to adopt options that cushion them from these shocks. If the existing poverty levels are not eliminated, climate change will worsen these problems. Environmental degradation in the form of soil erosion, deforestation and water pollution

Table 6 Cyclone activity and associated estimated cost for southern Africa

Cyclone	Year	Countries affected	Estimated cost in damages
Idai	2019	Mozambique, Zimbabwe, Malawi	1 billion
Ava	2018	Angola, Malawi, Mozambique, Tanzania, Zambia and Zimbabwe	156 million
Dineo	2017	Mozambique, Madagascar, South Africa	267 million
Bansi	2015	Mozambique, Malawi, Angola, Tanzania, Zambia, Northern and Central Malawi, Madagascar and Mauritius.	46.3 million
Chedza	2015	Malawi, Madagascar, Mozambique	40 million
Dando	2012	Mozambique, South Africa	Unknown
Funso	2012	South Africa, Mozambique, Madagascar	30.5 million
Irina	2012	Madagascar, Mozambique and South Africa	309 million
Favio	2007	Mozambique, Madagascar, Zimbabwe	71 million
Japhet	2003	Mozambique, Zambia, Zimbabwe, South Africa, Madagascar	200 million
Eline	2000	Zimbabwe, Mozambique, South Africa	309 million
Leon-Eline	2000	Namibia, Mozambique, South Africa, Zimbabwe	170 million
Huda	2000	South east Africa and Madagascar	Unknown
Domoina	1984	South Africa, Swaziland, Madagascar Mozambique	200 million
Imboa	1984	Mozambique, South Africa	496 million
Justine	1982	South Africa, Madagascar, Mozambique	250 million
Kolia	1980	South Africa	342 million
Emilie	1977	South Africa Mozambique	Unknown
Danae	1976	South Africa, Mozambique	Unknown
Caroline	1972	South Africa, Mozambique	Unknown
Eugenie	1972	South Africa, Madagascar	Unknown
Claude	1966	Swaziland, South Africa, Mozambique	Unknown
Astrid	1958	Swaziland, South Africa, Mozambique	1 million

Source EM-DAT: The Emergency Events Database—Universite Catholique de Louvain (UCL)—CRED, D. Guha-Sapir—www.emdat.be, Brussels, Belgium

adversely affects the available water resources. Climate change will worsen these problems increasing the vulnerability of the affected communities.

Southern African communities are often caught unaware when floods and droughts exist. This is partly due to poorly developed early warning system which contributes to increased exposure to these shocks. A contributory factor is that large parts of southern Africa have inadequate hydro meteorological monitoring system (Mazvimavi (2010a). This has led to limited location-specific knowledge about the nature and effects of climate variability and potential effects climate change. Lack of locally relevant data constrains the development of adaptation measures appropriate for different environmental settings and communities. This often leads to the recommendation of measures based on "one-size fits all". The limited financial resources are invested in ineffective adaptations measures.

6.4 Hydrological Regimes

Due to the altered rainfall patterns (increase in flooding and droughts in addition to increased variability and changes in rainfall characteristics) flow regimes of rivers are likely to change. For example, dry season flows are likely to be reduced significantly during dry seasons, that which will not only affect irrigation but also hydro electricity generation. Ecologically, several observations can be made from this of which the following are key: Firstly, river flows are a major determinant of physical habitat in streams, which in turn is a major determinant of biotic composition; Secondly, aquatic species have evolved life history strategies primarily in direct response to the natural flow regimes; Thirdly, maintenance of natural patterns of longitudinal and lateral connectivity is essential to the viability of populations of many riverine species; Finally, the invasion and success of exotic and introduced species in rivers is facilitated by the alteration of flow regimes. Thus the impacts of flow change (naturally or due to climate change) are manifest across broad taxonomic groups including riverine plants, invertebrates and fish. However, despite growing recognition of these relationships, there is still dearth in literature for Southern Africa on prediction and quantification of biotic responses to altered flow regimes. One obvious difficulty is the ability to distinguish the direct effects of modified flow regimes from impacts associated with land use change and climate change that often accompanies water resource development.

7 Conclusion

From this discussion, a picture of the availability of the water resources in the future can be constructed and the implications of these findings are as follows: (1) Water availability is likely to be compromised, since, for example, rainfall for a season may now fall within a few days. Thus altered rainfall characteristics (intensity, duration, etc…) will eventually impact on flow regimes, hence available water resources—spatially and temporally. (2) Water management measures are likely to become strict, since water shortages as experienced in Cape Town, South Africa from 2017 to 2018, are a certain possibility. Hence management of water resources, i.e. implementation of Integrated Water Resources Management Systems have to be expedited. (3) Most of the population in southern Africa depends on rain-fed agriculture and most of this population lives in rural areas where mean annual rainfall is approximately below 600 mm. Thus the impacts of climate variability are likely to be severest. In light of this, completion of overdue dams, construction of other water infrastructure and establishment of irrigation infrastructure to support agriculture need to be seriously considered in order to support small holder irrigation schemes, so as to ensure food security. (4) Similarly, for electricity generation, due to rainfall variability and increased temperatures, available water for electricity generation is likely to be reduced. This is particularly severe for dry season flows, since for hydroelectric power generation, a low threshold of water levels must not be exceeded for continual power generation. When exceeded, power generation becomes unsustainable. A typical implication associated with this deduction is faltering electricity production, as is the current case in Zimbabwe- 2019. The Zimbabwe Electricity Transmission and Distribution Company (ZETDC) issued a statement on their social media platform acknowledging the high electricity demand in the Zimbabwe and warning citizens of potential power cuts. From whence, a power cut schedule has since been developed and followed strictly. Thus alternative power sources such as solar, need to be seriously considered. (5) Lastly, the implications of these changes on agriculture and food security must be emphasised. Planting dates become unsteady due, for example, to the shift in the rainfall season onset. Thus, when these underlying fundamental processes for agriculture are subject to change, food security is heavily compromised and food shortage is probable. Thus, changes in climatic patterns may be a basis for changes in growing patterns, that is, moving from growing a particular crop in a region which it was suitable before to another area which may seem to possess the suitable conditions. As such new crop varieties (genetically modified) have to be seriously considered.

References

Aguilar E, Aziz Barry A, Brunet M, Ekang L, Fernandes A, Massoukina M, Mbah J, Mhanda A, do Nascimento DJ, Peterson TC, Thamba Umba O, Tomou M, Zhang X (2009). Changes in temperature and precipitation extremes in western central Africa, Guinea Conakry, and Zimbabwe, 1955–2006. J Geophys Res Atmos 114(D02115):1–11

Allen MR, Ingram WJ (2002) Constraints on future changes in climate and the hydrologic cycle. Nature 419(6903):224–232

Arnell NW (1999) Climate change and global water resources. Glob Environ Chang 9(1):S31–S49

Ashton PJ, Love D, Mahachi H, Dirks P (2001) An overview of the impact of mining and mineral processing operations on water resources and water quality in the Zambezi, Limpopo and Olifants Catchments in Southern Africa. In: Contract report to the mining, minerals and sustainable development (Southern Africa) Project, by CSIR-Environmentek, Pretoria and Geology Department, University of Zimbabwe-Harare. Report No. ENV-PC, vol 42

Banze F, Jiali G, Shi X (2017) Variability and trends of rainfall, precipitation and discharges over zambezi river basin, Southern Africa: review. Int J Hydrol 2:2

Bathiany S, Dakos V, Scheffer M, Lenton TM (2018) Climate models predict increasing temperature variability in poor countries. Sci Adv 4(5):eaar5809

Bullock A (1992) Dambo hydrology in Southern Africa—review and reassessment. J Hydrol 134(1–4):373–396

Chamaillé-Jammes S, Fritz H, Murindagomo F (2007) Detecting climate changes of concern in highly variable environments: quantile regressions reveal that droughts worsen in Hwange National Park, Zimbabwe. J Arid Environ 71(3):321–326
Collins JM (2011) Temperature variability over Africa. J Clim 24(14):3649–3666
Collins M, An SI, Cai W, Ganachaud A, Guilyardi E, Jin FF, Jochum M, Lengaigne M, Power S, Timmermann A (2010) The impact of global warming on the tropical Pacific Ocean and El Niño. Nat Geosci 3(6):391–397
Conway D, Persechino A, Ardoin-Bardin S, Hamandawana H, Dieulin C, Mahé G (2009). Rainfall and water resources variability in sub-Saharan Africa during the twentieth century. J Hydrometeorol 10(1):41–59
Crane TA, Roncoli C, Hoogenboom G (2011) Adaptation to climate change and climate variability: the importance of understanding agriculture as performance. NJAS-Wageningen J Life Sci 57(3):179–185
Desanker P, Magadza C (2001) Africa. In: McCarthy JJ, Canziani OF, Leary NA, Doken DJ, White KS (eds). Climate change 2001: impacts, adaptation and vulnerability. IPCC working group II: third assessment report. Cambridge University Press, Cambridge, UK
Desanker P (2002) The impact of climate change of life in Africa: climate change and vulnerability in Africa. WorldWide Fund for Nature, Washington, DC, USA
Easterling DR, Meehl GA, Parmesan C, Changnon SA, Karl TR, Mearns LO (2000) Climate extremes: observations, modeling and impacts. Sci Compass 289(5487):2068–2074
Eriksen S, Watson H (2009) The dynamic context of southern African savannas: investigating emerging threats and opportunities to sustainability. Environ Sci Policy 12(1):5–22
Falkenmark, M. 1989. The massive water scarcity now threatening Africa: why isn't it being addressed? Ambio 1(1):112–118
Fanta B, Zaake BT, Kachroo RK (2001) A study of variability of annual river flow of the Southern African region. Hydrol Sci J 46(4):513–524
Fauchereau N, Trzaska S, Rouault M, Richard Y (2003) Rainfall variability and changes in Southern Africa during the 20th century in the global warming context. Nat Haz 29(2):139–154
Field CB, Barros V, Stocker T, Qin D, Dokken D, Ebi K, Mastrandrea M, Mach K, Plattner GK, Allen S (2012) IPCC, 2012: managing the risks of extreme events and disasters to advance climate change adaptation. A special report of Working Groups I and II of the Intergovernmental Panel on Climate Change, vol 30(11). Cambridge University Press, Cambridge, UK, and New York, NY, USA, pp 7575–7613
Garderen EAV (2017) Chapter 13: regional climate change strategies and initiatives. In: Davis-Reddy CL, Vincent K (eds) 2017: climate risk and vulnerability: a handbook for Southern Africa (2nd Ed), CSIR, Pretoria, South Africa
Gardner-Outlaw T, Engelman R (1997) Sustaining water. Easing scarcity: a second update, Population Action International (PAI), Washington DC, US. http://www.populationaction.org/Publications/Reports/Sustaining_Water_Easing_Scarcity/Sustaining_Water_Easing_Scarcity_-_Full_Report.pdf
Githeko AK, Lindsay SW, Confalonieri UE, Patz JA (2000) Climate change and vector-borne diseases: a regional analysis. Bull World Health Organ 78(9):1136–1147
Gizaw MS, Gan TY (2017) Impact of climate change and El Niño episodes on droughts in sub-Saharan Africa. Clim Dyn 49(1–2):665–682
Goodess CM (2013) How is the frequency, location and severity of extreme events likely to change up to 2060? Environ Sci Policy 27:S4–S14
Groisman PY, Knight RW, Easterling DR, Karl TR, Hegerl GC, Razuvaev VN (2005) Trends in intense precipitation in the climate record. J Clim 18(9):1326–1350
Gyamfi C, Ndambuki JM, Salim RW (2016) Hydrological responses to land use/cover changes in the Olifants Basin, South Africa. Water 8(12):588
Harrison L, Michaelsen J, Funk C, Husak G (2011) Effects of temperature changes on maize production in Mozambique. Clim Res 46(3):211–222
Hewitson B, Crane R (1996) Climate downscaling: techniques and application. Clim Res 7(2):85–95

Hughes WS, Balling RC (1996) Urban influences on South African temperature trends. Int J Climatol 16(8):935–940
Hulme M, Doherty R, Ngara T, New M, Lister D (2001) African Climate Change: 1900–2100. Clim Res 17(2):145–168
Huntingford C, Hemming D, Gash JHC, Gedney N, Nuttall PA (2007) Impact of climate change on health: what is required of climate modellers? Trans Roy Soc Trop Med Hyg 101(2):97–103
Huntington TG (2006) Evidence for intensification of the global water cycle: review and synthesis. J Hydrol 319(1–4):83–95
IPCC (2014) Climate change 2014: synthesis report. Contribution of working groups I, II and III to the fifth assessment report (AR5) of the intergovernmental panel on climate change. [Pachauri RK, Meyer LA (eds.)]. IPCC, Geneva, Switzerland
Jury MR (2013) Climate trends in southern Africa. South Afr J Sci 109(1–2):1–11
Kabat P, Schulze RE, Hellmuth ME, Veraart JA (2002) Coping with impacts of climate variability and climate change in water management: a scoping paper. International Secretariat of the Dialogue on Water and Climate Wageningen, The Netherlands
Kadomura H (2005) Climate anomalies and extreme events in Africa in 2003, including heavy rains and floods that occurred during northern hemisphere summer. Afr Study Monograph 30(1):165–181
Kane R (2009) Periodicities, ENSO effects and trends of some South African rainfall series: an update. South Afr J Sci 105(5–6):199–207
King'uyu S, Ogallo L, Anyamba E (2000) Recent trends of minimum and maximum surface temperatures over Eastern Africa. J Clim 13(1):2876–2886
Kistin EJ, Ashton PJ, Earle A, Malzbender D, Patrick MJ, Turton AR (2009) An overview of the content and historical context of the international freshwater agreements that South Africa has entered into with neighbouring countries. Int Environ Agreements Polit Law Econ 9(1):1–21
Klinman M, Reason C (2008) On the peculiar storm track of TC Favio during the 2006–2007 Southwest Indian Ocean tropical cyclone season and relationships to ENSO. Meteorol Atmos Phys 100(1–4):233–242
Knox JW, Rodríguez Díaz J, Nixon D, Mkhwanazi M (2010) A preliminary assessment of climate change impacts on sugarcane in Swaziland. Agric Syst 103(2):63–72
Kruger A, Shongwe S (2004) Temperature trends in South Africa: 1960–2003. Int J Climatol 24(15):1929–1945
Kruger A (2006) Observed trends in daily precipitation indices in South Africa: 1910–2004. Int J Climatol 26(15):2275–2285
Kruger A, Sekele S (2013) Trends in extreme temperature indices in South Africa: 1962–2009. Int J Climatol 33(3):661–676
Kundzewicz Z, Mata L, Arnell NW, Döll P, Jimenez B, Miller K, Oki T, Şen Z, Shiklomanov I (2008) The implications of projected climate change for freshwater resources and their management. Water Resour Res 2(1):32–45
Kusangaya S, Warburton ML, Archer van Garderen E, Jewitt GP (2014) Impacts of climate change on water resources in southern Africa: A review. Phys Chem Earth, Parts A/B/C 67–69(1):47–54
Kusangaya S, Shekede MD, Mbengo I (2017) Chapter 8: water resources. In: Davis CL, Vincent K (eds.) Climate risk and vulnerability: a handbook for Southern Africa, 2nd ed. CSIR, Pretoria
Kusangaya S, Warburton Toucher ML, Garderen EAV (2018) Evaluation of uncertainty in capturing the spatial variability and magnitudes of extreme hydrological events for the uMngeni catchment, South Africa. J Hydrol 412–413:233–245
Lenton TM, Dakos V, Bathiany S, Scheffer M (2017) Observed trends in the magnitude and persistence of monthly temperature variability. Sci Rep 7(1):5940
Love D, Uhlenbrook S, Twomlow S, Zaag PVD (2010) Changing hydroclimatic and discharge patterns in the northern Limpopo Basin, Zimbabwe. Water 36(3):335–350
MacKellar N, New M, Jack C (2014) Observed and modelled trends in rainfall and temperature for South Africa: 1960–2010. South Afr J Sci 110(7–8):1–13

Magadza C (1996) Climate Change: Some likely multiple impacts in Southern Africa, Climate Change and World Food Security. Springer, Netherlands, pp 449–483

Malzbender D, Earle A, Deedat H, Hollingworth B, Mokorosi P (2009) Review of regulatory aspects of the water services sector. Water Research Commission, WRC, Pretoria, South Africa

Manatsa D, Chingombe W, Matarira C (2008). The impact of the positive Indian Ocean dipole on Zimbabwe droughts. Int J Climatol 28(15):2011–2029

Manatsa D, Matarira C, Mukwada G (2011) Relative impacts of ENSO and Indian Ocean dipole/zonal mode on east SADC rainfall. Int J Climatol 31(4):558–577

Manatsa D, Behera SK (2013) On the epochal strengthening in the relationship between rainfall of East Africa and IOD. J Clim 26(15):5655–5673

Manatsa D, Reason CJC (2016) ENSO-Kalahari desert linkage on Southern Africa summer surface air temperature variability. Int J Climatol 10(5)

Mathews R, Richter BD (2007) Application of the indicators of hydrologic alteration software in environmental flow setting. J Amer Water Resour Assoc 43(6):1400–1413

Matondo JI (2010) Adaptation options to climate change and variability on the water resources in Africa. In: Proceedings of the second science with Africa (swa) conference Addis Ababa, Ethiopia. Economic Commission for Africa

Mazvimavi D (2010a) Climate change, water availability and supply. In: Kotecha P (ed) Climate change, adaptation and higher education: securing our future. SARUA Leadership Dialogue Series, South Africa, pp 81–97

Mazvimavi D (2010b) Investigating changes over time of annual rainfall in Zimbabwe. Hydrol Earth Syst Sci 14(12):2671–2679

McMichael AJ, Woodruff RE, Hales S (2006) Climate change and human health: present and future risks. The Lancet 367(9513):859–869

Mekonnen MM, Hoekstra AY (2016) Four billion people facing severe water scarcity. Sci Adv 2(2):e1500323. https://advances.sciencemag.org/content/2/2/e1500323/tab-pdf

Monyela, BM (2017) A two-year long drought in summer 2014/2015 and 2015/2016 over South Africa. PhD dissertation, University of Cape Town. http://hdl.handle.net/11427/27111/. https://open.uct.ac.za/handle/11427/27111

Morishima W, Akasaka I (2010) Seasonal trends of rainfall and surface temperature over Southern Africa. Afr Study Monogr 40(1):67–76

New M, Hewitson B, Stephenson DB, Tsiga A, Kruger A, Manhique A, Gomez B, Coelho CA, Masisi DN, Kululanga E (2006) Evidence of trends in daily climate extremes over Southern and West Africa. J Geophys Res 111(D14102):1–11

Ngigi SN (2009) Climate change adaptation strategies: water resources management options for smallholder farming systems in sub-Saharan Africa. MDG Centre for East and Southern Africa, The Earth Institute, Columbia University, USA

Nicholson SE (2001) Climatic and environmental change in Africa during the last two centuries. J Clim Res 17(1):123–144

Nicholson SE, Funk C, Fink AH (2017) Rainfall over the African continent from the 19th through the 21st century. Glob Planet Change 165:114–127

Osbahr H, Twyman C, Adger WN, Thomas DS (2008) Effective livelihood adaptation to climate change disturbance: scale dimensions of practice in Mozambique. Geoforum 39(6):1951–1964

Patt A, Schröter D (2008a) Perceptions of climate risk in Mozambique: implications for the success of adaptation strategies. Glob Environ Change 18(3):458–467

Patt AG, Schröter D (2008b) Perceptions of climate risk in Mozambique: implications for the success of adaptation strategies. Glob Environ Change 18(3):458–467

Pietersen K, Beekman H (2016) Groundwater management in the Southern African development community. Position paper prepared for SADC Groundwater Management Institute. https://sadcgmi.cyberdevs.co.za/wp-content/uploads/2020/01/SADC-GMI-Position-Paper-Final.pdf

Pielke RA (2007) Future economic damage from tropical cyclones: sensitivities to societal and climate changes. Philos Trans Series A Mathe Phys Eng Sci 365(1860):2717–2729

Rateb A, Chung-Yen K, Moslem I, Kuo-Hsin T, Wen-Hau L, Kuo-En C, Tzu-Pang T (2017) Terrestrial water storage in African hydrological regimes derived from GRACE mission data: intercomparison of spherical harmonics, mass concentration, and scalar slepian methods. Sensors

Reason C, Keibel A (2004) Tropical cyclone Eline and its unusual penetration and impacts over the Southern African mainland. J Weather Forecast 19(5):789–805

Richard Y, Fauchereau N, Poccard I, Rouault M, Trzaska S (2001) 20th century droughts in southern Africa: Spatial and temporal variability, teleconnections with oceanic and atmospheric conditions. Int J Climatol 21(7):873–885

Richter BD, Baumgartner JV, Powell J, Braun DP (1996) A method for assessing hydrologic alteration within ecosystems. Conserv Biol 10(4):1163–1174

Rouault M, Richard Y (2005) Intensity and spatial extent of droughts in southern Africa. Geophys Res Lett 32:15

Saraiva OAML, Masih I, Uhlenbrook S, Jewitt GPW, Van der Zaag P, Riddell E (2015) Drivers of spatial and temporal variability of streamflow in the Incomati River basin. Hydrol Earth Syst Sci 19(2):657–673

Sarron C (2005) Effects of wetland degradation on the hydrological regime of a quaternary catchment: Mohlapitse River. Ga Mampa valley, Limpopo Province, South Africa

Schulze R (2005) Climate change and water resources in Southern Africa: studies on scenarios, impacts, vulnerabilities and adaptation. WRC Report No. 1430/1/05. Water Research Commission, Pretoria, South Africa

Schulze R (2011) Approaches towards practical adaptive management options for selected water-related sectors in South Africa in a context of climate change. Water 37(5):621–646

Schulze RE (2000). Modelling hydrological responses to land use and climate change: a Southern African perspective. Ambio 29(1):12–22

Schulze RE, Pike A (2004) Development and evaluation of an installed hydrological modelling system. Water Research Commision Report, vol. 1155/1. pp: 04

Schulze RE, Kunz RP, Lumsden TG, Davis NS (2014) On model selection, downscaling and bias correction of GCM output for design hydrological applications, with emphasis on the CSIR GCMs. In: DEA (ed.) Long-term adaptation scenarios flagship research programme (LTAS) for South Africa Phase Two. Report No. 7. Department of Environmental Affairs (DEA), Pretoria, South Africa

Shongwe M, Van Oldenborgh G, Van Den Hurk B, De Boer B, Coelho C, Van Aalst M (2009) Projected changes in mean and extreme precipitation in Africa under global warming part I: Southern Africa. J Clim 22(13):3819–3837

Solomon S, Qin D, Manning M, Chen Z, Marquis M, Averyt K, Tignor M, Miller H (2007) IPCC, Climate Change 2007: The physical science basis. Contribution of working group I to the fourth assessment report of the intergovernmental panel on climate change. Cambridge University Press, Cambridge, UK and New York, USA

Syed TH, Famiglietti JS, Chambers DP, Willis JK, Hilburn K (2010) Satellite-based global-ocean mass balance estimates of interannual variability and emerging trends in continental freshwater discharge. Proc Nat Acad Sci 107(42):17916–17921

Tadross M, Jack C, Hewitson B (2005) On RCM based projections of change in southern African summer climate. Geophys Res Lett 32(1):23–35

Taylor V, Schulze RE, Jewitt GP (2003) Application of the indicators of hydrological alteration method to the Mkomazi River, KwaZulu-Natal, South Africa. Afr J Aqua Sci 28(1):1–11

Timmermann A, Oberhuber J, Bacher A, Esch M, Latif M, Roeckner E (1999) Increased El Nino frequency in a climate model forced by future greenhouse warming. Nature 398(6729):694–697

Tirado MC, Clarke R, Jaykus LA, McQuatters-Gollop A, Frank JM (2010) Climate change and food safety: a review. Food Res Int 43(7):1745–1765

Todd MC, Andersson L, Ambrosino C, Hughes DA, Kniveton DR, Mileham L, Murray-Hudson M, Raghavan S, Taylor R, Wolski P (2011) Climate change impacts on hydrology in Africa: case studies of river basin water resources, African Climate and Climate Change. Springer, pp 123–153

Trenberth K (1999) Conceptual framework for changes of extremes of the hydrological cycle with climate change. In: Karl T, Nicholls N, Ghazi A (eds) Weather and climate extremes. Springer, Netherlands, pp 327–339

Troy B, Sarron C, Fritsch JM, Rollin D (2007) Assessment of the impacts of land use changes on the hydrological regime of a small rural catchment in South Africa. Phys Chem Earth Parts A/B/C 32(15–18):984–994

Tshiala FM, Olwoch JM, Engelbrecht FA (2011) Analysis of temperature trends over Limpopo Province, South Africa. J Geograp Geol 3(1):13

Tshimanga RM (2012) Hydrological uncertainty analysis and scenario-based streamflow modelling for the Congo River Basin. PhD Thesis, Hydrology, Rhodes University, South Africa

UNESCO (1997) International Standard Classification of Education-ISCED 1997: November 1997. UNESCO

Unganai L (1996) Historic and future climatic change in Zimbabwe. J Clim Res 6:137–145

Vermeulen SJ, Aggarwal PK, Ainslie A, Angelone C, Campbell BM, Challinor A, Hansen JW, Ingram J, Jarvis A, Kristjanson P (2012) Options for support to agriculture and food security under climate change. Environ Sci Policy 15(1):136–144

Vörösmarty CJ, Douglas EM, Green PA, Revenga C (2005) Geospatial indicators of emerging water stress: an application to Africa. Ambio 34(3):230–236

Warburton M, Schulze R (2005a) Detection of climate change: A review of literature on changes in temperature, precipitation and streamflow, on detection methods and data problems, Climate Change and Water Resources in Southern Africa: Studies on Scenarios, Impacts, Vulnerabilities and Adaptation. WRC Report 1430/1/05. Water Research Commission, Pretoria, South Africa

Warburton M, Schulze R, Maharaj M (2005b) Chapter 16: Is South Africa's temperature changing? An analysis of trends from daily records, 1950–2000. In: Schulze R (ed.) Climate change and water resources in southern Africa: Studies on scenarios, impacts, vulnerabilities and adaptation. WRC Report 1430/1/05, pp 275–295. Water Research Commission, Pretoria, South Africa

Warburton M, Schulze R, Jewitt G (2010) Confirmation of ACRU model results for applications in land use and climate change studies. Hydrol Earth Syst Sci 14(12):2399–2414

Wuebbles DJ, Hayhoe K, Parzen J (2010) Introduction: assessing the effects of climate change on Chicago and the Great Lakes. J Great Lakes Res 36(2):1–6

Yamba FD, Walimwipi H, Jain S, Zhou P, Cuamba B, Mzezewa C (2011) Climate change/variability implications on hydroelectricity generation in the Zambezi River Basin. Mitig Adapt Strat Global Change 16(6):617–628

Yanda PZ (2010) Climate change impacts, vulnerability and adaptation in Southern Africa. In: Kotecha P (ed.), Climate change, adaptation and higher education: securing our future. SARUA Leadership Dialogue Series, . pp 11–30. Cape Town, South Africa

Zinyowera MC, Unganai SL (1992) Drought in Southern Africa: An update on the drought: 1991–92. Drought Netw News 4(34):3–4

Chapter 6
Historic Climatic Variability and Change: The Importance of Managing Holocene and Late Pleistocene Groundwater in the Limpopo River Basin, Southern Africa

Tamiru A. Abiye and Khahliso C. Leketa

Abstract The isotopic signature in local precipitation is driven by factors that control fractionation such as temperature and season, where stable isotopes in groundwater retain the signature of the effects of physical changes at the time of recharge. High ambient temperature and decrease in rainfalls are characteristic features of the El Niño event in southern Africa that has resulted in tremendous pressure on water resource management. Based on the environmental isotope records from the Limpopo River Basin in South Africa and Botswana, this study determines ambient temperature during rainfall at the time of recharge and correlates it with the ^{14}C Mean Residence Time to observe temperature variability in the Holocene and Late Pleistocene. Its implication for the need to manage such historic groundwater is discussed. Temporal variation in stable isotopes and ambient temperature in relation to ^{14}C data revealed the presence of evaporation and recharge at different times due to long-term climatic variability in the areas located within the arid and semi- arid climatic settings in the basin. Evidence of variability in ambient temperature was found in geological times, with increasing trends in the Holocene and Late Pleistocene. Therefore, increasing water demand by various economic sectors requires planned intervention in groundwater resource management to achieve sustainable development.

Keywords Groundwater management · Environmental isotopes · Ambient temperature · Holocene-Late pleistocene · Impact on groundwater resources

---

T. A. Abiye (✉) · K. C. Leketa
School of Geosciences, University of the Witwatersrand, Private Bag X3, P.O. Box Wits 2050, Johannesburg, South Africa
e-mail: tamiru.abiye@wits.ac.za

K. C. Leketa
e-mail: kleketa@gmail.com

S. Diop et al. (eds.), *Climate Change and Water Resources in Africa*,
https://doi.org/10.1007/978-3-030-61225-2_6

1 Introduction

Groundwater, often considered a strategic resource in Africa, is not well understood, despite its role in addressing clean water shortage for the vast population. Compounding this, groundwater is under pressure from uncontrolled waste disposal, mining effluents, excessive pumping for irrigation and industrial activities (Abiye et al. 2011; Abiye 2013a, b, 2015). Groundwater plays a vital role, predominantly in arid and semi-arid regions, fulfilling an increasing water demand due to its resilience to short-term climatic variation. In arid and semi-arid regions, aquifers receive limited recharge from sporadic rainfall (Selaolo 1998; Abiye 2016). Therefore, it is necessary to dedicate attention to sustainable development policy in the region. In the Limpopo River Basin, groundwater exploitation is unregulated and in most cases storage and recharge history are unknown. The basin hosts about 18 million inhabitants within the 408,250 km^2 catchment, including portions of Botswana, South Africa, Mozambique and Zimbabwe, with a total average rainfall ranging from 300 to 1050 mm/yr across the catchment (Owen 2011).

The potential impact of climate change on groundwater has been researched in depth globally, with a dominant interest in arid and semi-arid regions due to the scarcity of surface water. High evaporation due to an increase in temperature (IPCC 2018) and low rainfall obviously have a tremendous impact on surface water availability. Constant improvement in the modeling of global climatic variables also contributes to understanding the extent of climate change at different levels, even though the capacity to utilize data at the local level is very limited.

In order to investigate the isotopic characteristics of groundwater that is in great demand by local communities and the industrial sector, different sites with a known groundwater history in the Limpopo River Basin have been considered in this study (Fig. 1).

Groundwater recharge originates from rainfall with a buffer zone that serves as ambient impoundment, which either directly or indirectly replenishes aquifers that later release water onto the surface through wells, springs and base flow. Therefore, it is essential to understand the status of climate at the time of rainfall event and subsequent recharge into the aquifer. Naturally, groundwater recharge is a slow process that can be subjected to extreme climatic variation. Particularly in semi-arid and arid regions, rainfall occurs in an unpredictable manner, which restricts regular aquifer recharge. Hence, people rely on the available groundwater without knowing the recharge history that is necessary for knowledge-based groundwater management.

It is known that an increasing global ambient temperature leads to a change in precipitation and atmospheric moisture, which ultimately affects the amount of recharge to aquifers (Négrel and Petelet-Giraud 2011). Groundwater provides an archive of past climate variations by recording changes in the recharge amount or the chemical and isotopic evolutionary history of a groundwater system (Hughes et al. 2011). Négrel and Petelet-Giraud (2011) investigated the recharge conditions and groundwater characteristics of big aquifers in Europe at a large scale, involving recent, Holocene and Pleistocene components and their eventual mixing. If isotopic

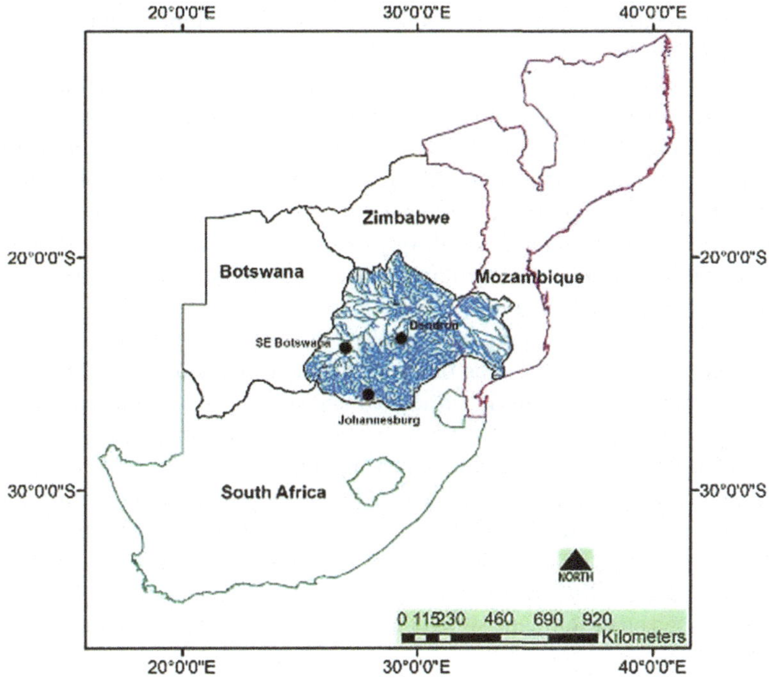

Fig. 1 Location map of the Limpopo River basin with three localities considered in this study

signatures are not affected by any process from surface water to groundwater, the isotopic ratio reflects the origin of the water such as location, period and recharge process (Négrel and Petelet-Giraud 2011). This suggests that, if recharge takes place in a cold climate, other than from high latitude moisture sources such as from the polar region, a depleted isotope signature is expected in the recharging water. On the other hand, isotopic enrichment is linked to water that experienced evaporation prior to recharge. The spatial and temporal variability of the isotopic composition of meteoric water is controlled by factors such as ambient temperature, relative humidity, precipitation, latitude, distance from the coast, and elevation of the given area above sea level (Rozanski et al. 1997). Consequently, the isotopic composition of atmospheric water vapor and precipitation exhibits broad temporal variations (Rozanski et al. 2000). The most important paleoclimate application of isotopes in global rainfall is an apparent relationship between the isotopic composition of precipitation and surface temperature (Rozanski et al. 1997). Once recharge with a distinct isotope signature joins an aquifer, the stable isotope values of recharging water are retained in groundwater. In this regard, the presence of paleo-recharge under cold climatic conditions with depleted stable isotope values can be easily associated with the last glaciation period in the Late Pleistocene. The widespread occurrence of C_4 grasses in the warmer summer rainfall areas of southern Africa reveals temporal shifts of climatic boundaries, which indicates that temperatures and also precipitation in southern Africa have changed since the Last Glacial maximum, about 18,000 years ago (Vogel 1983).

The stable isotope signature of groundwater retains the effect of climatic conditions through which rainfall occurred. Therefore, environmental isotopes are suitable to characterize climatic factors that alter the isotopic composition of water. In groundwater, stable isotopes help to explore the presence or absence of evaporation, which could be directly linked to ambient temperature. Groundwater that receives indirect recharge from surface water systems through ponds, channels, wetlands, streams and lakes tends to show more enrichment in stable isotopes relative to local rainfall. This is because of exposure of water to evaporation prior to infiltration, whereby the lighter isotopes preferentially evaporate, leaving heavier isotopes behind, which eventually percolate into groundwater. On the other hand, water that is directly recharged into groundwater from rainfall through fractures and pore spaces could have a similar isotopic signature as the rainfall, with a very limited effect of evaporation. In most cases, surface water acts as a buffer between precipitation and groundwater so that the effect of ambient temperature can be traced within the groundwater archive through the identification of the stable isotope signature. Varying rates of infiltration during summer and winter seasons could also result in an isotopic shift in rainfall and/or temperature, besides climatically induced changes in the plant cover. These effects are superimposed on the climatic signal imprinted in regional precipitation (Rozanski et al. 2000).

For several decades, stable isotopes have been widely used as environmental tracers in groundwater studies, catchment water balance, infiltration through soil horizon, surface water and groundwater interaction, and dam leakages across the world. Particularly in arid and semi-arid regions that are characterized by high temperature, the isotopic composition of groundwater is dependent on the extent of evaporation that is affected by the occurrence of effective rainfall to replenish aquifers and hence, maintain the signature of atmospheric effect.

The spatial and temporal variability in the δ^2H and $\delta^{18}O$ values of meteoric water results from the isotope-fractionation effect that accompanies evaporation and condensation (Yurtsever and Gat 1981; Clark and Fritz 1997; Edmunds et al. 2005). Variability could be long term, based on the change in climatic variables. On a millennial time-scale, changes in the $\delta^{18}O$ values of ice in the tropics represent large-scale climatic variations (Thompson and Davis 2005; Hoffmann et al. 2005). Past rainfall stored as groundwater and ice provides evidence of former climatic conditions (Rozanski et al. 1992; Edmunds 2005; Dragoni et al. 2008). The $\delta^{18}O$ distribution in the Holocene and Late Pleistocene in large sedimentary aquifers in North Africa has been documented by Edmunds 2005, and revealed the presence of highly depleted $\delta^{18}O$ values within the Nubian sandstone aquifer in the Late Pleistocene as compared to the West African region, which indicates a cooler climatic condition at the time of recharge in North African region.

The Pleistocene climate is marked by major glacial and interglacial oscillations, where the most recent glacial period culminated about 21,000 years ago (the last glacial maximum) with vast ice sheets that extend into the mid-latitude of North America and Europe (Jones and Mann 2004). In the Holocene and Late Pleistocene in northern Africa and Europe, significant groundwater replenishment took place during the pluvial periods in the late glacial maxima, where general cooling of around 5 to 6 °C was estimated through the noble gas method (Edmunds 2005). Jones and Mann (2004) indicated that in the Late Pleistocene, global temperatures were about 4 °C cooler than in the mid-twentieth Century, confirming unprecedented warming at a global scale. Evidence also suggests the existence of ancient lakes in the Sahara (Street and Grove 1979), which dried up due to high ambient temperature. This, obviously, has implications for the absence of active groundwater recharge in the region, as well as on extreme desertification, which impacted the availability of freshwater.

In recent times, southern Africa experienced an El Niño event, which is characterized by high surface temperature and a decrease in rainfall amount (Kogan and Gu 2017) resulting in increased evaporation, while causing tremendous pressure on water resources management (Abiye 2016). Based on the available environmental isotope record from South Africa and Botswana (Limpopo River Basin, Fig. 1), this study tries to determine the ambient temperature during rainfall events that generated recharge to groundwater and correlate them with ^{14}C Mean Residence Time (MRT) to observe temperature change in line with long-term climate variability and change. In order to estimate the groundwater MRT, radiocarbon (^{14}C) with a half-life of 5,730 years (Clark and Fritz 1997) was used. ^{14}C is naturally produced in the atmosphere to form $^{14}CO_2$, which is assimilated into plants through plant respiration. ^{14}C is measured in "percent Modern Carbon" (pMC), which corresponds with 95% of the Oxalic acid standard for the ^{14}C activity of 1950 (Clark and Fritz 1997; Suckow et al. 2013). In water bodies, CO_2 exchange occurs through carbonate formation that retains ^{14}C in its molecular structure, where the infiltrating water picks up $^{14}CO_2$ in the root zone from respiring roots and decomposition of dead organic matter. According to Leaney and Allison (1986) and Suckow et al. (2013), water that was recharged pre-1952 when the atmospheric ^{14}C was 100 pMC for deep groundwater, is not affected if a closed system is assumed for carbonate dissolution. Values of ^{14}C close to 100 pMC could suggest that the recharge rate is fast or it is restricted to certain areas of the aquifer. Therefore, the temporal variation in stable isotopes has been thoroughly assessed based on ^{14}C data in order to gain insight into long-term climate variability in the Limpopo River Basin. With the current high rate of population growth and an increase in water demand, this analysis is intended to assist water managers to plan and an implement intervention to protect groundwater depletion due to recurrent drought.

2 Methodology

Three sites were identified within the Limpopo River Basin in South Africa and Botswana (Fig. 1), where groundwater samples were collected for environmental isotope analysis. These sites are located in southeast Botswana, Dendron (Limpopo Province, South Africa) and the Johannesburg region (Gauteng Province, South

Africa). Pertinent data representative of annual averages for stable isotopes was systematically collated. Groundwater samples were analyzed for stable isotopes of water molecules (^{18}O and ^{2}H), and ^{14}C dating; 14 samples were taken from southeast Botswana, six from Dendron and 18 from the Johannesburg region. Analysis for stable isotopes of water was conducted at the Hydrogeology Lab, University of the Witwatersrand, South Africa, using an LGR Liquid Water Isotope Analyzer, while ^{14}C analysis was conducted at iThemba Labs in Gauteng, South Africa.

To estimate ambient temperature, regression equations from various published studies were considered. Cuffey et al. (1995) proposed a linear regression for the paleo-thermometer based on ^{18}O values as $\delta^{18}O$ (‰) = 0.327T−24.8‰ for Greenland, using ^{18}O values of ice core and calibrated with borehole temperature. More important, the pioneering work of Dansgaard (1964), based on the global isotopic composition of precipitation and temperature, was applied. Based on the global relationship, the ^{18}O values in relation to temperature at the time of rainfall event was represented by $\delta^{18}O‰ = (0.695xT_{annual} - 13.6‰)$. Rearranging the relationship for temperature leads to: $T_{annual} = (\delta^{18}O‰ + 13.6‰)/(0.695)$, which was used to calculate the ambient temperature at the time of recharge in this study.

Based on the exponential function, groundwater mean residence time was estimated where: t(years) = −8267 ln ($a_t^{14}C/a_0^{14}C$) (Clark and Fritz 1997), and $a_t^{14}C$ is the measured ^{14}C activity and $a_0^{14}C$ is the initial ^{14}C activity (100 pMC).

3 Results and Discussion

Data for the environmental isotopes of ^{18}O, ^{2}H and ^{14}C and the calculated ambient temperatures are presented in Table 1. The stable isotopes of ^{18}O and ^{2}H have been plotted in Fig. 2, which reveals at least three clusters. Results from southeast Botswana indicate that the $\delta^{18}O$ values fall between −8.22‰ and −3.95‰, with $\delta^{2}H$ values in the range of −38.8‰ to −27.3‰. In this region, the lowest ^{14}C recorded is 1.4 ± 1.5 pMC (long Mean Residence Time) with the highest value of 96.5 ± 2.5 pMC (young water), and a corresponding ^{14}C age of 35,500 ± 50 years and 300 ± 5 years, respectively (Table 1). Obviously, the highly enriched $\delta^{18}O$ values of −3.95‰ (BH9), −4.56‰ (BH12) and −4.61‰ (BH11) could suggest the occurrence of recharge after evaporation (Fig. 2) where two distinct clusters of groundwater are identified based on the ^{18}O and ^{2}H distribution: one cluster falling above the GMWL (Cluster A), indicating recharge in a cold climatic condition affected by the condensation process. The other cluster plotted below the GMWL line (cluster B), from the Dendron area (Limpopo Province), indicates the impact of evaporation prior to recharge. The highly depleted values in $\delta^{18}O$ (Cluster A) could be related to moisture that emanated from high latitude sources with minimal evaporation effect. The cold south Atlantic air mass could also be a source for highly depleted moisture. Large variation in stable isotope values could be attributed to variation in temperature during rainfall events that generate recharge to the groundwater. Calculated temperature values during the rainfall event reveal variation in southeast Botswana

Table 1 Isotope data from southeast Botswana, Dendron area, and Johannesburg region, with calculated mean annual ambient temperature and groundwater mean residence times

Sample ID	$\delta^{18}O$ (‰)	$\pm\delta^{18}O$ (‰)	$\delta^{2}H$ (‰)	$\pm\delta^{2}H$ (‰)	^{14}C (pMC)	^{14}C (years)	T (°C) calculated	
BH1	−7.66	±0.11	−36.5	±0.4	96.5±2.5	300±5	8.5	SE Botswana
BH2	−5.41	±0.00	−34.0	±0.0	88.5±2.4	1000±10	11.8	
BH3	−5.10	±0.08	−32.0	±0.4	81.2±2.4	1700±10	12.2	
BH4	−5.38	±0.00	−34.3	±0.0	64.9±2.4	3550±10	11.8	
BH5	−8.22	±0.00	−38.8	±0.0	64±2.4	3700±10	7.7	
BH6	−7.65	±0.00	−35.7	±0.0	58.1±2.4	4500±10	8.6	Dendron
BH7	−5.23	±0.00	−33.3	±0.0	56.5±2.1	4700±10	12.0	
BH8	−7.37	±0.00	−34.8	±0.0	56.3±2.1	4750±10	9.0	
BH9	−3.95	±0.03	−27.3	±0.2	47.7±2.1	6100±10	13.9	
BH10	−5.11	±0.00	−30.8	±0.0	47.2±2.1	6200±10	12.2	
BH11	−4.61	±0.00	−29.9	±0.0	18±1.7	14200±10	12.9	
BH12	−4.56	±0.00	−28.4	±0.0	15.7±1.7	15300±10	13.0	
BH13	−7.75	±0.00	−35.7	±0.0	4.9±1.5	24950±10	8.4	
BH14	−8.06	±0.05	−37.9	±0.3	1.4±1.5	35300±50	8.0	
D1	−4.70	±0.28	−30.6	±1.7	85.4±2.5	1300±10	12.8	
D2	−4.60	±0.22	−32.0	±0.9	93.4±2.6	560±5	12.9	
D3	−5.22	±0.06	−35.8	±0.3	84.3±2.5	1400±10	12.1	
D4	−4.62	±0.19	−29.1	±1.1	100±2.6	0±5	12.9	Johannesburg
D5	−3.86	±0.24	−26.9	±3.1	100±2.7	0±5	14.0	
D6	−4.65	±0.06	−28.7	±0.2	92.9±2.6	600±5	12.9	
BH1	−1.39	0.13	−6.5	0.6	97.5±2.6	209±5	17.6	
BH2	−2.69	0.07	−12.2	0.9	91.3±2.6	752±5	15.7	
BH3	−2.87	0.11	−13.0	0.7	98.5±2.6	125±5	15.4	
BH4	−2.87	0.04	−12.6	0.8	99.7±2.6	25±5	15.4	
BH5	−4.52	0.07	−23.2	1.2	100±2.7	0±5	13.1	
BH6	−5.73	0.05	−29.1	0.4	88.2±2.6	1038±10	11.3	
BH7	−2.23	0.05	−12.3	0.3	95.7±2.6	363±5	16.4	
BH8	−3.55	0.05	−15.8	0.7	62.7±2.3	3859±10	14.5	
BH9	−2.62	0.27	−12.5	3.5	84.4±2.5	1402±10	15.8	
BH10	−5.08	0.36	−22.5	0.5	99.6±2.6	33±5	12.3	
BH11	−4.32	0.05	−21.6	0.4	95.1±2.6	415±5	13.4	
BH12	−3.27	0.08	−16.7	0.3	100±2.7	0±5	14.9	
BH13	−3.83	0.11	−17.3	1.0	92.9±2.6	609±5	14.1	
BH14	−3.83	0.17	−18.2	1.5	98.1±2.6	159±5	14.1	
BH15	−5.15	0.03	−25.3	0.4	80.7±2.5	1773±10	12.2	

(continued)

Table 1 (continued)

Sample ID	$\delta^{18}O$ (‰)	$\pm\delta^{18}O$ (‰)	δ^2H (‰)	$\pm\delta^2H$ (‰)	^{14}C (pMC)	^{14}C (years)	T (°C) calculated
SP1	−4.97	0.10	−27.6	0.4	82±2.5	1641±10	12.4
SP2	−5.11	0.10	−27.5	0.5	79.9±2.5	1855±10	12.2
SP3	−2.81	0.15	−11.7	1.5	81.7±2.5	1671±10	15.5

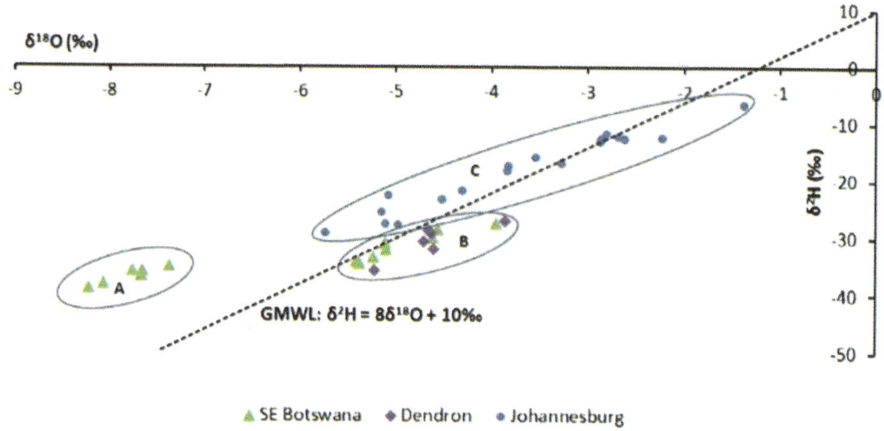

Fig. 2 The relative distribution of stable isotopes of water with respect to the global meteoric water line (GMWL, Craig 1961)

(Table 1), where the mean annual temperature ranges between 7.7 °C and 13.9 °C (Fig. 3b) with high temperatures limited to the Holocene (Fig. 3b).

The stable isotope data in Table 1 from the Dendron area consists of enriched $\delta^{18}O$ values (−3.86‰ to −5.22‰) with the corresponding δ^2H values ranging from −26.92‰ to −35.78‰. These data were plotted along with Cluster B, with high-temperature recharge condition similar to the southeast Botswana samples indicating similar atmospheric condition and moisture source. The ^{14}C values are relatively higher than those of the southeast Botswana, with a range of 84 ± 2.5 pMC to 100 ± 2.0 pMC, resulting in a MRT ranging from 1300 ± 10 years to recently recharged water. In the Dendron area, the calculated ambient temperatures are substantially high, varying between 12.1 °C and 14.0 °C (Table 1).

Based on the globally adopted geological time scale, the boundary between the Holocene and Pleistocene is set at 12,500 years from the present, which is equivalent to the ^{14}C values of 22 ± 1.7 pMC. Even though the areas in this study are all located within the Limpopo River Basin, there is a variation in temperature, which could be attributed to extensive desertification in the region, probably linked to the expansion of the Greater Kalahari Desert. This might have an impact on selective isotope enrichment based on the season (Fig. 3a). The MRT (Fig. 3b) depicts two geological times where the recorded low temperature occurred about 24,950 years

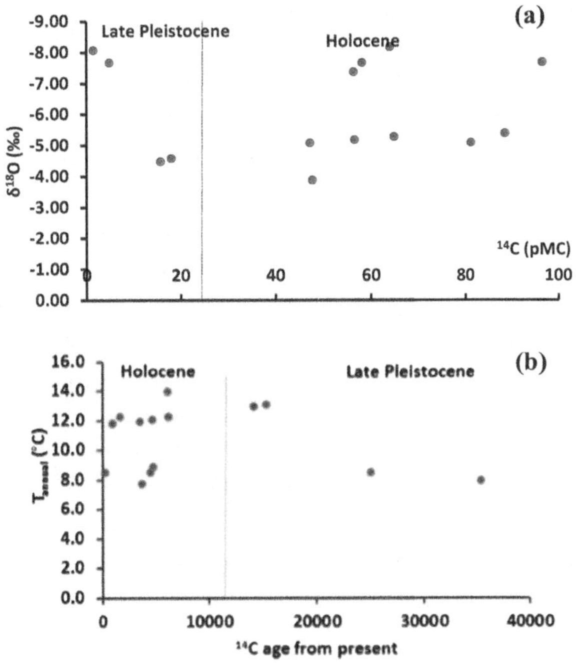

Fig. 3 **a** The $\delta^{18}O$ variation with ^{14}C activity (22 ± 1.7 pMC as the Holocene and Late Pleistocene boundary), **b** Annual temperature variation with respect to ^{14}C ages during Holocene and Late Pleistocene (12500 years as a boundary) in southeast Botswana

ago, which could correspond to the glacial period. The depletion of $\delta^{18}O$ in most of the groundwater samples can be explained by a significant contribution of Late Pleistocene recharge from cold climatic conditions. On the other hand, warm temperature was linked to isotope enrichment in recharging water in the Holocene (Fig. 3b). This signifies the loss of rainwater through evaporation and less availability of water for recharge due to an increase in temperature.

The stable isotope data from the Dendron area occupy the same cluster (Cluster B) as do the evaporated southeast Botswana samples that fell below the GMWL (Fig. 2), perhaps signifying a similar source of recharge after evaporation. The three clusters, which were identified based on the $\delta^{18}O$ and ^{14}C data imply the progressive enrichment of isotopes into recent times due to increased ambient temperature (Fig. 4a and b). This is the estimated temperature using the Dangaard 1964 temp versus $\delta^{18}O$ relationship. Combining this analysis with the ^{14}C, it is deduced that the period was in the Holocene. The change in annual temperature was in the range of 1 to 2 °C where the lowest temperature was 12.1 °C and the highest was 14.0 °C (Table1 and Fig. 4b). Based on the ^{14}C data, groundwater recharge took place in the Holocene, with relatively warm climatic conditions. The plot in Fig. 4a further displays at least three pairs with relatively enriched $\delta^{18}O$ values with respect to recent recharge. This observation is supported by low temperatures, between 12 and 13 °C, recorded between 1000 and 1500 years ago (Fig. 4b).

The stable isotope data from the Johannesburg region contain highly enriched $\delta^{18}O$ values (−1.39‰ to −5.73‰) with $\delta^{2}H$ ranging from −29.1‰ to −6.5‰. The

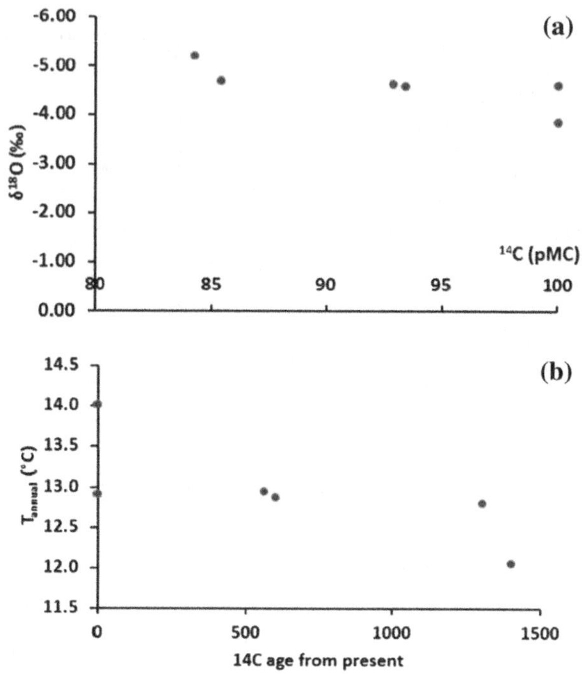

Fig. 4 The distribution of $\delta^{18}O$ with respect to ^{14}C (pMC) **a** with decreasing in ambient temperature in Holocene **b** in the Dendron area, Limpopo Province, South Africa

^{14}C values range from 62.7 ± 2.3 pMC to 100 ± 2.0 pMC, resulting in the MRT of 3,859 ± 10 years to recently recharged water (Fig. 5a). In this region, the calculated annual temperature varies between 11.3 °C and 17.6 °C (Table 1 and Fig. 5b), which is higher relative to the other two areas described above.

4 Managing Historic Groundwater

Groundwater represents the largest reserve of accessible freshwater for arid and semi-arid countries that are less dependent on rainfall and surface water (Abiye, 2016). The challenge to the availability of groundwater is aggravated by climate change and widespread pollution linked to uncontrolled disposal into water supply aquifers from urban, industrial, agricultural and mining wastewater. Since groundwater plays an essential role in the economic development of communities in addition to its vital role as a source of water supply through decentralized provision for the majority of the population, and because of the impacts of its complexities, groundwater must be managed through a knowledge-based approach. In the Limpopo River Basin, which is characterized by low rainfall and high temperature, the scarcity of surface water drives communities to depend on groundwater for various developmental activities. The current study reveals that groundwater recharge is not regular, affected by increasing ambient temperature in addition to the complex nature of crystalline

Fig. 5 The change in $\delta^{18}O$ with respect to ^{14}C (pMC) (**a**) and distribution of the calculated temperature during with ^{14}C age of groundwater in the Johannesburg area, South Africa

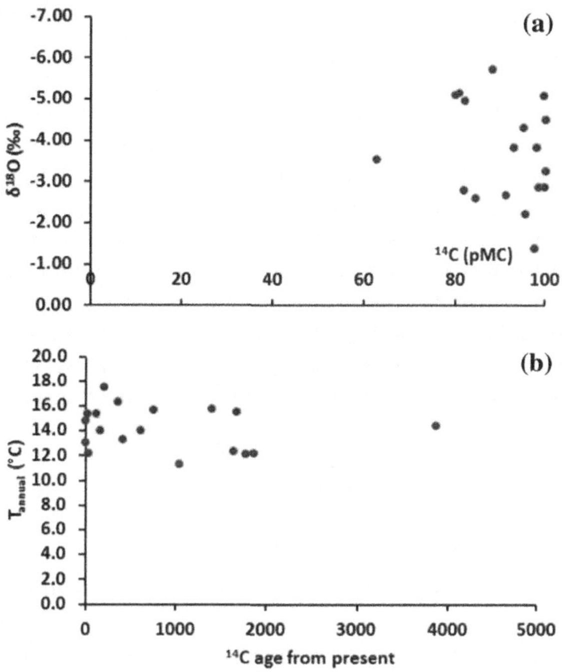

aquifers. Groundwater in the region is a life line for irrigated agriculture in addition to cattle watering, domestic-use mining and industrial activity, all of which can potentially cause over-abstraction due to increasing demand. In this basin, numerous boreholes are actively serving different economic sectors. However, it is essential to gather data on spatial and temporal characteristics of recharge through geological time in view of the increasing impact of climate change.

Currently, there is no clear understanding of the nature of recharge in the area, except the estimation of recharge amount that accounts for 3 to 5% of rainfall in arid and semi-arid regions of southern Africa (Abiye 2016 and references therein). With the increasing impact of climate change on water resources, it is naïve to expect an increase in recharge in the region to sustain current groundwater demand and abstraction. ^{14}C values reveal that groundwater recharge has large temporal variation over several thousand years. In southeast Botswana, for example, the time span for recharge is highly variable. Out of 14 groundwater samples, 14% were recharged 1,000 years ago, 57% were recharged between 1,000 and 10,000 years ago, and 29% were recharged more than 10,000 years ago. In the Dendron area, out of six groundwater samples, 67% were recharged less than 1,000 years ago, while 33% were recharged between 1,000 and 1,400 years ago. In the Johannesburg region, out of 18 groundwater samples, 56% were recharged less than 1,000 years ago, while 44% were recharged between 1,000 and 3,900 years ago. Such diverse recharge episodes could indicate the heterogeneity of aquifers and the strong influence of atmospheric conditions that dictate groundwater recharge through geological time into aquifers upon which the community depends on for day-to-day activities. Therefore, recharge rarely reaches water supply aquifers except in the shallow and weathered aquifers with modern water. The increase in annual temperature from the Late Pleistocene to the Holocene, and even in the recent times observed in this study, could encourage policymakers and water managers to introduce integrated assessments of groundwater storage for resource allocation. Often, the extent of aquifer and storage for water supply is not known and it can hinder sustainable management of the groundwater resource in the region. In many areas, well fields that have been supplying water for big cities and irrigation schemes should also be accompanied by managed aquifer recharge from available runoff and treated wastewater return into catchments. In the Johannesburg region, on average, about 5 m^3/s (Leketa and Abiye 2019) of treated wastewater flows into streams from the wastewater treatment works. This amount can be sustainably used to recharge the weathered and alluvial aquifers.

5 Conclusions

The integrated results from $\delta^{18}O$ and ^{14}C records reveal an alarming temperature variation over millennia within the southern African region that requires a thorough reflection on the management of the available groundwater resource. Through geological times, temperature showed an increase based on the $\delta^{18}O$ record that has an impact on the availability of recharging water from rainfall. The results also reveal the presence of historical groundwater in the region as a critical resource, especially because of the lack of active groundwater recharge, which is neither consistent nor reliable. So, in order to plan large-scale exploitation of groundwater that was recharged over a thousand years ago, a contingency plan to include managed aquifer recharge through runoff harvesting, treated wastewater storage and use of water-efficient irrigation technologies, must be designed.

References

Abiye TA (ed) (2013a) The use of isotope hydrology to characterize and assess water resources in Southern Africa. WRC Report No TT 570/13, Pretoria, p 211. ISBN 978-1-4312-0480-9

Abiye TA (2013b) Surface water and groundwater interaction in the upper Crocodile River basin. In: Abiye T (ed) The use of isotope hydrology to characterize and assess water resources in South (ern) Africa. WRC Report No TT 570/13, Pretoria, pp 141–164. ISBN 978-1-4312-0480-9

Abiye TA (2016) Synthesis on groundwater recharge in Southern Africa: A supporting tool for groundwater users. Groundwater Sustain Dev 2:182–189

Abiye TA, Mengistu H, Demlie MB (2011) Groundwater resource in the crystalline rocks of the Johannesburg area, South Africa. J Water Resour Protect 03:199–212

Abiye TA, Mengistu H, Masindi K, Demlie MB (2015) Surface water and groundwater interaction in the upper Crocodile River Basin, Johannesburg, South Africa: Environmental Isotope Approach. South Afr J Geol 118(2):109–118

Clark ID, Fritz P (1997) Environmental isotopes in hydrogeology. Lewis Publishers, Boca Raton, Fla

Craig H (1961) Isotopic variations in meteoric waters. Science 133:1702–1703

Cuffey KM, Clow GD, Alley RB, Stuiver M, Waddington ED, Saltus RW (1995) Large Arctic temperature change at the Wisconsin-Holocene Glacial transition. Science 270:455–458

Dansgaard W (1964) Stable isotopes in precipitation. Tellus 16:436–468. https://doi.org/10.1111/j.2153-3490.1964.tb00181.x

Dragoni W, Sukhija BS, Dragoni W,. Sukhija BS (ed) (2008) Climate Change and Groundwater, vol. 288. Geological society. Special Publ., London, p 1

Edmunds WM (2005) Groundwater as an archive of climatic and environmental change. In: Aggarwal PK, Gat JR, Froelich KFO (eds) Isotopes in the water cycle; past, present and future. IAEA, pp 341–352

Edmunds WM, Aggarwal PK, Gat JR, Froelich KFO (eds) (2005) Isotopes in the water cycle; past, present and future of a developing science. Springer, Heidelberg, Germany, pp 341–352

Hoffmann G, Cuntz M, Jouzel J, Werner M (2005) How much climatic information do water isotopes contain? In: Aggarwal PK, Gat JR, Froehlich KFO (eds) Isotopes in the water cycle: past, present and future. IAEA, pp 303–320

Hughes CE, Cendón DI, Johansen MP, Meredith MT (2011) Climate change and groundwater. In Jones JAA (ed) Sustaining Groundwater Resources, International Year of Planet Earth, pp 7–117

IPCC (Intergovenmental panel on Climate change) (2018) An IPCC Special Report on the impacts of global warming of 1.5 °C, p 32

Jones PD, Mann ME (2004) Climate over past Millennia American Geophysical Union. Rev Geophys 42:1–42

Kogan F, Gu W (2017) Strong 2015–2016 El Niño and implication to global ecosystems from space data. Int J Remote Sens 38(1):161–178

Leaney F, Allison G (1986) Carbon-14 and stable isotope data for an area in the Murray Basin: Its use in estimating recharge. J Hydrol 88(1–2):129–145

Leketa K, Abiye T (2019) Modelling the impacts of changes in climate variables on the hydrology of the Upper Crocodile River Basin, Johannesburg South Africa. Environ Earth Sci 78:358

Négrel P, Petelet-Giraud E (2011) Isotopes in groundwater as indicators of climate changes. Trends Anal Chem 30(8):1279–1291

Owen R (2011) Groundwater needs assessment, Limpopo River Commission. Africa Groundwater Network, p 43

Rozanski K, Araguás-Araguás L, Gonfiantini R (1992) Relation between long-term trends of oxygen 18 isotope composition of precipitation and climate. Science 258:981–985

Rozanski K, Johnsen SJ, Schotterer U, Thompson LG (1997) Reconstruction of past climates from stable isotope records of palaeo-precipitation preserved in continental archives. Hydrol Sci J 42:725–745

Rozanski K, Froehlich KW, Mook G (2000) Environmental in the hydrological cycle: principles and applications, vol-III. Surface Water. International Atomic Energy Agency

Selaolo ET (1998) Tracer studies and groundwater recharge assessment in the eastern fringe of the Botswana Kalahari: the Letlhakeng-Botlhapatlou area. PhD, Free University, Amsterdam

Street AE, Grove AT (1979) Global map of lake level fluctuations since 30,000 years B.P. Quat Res 12:83–118

Suckow A, Aggarwal PK, Araguás Araguás L (2013) Isotope methods for dating old groundwater, Vienna, International Atomic Energy Agency

Thompson LG, Davis ME (2005) Stable isotopes through the Holocene as recoded in Low-latitude, high-altitude ice cores. In: Aggarwal PK, Gat JR, Froehlich KFO (eds) Isotopes in the water cycle: past, present and future. IAEA, pp 321–339

Vogel JC (1983) Isotopic evidence for the past climates and vegetation of southern Africa. Bothalia 14(3 & 4):391–394

Yurtsever Y, Gat JR (1981) Atmospheric waters. In Stable isotope hydrology: Deuterium and oxygen in water cycle; Tech. Rep 210, IAEA, Vienna, pp 103–142

Chapter 7
A Framework for IWRM in the Water-Energy-Food Nexus for the Senegal River Delta

Mor Talla Sall, Penda Diop, Joost Wellens, Mamoune Seck, and Jean Louis Chopart

Abstract The Delta of the Senegal River is a complex ecosystem that serves as a source for various uses of fresh water: fishing, agriculture, grazing, ecotourism and drinking water. Climate variability and change lead to changes in the hydrological regime that impact the activities of both populations and ecosystems. Faced with this vulnerability, agricultural models must be improved to achieve integrated and sustainable water resource use. A Water-Energy-Food nexus approach is being implemented by the Senegalese Sugar Company (CSS), which produces irrigated sugar cane crops (11,000 ha). A milling by-product, bagasse (420 megatons/year), is used for power production, replacing fossil fuel and realizing the reduction of CO_2 emissions by 43,000 tons/year. The current annual production of energy from bagasse allows the plant to reduce fossil fuels by 61 megatons/year and also provides energy for irrigation at CSS and surrounding smallholder fields. Next steps at CSS are to: (i) transfer 28 GWh/year from CSS into the national electric network; (ii) explore the feasibility of using high fiber energy cane cropped by smallholders on poor soils surrounding CSS farmland to be used for electricity production when bagasse stock is exhausted; (iii) promote small power station projects to run on high fiber and rustic sugarcane varieties cropped by smallholders for local electricity production in villages where electricity is not available. It is anticipated that this integrated approach will contribute to a reduction in climate change effects.

Keywords Senegal river delta · Climate change · Agricultural water · Bagasse · Energy crop

---

M. T. Sall (✉) · M. Seck
Senegalese Sugar Company, Richard-Toll, Senegal
e-mail: mortalla.sall@css.sn

M. T. Sall · J. Wellens
University of Liege, Liege, Belgium

M. T. Sall · P. Diop
Anta Diop University, Dakar, Senegal

J. L. Chopart
AgerConsult, Montpellier, France

S. Diop et al. (eds.), *Climate Change and Water Resources in Africa*,
https://doi.org/10.1007/978-3-030-61225-2_7

1 Introduction

According to the Intergovernmental Panel on Climate Change (IPCC) 2013 scenarios, temperature increase over the entire African continent could range between 2 and 3 °C over 50 years.

Climatic conditions in the Sahel and, in particular, in the Senegal River Delta are characterized by high evaporative demand (with potential evapotranspiration of up to 17 mm/day in April), high temperatures, and, above all, low as well as spatial and temporal heterogeneous rainfall.

Climate variability in the Sahel is not new. Climate changes have had an impact on social ecosystems and agricultural production systems for some time. The transition from wet conditions in the 1950s–1960s to drier conditions in the 1970s–1990s—with disastrous consequences—in the Sahelian strip of West Africa was one of the most significant inter-decade climate signals on the planet during the 20th century.

Between 1968 and 1995, there was a historical decrease in rainfall in West Africa of 15 to 25%, with a shortening of the rainy season. Inter-annual variability has also increased over the past 23 years (Hulme 2001).

For the future, according to TF Stocker et al. in the IPCC 2013 report, rainfall will remain at the current level until 2030–2035 and then decrease. Whatever the scenario, even if the results of the 21st Conference of the Parties (COP21) are applied, a decrease in yields is to be expected for rainfed crops in the region (FAO 2009).

In sub-Saharan Africa, the consequences of climate change are already visible in hydrological regimes, coastal erosion and sea-level rise. We are witnessing a decline in rainfed agriculture and an acceleration in deforestation, the latter linked to high population growth resulting in increasing food and energy deficits.

Irrigation is one of the possible tools to redress this situation. But there are multiple pressures by various actors on natural resources, mainly on fresh water and soils best suited for irrigated or intensive farming.

The economies of most sub-Saharan African countries that depend on agriculture are currently weakened. Their agricultural sectors are dependent on climatic hazards (low precipitation, high seasonal variability, rising temperatures, floods); rural populations are the most vulnerable. Senegal, in the Sahel, is one of the most sensitive and affected countries (Faye et al. 2015).

The Senegal River Valley, along with the Niayes area, have been main frontiers in agricultural intensification in Senegal, thanks to the exploitation of surface and groundwater resources, respectively. Agricultural production in the Senegal River Delta (Fig. 1) is closely linked to the practice of irrigation. Research on irrigated agriculture has been ongoing for more than 100 years (Dia 2004).

The droughts of the 1970s prompted Senegal to partner with three neighboring countries (Mali, Guinea and Mauritania) to construct the Diama and Manantali dams between 1984 and 1988. One of the objectives was to block seawater intrusion in the Delta, reducing surface and groundwater salinity, and thereby their negative impacts on agricultural lands, thus increasing hydropower production. The main objective was to raise and stabilize river and lake water levels for irrigation; however, this

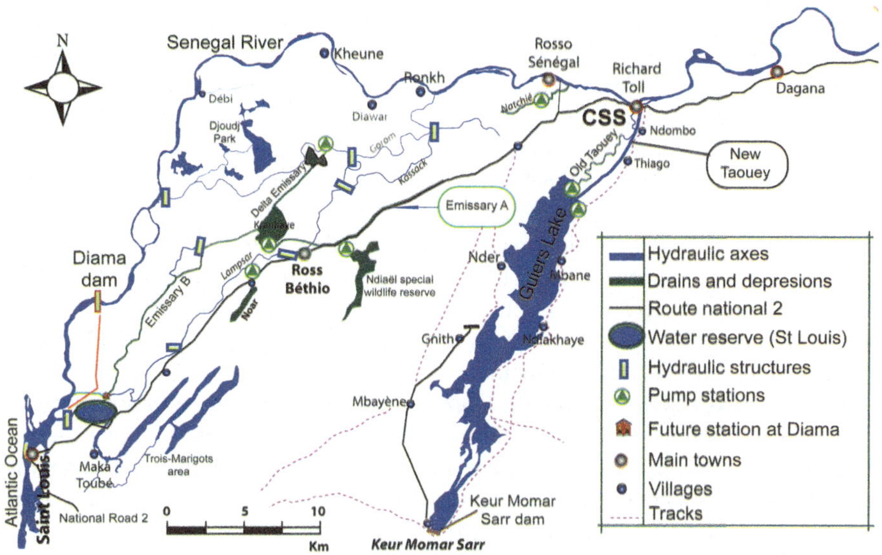

Fig. 1 Map of the area with hydraulic structures (Cissé 2011)

stimulated other harmful phenomena: eutrophication and proliferation of invasive aquatic plants (Kane 1997; Mietton et al. 2008).

The left bank of the Senegal River delta, discussed in this chapter, lies between 15° and 16° North and covers an area of 5,000 km^2 from Dagana to Saint-Louis (Fig. 2). Although it occupies only about 3% of the total Senegal River catchment

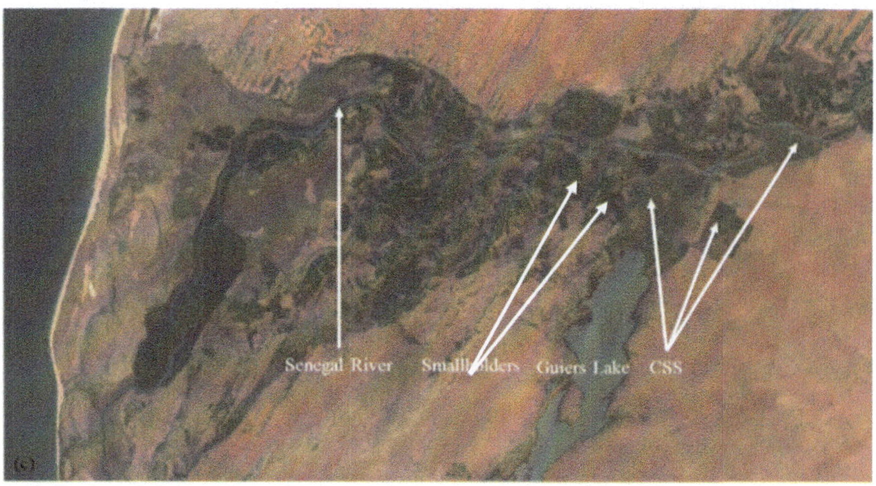

Fig. 2 Delta of the Senegalese river and area of study: (Richard Toll-Saint Louis-Lac de Guiers)

area (Kamara 2013), it is of particular value for its diversified and rich natural environment, its past and current developments (Diop, 1995 cited by 2017) and its strong potential for Integrated Water Resources Management (IWRM), thanks to its rich but vulnerable water resources.

The area has important potential for economic development, which in turn offers the potential to contribute significantly to the country's water and food security. But it is a highly anthropized environment where fresh water mixes with saltwater on very heterogeneous soils, which makes it highly vulnerable. Environmental degradation affects agriculture, fishing, hydropower production, shipping and drinking water production (Diop 2017). For these reasons, local public and private actors are calling for the implementation of an IWRM approach to address these challenges in a participatory, coordinated and inclusive manner.

In the context of climate change adaptation, the time has come to reflect on the state and evolution of current agricultural production models and to move towards integrated and sustainable development. In this respect, a new approach through the Water-Energy-Food nexus (i.e., the connection among these several elements) seems relevant. Since the Bonn Summit in 2011 (The Water, Energy and Food Security Nexus, November 16, 2011), this concept has been the subject of numerous studies, especially in southern nations.

After describing the natural and cultivated environments and the role of the main agricultural and agro-economic stakeholders (Table 1) on the left bank of the Senegal River Delta, we will present an embryonic IWRM and Water-Energy-Food nexus plan implemented around the CSS sugar state. The ultimate objective is to make innovative recommendations for sustainable adaptation of agricultural practices to climate change in the Senegal River Delta, to depart from the natural link between

Table 1 Brief typology of agriculture in the delta

Type	Example	Main threats
Productivist-intensive	CSS-GDS-CASL…	Land tenure stability—Labour costs—World food prices—Conflicts with local residents—Pollutants and regulations
Intermediary SME	Integrated Polyculture	Profitability—Water accessibility—Financing of production—Maintaining sectors—Competition of global products—Agro-food processing—Conflicts with pastoralists
Fragile livelihood	Rice or market gardening on small areas	Access to land—Accessibility and availability of water—Food insecurity and energy-accentuated vulnerability

the two approaches in order to transfer best practices from the agro-industry to smallholders.

The energy aspect of the nexus will be discussed in relation to the residual or dedicated biomass of crops and its possible adaptation in the village environment. Our summary will focus on prospective potential links among irrigated agriculture, non-food biomass and sustainable energy.

2 Natural and Cultivated Environments: Constraints and Assets

The Senegal River Delta is a vast set of lowlands and forms a kind of oasis between two dry Sahelian countries (Fig. 2). Its dense hydrographic network is composed of marigots (basins fed by several rivers: the Gorom, Djeuss, Lampsar), and lagoons (Fig. 1), separated from the sea by the coastal zone (Cissé 2011). The main basin, Guiers Lake, is connected to the Senegal River by the Taouey channel. It is a natural depression, where the flow is now partially controlled. It plays a role in regulating floods in the Senegal River thanks to its high storage capacity (Kane 1997), and has an important store of fresh water. Annual rainfall is highly variable (Fig. 3).

This inter-annual rainfall variability is a determining factor in the great dependence of agriculture on irrigation and thus the great demand for agricultural water for food production to the detriment of other uses.

The soils of the delta were formed in connection with the Senegal River and the nearby Sahara. Soils characterized by the clay of alluvial origin contrast with sandy soils deposited by wind. Shallow groundwater tables are often found along the river

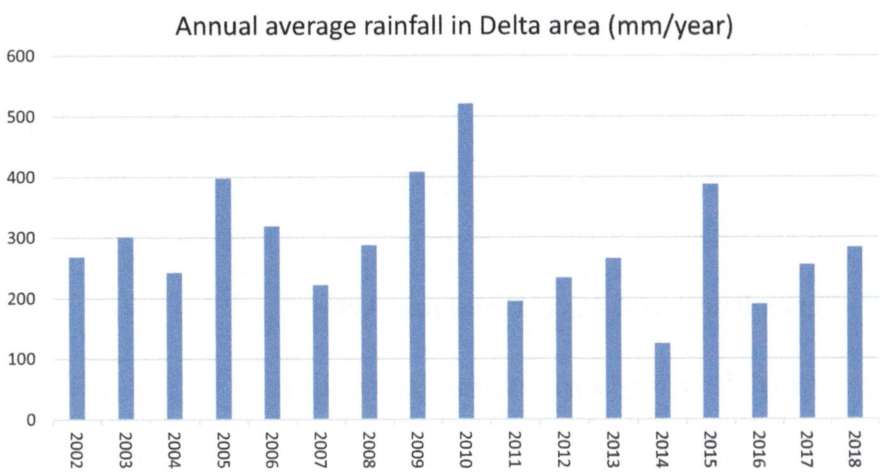

Fig. 3 Annual rainfall in the Senegal River Delta region in mm per year (average of several weather stations of CSS, 2019)

banks and around the lake. The variable salinity of the water is a major constraint on agricultural production in the region.

Wetlands are omnipresent in the Delta. The Ndiaël basin near Guiers Lake, Diawling National Park (Mauritania) and the Djoudj Bird Park (Senegal)—with its plant and animal biodiversity—are listed as Ramsar wetlands (treaty adopted on February 2, 1971 on Wetlands of International Importance). This designates the region with special heritage status and proven ecotourism potential. These natural wetlands need water in quantity and quality to maintain their rich biodiversity. They also have a role in flood control and as a natural filter. In the past, the lake was full of fish, but today there are few because of pollution and the proliferation of invasive plants.

The Senegal River Delta is also a region with great agricultural potential, thanks to the presence of high-quality fresh water (class C1-S1 according to the Riverside classification), fertile soils and strong solar radiation conducive to photosynthesis. The Delta has been the subject of several hydraulic and hydro-agricultural developments. For a long time, it has been accessed by water resource users from the agriculture, livestock and fishing sectors.

Traditionally, subsistence farming was practiced mainly on the river banks or its tributaries, benefiting from soil moisture following receding floods. However, rainfed agriculture is highly vulnerable due to the very low rainfall. Hydro-agricultural developments have gradually led to the emergence of more profitable, but also more water-consuming, irrigated areas. Agro-industry in the Delta currently consumes about 325 million m^3/year over an estimated 22,000 ha, and private farmers consume 400 million m^3/year over an estimated 35,000 ha. The Senegal River has an annual volume of about 18 billion m^3/year, with an average flow rate of 700 m^3/s (OLAC 2017).

Agro-industry and smallholder farmers alike take their water from and discharge wastewater into, the Senegal River or Lake Guiers (Tandian 2008).

Climate change will, by definition, have less impact on irrigated agriculture compared to rainfed agriculture, particularly in our case study (abundance of water). However, irrigated agriculture in the Senegal River Delta is and will remain vulnerable, especially because there are so many negative externalities. More than the agricultural performance itself, profitability, sustainability and resilience of irrigated agriculture should be the most relevant decision-making criteria in order to address climate change, demographic pressure and environmental risk.

3 Agricultural and Agro-Economic Actors

3.1 Brief Review of Agricultural Stakeholders

Currently, many different actors are working in the region independently or semi-independently. There are the autonomous agro-industrial and industrial complexes, such as the *Compagnie Sucrière Sénégalaise* (CSS) for sugar production, the *Société de Conserves Alimentaires du Sénégal* (SOCAS) for tomatoes, the *Société de Cultures Légumières* (SCL) for fresh vegetable market gardening and the more general consortia *Grands Domaines du Sénégal* (GDS) and *Compagnie Agricole de Saint Louis* (CASL). At the other end of the spectrum are smallholder subsistence farmers and, in between, more-or-less efficient agricultural and agri-food small and medium-sized enterprises (SMEs).

Large agro-industries benefit from directly managed hydro-agricultural development. The largest and oldest is the CSS: with 11,300 ha, it is the only sugar producer in Senegal. CSS was created at the beginning of the 1970s and exploits and maintains the Taouey canal and drainage structure. Its highly developed agricultural operation (traffic lanes, irrigation and drainage canals) is integrated into the hydraulic canvas of the Delta. Its technical and human resources have shaped the northern landscape of the lake, making it the "economic lung" of the region.

Table 1 presents a brief typology of the irrigated agriculture of the area. The multitude of users and actors, with diverse and sometimes conflicting interests, inevitably leads to a real problem of natural resource (i.e. water and soil) management (Kane 1997; Dia 2004; Kamara 2013, Diop 2017). Every aspect of management is aggravated by high population growth.

The threats listed in Table 1 result from the current vulnerability of the agricultural and agro-industrial sectors in Senegal. Productivist and intensive agriculture consumes large amounts of input and suffers from land insecurity, which makes investment and production financing complicated.

3.2 Competition and Complementarities

Several public and private projects concentrate on inclusive and sustainable development of agribusiness and small-scale agriculture in the area. PDMAS (*Projet de Développement des Marchés Agroalimentaires et Agricoles au Sénégal*), on 30,000 ha including 14,000 ha in the Delta, and PDIDAS (*Projet de Développement Inclusif et Durable de l'Agrobusiness au Sénégal*), on 10,000 ha, are currently involved in the development of horticulture around Lake Guiers. In recent years, irrigated farms and water-consuming enterprises in the region experienced various problems:

- Abandoned, inefficient and unprofitable hydro-agricultural developments, in turn, increasing the diffusion of pollution;
- Exacerbated tensions between those with a primary interest in fresh water and fertile soils;
- Insufficient food production, leading to food insecurity in a context of high population growth; and
- Increasing differentiation and competition between agro-economic models that are developing modern agricultural or agro-industrial enterprises rapidly and declining traditional family farms.

There are rivalries and conflicts over the use of land and water resources. The oldest and most recurrent one is the conflict between sedentary agriculture and nomadic herders. Everyone seeks, naturally, to have or maintain easy access to fresh water. Livestock, abundant in the area, needs to find drinking places easily, associated with clear routes to move around. The difficulty of water access for livestock is further complicated by the current rise of invasive species, such a Typha, in watering points (Mboup 2014).

In the Delta, land ownership remains a central development constraint (Diop 2017). A conflict in 2012 pitted local populations against the agrobusiness project SenHuile following the allocation of 20,000 ha of land to the latter in the village of Ndiaël. In Keur Momar Sarr, conflicts are also recurrent among small-scale farmers practicing traditional rainfed agriculture. However, according to Diop (2017), "these conflicts tend to evolve into a certain complementarity and solidarity, offering the prospect of important progress towards the development of a high-performance system." For example, agro-industries such as CSS, *Société des conserves alimentaires du Sénégal* (SOCAS) and West Africa Farm provides free water for irrigation to local small farmers. These companies also employ local people in casual or year-round paid labor and contribute to their well-being by the construction of corporate health centers.

The efforts of local actors are real, but there are still threats to the Delta's agricultural development. Several examples are:

(1) The extension and intensification of irrigated agriculture that leads to a rise in the water table (Kane 1997);
(2) Environmental pollution caused by the uncontrolled use of chemicals between Dagana and Saint-Louis (Diop 2017);
(3) Station X6 drainage water discharge containing chemicals on the axis of Lake Guiers, low Ferlo and Niety Yone village; and
(4) The practice of gravity irrigation on highly permeable sandy soils, leading to significant drainage losses.

The discharge of drainage water into the lake reduces its physicochemical water quality (Cissé 2011). For OLAC (*Office des Lacs et Cours d'eau*), the entity responsible for the management of the Lake, CSS and other agro-industrial companies are identified and considered to be polluting users. But this in isolation may be misleading: the main problem is the thousands of poorly-identified producers. It is, therefore, difficult to determine the origin and exact quantity of fertilizers and pesticides used around Lake Guiers. This situation gradually reduces the quality and quantity of water resources in the lake and Delta. Agro-industry, like small producers, is a large water consumer and contributes to water pollution; there is a complex tradeoff between this consumption and its contribution to food security and economic development.

3.3 Other Local Water Users

Another important and special water user of Lake Guiers is the Senegalese Water Company (SDE). The water collected from the lake does not serve the local population exclusively, but rather mainly the urban dwellers of Dakar, several hundred kilometers from the Delta. Water pumping and treatment plants supply almost all of Dakar and its suburbs with drinking water (projected to reach 500,000 m^3/day for five million inhabitants by 2030). Pumping the Delta to supply local populations with drinking water is, of course, necessary. However, it is also necessary to consider other and/or additional solutions to supply the city of Dakar, because the transport of water over several hundred kilometers through old pipelines, often subject to water 'pirates' who vandalize the supply, is highly inefficient.

3.4 Industrial Water Process Uses: Case of CSS

The CSS uses a lot of water in its industrial process to produce sugar and energy. Water is collected in the main gravity irrigation network and then treated in a drinking water treatment plant before being distributed to an internal domestic network to supply industrial needs. Every day, 12,000–15,000 m^3 of water is treated and reused in the CSS factory. The wastewater at the end of the process is stored in lagoons and partly reused in the irrigation system (for sugar cane) after monitoring and mixing with raw water.

4 Current and Suggested Strategies for Climate Change Adaption

As already indicated, the Senegal River Delta, like the entire Sahel-Saharan strip, is marked by a recurrent food and energy deficit, despite hydraulic and hydroelectric developments (Manantali Dam by OMVS for 200 MW and the Bokhol solar platform for the Senegal grid). The situation is due in part to climate change, and in part to other, aforementioned related factors:

- Highly contrasting soil types (sandy soils and salty clay soil) that require specific management;
- Periods of low water availability due to difficult access or water points infested by aquatic plants;
- Continued and significant growth of the urban population associated with increasing water needs;
- Structural weakness of local agricultural sectors marked by excessively high production costs, low protection and price volatility;

- Low cereal yields aggravated by significant post-harvest and avian losses, and
- Continued dependence on the national electricity grid, which loses money and is unable to adequately fulfill its public service role.

The area's strong agricultural potential has not yet been fully realized and hence has not yet had a significant impact on living standards. An overly segmented agricultural development policy lacking an integrated approach is thought to be a major culprit.

4.1 Towards an IWRM and Water-Energy-Food Nexus

Following the Rio Summit in 1992, the Global Water Partnership (GWP) was created in 1996 by the World Bank, United Nations Development Program (UNDP) and the Canadian International Development Agency (CIDA) to promote an Integrated Water Resources Management (IWRM) approach by bringing together public and private actors, and by assisting member countries in defining and implementing an IWRM approach.

According to an ECOWAS (Economic Community of West African States) statement in 1997, "IWRM is a process that promotes the coordinated development and management of water, land and associated resources in order to maximize the resulting economic and social well-being in an equitable manner, without compromising the sustainability of social ecosystems." The purpose of IWRM is to facilitate solutions to water resource-related problems through dialogue and trust-building among concerned stakeholders.

The social well-being mentioned in the declaration includes the availability of energy (including electricity) for lighting, cooking, health, information, etc. As a consequence, this study proposes to focus on the establishment of an IWRM-based consultation framework for various water stakeholders, and the attempt to implement a Water-Energy-Food nexus approach.

IWRM and WEF nexus approaches are closely linked. Both promote better resource use for sustainable development – socially, economically and environmentally. The key difference is that IWRM encompasses elements all around the water and their eventual interrelationships with food and energy. The nexus approach focuses on the three highly-interrelated factors.

Between 2011 and 2012, thirteen international scientific meetings discussed and developed the nexus approach for a variety of contexts and situations (IISD 2013). Several scientific organizations, including the International Institute for Sustainable Development (IISD), the International Food Policy Research Institute (IFPRI), the Stockholm International Water Institute (SIWI), the United Nations (UN), the World Bank (WB), the US Department of Energy, the Organization for Economic Co-operation and Development (OECD) and the World Economic Forum (WEF), have conducted or are currently conducting studies consistent with this approach. It is also currently used as a methodology for several integrated development projects in the West (IISD 2013). This approach frames agricultural production as a multi-dimensional agricultural system where several variables can be combined and result in multi-purpose products.

The Water-Food-Energy nexus approach is natural, recognizing how men have historically eaten and dressed, initially through the rational use of hunting and gathering, and later, by cropping food, textile and energy crops to eat, cook food, dress and heat. These local production models have been lost over time due to modernization and excessive specialization of agriculture, inherited from colonization and the agrarian revolution in the West.

The modern productivist agriculture model, which has succeeded in the West albeit sometimes imperfectly, has not been successfully applied to tropical Africa, especially the Sahel region. Reasons have been detailed elsewhere, including by Dumont (1962), Mazoyer and Roudard (2002) and Stiglitz (2004).

Methodology, as it applies to the nexus approach, will also not be detailed here; Burnett and Christopher (2018) lists three types of links among nexus components:

- Dependent and interconnected relationships for inputs/outputs (e.g., water used to produce food or energy, plants used to produce energy, and/or energy used to produce water or food);
- Competitive relationships, where the use of the same water source to produce food, may compete with the use of water to produce energy (biofuels and/or hydropower). Choices must therefore not be made exclusively on the basis of financial criteria, but also account for social and environmental criteria, with priority to achieving complementarity of uses, and
- Negative or positive external relations, wherein the use of one component risks an impact on the quality and fate of another component.

The nexus approach must focus on these three types of relationships and seek to minimize negative or regressive relationships, and has the potential to be an innovative way to foster agricultural development in the Delta.

First attempts are being made to combine IWRM and the W-E-F nexus approach among stakeholders. For example, the CSS, one of the first economic actors in the Delta, and one of the most criticized for its sectoral approach and environmental pollution, has implemented a research and development department that allows for the progressive exploration of new integrated productions.

4.2 *First Achievements: Example of the CSS*

Several authors recognize sugar cane as an efficient renewable energy source, especially in developing countries (Hess et al. 2016). The Water-Food-Energy nexus approach is now being implemented at CSS through the operation of a large, irrigated 11,300 ha sugar cane farm in the Delta (Fig. 4).

Agricultural production is intensive, fully irrigated and highly mechanized (Fig. 5). Water needs are high (16,000 m^3/ha/year) because of climate (average

Fig. 4 Presentation from Google map of the area study (CSS and smallholders)

Fig. 5 Perspective of the CSS Farm: on the front sugarcane fields, irrigation channel and at the back: heap of bagasse and boiler (@Wellens)

annual evaporation of 2,000 mm, very low rainfall) and the length of the cultivation cycle (12 months). Annual production is around 1,400,000 tons of sugar cane for an average gross yield of 134 TC/ha.

The industrial process of CSS seeks to optimize the production of food, energy products and fertilizers. It consists of a sequence of different industrial operations (Fig. 6), from the entry of sugar cane into the factory to final products and by-products (Table 2).

System innovation gives about equal importance to products (sugar, alcohol) and by-products (bagasse, scum, vinasse). Bagasse is an organic residue obtained after crushing cane stalks, the first step in the industrial transformation process. It is the fibrous part (45% to 51%) of the sugar cane stalk, representing about 30% of the stalk biomass, with good caloric value (Low Heating Value LHV of 1850 kcal/kg

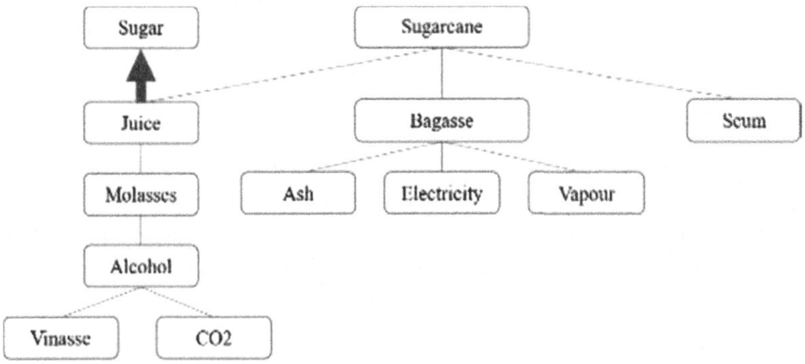

Fig. 6 Industrial sugar cane production diagram at CSS

Table 2 Sugar cane products and by-products at the CSS (direct products are in bold; derived by-products are in italics)

Products and *by-products*	Use	Product/year quantity
Sugar	**Food**	**145,000 tons**
Bagasse	**Energy**	**420,000 tons**
Molasses	**Food and raw materials**	**53,000 tons**
Pure alcohol	Food, Adjuvant, pharmaceutical	10, 700 000 L
Fuel oil	Chemical adjuvant	46,000 L
Vinasse	Organic fertilization	130,000 m^3
Water Vapor	Energy	840,000 tons
Electricity	Energy	100 GWh
Scum and sludge	Organic fertilization	27,700 tons
Ash and smoke	Organic fertilization	20,000 tons
CO2	Chemical adjuvant	1912 tons

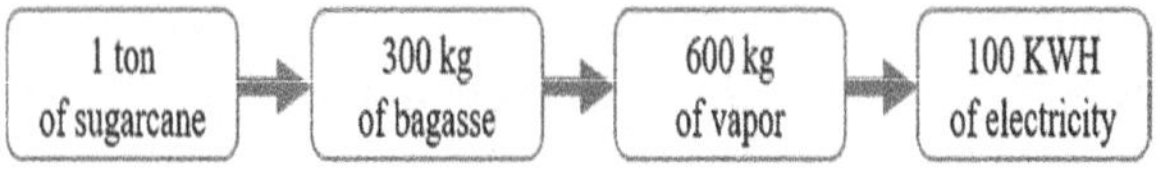

Fig. 7 Connection from bagasse to electricity at CSS

at 50% humidity). The CSS produces 420,000 tons/year of bagasse (Table 2); the combustion of one ton of bagasse generates about 330 kWh of electricity (Fig. 7).

Bagasse is used as fuel for the boiler, the industrial unit of the CSS, the role of which is to supply water vapor at a flow rate of 120 to 150 t/h, at 400°C and under 40 bars of pressure, necessary for the operation of the other units (refinery, agglomeration, power plant, etc.). The boiler room consists mainly of:

- Four mixed boilers that can operate on bagasse or fossil fuel, with a total production capacity of 30 tons of steam per hour, and
- A new bagasse boiler IJT (ISGEC JOHN THOMPSON). This unit alone can satisfy all the plant's steam demand and has a capacity of 150 tons of steam per hour.

CSS has two power plants:

- An old power plant with a capacity of 22 MW. This plant is currently used by a single 9.5 MW turbo-alternator, and can only run on fossil fuel oil.
- A new 25 MW thermal power plant, built as part of the KT150 project. This plant is cogenerative and can operate on fossil fuel or bagasse (the five boilers mentioned above).

The 420,000 tons of bagasse produced in this way are used exclusively in the new plant and can satisfy CSS's energy demand, which provides for annual production of 100 GWh. In the event that power plants become unavailable, CSS has more than ten back-up generators.

Electricity production is distributed among the various units of the CSS agro-industrial farm (Table 3); the company is entirely self-reliant on energy and is not connected to the national network.

The agricultural sector (plantation) uses 23% of the produced energy, mainly to pump water irrigation (260 billion m3/year) and drainage water. 50.5% of produced

Table 3 Amount of energy used per activity at the CSS

Energy Used at the CSS							
Sector	Turbo alternator units and boilers	Factory Production	Factory Workshop	Fields	CO2 Power plant	CSS village	Total energy consumed
Consumption (MWh)	2,620	6,233	461	2,763	65	124	12,266
Consumption (%)	21	50.5	4	23	0.5	1	100

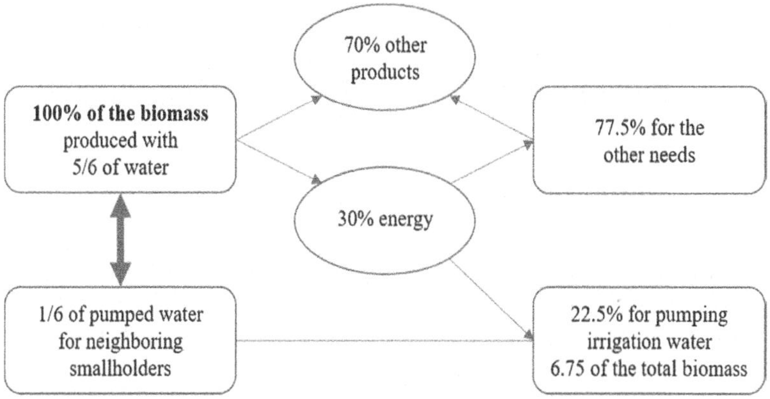

Fig. 8 Diagram of produced biomass use by CSS

energy is used to process sugar (food) and bagasse (energy). The residual energy is consumed by various factory workshop units and domestic needs. Overall, around 74% of the energy produced by the plants impacts the nexus relationship.

CSS reuses only 23% of the energy it produces, that is, 6.75% of the total produced biomass, to irrigate the entire production of CSS sugar cane production and 4,500 ha for smallholders. These village fields consume one-sixth of the pumped irrigation water, and CSS consumes the remaining five-sixths (Fig. 8).

These technical contributions are complemented by economic and environmental improvements:

- Clean, renewable and cheaper energy, saving several billion dollars per year;
- Reduction in CO_2 emissions of 43,000 tons/year eligible for carbon credits (third-most in West Africa), attributable to weaning from fossil fuel oil, eliminating around 61,000 tons/year;
- Industrial production of 1,962 tons/year of CO_2;
- Free water supply for the 4,500 ha of small riverside producers, with an annual production of 16,000 tons of paddy rice and 4,000 tons of onions (representing gross income of more than 2 billion CFA), and
- Plant or distillery by-products (ash, recovered smoke filters, scum, sludge and vinasse) reused as organic soil enhancements in the fields, thus reducing demand for synthetic fertilizer.

Current operations do not achieve an ideal Water-Energy-Food nexus, nor are they perfect integrated production models. Even though they use a highly photosynthetic-efficient C4 crop (sugar cane) for biomass and energy production, the crop is demanding in terms of inputs (between 700 to 800 kg/ha/year of fertilizer) and water (22,000 m^3/ha/year, from which 16,000 net).

This case study is intended to model optimization of plant production, in this case, sugar cane, through food, energy, organic fertilizers and even chemical products (fuel oil), leaving very little unused biomass.

Although this example is modest compared to demand at the Senegal River catchment scale, it shows the willingness of one of the most important companies in the delta to move towards greater IWRM and an improved Water-Energy-Food nexus.

4.3 Proposed Water Energy Food Nexus Approach

4.3.1 At the CSS

At the CSS, the next steps are to better link to the Water-Energy-Food nexus. Further research and investment required, starting in about 2020, is to:

- Transfer 28 GWh/year of renewable electricity from bagasse to the benefit of the neighboring population through a connection between the CSS network and the national network;
- Explore the possibility of cropping a high fiber energy cane on poor soils around the CSS estate, with the potential to increase renewable electricity production by CSS, and
- Promote small-scale power station projects to run on high fiber and rustic sugar cane varieties cropped by smallholders for local electricity production in villages where electricity is not yet available or too expensive.

In addition to the bagasse already fully utilized at the CSS to produce electricity, the company will explore the possibility of using all the sugar cane biomass produced for electricity output. This could exploit surrounding soils that have constraints (poor quality, difficult to irrigate and/or drain, weedy) that currently make them inefficient for sugar production.

Within the CSS, the fiber cane (dedicated to energy production) harvesting season would be outside the sugar cane harvesting season. This way, sugar cane and high-fiber production are complementary, optimizing the use of local agricultural labor, harvesting and mechanized transport equipment. It would also allow the CSS thermal power plant to be supplied with biomass produced locally, eliminating the demand for the imported fossil energy required when the bagasse stock is depleted.

This would require the use of appropriate plant varieties, characterized by their high total biomass and their adaptation to poor or slightly salty soils. The cropping system would also have to be adapted to minimize the use of imported herbicides and fertilizers.

Reduction in the use of fertilizers, in particular nitrogen, by reusing potassium-rich combustion ash, can lead to a slight reduction in biomass yield. This can be compensated by reducing production costs and improving the energy balance of this biomass production for energy production.

These fields, if located near drainage discharge areas, could, at least in part, reuse drainage water as irrigation water, enabling fuel cane to work as a filter for pollutants, particularly nitrates, that can be reused as fertilizer.

In the framework of using sugar cane for energy production without first extracting sugar, sugar is considered to be a fuel contained in biomass. This concept is new but seems to be gaining traction in some environments and contexts even where the main use of sugar cane is and will remain sugar production.

As indicated previously, this could be of interest to CSS in a Water-Energy-Food nexus approach by promoting the production of energy from a local agricultural product using local water and solar energy efficiently for photosynthesis. Reducing the use of fossil energy for electricity production when bagasse stock is depleted would allow the CSS to participate in reducing the use of fossil energy products (oil, coal) in the current context of climate change.

Research has been conducted in this area with encouraging results. For example, a five-year project in Guadeloupe, an island country in the Lesser Antilles, as part of the REBECCA research and development project (*'REcherche Biomasses Energie Canne à CApesterre'*, Biomass for Energy Research from Cane in CApesterre) is three years into a partnership with an industrial company (QUADRAN) already producing electricity in Guadeloupe. The research was first conducted to identify the best varieties of high-fiber cane for biomass production to be dedicated to electricity production and to characterize the performance of these varieties, in particular, their total dry matter yield and moisture content (Fig. 9), showing that under conducive growing conditions with rain and without water stress, the two best varieties (WI81456, WI79460) from WICSBS (Barbados) can produce 81 tons/ha/year of total dry matter (average result obtained over 3 years, plant cane and two ratoons) (Chopart 2016). These high-fiber cane varieties were cultivated using the same conventional practices as those used for local commercial sugar cane, but yields increased by 33% for stalks and 71% for total dry shoot biomass (Chopart 2016). Despite their significant height and biomass, the two best varieties were not more sensitive to lodging than were commercial varieties (Chopart 2016). The adaptation of crop systems for this new use of sugar cane (Chopart et al. 2015) has also been studied, and show that, under local conditions in Guadeloupe, it is possible to opt either for conventional 12-month cropping cycles or 8-month cycles, which allow

Fig. 9 High fiber cane variety cropped in Guadeloupe as part of the REBECCA project in a low fertility soil grows under high water stress conditions. The aboveground dry matter is 56 t ha^{-1} (@JL Chopart)

three harvests in 24 months. After two years, the dry matter yields obtained with the two cropping systems (8-month or 12-month cycles) are very similar (Chopart et al. 2015).

Even if it is unlikely that this approach can be adopted directly by the CSS, it demonstrates a model for cane harvesting outside the sugar cane harvesting period. This would allow the power plant to be supplied with renewable energy produced on-site (bagasse and then a high-fiber cane) almost all year round, with a very significant reduction in the use of fossil fuel and no net impact on the annual water consumption of the crop and therefore on IWRM. In this scenario, water from the Senegal River would be used to produce electricity for the needs of inhabitants (including food cooking) by optimizing existing industrial equipment (Water-Energy-Food nexus).

It has also been shown that this fibrous sugar cane is not very sensitive to the stemborer moth *Diatraea saccharalis* (Chopart et al. 2015), that it is possible to reduce fertilization and herbicides compared to conventional sugar cane and that these canes can be mechanically cut with the same equipment as sugar cane (Chopart, personal communication, Goebel et al. 2016). They are also more drought-tolerant (Fig. 9), which could be an advantage if used by small producers who may have limited or irregular irrigation water availability. Studies have focused on organic matter management in fibrous cane cultivation systems for biomass, showing that it is possible to maintain the organic matter stock in these systems (Sierra et al. 2016). The fact that fiber canes produce more roots (Chopart and Sergent 2015) contributes to the maintenance of organic soil stock despite higher above-ground biomass withdrawals than in conventional green sugar cane systems without pre-harvest burning.

Finally, it has been shown that the low heating value of dry above-ground biomass (LHVd, MJ/kg) is the same in each part of the fiber cane (stems, leaves, green, dry leaves), about 16.65 MJ/kg. A tight linear relationship between the LHVd and its energy yield (MJ/m^2), regardless of cultivar, age and environment, was found. Sugar cane energy content assessment could thus be simplified by measuring the dry above-ground biomass (DAB, kg/m^2) and its water content (percentage).

Using results described from the Guadeloupe study on this concept of biomass cane dedicated to energy production, it would be possible to produce more than 70 tons/ha/year of total dry matter (stems and leaves) with simplified and economical cultivation practices that reduce consumption of fertilizer and herbicide compared to conventional sugar cane for sugar production. It has been shown that, unlike sugar production, it is possible to harvest fiber cane after eight- to fourteen-month cropping cycles. Compared to conventional sugar cane, this would allow planting and harvesting dates to be spread over a larger part of the year and thus enable more efficient irrigation management.

The Guadeloupe findings could potentially be adapted for the environmental and technical conditions of the CSS with few modifications. Research on fibrous cane varieties dedicated to energy or for multiple purposes is, or has been, undertaken in some other countries, such as the southern USA (Sandhu 2018), Brazil and the West Indies (Matsuoka et al. 2014).

4.3.2 Small Producers

According to Smeets et al. (2012), any variety of sugar cane bagasse pellets, including energy cane biomass, can serve as an alternative to firewood in developing countries, especially in sub-Saharan Africa.

The model currently being tested at the CSS and its potential extensions will not be directly applicable to small producers in the area before developing it with various stakeholders; such partnership will result in appropriate and undoubtedly diversified solutions for local biomass production of energy.

A- Small producers on areas adjacent to CSS

Small producers bordering the CSS produce a rustic fibrous cane with very few inputs, cut manually outside the sugar cane cutting period. Under the envisioned model, irrigation water will be provided by the CSS, partly from drainage water. Sale of this total biomass (stems and leaves) to CSS would be negotiated between producer and buyer. Pricing variables would include fresh weight of the total biomass during loading and its water content, which are objective and easily measured parameters. Indeed, the facility of estimating the energy yield of a sugar cane crop from the wet above-ground biomass and its dry matter content (Chopart et al. 2013, see par. 331) will serve as a guide for simple, fair and transparent payment methods for growers based on simple wet matter and dry matter content measurements, and constant low heating value of the dry above-ground biomass (LHVd, MJ/kg), which in turn will incentivize the development of this new crop.

B- Small producers further away

It seems plausible to extend the concept of energy cane production to small producers already settled or to populate irrigated or irrigatable areas near the river, its tributaries, or on the periphery of Lake Guiers. Rather than selling the product to the CSS, which is too far from the production area to make such an exchange feasible, it would be used to produce local electricity through rustic micro-power plants located near the production area and provide electricity for one or a few villages. This "rural electricity" would be a contributor to the development and reduce the use of expensive fossil fuels. It will also contribute, modestly but tangibly, to reducing climate change.

Most of these outer farmers will probably be new to sugar cane cultivation. They will therefore require training and supervision, at least initially. The CSS, with its skilled technical staff, could contribute to this training and supervision, alongside local authorities, thus strengthening its social and CSR (Corporate Social Responsibility) function, a priority of the proposed IWRM.

Well-dried fiber cane can even more directly replace wood for cooking food, with the additional benefit of preserving the local environment. It would also contribute as a processor of non-digestible raw food products (such as rice) into digestible cooked foods. It may be worthwhile therefore to test the value of sugar cane to the benefit of populations, local industry and the environment in the interest of building the Water-Energy-Food nexus.

Fig. 10 Typha harvested by women near the Lake (@Diop)

Another local plant, the Typha (*Typha australis*), can be used for local energy production. It is a wetland aquatic plant that proliferates around Lake Guiers from its tributaries as well as the open drainage channels of the CSS (Fig. 10); currently, it is disrupting their hydraulic operation and use. This invasive species could, a priori, be used as a biofuel, such as fiber cane, to protect the environment in addition to reducing greenhouse gas emissions as with fuel cane.

The potential for the use of this species as a biofuel has not been well-studied, in contrast to fiber cane (there is only one study cited, referenced by the ACT4SSAWS (an African Union project) in reports N°8, 2015 and N°13, 2016). A new project, TyCCAO (*Typha Combustible et Construction en Afrique de l'Ouest)*, has been initiated by OLAC. It is worth keeping in view the potential for the use of this biomass because it co-exists with the fiber cane produced by farmers in micro-thermal power plants.

The transformation of fiber cane or Typha biomass into energy is achievable by combustion, making steam and then electricity, as is currently the case for bagasse or to cook food directly. It is also possible to transform this same plant biomass into energy through other technologies such as gasification or pyrolysis; alternatives may be studied. Action plans are made, of course, by local decision-makers according to various technical, economic and social criteria, which must include simple, low-cost, lightweight technologies and equipment suitable for small area production. It is indeed desirable that this production of bioenergy from local biomass be managed by local companies and technicians, according to the principle of economists such as Léopold Kohr: "Small is beautiful."

One option is to train charcoal cutters to harvest or purchase Typha to be used as an alternative fuel to charcoal (Fig. 11). Currently the harvesting of charcoal deforests

Fig. 11 charcoal from the transformation of Typha (@Cissé)

the protective trees along lake banks on embankments with a risk of dam failure. This would thus contribute to the nexus approach, enabling small producers in the Senegal River Delta to promote integrated production systems that exploit all the potential uses of agricultural products as well as some self-propagating species such as Typha.

Notwithstanding the promise of these approaches, IWRM via the Water-Energy-Food nexus remains complex to implement. Traditional extensive agricultural production systems are not ready to adopt this sort of innovation in a uniform and comprehensive way. Yet it is these systems that would benefit most in making water management and agricultural practices efficient and sustainable. As a first step, the development and production support agents of these small growers must adopt an IWRM approach and strengthen their training capacity by integrating IWRM into the Water-Energy-Food nexus to reduce the gap between them and agro-industries, making them less vulnerable to ongoing climate change.

4.4 Proposed IWRM Approach

To promote and implement innovative solutions among agro-industrial companies and especially among small producers in the Delta, an Integrated Water Resources (and Land) Management approach is desirable to help stakeholders apply the Water-Energy-Food nexus to their production systems. This will require optimal and participatory governance of this complex and diverse natural and human environment. The diagram in Fig. 12 suggests a framework that could help to promote sustainable water and soil management of irrigated agriculture in the Delta, requiring interaction among various stakeholders, including policy makers, trainers, development agents, technical experts and, of course, concerned producers.

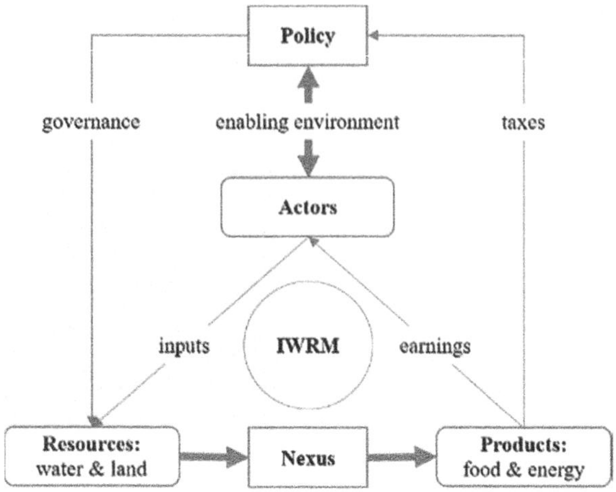

Fig. 12 Conceptual model of how to govern and to implement integrated water resources management methods in the delta

In Fig. 12, governance is directly related to land and water resources. Whenever the term "Water" in "IWRM" is used at the nexus level, the notion of soil to produce food and energy (with products and/or by-products) is implied.

Governance of these resources should be under the supervision of the State or its decentralized administration. Indeed, it is desirable that control, pollution management and associated rules be formalized by a neutral and unifying entity, to include a land and water use planning component to safeguard and allocate water and local energy resources to reduce local conflicts.

To achieve this, government institutions must, therefore, implement a regulatory and enabling framework for IWRM approaches that is outside the operational framework so as not to neutralize the dynamics of the project's stakeholders. A feature of governance must be strong awareness and supervision so that there is a common understanding of rules and regulations. This component could be financed through water charges paid by agricultural and agro-industrial producers and would be expected to result in broad awareness, be a major focus of IWRM and promote exchange among local and foreign private actors.

The suggested IWRM approach would no longer be under State control, as with initial attempts in the early 1990s. Rather, we recommend that they be driven by "champions" such as the CSS, and include all relevant local actors (agro-industry, small producers, local decision-makers). Through the establishment of a consultation framework, stakeholders will promote their self-interest in developing natural resource management within cooperative models centered on water resources, thus minimizing conflicts and rivalries. Local actors themselves are best placed to solve problems at their level, with minimal top-down State intervention. The State, on the other hand, will continue to assume responsibility for the establishment of the enabling environment that will specify the "rules of the game," that is, the laws constraining and obligations of each party.

Such an IWRM model is currently being built around the irrigation canals of the CSS, a relatively small area that is nevertheless quite representative of the challenges in the Delta related to soil and water, agro-economic models, types of actors, etc.). In recent years, local stakeholders have initiated a cooperative structure for the use of water from CSS irrigation canals and adjacent land by farmers, granting rights to cane harvest residues and artificial pools for livestock farmers. The challenge now is to strengthen and improve the existing system to achieve better agricultural and economic results, and participatory and conservatory management of natural resources, integrating other stakeholders and all the dimensions selected in the IWRM framework. Therefore, a consultation framework is being implemented to integrate water resource management in an environment of trust.

5 Conclusion and Prospects

The proposed nexus and IWRM approaches outlined are complementary, as indicated by the United Nations (Mohtar and Lawford 2016). Both contribute to sustainable development and environmental protection. In the case of the Senegal River Delta, the availability of natural resources (fresh water, soil, transformable aquatic plants and sunlight) is favorable to produce models using these approaches. Local populations already consume electrical energy from river water (the Manantali Dam) and the sun (the Bokhol solar platform). These renewable electricity producers are indicators of the reliability and sustainability of solar and water energy production in the area. However, they are disadvantaged by their centralization and, therefore, disconnection from rural populations and agricultural systems.

The many options for food (cereals, industrial crops, horticulture) and energy (residual biomasses, energy cane, harvested Typha, solar energy) production are still underexploited. Intensive and industrial agricultural production certainly consumes water, as is the case of sugar cane at the CSS. But with an average sea discharge flow of 13 billion m^3/year, compared to five billion used, the volume of unused water remains enormous, even considering the needs of spontaneous vegetation and natural ecosystems. From this perspective, fresh water is not currently a constraint in the Delta. The actual availability of this water at the local level, however, may be reduced or constrained, especially for small producers, due to physical (invasive aquatic plants) and economic (e.g., pumping costs) limits.

Despite the diversity and heterogeneity of these actors, synergy that benefits everyone must be reached. For example, the by-products and technical knowledge of industrial or agro-industrial actors could benefit small producers by supporting diversification of their production systems, integrating, in particular, the production of biomass for energy purposes alongside food production. These interactions set the stage to promote a productive dialogue among stakeholders. In the coming years, this will lead to better food and energy security in the river delta. Further Research & Development, conducted with local actors to assist in policy and technical choices

(Hoolohan et al. 2018), is needed for successful adoption, which would strengthen the capacity of local actors to take ownership of the approach.

In the current context of global and local climate change with high population growth, it is, of course, difficult to predict the future, especially in the Senegal River delta. The environment is rich in water resources and agricultural development potential but is also fragile, on the border of very poor and already semi-desert areas. IWRM within a Water-Energy-Food Nexus approach, applied to the river delta, will support decision-makers and stakeholders in the area to optimize the use of water and energy resources.

In the ideal case, IWRM and the Water-Energy-Food nexus approaches will be adopted through appropriate investments, making it possible to consider sustainable agro-economic development for the benefit of all. If this fails, it could lead, in the context of climate change, to greater vulnerability of small farmers as well as agro-industry. The latter, although technically better equipped, would be subject to social and environmental constraints that threaten the sustainability of their production models. Considering the complexities, it may be most realistic to imagine an initial intermediate scenario consisting of partial implementation. This could lead in turn to further development at the expense of optimizing progress agriculturally, economically and socially. Smaller and more fragile producers are more vulnerable to negative impacts than are modern and autonomous producers.

Acknowledgements The authors wish to thank

The CSS Executive Director, Mr Vincent Leroux, for his agreement to use internal CSS data

Prof Bernard Tychon for the appreciated contribution on several topics of discussion.

Prof Alioune Kane and Prof Awa Niang for the appreciated opportunity to work with them on IWMR in the Delta area.

References

Burnett K, Christopher AW (2018) Accounting for externalities in the water energy food Nexus Wada from the Water-Energy-Food Nexus. Global Environ Stud :261–272

Chopart JL (2016) Two high-fibre sugarcane varieties adapted for energy use in Guadeloupe In: ISSCT (edit) 29th ISSCT congress, Chiang Mai, Thailand, vol. 29, pp 34–37

Chopart JL , Bonnal L et al (2013) Functional relationships between dry above-ground biomass and the energy yield of sugarcane, In: STAB & ISSCT (edits) 28th ISSCT congress, vol. 28, p 39

Chopart JL, Sall MT et al (2018) Advances in sugarcane irrigation for optimisation of water supply. In: Rott P (ed) Achieving sustainable cultivation of sugarcane, vol. 1. Cultivation Techniques, quality and sustainability, Burleigh Dodds, London, pp 141–160

Chopart JL, Sergent G (2015) Root biomass quantification of sugar and multipurpose cane varieties for sustainable production, In: Abstracts of the second ISSCT Agronomy Workshop, Durban South Africa, 1 p

Chopart JL, Sergent G, Goebel FR (2015) Cropping systems for energy cane grown on volcanic soil in a tropical climate: initial results on planting dates, cycle duration and pest pressure. In: Abstracts of the second ISSCT Agronomy Workshop, Durban, South Africa, 1 p

Cissé B (2011) Les eaux de drainage des périmètres irrigués du Delta du Fleuve Sénégal : systèmes d'évacuation et qualité des eaux, Dissertation, University of Dakar and University of Bordeaux, 301 p

Dia S (2004) Nature et Métropole : évolution des politiques publiques dans le Delta du Sénégal: les réponses organisationnelles et spatiales au désengagement de la SAED. VertigO 12(2):19

Diop P (2017) Vers une gestion participative de la ressource en eau multi-usage dans le Delta du fleuve Sénégal. Processus de décision et outils de régulation autour du Lac de Guiers, Dissertation, University of Paris Saclay and University of Dakar, 370 p

Dumont R (ed) (1962) L'Afrique noire est mal partie (Le Seuil, collection « Esprit » , Paris), p 287

Faye C, Diop ES, Mbaye I (2015) Impacts des changements du climat et des aménagements sur les ressources en eau du fleuve Sénégal: caractérisation et évolution des régimes hydrologiques de sous-bassins versants naturels et aménagés, Belgeo (4) 26 p

Food and Agriculture Organization FAO (2009) The state of the food and agriculture. Rome

Goebel FR, Chopart JL et al (2016) Intégrer les nouvelles valorisations de la biomasse dans la conception des systèmes de culture. In: Biénabe E, Rival A, Loeilllet D (eds). Développement durable et filières tropicales. Quae, Paris, pp 183–196

Hess TM, Sumberg J et al (2016) A sweet deal? Sugarcane, water and agricultural transformation in sub-Saharan Africa. Global Environ Change 39:181–194

Hoolohan C et al (2018) Engaging stakeholders in research to address Water–Energy–Food (WEF) Nexus challenges Sustainability Science. https://doi.org/10.1007/s11625-018-0552

Hulme M (2001) Climatic perspectives on Sahelian desiccation: 1973-1998. Glob Environ Change 11:19–29

International Institute for Sustainable Development Report, The Water–Energy–Food Security Nexus (2013) Towards a practical planning and decision-support framework for landscape investment and risk management. https://www.iisd.org/pdf/2013/wef_nexus_2013.pdf

Kamara S (2013) Développements hydrauliques et gestion d'un hydro-système largement anthropisé: Le delta du fleuve Sénégal. Dissertation, University of Avignon and University of Saint-Louis 472 p

Kane A (1997) L'après-barrage dans la vallée du Fleuve Sénégal : modifications hydrologiques, morphologiques; conséquences sur le milieu naturel et les aménagements hydro agricoles. Dissertation, University of Dakar

Matsuoka S, Kennedy EGD et al (2014) Energy cane: its concepts, development, characteristics and prospects. Adv Bot. https://doi.org/10.1155/2014/597275

Mazoyer M, Roudart L (2002) Histoire des agricultures du monde, du néolithique à la crise contemporaine. Editions du Seuil, Paris, p 705

Mboup M (2014) Changements socio-environnementaux et dynamique de la végétation aquatique envahissante dans le delta du Fleuve Sénégal. Dissertation, University of Dakar

Mietton M, Dumas D et al (2008) Le delta du fleuve Sénégal : gestion de l'eau dans l'incertitude chronique. In Allard P, Fox D, Picon B (eds) Incertitude et environnement. La fin des certitudes scientifiques (Editions Sud) pp 321–336

Mohtar R, Lawford R (2016) Present and future of the Water-Energy-Food Nexus and the role of the community of practice. Environ Stud Sci 6:192–199

Sandhu SH (2018) Cultivating sugarcane for use in bioenergy applications: key issues. In Rott P (eds) Achieving sustainable cultivation of sugarcane vol. 1. Burleigh Dodds Science Publishing, London, pp 235–250

Sierra J, Chopart JL et al (2016) Optimization of biomass and compost management to sustain soil organic matter in energy cane cropping systems in a tropical polluted soil: a modeling study. Bioenergy Res 9:798–808)

Stiglitz JE (2004) La grande désillusion. Edition Fayard, Paris, p 407

Stocker TF et al (2013) Climate Change 2013, The physical science basis. contribution of working group i to the fifth assessment report of the Intergovernmental Panel on Climate Change IPCC, ed. Cambridge University Press, Cambridge, United Kingdom and New York, NY, USA, 1535 p

Smeets EMW, Johnson FX, Ballard-Treemer G Traditional and improved use of Biomass for energy in Africa. In: Janseen Rand Rutz D (eds) Bioenergy for sustainable development in Africa. Springer, Berlin, pp 3–12

Tandian H (2008) Les organisations paysannes du Delta du Fleuve Sénégal dans la gestion des aménagements hydro agricoles : Problématique sociologique du transfert des aménagements hydro agricoles de la gestion publique (Etat) à la gestion privée par les organisations paysannes. Dissertation, University of Saint-Louis (Sénégal

Chapter 8
Cumulative Impacts of Climate Change Variability Around the Goronyo Dam in the Iullemmeden Basin, Northwest Nigeria

Martin O. Eduvie and I. Francis Oseke

Abstract This chapter considers studies of the persistent multi-dimensional impact of climate change variability around the Goronyo Dam and across the Rima River (Latitude 13° 25' 56"E) in the Lullemmeden Basin of Sokoto state, North West Nigeria. With an initial designed storage capacity of 942 million cubic meters, mainly for water supply and irrigation to Sokoto and Birnin Kebbi states with a total population of over 8.5 million people, recent water level and storage in the dam has been depleted by almost 90% in the past decade of its installed capacity, the biggest decrease since the construction of the dam over 30 years ago. The peak of climate change variability was observed as a consequence of a rainfall shortage and the siltation of the dam in 2017, which together resulted in significant socioeconomic impact on the beneficiaries of these water infrastructures. This has, in turn, resulted in inadequate water supply to state water agencies that depend on the dam to supply water for drinking, irrigation and other uses. The water shortage due to this climate change variability has compelled the dislocation of the socioeconomic life of people, families and livestock, forcing migration of human and animal life, thereby igniting social tension and poor personal and communal hygiene, resulting in epidemics including cholera and dysentery. This study proposes immediate remedial solutions, including the construction of tubewells and effective integrated water management of the entire water resource in order to improve and mediate these multiple and widely varying impacts on the settlements of Goronyo infrastructures, including on their dependents and beneficiaries.

Keywords Climate change · Goronyo dam · Precipitation water resources · Conflict and tube wells

M. O. Eduvie (✉) · I. F. Oseke
National Water Resources Institute, Kaduna Nigeria, Nigeria
e-mail: martineduvie@gmail.com

S. Diop et al. (eds.), *Climate Change and Water Resources in Africa*,
https://doi.org/10.1007/978-3-030-61225-2_8

1 Introduction

A dam is a man-made structure built across sections of a river or stream to retain water, which is generally used for agricultural purposes such as farmland and garden irrigation and the watering of livestock (Lodha 2007). It also can serve as a constructed barrier across a stream or river to impound water and raise its level for various purposes, such as drinking water supply and irrigation systems, and also increase river depth for navigation, generating electricity, control of water flow during times of floods and droughts, and the creation of artificial lakes for fisheries and recreational use. Many dams serve multiple of these purposes (Uyigue 2006). However, in the past twenty (20) years especially in 2017/2018, there has been upsurge in extreme events which negatively impacted the dams as a result of changes in climate (NASRDA 2012). In this particular event of 2017/2018, the rainfall period was not as long as anticipated. On the average, rainfall period always occurs between March/April to September/October but in the 2017/2018, the rains only occurred between April to early August. Climate and water resource systems have a special relationship insofar as water resources depend on the hydrological cycle which itself is part of the climate system (Stainforth et al. 2005). The impact of climate change on the function and operation of existing water resource systems and infrastructures, including hydropower, structural flood defenses, drainage and irrigation systems, as well as water management practices (Alvarez et al. 2014) are critical. As effects of climate change on water resource systems accumulate, there is corresponding impact of other stresses, such as population growth, changing economic activity, land-use change and urbanization (Thornton et al. 2008).

The concept of climate change revolves around the atmospheric concentrations of greenhouse gases (GHG) which have increased since the pre-industrial era due to human activities, primarily the combustion of fossil fuels and changes in land use and land cover (Gwary 2008). These, together with natural forces, have contributed to changes in the Earth's climate over the twentieth century. Land and ocean surface temperatures have warmed, the spatial and temporal patterns of precipitation have changed, sea level has risen and the frequency and intensity of cumulative pressure from climate change events have increased. These changes have affected dams and reservoirs and other water infrastructures (Gwary 2008).

The importance of water resources to man cannot be overemphasized. These include provision of water for domestic use, agricultural production, fishing, transportation, industrial use, hydro-electric power, recreation, tourism and mining. Furthermore, according to the World Bank (2014), every development challenge of the twenty-first century, such as food security, management of rapid urbanization, energy, security, environmental protection and adaptation to climate change requires urgent attention to the infrastructure management of water resources. Available information and statistics indicate that in recent times, climate variation or change is becoming more extreme in certain areas. This dynamic is not specific to Nigeria of course, but rather is represented by a rapidly growing global catalogue of storms, floods and drought in all parts of the world, which together make climate variability headline news.

Flood and drought as consequences of climate change cause tremendous losses to infrastructure and industry, including reducing the quality of personal and communal hygiene, with the consequence of epidemics like cholera and dysentery. In addition, losses to agriculture caused both by floods and droughts are staggering and affect a vast majority of the population directly dependent on water resources for their livelihood (Azman 2007). A recent example of this is the horrendous hardship inflicted in 2017 on the people of the town of Goronyo in the Sokoto state, located in north central Nigeria, caused by drought as a result of the shortage of rainfall in 2017 and the related siltation of the Goronyo Dam. This devastated socioeconomic activities and cost lives.

Cumulative climate change, water resources and socioeconomic systems are interconnected in complex ways (Alex et al. 2005), so a change in any one of these variables causes change in one or more others. Persistent drought and flooding, off-season rains and dry spells have interrupted the growing season of areas dependent on rain-fed agriculture. Plants that require low or high temperature at some stage of their life cycle may adapt and survive in the short term, but could become extinct in the long term.

As available water resources decrease and concern rise over the influence of climate change, the reliability of estimates of current and future water resources in reservoirs are becoming increasingly important to water managers. We, therefore, consider the cumulative pressure from climate change that encompasses all pressures from human activities, specifically examining impacts on the Goronyo Dam of the Iullemmeden basin in northwest Nigeria.

1.1 Aim and Objectives

The aim of this study is to consider climate variability in the context of global change that encompasses all pressures from human activities on water resource systems and infrastructures. Specific objectives are to analyse and review the impacts of climate change variability on the Goronyo Dam and to evaluate possible response options (or adaptation options) through an integrative approach covering

i. Climate variability and change, and its impact on dams;
ii. Observed changes in the spatial distribution of water resource systems and infrastructure in response to climate variability, especially change in temperature and rainfall and
iii. Development of responses that address the impacts of cumulative pressures, including climate change on dams.

2 Literature

2.1 Climate Change and the Impact on Dams

Climate change and its impact of dams are characterized by

i. Observed changes in the hydrological circle (atmosphere, oceans) and the role of human activity in those changes;
ii. Observed impacts of these changes on water resource systems, especially dams and reservoirs and
iii. The role that adaptation can play in responding to these changes.

2.2 Projected Changes in the Climate System

Future emission of greenhouse gases and aerosols are driven by forces such as changes in human population, socioeconomic development, technological and policy with necessary action plans by Governments as well as other stakeholder interventions. In 2014, the Inter-Governmental Panel on Climate Change (IPCC) presented in its special report on emissions six groups of scenarios or plausible futures which are based on narrative storylines and account for a wide range of driving forces. The scenarios project future emissions of the greenhouse gases, carbon dioxide, methane and nitrous oxide, and the aerosol sulphur dioxide. For gases that stay in the atmosphere for a long period, such as carbon dioxide, the atmospheric concentration responds to changes in emissions relatively slowly, whereas for short-lived gases and aerosols, such as sulphate aerosols, the atmospheric concentration responds much more quickly.

This is due to the length of the half-life of carbon dioxide in the atmosphere as well as inertia in the system. Even if emissions were to halt today, the Earth's surface temperature could continue to rise for a few centuries, and sea level could continue to increase for several millennia, due to thermal expansion and melting of ice. According to the IPCC, the following effects are likely to occur to surface temperatures, precipitation and climate-related extreme events:

2.2.1 Projected Changes for Surface Temperatures, Precipitation

i. The globally averaged surface temperature is projected to increase by 1.4 to 5.8 °C between 1990–2100 and
ii. Some areas are projected to become wetter and others drier, with a net increase in precipitation projected.

2.2.2 Projections for Climate-Related Extreme Events (E.G. Floods, Heat Waves). Include

i. Higher maximum temperatures: more hot days and heatwaves over nearly all land areas;
ii. Higher minimum temperatures: fewer cold days, frost days and cold spells over nearly all land areas;
iii. More intense precipitation events over many areas;
iv. Increased summer drying over most mid-latitude continental interiors and associated risk of drought and
v. Increase in peak wind intensity and mean and peak precipitation intensities in tropical cyclones.

2.3 Current Concern

2.3.1 Cumulative Impact of Climate Change on the Goronyo Dam

Climate change is likely to affect dams directly (e.g. through changes in temperature) and indirectly (e.g. through impacts on hydrology). Dams have typically been designed using historic periods of streamflow record to determine storage capacity and related dam infrastructure features, including spillways and hydroelectric turbine capacities. Understanding climate change will likely impact the quantity and timing of water availability. Specific impacts are projected to include

i. *Decreased water availability in the Sokoto River Rima Basin (Goronyo Dam catchment).*

The decrease of water availability is attributed to poor watershed management, deforestation, irrigation systems, waste and poor management of hydrology in the catchment area and shortage of rainfall as well as siltation of the dam (Plates 1 and 2).

Developing countries are projected to be the most vulnerable to climate change: many are already more prone to water shortages. In the area under study, indigenous communities that depend on dam water resources are already vulnerable and will become more vulnerable as a result of these projected impacts; current conditions are the worst seen in the dam in over 25 years. Compounding the chronic problems is the shortage of rainfall in 2017 and siltation of the dam, which is directly related to climate change. In addition, increasing sea surface temperatures were recorded in most of the region during the last three (3) decades 1980–2019 (Plate 3, 4 and 5).

Plate 1 Section of Goronyo Dam from the northern side before 2017

Plate 2 Goronyo Dam Showing Upstream (A–B) and Downstream (C–D)

2.4 Study Area

2.4.1 Characteristics of Goronyo Dam

The Goronyo Dam in Sokoto is located in the Sokoto Rima basin, situated within the Niger north hydrological area (HA-1) in the semi-arid region of the Northwestern

Plate 3 Goronyo Dam water shrinkage from spillway in 2017

Plate 4 Sand Dome presence in Goronyo Dam

part of Nigeria (Fig. 1), between latitude 11°N to 16°N and longitude 3.3°E to 10°E. The entire basin covers a total catchment area of 131,600 km^2 and is drained by many rivers, including the Tarka (upstream of Goronyo Dam), Kaba, Tagwai, Maradun, Bunsuru, Gagare, Sokoto, Zamfara and Ka rivers. The Sokoto-Rima extends from parts of Katsina and the entirety of Sokoto, Kebbi and Zamfara, and covers a small portion of Kaduna and Niger States. Major dams in the basin are the Bakolori and Goronyo dams. The capacities of the dams are 942 and 450 million m^3 with planned irrigation area of 69,000 ha and 30,000 ha, respectively. The actual developed irrigation area is just 40,000 ha.

Plate 5 Goronyo Dam showing water resource shrinkage

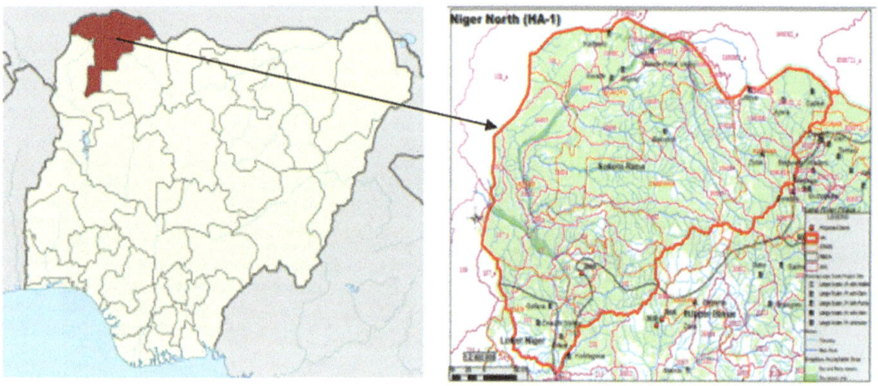

Fig. 1 The Study Area Showing Goronyo Dam; *Source* NASRDA 2012

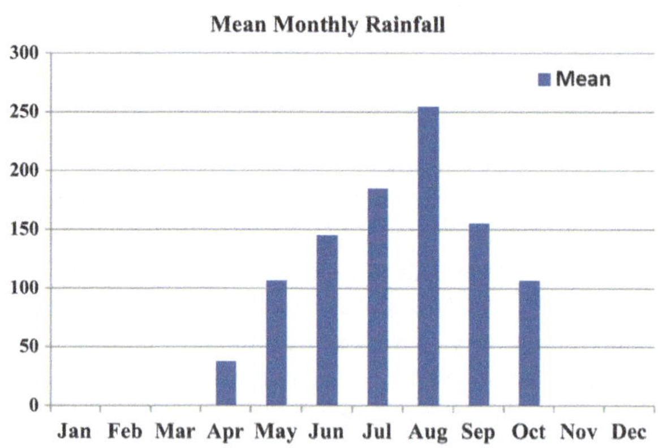

Fig. 2 Average Rainfall Amount (mm) in Sokoto from 2007 to 2017

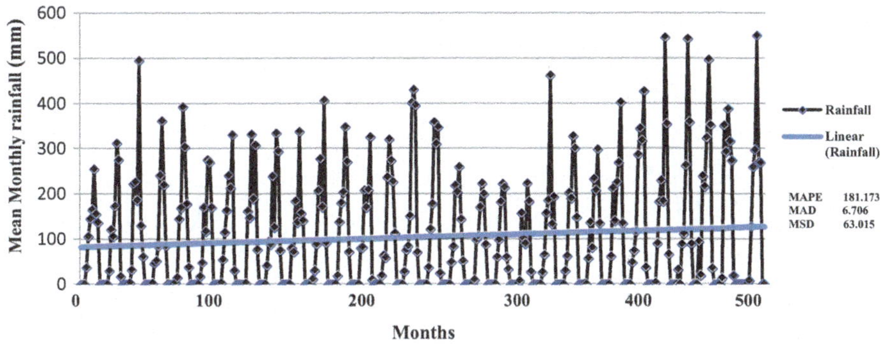

Fig. 3 Time series of total rainfall; 1920–2007 at Gusau Aerodrome

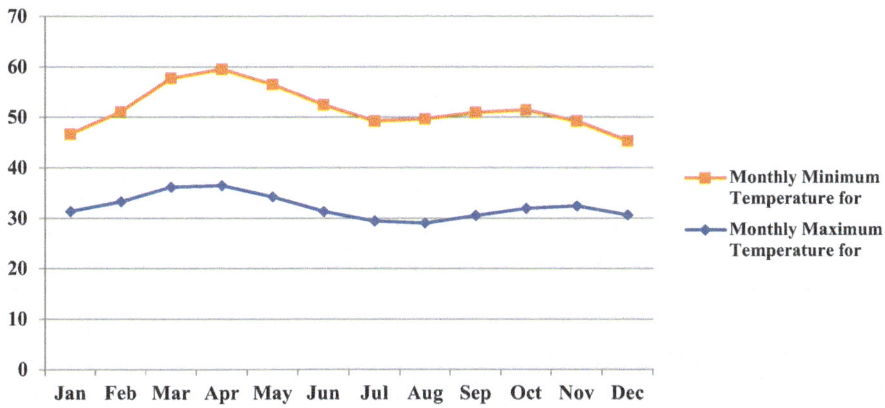

Fig. 4 Maximum, Minimum and Average Temperature (°C) in Sokoto

The Goronyo dam is classified as large dam in Nigeria and it is multipurpose in nature. The Goronyo dam was completed in 1984 but commissioned in 1992 and is to store 942 millioncubic meters for irrigation and development of downstream areas from Goronyo to Argungu, covering an area of about 200 km^2. The dam, which is earth-filled and built across the Rima River on the Sokoto Rima basin, is located near Keta Village, about 25 km east of the town of Goronyo and 90 km northeast of Sokoto City; it is the largest dam under the jurisdiction of the Sokoto Rima River Basin Authority (SRRBDA) (Figs. 2, 3, 4 and 5).

2.4.2 Hydrology of the Goronyo Dam

The Goronyo Dam is located within the Sokoto Rima River Basin, which is essentially drained by the Sokoto River, a prominent part of the Niger River drainage system.

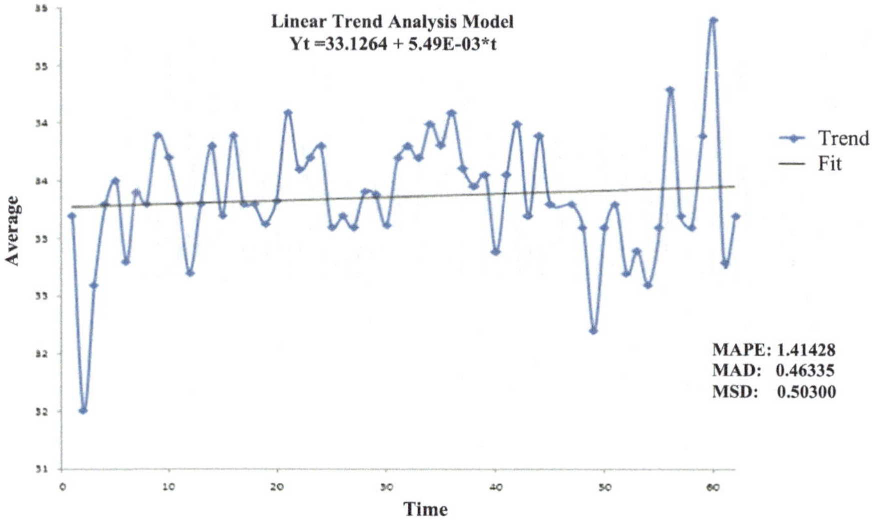

Fig. 5 The trend analysis for temperature 1949–2009 (*Source* Sokoto New Airport)

The Sokoto River rises with its main tributaries, the Ka, Zamfara and Rima, from the 600 to 900 meter-high Mashika and Dunia highland areas bordering the eastern part of the basin, and flows sluggishly down a gentle slope towards the northwest around Sokoto town. It is joined by the Rima in the north, making a southward swing, collecting the rivers Zamfara and Ka before entering into the Niger River. The river systems thus effectively drain the whole basin. At the source areas in the east, the Sokoto River System is only seasonal, but in the western parts of the basin, the river becomes perennial as substantial groundwater contributes to its flow.

The dam comprises three earth-filled dams; the main, secondary and saddle dyke with a total length of 12 km, a concrete intake and outlet structure including a spillway of 1,540 cubic meters and other infrastructure.

2.4.3 Goronyo Dam Capacity

The dam is designed to provide water for irrigation and development of downstream areas including the middle Rima Valley and the Zauro Polder Project, covering a total of 17,000 hectares. It provides an annual regulated flow of 425 million cubic meters to double the present rice cultivated fadamas from 40,000 to 80,000 hectares, and supplies 80 million cubic meters of water annually to the Sokoto, Argungu and Birin Kebbi water supply scheme, while the 200 km^2 lake formed by the dam is expected to boost the fishing industry in the area.

2.4.4 Climate Around Goronyo Dam

Like the rest of West Africa, the climate of the region is controlled largely by the two dominant air masses affecting the sub-region. The regions are also associated with the movement of air masses of the Inter Tropical Convergence Zone north and south of the equator.

3 Methodology

3.1 Data Collection and Analysis

Data for this study was generated through field visits to the dam site on 30 November 2018, and geographical review of the Goronyo Dam by comparing historical data from imagery, field observations and assessment of the dam and water shortage evidence, and field assessment of the effects of climate variability on settlements and farmlands. Using a digital camera, pictures were taken of the dam and of the farmland that was once irrigated using the dam water. Farms using dam water for irrigation were also investigated. The historical climate records up to 2018 were compared to determine the trends in various climate parameters, including temperature and precipitation in the hydrological area. The catchment for analysis, chosen as a basis for the general climate conditions of the area, was Sokoto New Airport and Katsina and Gusau aerodromes. Secondary sources of data were collected through desk research from subject-matter Internet sources, including published journal articles, discussion papers, environmental reports and presented conference papers (National Water Conference and National Water Resources Council) (Plate 6).

3.2 Climate Change Variability Analysis

Stream flows are generally affected by the cumulative effect of climate change. Hence, study of stream flow over time is essential, especially to determine the sustainability of water resource systems. The most significant observed impacts include changes in temperature and precipitation, which in turn have varying consequences for water availability. For this reason, trend analysis was used to define climate change with respect to rainfall, evaporation, temperature and humidity in the hydrological area, using Microsoft Excel (Plate 7).

Plate 6 Goronyo residents queuing for water from alternative sources due to water shortage from Goronyo dam to the water treatment plant

Plate 7 Stream drying up and fishermen abandoning fishing net as a result of impact of climate change on the dam

4 Results and Discussion

Results have established the impacts of climate variability around the Goronyo Dam located in Hydrological Area I. However, all the stations considered in the analysis for stream data, i.e. Wamakko on the River Rima, Jega on the River Sokoto and Gidan Doka on the River Zamfara, have demonstrated the impact of climate variability on their discharges over the duration under study. Moreover, the stations, i.e. Sokoto New Airport and the Katsina and Gusau aerodromes, were selected for the analysis of temperature and precipitation and have also shown the impacts of climate variability.

4.1 Trend in Climate Variability Analysis

The intensity, frequency and seasonal distribution of rainfall have particular impact on the amount of water available for various uses in the study area. The dry season, from early November until March, has no trace of rainfall at all. The mean monthly rainfall for the study area is depicted in Fig. 2, with the historical data in Fig. 2. This shows a reduced rainfall of 1.2% in the period 1949–1976 compared to 1976–2008. In addition, the historical trend shown in Fig. 3 at Katsina and Gusau aerodromes also shows an increase in annual rainfall in 1936 and 1965, reaching a maximum of 1,025 mm and 1039.4 mm, respectively, an increase of 1.4%. The average annual rainfall in Sokoto from the end of the 1970s through the 1980s was 632 mm, a decreasing trend, with the lowest recorded year at 325 mm in 1987.

The mean monthly maximum temperature of the study area is between 30.6 and 36.5 °C, with the hottest period in the months of March and April. The highest mean monthly temperature is observed in April with 36.5 °C, while the lowest mean monthly temperature is in December with 14.7 °C. Seasonal and latitudinal variations affect the extremes and the diurnal and seasonal ranges of the temperature in all areas. In addition, the daily minimum temperature by monthly average is lowest in August at 30.5 °C. From Fig. 4, it is evident that there was a significant rise in temperature at Sokoto New Airport between 1989 and 2009. A decrease of 1.8% was observed between the periods 1949–1976 and 1976–2008 (see Fig. 5).

4.2 Cumulative Effect on Water Supply

The communities of Goronyo that consist mostly of farmers depend largely on dam water for domestic purposes, especially during the dry season. The importance of dam management is most evident during times of drought. The 2017 drought in the Goronyo dam affected the lives and livelihoods of almost two million people as a result of the shortage of water. Levels in the Goronyo dam basin, which provides water for farmers, fishermen and families, have dropped dramatically in the 2017/2018 to about 10% of its capacity, forcing authorities to ration water to homes and seek alternative water sources (Plate 6).

The impacts of climate change are being felt by both developed and developing countries. To the extent that they are felt more by developing countries, it is not necessarily because they are the highest contributors of climate variation, but rather because these countries lack economic, social and political infrastructures to respond adequately to the effects of climate change. The Sokoto state of Northern Nigeria, where Goronyo dam is located, contains a significant portion of the Sudan-Sahel ecological zone of West Africa. Since the early 1970s, climate anomalies in the form of recurrent droughts and numerous dust storms have threatened access to clean drinking water.

4.3 Loss of Fishing Grounds

Fishermen set traps in dam water bodies for home consumption and sale. Shortage of water from the dam has resulted in the disappearance of fish on the dried neighbouring streams. According to Tambuwal (2010), fishermen were negatively affected by the extreme event of fish stocks being carried away with the escape of the water, which devastated fishing activity.

4.4 Effect on Agricultural Farmlands

Climate change compounds the linkage of agricultural farmlands to poverty; the adverse impacts of climate change on farmland exacerbate the incidence of rural poverty. This is likely to be most severe in Nigeria, where agriculture is a major livelihood for the majority of the rural population. If nothing proactive is done to replenish the shrinking water supply, water resources in the Goronyo dam will become insufficient to support its communities.

In communities surrounding the dam, over 80% of the population depends on the dam for most of its survival, including from livelihoods from agriculture and fishing. The severe drought in 2017 brought into dramatic relief the adverse effects of variability in timing and amount of rainfall, and lack of water for the irrigated farmlands (Plate 8).

4.5 Migration and Conflict

Nigeria is home to more than 15% of the entire African population, with a median age of just 19 years. Its total population is projected to double to roughly 320 million by 2040 (Red Cross 2013). Climate change is already affecting Nigeria in significant ways. The most noticeable is in the 1,350 square miles of Nigerian land that turns to desert each year, driving farmers and herders south from the Sahel and into cities. Lake Chad, which marks Nigeria's northeast border, was reduced to one-twentieth of its size during the 1960s due to a drier climate and changing water management. These effects have exacerbated additional destabilization. More recently, in 2017, the Goronyo dam saw its water level depleted to about 10% of its installed capacity due to the cumulative pressure of climate change. This has devastated the livelihood

Plate 8 Farmlands showing symptoms of stress downstream of the dam

of many farmers in Goronyo and its surrounding communities, and resulted in some farmers migrating to other areas, in particular the Bakolori Dam in neighboring Zamfara state.

The Bakalori Dam was the source of water for cattle farmers living within Goronyo, as well as for nomadic herdsmen who brought their cattle to drink at the dam. Herdsmen also graze their cattle, because grasses grow along the banks of the dam. During the study visit to Goronyo dam in 2017 and 2018, only a few small herds of cattle were observed to be moving on the plain of the dam, compared to the thousand observed previously in the 1980–2000 (Plate 9).

Plate 9 Few cattle with herdsmen and a fisherman behind them (showing decline in cattle numbers)

It is critical to consider the resulting effect with regard to migration and conflict associated with the 2017/2018 drought due to shortage of water in the Goronyo Dam, because it includes migration, leading to tension, violence, conflict and loss of livelihood in the affected communities or areas of new settlement. The majority of those affected by the migration lost their farm land, the source of pasture and water for their livestock, compelling them to leave their familiar rural environments to reside in an unfamiliar urban setting.

4.6 The Role of Adaptation (Interventions)

Adapatation to climate change has the potential to mitigate its adverse effects, but cannot prevent all impacts. Numerous projects and processes designed to reduce the impact of climate change have been identified that can reduce adverse and enhance beneficial impacts of climate change, but all come at a cost (Gwary 2008). Adaptation is necessary to complement efforts to take deliberate action to reduce the sources of and/or enhance the sinks of greenhouse gases. Together, adaptation and mitigation contribute to sustainable development objectives for dams and reservoirs.

Adaptation activities can promote conservation and sustainable use of man-made and natural solutions to reduce the impact of changes in and extremes of climate on cumulative pressures on dams and reservoirs. These activities include the establishment of interconnected, multiple-use reserves designed to take into account projected changes in climate, and integrated land and water management activities that reduce non-climate pressures on dams and reservoirs and hence reduce structural and human vulnerability to changes in climate. The effectiveness of adaptation and mitigation activities are enhanced when they are integrated within broader strategies for sustainable development.

4.7 Management Approaches for Addressing the Impacts of Climate Change on Dams (the Goronyo Experience)

The Nigerian government should as a matter of urgency facilitate a sustainable program to mitigate the upsurge and cumulative effects of climate change. Within such a strategy, the following approaches for the case of Goronyo Dam are suggested.

4.7.1 Use of Wetlands as Improved Watershed Infrastructures

Wetlands can be used to trap sediment, retain water during high flow periods and attenuate nutrient loads. Strategically placed constructed or restored wetlands in watershed headwaters or near dams have the potential to ameliorate the impacts of large precipitation events (World Commission on Dams 2015). The upstream debris dams and sediment basins can help trap coarse-grained sediment before it reaches the dam. River basins can be dredged periodically at less cost than large reservoir dredging.

4.7.2 Integrated Watershed Management (Sokoto River Rima Basin Development Authority-SRRBDA)

Sokoto Rima River Basin Development Authority should be encouraged to develop and implement a sophisticated Watershed-based Management Program for Restoration and Protection Strategy (WRAPS), based on the dams within its jurisdiction, involving collaboration among several state agencies. The program should be built on a planning and management framework with stakeholder involvement, which is essential because it is the stakeholders who are responsible for developing watershed assessment, establishing goals, identifying necessary actions and costs, preparing a watershed plan, and securing resources needed to execute on it. This process must be closely monitored by the SRRBDA, under the authority of the Federal Ministry of Water Resources in the capital of Abuja, Nigeria. Relevant stakeholders for the Goronyo Dam include

Bilateral and Multilateral Organizations

i. Niger Basin Authority (NBA)
ii. IAEA-UNESCO-GEF Regional Project on Development of Water Resources of the trans-boundary Iullemeden Aquifer System
iii. UNICEF: Water Supply and Sanitation Projects
iv. World Bank: Water Supply and Irrigation Projects
v. African Development Bank (AfDB)

Nigeria Federal Level:

i. Federal Ministry of Agriculture
ii. Federal Ministry of Environment
iii. Federal Ministry of Water Resources—Implementing and Coordinating Capacity.

State Level:

State Water Boards (SWBs)/Corporations or Utility Boards for water supply schemes, including the development and Operation & Maintenance (O & M) of the schemes.

Areas of Collaboration:

i. Various stakeholders should provide relief material and supports (which include seedlings for agricultural purposes, technological support with information from extreme weather events, water pumps and credit facility, etc.) to the farmers affected by the shortage of water from Goronyo Dam who have suffered loss of farmlands and food crops with a view towards rehabilitating their shattered lives;

Table 1 Dam services and possible impacts due to climate change

Reservoir-Derived Ecosystem Services	Possible Effects due to Climate Change
Flood Control	Overwhelm flood control capacity
Water Supply: Municipal	Sedimentation diminishes water supply capacity; uncertainty in drought adds water supply stress; increased nutrient loading will increase eutrophication
Water Supply: Industrial	Sedimentation diminishes water supply capacity; uncertainty in drought adds water supply stress and decrease in water quality
Water Supply: Agriculture	Sedimentation diminishes water supply capacity; uncertainty in drought adds water supply stress
Power Generation	Decreased inflow may bring water levels below turbines and decrease power generating potential

ii. Every stakeholder should work towards ensuring that water resources, especially small and large-scale dams, are properly managed: mismanagement leads to the loss of water resources as well as the destruction of live sand properties and
iii. Technical personnel should be hired to supervise the dam and its condition during the rainy season so that excess water is released in a timely manner.

4.7.3 The Need for Flexible Financial Mechanisms to Cope with Climate Change

Another strategic management approach to address the impacts of climate change on dams is the consideration and development of flexible financial mechanisms to manage climate change adaptation. The most obvious option is to provide agricultural subsidies to farmers and herders during floods and drought. This measure has the potential to create a highly enabling macroeconomic environment in which approaches for addressing climate change adaptation becomes the responsibility of individuals. The challenge is to secure equitable wealth distribution to affected individuals and to target support to those most likely to be affected by climate change (Table 1).

4.7.4 Innovative Technology Through Use of Tubewells

Tubewells (also known as sand point wells) are an important adaptation technology measure for providing a domestic water supply during times of water shortage and drought. They extract freshwater from subsurface or deeper groundwater aquifers. The approach includes both creating new tubewells as a response to drought, and/or deepening and rehabilitating existing ones. Construction is relatively easy, and tubewells are most often installed using a hand auger. However, for a tubewell to serve as a source of potable water, precautions must be taken to ensure that water quality

Table 2 Major factors in tubewell management

Consideration	Description
Environmental benefits	Relieves pressures on surface water sources, reducing risks of pollution and degradation
Socioeconomic benefits	i.Provides freshwater for domestic and other uses in times of drought ii.Produces high-quality water, reducing health risks that may occur from use of surface water sources iii.Helps avoid interruption of significant socioeconomic activities during dry periods
Opportunities	Increased diversification of water sources provides for more water but also increases water supply resiliency
Major Barriers	i.Requires pumping and associated energy supply (and costs) for larger volumes, ii.Requires knowledge of local geological conditions and assessments of chosen drilling sites iii.Poorly coordinated well development can cause a groundwater table decrease and create risk of over-abstraction iv.In areas with high climate variability (floods and droughts), tubewells and boreholes are at risk of contamination during flood events

is acceptable. This is done through groundwater surveys. Major factors in the choice of tubewells include Table 2.

4.7.5 Improvement of Disaster Preparedness and Management Skills

Given the general lack of disaster preparedness in the region, community-created emergency structures, including procedures and training for potential disaster scenarios, must be created, institutionalized and expanded to include stabilization or livelihood programs, to build a systematic structure for redeploying labour after drought or other extreme events of climate change (Plate 10).

4.8 Integrated Approach to Reservoir Management with Climate Change

The sustainability of dams is critical, but long-term sustainability efforts are complicated by the uncertainties of climate change. Even if past climate data can be used as a proxy for possible future droughts and floods, it may not be sufficient for planning into the future (Uyiguie, 2006). In an era of uncertainty, water resource managers must employ flexible methods to adapt to a changing climate. Adaptive policies and strategies can be informed and developed through simulation modelling. The most common approach is to combine a series of climate, hydrologic, reservoir and/or

Plate 10 Tubewells for Agriculture and water supply during drought

ecological models. A variety of mathematical and statistical models used to study the impacts of climate change on reservoirs is presented in Table 2.

Hydrologic tools (Table 3) are currently used to conduct climate change impact analyses for dams, and are ideal for examining a variety of climate simulations to inform management decisions.

Table 3 Mathematical and statistical tools used to study impacts of climate change on reservoirs

Tool	Description	Developer
Soil and Water Assessment Tool (SWAT)	Simulates water quality and quantity of surface water and can test scenarios related to land use, land management practices and climate change	USDA-ARS and Texas A&M Agri-Life Research
Hydrologic Engineering Center—Reservoir System Simulation (HEC-ResSim)	Uses rule-based approach to mimic decision-making process	Army Corp of Engineers
Integrated Adaptive Optimization Model (IAOM)	Contains three modules: weather generator, hydrological simulator and multipurpose reservoir optimization to develop optimal operating rule curves under climate change	Y. Zhou and S. Guo
Dynamic Hydroclimatological Assessment Model(DYHAM)	Utilizes system dynamics theories and feedback causal loops to simulate dynamic processes within watershed and reservoir	SP Simonovic and LH Li
Phytoplankton Responses to Environmental Change: PROTECH	Simulates the daily change in Chl-a concentration for up to 10 algal species in response to environmental variability in lakes and reservoirs	Alex Elliott, Colin Reynolds, Tony Irish

5 Conclusion

Reservoirs are established to provide critical services; they represent a large financial legacy from previous generations. With often limited locations appropriate to site new reservoirs, it is imperative that current reservoirs are analyzed and studied to improve and enhance their overall sustainable management.

In Nigeria, there are many dams that have been built and utilized for various purposes, particularly in semi-arid northern Nigeria. These dams are vital resources because they enhance the socioeconomic life of the people in areas where they are established.

The purpose of Goronyo Dam is to regulate flood from watershed areas in Kastina and Zamfara, including parts of the Niger Republic. However, in exceptionally dry years such as 2017 and 2018, there was an unanticipatedly low flow, siltation problems and little volume of rainwater, coupled with over-abstraction for irrigation agriculture from the upstream area of the dam.

Climate change in Nigeria is now devastating and destructive: the promotion of concerted effort and action is necessary to decelerate its impact on society.

Maintenance, including de-silting which has not been undertaken in over twenty years, is essential in order for Goronyo residents in Sokoto State to reduce the risk of even more water shortage challenges. To achieve this, collaboration among reservoir managers and climate scientists in Nigeria must result in the development of simulation modeling platforms to explore and test adaptive management strategies for altered climate patterns through careful review of available tools and thorough consultation with resource managers.

References

Alex Elliott SJ, Thackeray J, Huntingford RG Jones (2005) Combining a regional climate model with a phytoplankton community model to predict future changes in phytoplankton in lakes. Freshwater Biol. 50:1404–1411

Alverez M, UFH R, Trudel L, Leconte (2014). Impacts and adaptation to climate change using a reservoir management tool to a northern watershed: application to Lievre River watershed, Quebec, Canada. Water Resource Manage. 28:3667–3680

Azman B (2007) Impact of Climate Change on Water Resources Availability in the Komati River Basin Using WEAP21 Model. At the UNESCO-IHE Institute for Water Education, Delft, the Netherlands

Calebwura (2014) Impact of Climate Change That Causes Dryness on Water-Coast in Kware Stream, Sokoto State, Nigeria. https://calebwura.wordpress.com/2014/02/19/impact-of-climate-change-that-causes-dryness-on-water-coast-in-kware-stream-sokoto-state-nigeria/. (Date Accessed 9 September 2018)

Inter-Governmental Panel on Climate Change (2014) Synthesis Report. A Contribution of Working Groups I, II, and III to the Third Assessment Report of the Intergovernmental Panel on Climate Change. R.T. Watson and Core Writing Team. Cambridge University Press, Cambridge, U.K. and New York, USA

World Commission on Dams (WCD) (2015) Dams and Development, www.dams.org

International Federation of Red Cross and Red Crescent Societies (2013) Information Bulletin Nigeria; floods. www.frc.org/docs/Appeals/13/IBNG

Gwary D (2008) Climate change, food security and Nigerian agriculture. In Challenges of climate change for Nigeria; a multi-disciplinary perspective pp 179–202. Ibadan: Nigerian Institute of Social And Economic Research (NISER)

JICA (2013) The project for review and update of Nigeria national water resources master plan. Japan International cooperation Agency Yachiyo Engineering Co., Ltd. CTI Engineering International Co., Ltd. Sanyu Consultants Inc

Stainforth DA, Aina T, Christensen C, M., Collins, N., Faull, D., Frame, J. A., Kettleborough, S., Knight, A., Martin, J, M., Murphy., Piani, D., Sexton., Smith, R.A., Spicer, A, J., Thorpe M. R., Allen (2005) Uncertainty in predictions of the climate response to rising levels of greenhouse gases. Nature, 433, 403–406

Tambuwal UA (2010). Climate change: Threat to food security and livelihoods in selected states of Northern Nigeria. Paper presented at the 51st annual conference of the association of Nigerian Geographers at Department of Geography and Planning, Kogi State University, Anyigba

Thornton PK, Jones PG, Owiyo T, Kruska RL, M, Herrero, V., Orindi., S., Bhadwal, P., Kristjanson, A, Notenbaert, N, Bekele, A et al (2008) Climate change and poverty in Africa: Mapping hotsports of vulnerability. AFJARE, 2(1), 24–44 (Department for International Development, ILRI, Nairobi, Kenya)

Uyiguie E (2006). Dams are Un-renewable: A Discussion Paper. www.credcentre.org/publications/dams

World Weather Report (2018). https://www.worldweatheronline.com/kanoweather/kano/ng.aspx/wwo_r=srch (Date Accessed 5th August 2018)

Chapter 9
Lentic-Lotic Water System Response to Anthropogenic and Climatic Factors in Kenya and Their Sustainable Management

Daniel O. Olago, Jackson Raini, Christine Omuombo, Godfrey Ogonda, Jones Muli, Cornelius Okello, Willis Memo, and Obiero Ong'ang'a

Abstract Kenya has many lakes which have evolved through geological time; most of the modern ones were created during the Lower to Middle Pleistocene. While all of them react to climate changes over geological timescales, a number of them are highly sensitive to even seasonal climate variations, fluctuating quite dramatically in surface area and level, and are, therefore, commonly referred to as "amplifier" lakes. There is a range of biological diversity in the lakes, and the rivers and wetlands within the lake basins, in diverse ecosystems that provide various goods and services to the local communities around them. However, there is mounting evidence that the impacts of human activity (e.g., deforestation, agriculture, water abstractions) and hydrological variability related to global warming effects on climate (rainfall, temperature) are already affecting these natural aquatic resources, leading to changes in fauna and flora distributions and in their overall resource values. The consequences are mostly adverse, but there are also beneficial ones. Climate, environment, and society interact in complex ways in lake basin ecosystems. For example, there has been no clear understanding of the cause(s) of the striking rise in lake levels in the

D. O. Olago (✉) · C. Omuombo
Institute for Climate Change and Adaptation & Department of Geology, University of Nairobi, P.O. Box 30197-00100, Nairobi, Kenya
e-mail: dolago@uonbi.ac.ke

J. Raini
FlamingoNet, P.O. Box 13493-20100, Nakuru, Kenya

G. Ogonda · O. Ong'ang'a
OSIENALA (Friends of Lake Victoria), P.O. Box 4580-40103 Kisumu, Kenya

J. Muli
Kenya Marine and Fisheries Research Institute (KMFRI), Baringo Field Station, P.O. Box 31-30406, Kampi Ya Samaki, Kenya

C. Okello
Department of Environmental Sciences, Machakos University, P.O. Box 136-90100, Machakos, Kenya

W. Memo
Rift Valley Catchment Region, Water Resources Authority, P.O. Box 1600-20100, Nakuru, Kenya

S. Diop et al. (eds.), *Climate Change and Water Resources in Africa*,
https://doi.org/10.1007/978-3-030-61225-2_9

central rift lakes of Kenya since 2010 which have resulted in submerged buildings and road infrastructure, and displacement and/or disruption of the socio-ecological system. Implementation of viable and sustainable management and use options is critical, but it is currently precluded by myriad factors, including lack of timely and adequate data for decision-making, siloed sectoral approaches, jurisdictional challenges at sub-national and regional scales, overlapping institutional mandates, and diverse and uncoordinated stakeholder groupings. A pathway for development of lake basin-specific management plans in Kenya is outlined, based on the Integrated Lake Basin Management (ILBM) approach, that can help to ensure the health and sustainability of the lakes and their basins and continued provision of goods and services to the people and wildlife that are dependent upon them.

Keywords Kenya lakes · Climate change · Aquatic systems · Anthropogenic pressures · Management

1 Introduction

Kenya has a number of lake basins, most of which, with the exception of Lake Victoria and some minor seasonal/quasi-perennial lakes such as Chalbi and Lake Kenyatta, are in a topographically closed depression (bordered by mountains and plateaus) known as the Kenya Rift Valley. This region has two end-member lakes: Lake Turkana to the north and Lake Natron to the south (Fig. 1). Several aspects of the development of Kenya's lakes through geological time and their associated climatic/hydrological evolution have been synthesized in a number of previous studies (Tiercelin and Lezzar 2002; Odada and Olago 2005; Olago et al. 2009; Olago 2013; Woldegabriel et al. 2016). Generally, it is the case that within the western branch of the East African Rift System (EARS), older lake basins that are not in existence today (palaeolakes) were initiated in the Neogene. The present-day lakes Baringo, Nakuru, Elementaita, and Naivasha are smaller representations of older large lakes that developed during Lower-Middle Pleistocene times (Tiercelin and Lezzar 2002; Woldegabriel et al. 2016). In and off the rift flanks, related volcanism has been associated with the development of craters, some of which host well-known crater lakes such as Sokorte Dika on Mount Marsabit, Sacred Lake on Mount Kenya, and Lake Challa on the border of Kenya and Tanzania. Lake Victoria, in contrast, was formed in the late Pleistocene (ca. 400,000 years ago) by uplift along the western branch of EARS and back-ponding of rivers that previously drained westwards (Johnson et al. 1996).

The present-day lakes in the Kenya Rift are generally small (a maximum of 30×20 km) and shallow (mean water depths ranging from 1 to 10 m), with the exception of Lake Turkana. The lakes are more-or-less saline-alkaline (with the exception of Naivasha and Baringo), mainly because they lie in areas under semi-arid climatic conditions and/or have no outlet. These rift lakes have been termed, "amplifier lakes" (Street-Perrott and Harrison 1985) because they are sensitive to climate shifts as a consequence of their closed basins and small sizes relative to their

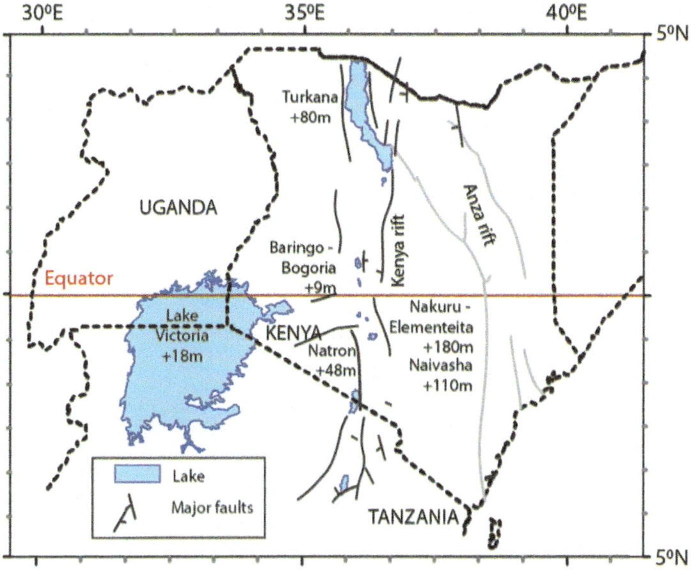

Fig. 1 Map showing the Kenya rift lakes and Lake Victoria and their early Holocene Lake level highstands (adapted from Olaka et al. 2010)

catchments, and their semi-arid setting which is characterized by a highly variable climate. Their relative sensitivities to climate fluctuations have been determined using basin morphology, the hypsometric index, and the aridity index (Olaka et al. 2010).

Kenya's lakes, and their inflowing/outflowing rivers, and wetlands (Fig. 2), range from mildly (e.g., Lake Turkana (Lowe-McConnell 1995)) to highly biologically diverse (e.g., Lake Victoria (Seehausen 1996)) ecosystems that provide various goods and services to the several million people living within their basins (e.g., Aloo 2002; Odada et al. 2004; Ong'ang'a 2010; Raini 2009). Some of them are major tourist attractions and have also been designated Wetlands of International Importance (Ramsar Sites) or World Heritage Sites, e.g., Lakes Baringo, Nakuru, and Elementaita. Their resource values are continually changing as a consequence of natural climatic and environmental changes (UNEP 2004, 2006), but they are also being degraded by multi-faceted anthropogenic impacts, including catchment degradation, pollution, unsustainable rates of resource extraction, and global warming (UNEP 2004, 2006). This chapter, therefore, evaluates the natural and anthropogenic factors that impact the country's aquatic systems (Fig. 2) and proposes, using Integrated Lake Basin Management (ILBM)[1], a pathway that is specific to lakes in Kenya which, if implemented, can ensure the sustainability of lake health and, therefore, their continued provision of goods and services to the people and other living organisms that depend on them.

[1]Conceptual framework to assist lake basin managers and stakeholders to achieve sustainable management of lakes and their basins [see ILEC 2007; ILEC and RCSE Shiga University 2012]

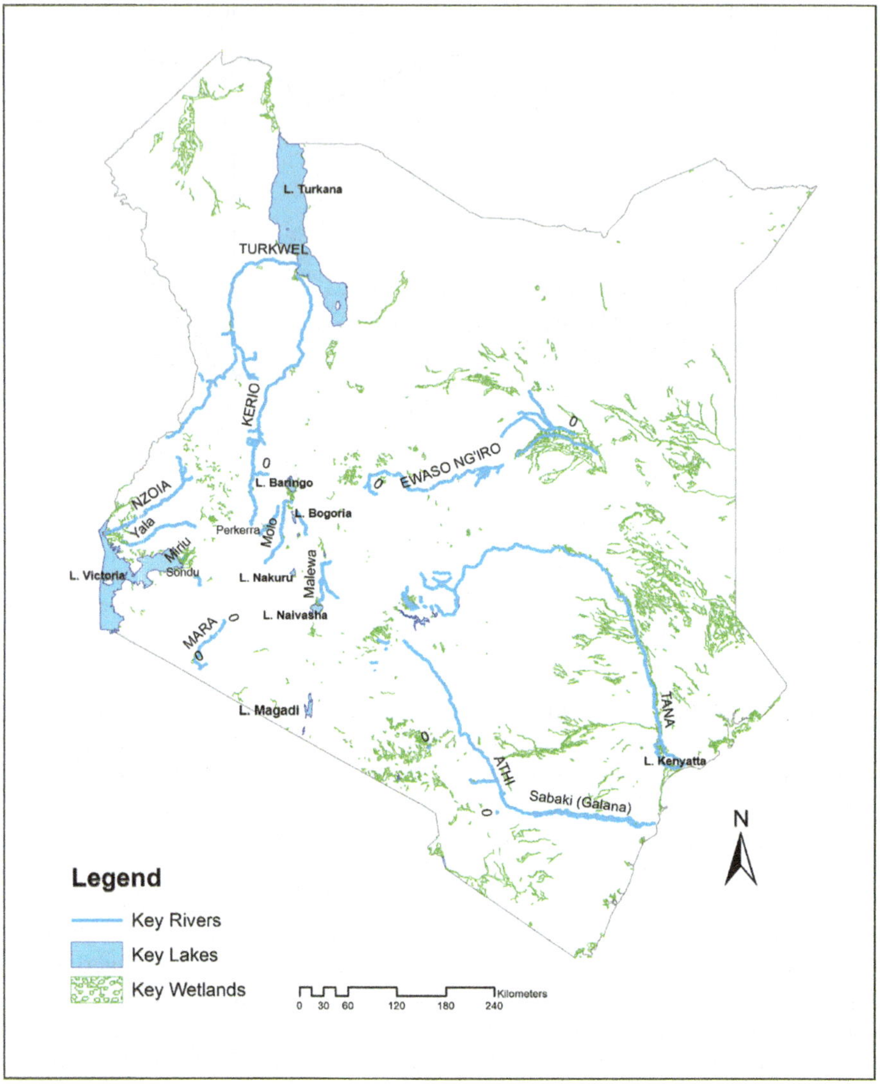

Fig. 2 The key lakes, rivers, and wetlands in Kenya. Note that the lake margins are associated with wetland ecosystems

2 Threats and Issues Facing the Lake Basins

2.1 Anthropogenic Factors

All lakes transform through time due to natural processes of aging caused by climatic, hydrologic, and ecosystem changes, but the greatest degradation impacts are caused by human intervention (ILEC 2007) (Table 1). The rapid rate of societal growth and change in modern times not only accelerates the types of changes that occur under natural conditions, but also introduces novel materials and substances that fundamentally, directly, and adversely impact aquatic ecosystem health and function. Since about the 1960s, Kenyan lakes and rivers have come under increasing and considerable pressure from a variety of inter-connected human activities on land and water. Deforested and degraded lands and wetlands are becoming increasingly characteristic features in the region as a consequence of the rapidly growing population (Waktola 1999; Shepherd et al. 2000; Swallow et al. 2002; Nyingi et al. 2013). Lake, river, and wetland ecosystems are being degraded by human activities through, for example, catchment degradation, pollution, siltation, bank encroachment, and over-abstraction (Thenya 2001; Moinde-Fockler et al. 2006; Xu 2008; UNEP 2009; SEI 2009; Onywere et al. 2011; Kiema 2013; Schagerl and Renaut 2016). Such stresses exacerbate vulnerability to current and future climate risks (SEI 2009; IPCC 2014). Continuously eroded lands are being cultivated, grazed, and/or settled, leading to rapid depletion of soil nutrients and forest resources.

Increasing deforestation and/or fragmentation of habitats for varied end-uses, such as settlement and agriculture, compromises the services of many ecosystems (Mogaka et al. 2005a, b ; Negishi and Nakamura 2006; Thenya 2001; UNEP 2009), and changes surface and groundwater regimes with resulting impacts on water availability for nature and human societies (Table 1). It also impacts the function and operation of existing water infrastructure (SEI 2009). For example, 100,000 ha of the Mau Complex was lost between 2000 and 2009, while on Mt. Elgon large tracts of the forest were excised for human resettlement in the 1970s (UNEP 2009); both of these water towers are catchments for rivers that drain into Lake Victoria Nyanza Gulf, the central rift lakes, and Lake Turkana. Between 1960 and 2000, for example, Lake Victoria Nyanza Gulf and its catchment underwent enormous ecological changes, which are linked to a number of inter-related problems, such as rapid population growth, poverty, land degradation and declining agricultural productivity and water quality (Odada et al. 2004; UNEP 2006; Odada and Olago 2007). Sedimentation, nutrient runoff, urban and industrial point source pollution and biomass burning induced the rapid eutrophication of Lake Victoria that was seen over the latter part of the twentieth century, resulting in invasion of water hyacinth, loss of endemic biodiversity and interelated and compounded problems for the lake environment and the welfare of its people (ICRAF 2000). Biomass burning is also a key source of phosphorus in Lake Victoria through atmospheric deposition (Tamatamah et al. 2005). These are still persistent issues today (Table 1).

Completed and planned development projects such as the construction and filling in of the Gibe series of dams in Ethiopia have raised concerns about the sustainability of Lake Turkana as a consequence of reduced inflows (Avery 2010, 2012). These are recently reported, through analysis of satellite data, to have translated to a lake level drop of approximately 1.5 m during the filling of the Gibe III reservoir whose construction was completed in December 2016 (Hodbod et al. 2018). Further, although a minimum of a 10-day flood has been proposed to sustain the ecological functioning of the lake, it is unknown whether this flood duration and size would

Table 1 Challenges and threats to the lakes and existing management structure

Challenge/Threat	Lake Victoria Nyanza Gulf	Lake Naivasha	Lake Nakuru	Lake Baringo	Lake Turkana	Some key references
Environmental	Land degradation Pollution (runoff/sewage) High sedimentation Introduced/Invasive species	Land degradation Pollution (runoff/sewage) High sedimentation Introduced/Invasive species	Pollution (runoff and sewage) Steep hydrological fluctuations	Land degradation Turbidity & high sedimentation in lake Nutrient loading Alkalinity	Land degradation Nitrate pollution from Omo River irrigation Turbidity and increased sedimentation Water quality	Hughes and Hughes 1992; Johnson and Odada (eds) 1996; UNEP 2004, 2006 Lowe-McConnell 1995; Paron et al. (eds), (2013)
Ecological	Changing food web structure Eutrophication—species extinction/invasion Anoxia Warming lake waters	Changing food web structure Eutrophication—species extinction/invasion	Changing food web structure due to hydrological changes, affects food chain	Biodiversity loss Habitat degradation, Land cover changes Algal blooms	Declining fish numbers Changes in the Omo River delta Biodiversity loss, Shoreline habitat degradation, Land-use changes Change in species contribution	Johnson and Odada (eds) 1996; Seehausen 1996; UNEP 2004, 2006; Sitoki et al. 2010; Aloo 2002; Hughes and Hughes 1992; UNEP 2004; Ojwang et al. 2016; Gownoris et al. 2017; Schagerl (ed) 2016; Njiru et al. 2017
Ecosystem Goods and Services	Unsustainable use of wetland resources Reduced livelihoods options	Decline in fish catch and biomass	Park & tourism	Decline in fish catch and biomass Aesthetics	Decline in fish catch Tourism Livelihood	UNEP 2004, 2006; Raini 2006; Ojwang et al. 2016

(continued)

Table 1 (continued)

Economic	Overfishing & pollution related to fisheries decline Poverty-environment trap Reduced land values	Overfishing & pollution related to fisheries decline Poverty-environment trap	Tourism affected by lake level changes and pollution from sewage/runoff	Decline in tourism income and fish revenue, livelihoods affected	Decline in fishing revenue	UNEP 2004, 2006; Omwega 2006; Abila et al. 2006a,b;
Social	Expanding settlements Displacements Conflict Loss of fish foods Gendered disparities in artisanal fisheries	Expanding settlements Displacements Conflict	Expanding settlements Water demand for multiple uses threaten inflowing surface/groundwater supply	Land/water-related conflicts Water demand for multiple uses threaten inflowing surface/groundwater supply	Growing coastal settlements Land/water conflicts Water demand for multiple uses	UNEP 2006; Lwenya et al. 2006; Ojwang et al. 2016

(continued)

Table 1 (continued)

Management	Several regional organizations/sectoral projects; Vision and Strategy Framework for the Management and Development of the Lake Victoria Basin, 2004; Protocol for Sustainable Development of Lake Victoria Basin, 2003; Lake Victoria Basin Commission	Environmental Management (Lake Naivasha Management Plan) Order, 2004 (Rev. 2012); Lake Naivasha Basin Integrated Management Plan, 2012–2022; Several sectoral & community-based management plans	Ramsar management protocols applied; Several sectoral and community-based management plans. Lake Nakuru Ecosystem Management Plan 2000–2012; Pilot set up of Lake Nakuru ILBM Platform, supported by Nakuru County Government—2014	Several sectoral and community-based management plans; No ILBM system in place	Several sectoral and community-based management plans; No ILBM system in place	Odada et al. 2004; Okia et al. 2006; Paron et al. (eds) 2013; Schagerl (ed) 2016; Njiru et al. 2017; Ramsar Regional Center—East Asia (2017)

be sufficient to do so (Avery 2012). Following the commissioning of the dam, the lake has undergone fluctuations affecting the productivity of fisheries (Gownaris et al. 2017). Changes in the lake littoral habitat could lead to a reduction of over two-thirds in fishery yields (Gownaris et al. 2017), and the analysis of fisheries data from Kolding (1995), MoLFD (2008), (KNBS 2012; 2013; 2014) comports with fishermen reports that the diversity of fish species catch has declined over the past five years due to littoral habitat changes.

Box 1: Ecological Impacts on Lake Nakuru

(a) **The bad**—In Lake Nakuru, dense suspensions of *Arthrospira fusiformis* are the main and preferred food for Lesser Flamingos (Jenkin 1957; Vareschi 1978; Kaggwa et al. 2013). This alkaliphilic cyanobacterium, which grows well even at pH 10 (Grant et al. 1990), is the key primary producer in soda lakes (Melack and Kilham 1974; Schagerl et al. 2016), forming the base of the food chain and supporting large concentrations of Lesser Flamingos (*Phoeniconaias minor*). Recently, cyanobacterial toxins, mainly from *Microcystis flos-aquae* and *Anabaena flos-aquae,* have been suggested as potentially lethal agents for Lake Nakuru Lesser Flamingos (Nelson et al. 1998; Raini 2009; Krienitz et al. 2003; Codd et al. 2003; Metcalf et al. 2006). These changes have had significant negative impacts on Lake Nakuru Park tourism, as flamingos are one of its major attractions.

(b) **The good**—The only fish species in Lake Nakuru is the filter feeding cichlid, *Sarotherodon alcalicum grahami*, introduced from Lake Magadi in 1953 to combat mosquitos. The introduction of fish substantially increased the diversity of the lake ecosystem by extending the food chain to over 30 species of fish-eating birds.

Evidence suggests that lakes are warming in response to anthropogenically-driven global warming; Lake Victoria, for example, was noted in the 1990s to be one-half of a degree (°C) warmer than it was in the 1960s (Hecky et al. 1994; Bugenyi and Magumba 1996), and is estimated to have warmed by >1°C between 1927 and 2009 (Sitoki et al. 2010), consistent with changes in surface temperature at tropical elevations above 1000 m worldwide. There appear not to be any new published data on long-term lake water temperature trends for other lakes in Kenya since the 1990s, except for the global analysis of nighttime surface (sometimes referred to as « skin ») temperatures of lakes, including Lake Turkana, over the period 1985–2009 by Schneider and Hook (2010). Their results show an average warming of 0.025 °C yr^{-1} for lakes in the tropics, with greater warming occurring in the mid- and high latitudes of the northern hemisphere than in low latitudes and the southern hemisphere. One consequence of the warming could be increased lake stratification that affects fishery productivity, as was observed in Lake Turkana in the 1980s (Kallqvist et al. 1988).

Thus, of immediate concern are the changes in nutrient dynamics and mixing regimes within the lakes that are a major fishery resource (Victoria, Turkana, and Baringo); these are in turn a result of increased thermal stability as they affect fishery production, as has been observed in Lakes Malawi and Tanganyika, and can completely alter the trophic structure of the food chain (O'Reilly et al. 2003; Verburg et al. 2003; ENSO Project 2003). Such concerns extend to the inflowing rivers and wetland buffers along the shores of the lakes.

2.2 Climatic Factors

The temporal distribution of precipitation over East Africa reveals daily to seasonal and inter-annual variability, and a simultaneous decline in the number of wet days in a year and increased frequency of intense rainfall (Kuya 2016). Rainfall in eastern Africa shows robust teleconnections to the rest of Africa and to the global tropics. It is strongly quasi-periodic, with a dominant timescale of variability of five to six years, influenced in part by the El-Niño-Southern Oscillation (ENSO) phenomenon and Sea-Surface Temperatures (SSTs) fluctuations in the equatorial Indian and Atlantic oceans. Further, rainfall in the region tends to be enhanced during ENSO years (Ropelewski and Halpert 1987; Ogallo 1989). An increase in the frequency of drought has been reported in the semi-arid Lake Turkana region (Waktola 1999; Hann et al. 2003) and more recently also in many other parts of the country, most notably in 2010/2011 and 2015/2016, strongly linked to La Niña, with the milder droughts attributed to low rainfall in typically arid to semi-arid areas that have a wide range of year-to-year rainfall variability (Uhe et al. 2018).

The water balance of Lake Victoria is dominated by rainfall on the lake and evaporation, with river inflow and outflow making minor contributions (Spigel and Coulter 1996). In contrast, in smaller lakes, river inflows, and evaporation are the dominant factors (e.g., Street-Perrott 1982; Finch and Hall 2006). As a consequence of climate variability, it has been observed that lake surface areas and levels fluctuate significantly, particularly for the smaller lakes. In addition, the very high potential evaporation which occurs throughout the year, which is in excess of 2000 mm per year over large tracts of Africa, significantly affects water resources (Schulz 2001). Thus, relatively small changes in rainfall and evaporation may lead to shifting between closed- and open-basin status, as was the case in historic times for some of the lakes in Kenya and eastern Africa as a whole (Beadle 1981; Richardson and Dussinger 1986; Verschuren et al. 2000; Bootsma and Hecky 1993; Owen et al. 1990; Spigel and Coulter 1996; Olago et al. 2007). For example, the surface area of Lake Nakuru (mean depth 2.5 m) normally fluctuates between approximately 40 and 60 km^2; this translates into significant water level fluctuations. Sometimes, the lake dries out almost completely, as happened between 1995 and 1997 (WWF 1998; McCord et al. 2009). On the other extreme, during prolonged sequences of wet years, the water level rises to form a lake with a surface area of 50–60 km^2 (Onywere et al. 2013). Extreme events such as the El Niño (1982–3 and the 1997–8) tend to rapidly raise lake levels and cause widespread flooding along lakeshores and rivers (Birkett et al. 1999; Conway 2002). For example, the El Niño phenomenon in 1997–1998 saw water level rise by 1.7 m (Lake Victoria), 2.1 m (Lake Tanganyika), and 1.8 m (Lake Malawi) (Birkett et al. 1999). The widespread heavy rainfall and flooding also produced adverse wide-ranging agricultural, hydrological, ecological, and economic impacts in East Africa (Conway 2002). Further, the consequences of aquatic ecosystem change and biodiversity resulting from naturally induced changes in pH, salinity, stratification, sediment, and nutrient load fluctuation due to lake level changes and rising water temperatures are still not known satisfactorily, yet it is clear that these factors are strongly compounded by anthropogenic activities.

East Africa is projected to become wetter in some GCMs (Endris et al. 2013; Masson-Delmotte et al. 2013), with projections of increased September–October–November (SON) and decreased March–April–May (MAM) rainfall in some models (Christensen et al. 2013). This is in contrast to observed (instrument-derived) data, which shows a drying trend. The mismatch between observed and model data in East Africa is known as the "East African Paradox": research is ongoing currently to explain this contradiction (e.g., Rowell et al. 2015). The frequency of intense rainfall is expected to increase as the planet warms, both over-lake (e.g., Thiery et al. 2016) and over land (e.g., Kuya 2016). The potential impacts of future climate changes on Kenya's water resources for the years 2030 and 2050, with a baseline year of 2010, have been evaluated against Kenya's National Water Master Plan (NWMP; Table 2) based on the Coupled Model Intercomparison Project Phase 3 (CMIP3) multi-ensemble model (11 GCMs) outputs based on the A1B scenario 3 and the Similar Hydrologic Element Response (SHER) model with the derived future climate conditions for 2030 and 2050 (MEWNR and JICA 2013). In terms of the latest IPCC

Table 2 Annual renewable water resources. Ten percent of groundwater recharge is adopted as sustainable yield, and maximum exploitable depth is set at 500 m (MEWNR and JICA 2013). The 2030 values are interpolated between the projected 2050 values and the 2010 baseline (MEWNR and JICA 2013)

Evapotranspiration Estimation	Aspect	2010 (MCM/yr)	2030 (MCM/yr)	2050 (MCM/yr)
Hamon's Method	Renewable Water Resources	76,610	80,474	83,583
	Surface Water Runoff	20,637	24,894	26,709
	Groundwater Recharge	55,973	55,580	56,874
FAO Penman-Monteith Method	Renewable Water Resources	42,107	44,301	45,996
	Surface Water Runoff	20,637	24,894	26,709
	Groundwater Recharge	21,470	19,407	19,287

scenarios, this would be in the range of RCP6.0 estimates (IPCC 2014). There are large uncertainties related to evapotranspiration estimation methods (Hamon and FAO Penman-Monteith) "since no data to properly estimate the evapotranspiration is available" (MEWNR and JICA 2013). Changes in evapotranspiration have knock-on impacts on the water balance of runoff, soil moisture, and surface and groundwater reservoirs (Bates et al. 2008). Given these uncertainties, the NWMP 2030 adopted the lower FAO Penman-Monteith method results as they offered a cautionary note to the development of the water resources. Despite these uncertainties, models generally predict that renewable water resources are likely to increase, and that, excluding factors such as water abstractions and a shift in the P:E ratio, the volume of water held in the lakes will increase in future.

The projected temporal and spatial distributions of the mean annual rainfall in 2030 and 2050 are similar to the climate of 2010, but the western part of Kenya is expected to be wetter with more rainfall in the MAM season, while the eastern part will be drier, due to potential increased evapotranspiration (MEWNR & JICA 2013). This is consistent with the findings of Shongwe et al. (2011), based on analysis of an ensemble of 12 CMIP3 GCMs (forced with A1B emissions), that show statistically significant widespread increases in short season (OND) rainfall, including extreme precipitation across East Africa. Stocker et al. (2013) also project increased short rains in East Africa due to the pattern of Indian Ocean warming, and increased rainfall extremes related to landfall cyclones on the east coast. Similar to the CORDEX analysis (Misiani 2013), the model ensemble used by Shongwe et al. (2011) was relatively poor in simulating the MAM season.

3 The Climate-Environment-Society Nexus in Lake Basins

The recent rise in water levels of the central rift lakes since 2010 (Figs. 3 and 4) and its effects have been a key subject of discussion in the country (see Box 1). Its impacts include submerged buildings, infrastructure, and displacement and/or disruption of the productivity of both humans and wildlife, including aquatic species. However, some of the infrastructures around the lakes, such as Lake Naivasha, have been built on what was likely previously riparian land because the water levels were that high in the first half of the twentieth century.

These discussions have taken center stage because there has been no clear change in rainfall amounts or patterns that is perceptible to humans, as occurs during ENSO events, and results of instrumental data analysis are ambiguous. For example, one early report suggests that the increase in Lake Nakuru's surface area from a low of 31.8 km^2 in January 2010 to a high of 54.7 km^2 in September 2013 was caused by an increase in the mean annual precipitation in the period 2009–2014 (e.g., Onywere et al. 2013). By July 2019, it had not recovered to average levels (Fig. 4). Indeed, all the lakes in Kenya's central rift valley (Nakuru, Bogoria, Baringo, Solai, and Logipi) have risen since 2010 to levels not seen in the last fifty years (Fig. 3). Moreover, some of the lakes exceeded the well-known and -documented historical high levels

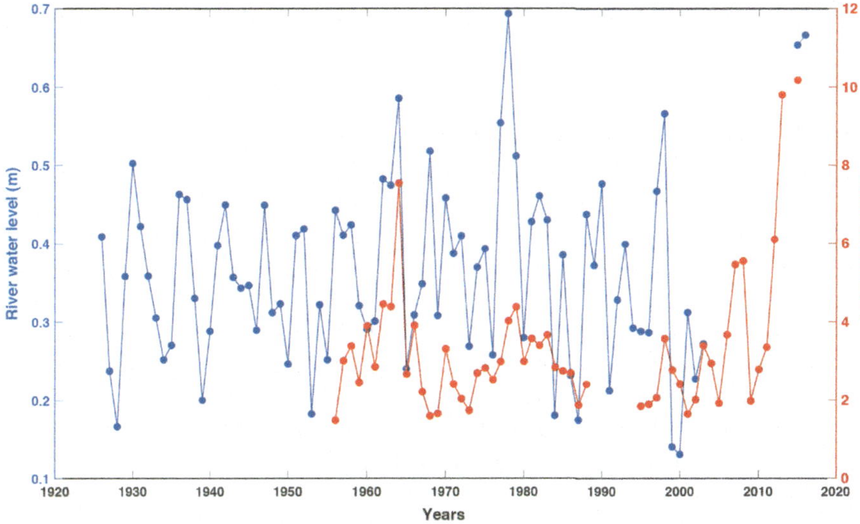

Fig. 3 Mean annual flows from in the Molo River (blue) that drains into Lake Baringo display a steep rise in water level between 2005 and 2015, although there is a gap in the dataset during this period. The annual lake water level in Lake Baringo (red graph) shows in contrast a recently reported rise in 2010: the area under water rose from 143.6 km^2 in January 2010 to a high of 231.6 km^2 in September 2013

Fig. 4 Left pane—Submerged road entry to the Park. The water mark on the trees in the background indicates that the lake level has receded by about 1 m from its highest level. Right pane—The flooded western wing of the Park offices at the Park gate. Both photos were taken on 24 July 2019. © D. Olago"

which occurred in 1961, and/or in the early 1900s (Schagerl and Renaut 2016, and references therein).

Box 2: Why the rise in water levels in Kenya rift lakes since 2011?

Several hypotheses from formal and informal discussions were proposed in 2013 to explain the rising lake levels in recent times—they are the opinions (not scientific findings) of several scientists who have researched the lakes. These hypotheses may have inter-related causes; exact cause(s) are currently being investigated.

Hypotheses for the rise in water levels in the lakes include

1. A primarily climatological effect affecting Rift lakes that are all situated in the same climate region. Their hydrological character as 'amplifier lakes' implies that relatively minor changes in the amount and seasonal distribution of (catchment-wide) rainfall and evaporation results in strong lake level response.
2. Intensified land use in recent decades and the resulting land degradation, which is severe in some areas, have resulted in higher rainfall runoff from land, and less percolation of it into groundwater systems, leading to larger volumes of water flowing directly and rapidly from the land surface into the lakes.
3. The rainfall runoff over degraded lands and enhanced soil erosion as well as degradation of the lakes' riparian zones has decreased the depths of the lakes through enhanced sediment inputs, shallowing them very rapidly. Thus, while volumes of water flow into the lakes and volume retention within them may remain essentially the same, there is an increase in the lake levels due to the shallowing effect of rapid sedimentation (similar to the siltation effect in man-made dams).
4. The impact of land degradation and related runoff flows to the lakes is enhanced because it causes more water to flow into the lakes via (faster) river inflow than via (slower) shallow groundwater input.
5. Geological factors such as tectonic uplift.
6. Inter/multi-decadal to centennial scale cyclical variability of the climate system in the lake region has been inferred from geological records, and cannot be captured from the short instrumental records that currently exist in the country.

Satellite monitoring of surface area fluctuations in Lakes Bogoria, Nakuru, and Elementaita reveals temporally congruent responses during the period 2000–2013, while Lake Logipi showed a response that was different to these lakes, but similar to Lake Natron (Emma Jayne Tebbs 2014). Whether or not the change in rainfall is the main driver of lake level changes is still, however, inconclusive, because the cause(s) of the rise in the other lakes has not been ascertained. It can be seen, for example, that from the mid-1950s up to the year 2005, discharge fluctuations in the

Molo River, which feeds into Lake Baringo, correspond to a lake level response of between 2 and 4 m even when the discharge was at its highest in the late 1970s. But between 2010 and 2013, the lake level response was +8 m with river inflow still not as high as it was in the late 1970s (Fig. 3). In addition, it has been demonstrated from palaeoclimate records and modeling that high amplitude climate variability and corresponding hydrological responses by the lakes also occur on multi-decadal and longer timescales which cannot be discerned from short instrumental records (e.g., Verschuren et al. 2000; Tierney et al. 2013; Shagerl and Renaut 2016) and should be taken into account when considering lake responses to climate variability.

Britton et al. (2008) demonstrated in Lake Baringo that water level is significantly and positively correlated with fish production. More generally, it has been noted that lake level changes in shallow lakes and man-made reservoirs are highly correlated with fish yield per unit area, and, therefore, the influence of water level changes on aquatic productivity should be considered in management of lake resources (Kolding and van Zwieten 2012). In a recent study of 13 African lakes, including Lakes Nakuru, Naivasha, Turkana, and Victoria, it was observed that there was a positive correlation between inter-annual water level fluctuations and primary and overall production, and a negative correlation to fish diversity, transfer efficiency, and food chain length. Also, seasonal water level fluctuation was positively correlated with biomass (Gownaris et al. 2017). In Lake Naivasha, Stoof-Leichsenring et al. (2011) noted that the chemistry and biology of the lakes were affected mainly by natural climate variations in the period 1820–1950, but that since 1950, anthropogenic activities have been the dominant factor. Awange et al. (2013) note, for example, that there was a significant correlation between the quantity of flowers produced and the level of Lake Naivasha during the period 2002–2010, which suggests that anthropogenic activities had some effect on the lake level drop that occurred during that period. Further effects on lake level changes could arise from changes induced by siltation that alters the lake littoral regimes (cf. UNEP 2004), bathymetry, and surface-volume ratios (cf. Shagerl and Burian 2016), but these effects have not yet been quantified.

4 Management

While lakes are essential habitats for diverse flora and fauna and also serve to support human livelihoods and economic development, they are much more vulnerable to stresses and more difficult to manage than river systems because they are easily impacted by complex land and water relationships (ILEC 2007), which are confounded by human activity (Fig. 5). It was conservatively estimated, in 2005, that water resource degradation cost the country about 30 million dollars (0.5% GDP) annually, while extreme events such as the El Niño-La Niña flood and drought in 1997–2000 cost the country about 14% of its GDP, with the drought exacting four times the cost of the flood (Mogaka et al. 2005). In addition, insufficient data for planning, management, and decision-making can lead to failure of development projects, frequent water supply disruption and rationing (Nyingi et al. 2013) and

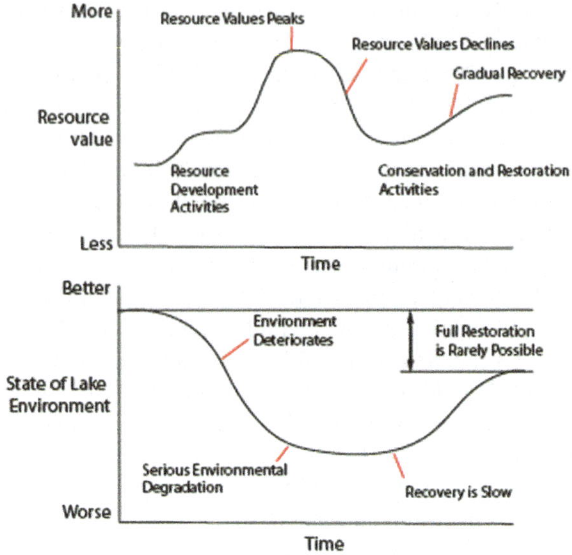

Fig. 5 Changing resource values and resulting environmental state of a hypothetical lake (from ILEC 2007)

sudden collapse of ecosystems (ILEC 2007) that have adverse knock-on effects on lake-dependent livelihoods.

Often, water abstractions and diversions of inflow waters for drinking, irrigation, and other uses can alter the delicate balance of lakes (Schagerl and Renaut 2016), while increasing temperatures, changing salinity, high levels of ultraviolet B radiation, lake fragmentation and salinization of habitable sections may impact fish (Kavembe et al. 2016) and other aquatic fauna and flora. Besides the significant revenue generation from lake tourism, some lake microorganisms have the potential to be used for biotechnological purposes (Shagerl and Renaut 2016). It has been correctly noted by Harper et al. (2016) that "several of the economic values of the saline lakes of East Africa are dependent upon their constancy, which is almost always assumed by non-scientists (such as most tourists and journalists), but rarely observed in reality." Looking to the future, Oduor and Kotut (2016) note that the increase in destructive human activities in the lakes and their catchments will lower their resilience to climate change-driven perturbations. Consequently, it has been proposed by various researchers, lake managers, governmental and non-governmental organizations and agencies that there is need for the different communities and interest groups that depend upon the lakes to formulate or to improve existing management plans to ensure their sustainable use (cf. Kiema 2013; Kavembe et al. 2016).

The monitoring of water resources has deteriorated greatly (Mogaka et al. 2005) due to non-maintenance and vandalization of river gauging stations, making it impossible to carry out in-depth water resource analyses and planning (Nyingi et al. 2013). Consequently, the symptoms of degradation often remain not only unknown, but also unnoticed for a long period of time because of their incremental nature and the delay in visible response of lakes to perturbations (ILEC 2007). Reactive conservation and remedial measures are often too little, too late (ILEC 2007). Protection of the lentic-lotic water systems will not work, therefore, if there is not proper land use planning and if water users do not fully understand the need for protection of such resources (cf. Mumma et al. 2011); this also extends to those who depend upon the lakes for their livelihoods and well-being. The vulnerability of the various aquatic systems to the cumulative effects of natural climate variability, human activities, land degradation, and climate change must be assessed and quantified in order to manage them sustainably.

The current lake management regime (see Table 1) is inadequate to respond to the multiple threats that have been outlined. The designation of some of the lakes as Ramsar sites, UNESCO World Heritage Sites, or National Parks/Reserves offers some protection, but often their catchments are excluded. In the large lake basins such as Victoria and Turkana, upper catchments which supply most of the water that flows into the lakes tend to be controlled by different administrative jurisdictions than are lower catchments. This makes it difficult to manage human activity in upper basin areas, and thus the consequences for downstream threats/impacts (Negishi and Nakamura 2006). Indeed, most of the lake basins lack a coordinated management approach or framework (Table 1).

The country's lakes and their associated or inter-linked rivers and wetlands, in the basin context, face major challenges of governance and management, because there are so many policies at national and county levels, as well as international treaties ratified by Kenya, which are applicable to their sustainable management. But these have not been coordinated on the level of specific lake management. Siloed sectoral approaches characterize most existing management plans, including those listed in Table 1. Further, conflicting or duplicated policies and strategies or mandates of institutions, inadequate institutional capacity both in terms of training and resource support, weak enforcement and implementation of policies, low public and stakeholder participation (including the private sector), inadequate financial resources, data and information, obsolete and inefficient technologies, emerging risks such as climate change, and the fact that most lake and many river basins fall within multiple political and administrative jurisdictions, all come together in a complex convergence that make it very difficult to successfully coordinate the proper sustainable management of lakes and their basins. Given the complexity of lake basin governance, we propose a pathway for adoption of the Integrated Lake Basin Management Approach (ILBM) in Kenya, fully incorporating Integrated Water Resources Management (IWRM) and Integrated River Basin Management (IRBM) principles and approaches, to ensure balance between conservation and development.

Policy: Kenya has recently undertaken policy and legislative reforms to align them with the Constitution of Kenya 2010, which embraces sustainability and participation principles that are critical for successful Integrated Lake Basin Management. There are several pieces of legislation and strategies that relate to water and environmental management, as well as a large number of applicable county government laws, programs, international frameworks, and conventions to which the country subscribes. These should all be carefully considered, including their interlinkages, in the development of fit-for-purpose sustainable management plans specific to each lake basin, in order to build holistic management of the lake basin and to manage conflicts of interest.

Institutions: Cross-sectoral coordination suffers from fragmentation of the many governmental and non-governmental institutions and agencies which adopt sectorally-oriented approaches to water and catchment management. This is compounded by the existence of multiple jurisdictions even within a single catchment. An institutional framework should be developed at the national and/or lake basin level that ensures that all relevant institutions work in a coordinated and effective manner under their various mandates to address specific issues related to the management of the lakes and basins within their jurisdictions. An established, credible institutional framework can also serve as an entry point for donor-funded programs to minimize duplication and enhance synergies and value for money.

Public participation: The requirement to enable participation of the public is firmly anchored in the Constitution of Kenya 2010 and offers to the communities and Civil Society Organizations (CSOs) an opportunity to better coordinate and effectively participate in environmental education, conservation, and management, and to reduce the duplication of effort which is currently prevalent in many of the lake basins.

Technology: Pollution from sewage and effluent treatment systems, solid waste, old and inefficient technologies, and unsustainable land use and management practices can be managed in part by the adoption of green technologies and approaches. The Kenya Green Economy Strategy and Implementation Plan 2015 (Republic of Kenya 2015) provides an ideal entry point under its *thematic Area 3: Sustainable Natural Resource Management* which focuses on the economy-environment nexus to optimize the contribution of Kenya's natural resources to the economy, industrialization, and livelihoods, while responsibly and effectively managing and conserving the country's natural capital.

Knowledge and information: There has been a long-term decline in environmental (e.g., Mogaka et al. 2005) and poverty (REACH 2015) monitoring in the country, though we recognize that agencies such as the Water Resources Authority are revamping their monitoring systems. Partial/incomplete, irregular, and often low-quality data provide insufficient scientific evidence for decision-making and management. It is, therefore, important to encourage relevant agencies to build the capacity to generate programs focused on carefully selected indicators of specific lake basin characteristics, using climatic, environmental, and socioeconomic data through cost-effective and sustained monitoring. These resources must be stored, analyzed, shared, and used to guide decision-making in the short- to long-term.

Finance: Lake-focused programs receive very little funding from national and county budgets. In addition, many programs are donor-driven and are, therefore, of short- to medium-term duration, often lacking the time needed to create the enduring infrastructure necessary for lasting impact. Revenues from existing levies, e.g., water use fees, fishing licenses, and tourism access fees, are not an adequate supplement to effectively support basin management activities. It is, therefore, critical to mobilize resources to create a shared commitment and implement sustainable management of the lake basins and their ecosystems. Lake-specific financial plans must also be instituted as part and parcel of lake basin management plans.

Cooperation and cross-cutting issues: There is no definitive institution responsible for holistic lake and catchment management nor cross-sectoral coordination at the national level. The proposed institutional framework must guide cooperation efforts from the local through to the national level. At the regional level, transboundary lake management organizations exist for some lakes, e.g., Lake Victoria, but not for others, e.g., Lake Turkana. Lake basin management plans must, therefore, also incorporate such platforms in their design. Cross-cutting issues such as poverty, environment, gender, inclusivity, human rights, population, health as it relates to the environment, climate change, and other socio-ecological issues must also be thoroughly incorporated into lake basin management plans.

5 Conclusion

Kenya's lentic-lotic water systems are highly sensitive to natural climatic and environmental changes from seasonal to geological timescales. This is compounded by anthropogenic threats to the water systems, including (1) Water quantity effects: deforestation, excessive water abstraction from lakes, rivers, wetland systems, and groundwaters; (2) Water quality effects: pollution from the use of agro- and industrial chemicals, siltation from soil erosion due to poor farming methods, livestock pressure, and destruction of riparian vegetation; (3) Ecological restructuring related to increased intensity of human exploitation of the resources such as fisheries and wetland plants, erosion and sedimentation, and invasive and introduced species. Thus, human activity in lake basins adversely impacts the ecosystem health and resource values of the lakes, rivers, and wetlands within them. While the problems facing each of the basins are multi-faceted, they are similar in terms of drivers (e.g., socioeconomic conditions, demography, climate variability and change) and pressures (e.g., nutrient enrichment from agriculture, wetland resource harvesting, water abstractions), varying only in the relative importance of the drivers/pressures between basins.

The socio-ecological complexity of the aquatic systems, diverse stakeholder groups, and uncertainties of factors such as climate change, population growth, and development trajectories make the management of these systems complex, but not impossible. Beneficial impacts can be achieved—some in the short-term, others in the medium-term, and others in the long-term—on aquatic ecosystems and their biophysical and socio-economic settings. A trans-generational view is, therefore, important when managing lentic-lotic water systems (lakes, rivers, and wetlands) in a basin context. The ILBM approach acknowledges that scientific knowledge of the lentic-lotic water systems in relation to natural climate variation and human influences, and the protection of the biophysical system associated with lakes, rivers, wetlands, and groundwater-dependent ecosystems, are of prime importance for their sustainable utilization and management in the long-, medium-,and short-term. The holistic and participatory ILBM approach proposed for the management of Kenya's lakes will, if implemented, enhance diverse livelihoods and accrue benefits to future generations by balancing conservation and development, and contributing to build resilience and reduce poverty in areas and communities where it is implemented.

References

Abila R, Rasowo JO, Manyala J (2006a) Biodiversity and sustainable management of a tropical wetland lake ecosystem: A case study of Lake Kanyaboli, Kenya. In: Odada EO, Olago DO, Ochola W, Ntiba M, Wandiga S, Gichuki N, Oyieke H (eds) 11th World lakes conference Nairobi, Kenya, 31 October to 4th November 2005, Proceedings Volume I: Socioeconomic aspects of lakes and lake management. ministry of water and irrigation, Kenya and International Lake Environment Committee (ILEC), Japan, pp 205–209

Abila RO, Odongkara KO, Onyango PO (2006b) Distribution of economic benefits from the fisheries of Lake Victoria. In: Odada EO, Olago DO, Ochola W, Ntiba M, Wandiga S, Gichuki N, Oyieke H (eds), 11th World lakes conference Nairobi, Kenya, 31 October to 4th November 2005, Proceedings volume i: socioeconomic aspects of lakes and lake management. Ministry of Water and Irrigation, Kenya and International Lake Environment Committee (ILEC), Japan, pp 87–92

Avery S (2010) Hydrological impacts of Ethiopia's Omo Basin on Kenya's Lake Turkana water levels and fisheries, final report. Prepared for the African Development Bank, Nairobi

Avery S (2012) Lake Turkana & the lower omo: hydrological impacts of major dam and irrigation developments. African Studies Centre, the University of Oxford, Oxford

Awange JL, Forootan E, Kusche J, Kiema JBK, Omondi PA, Heck B, Fleming K, Ohanya SO, Gonçalves RM (2013) Understanding the decline of water storage across the Ramser-Lake Naivasha using satellite-based methods. Adv Water Resour 60:7–23

Beadle LC (1981) The inland waters of tropical Africa: an introduction to tropical limnology, 2nd edn. Longman, London

Birkett CM, Murtugudde R, Allan JA (1999) Indian ocean climate event brings floods to East Africa's lakes and Sudd Marsh. Geophys Res Lett 26:1031–1034

Bootsma HA, Hecky RE (1993) Conservation of the African great lakes: a limnological perspective. Conserv Biol 7(3):644–656

Britton JR, Jackson MC, Muchiri M, Tarras-Wahlberg H, Harper DM, Grey J (2008) Status, ecology and conservation of an endemic fish, *Oreochromis niloticus baringoensis*, in Lake Baringo, Kenya. Aquatic Conserv Freshwat Ecosyst. Published online in Wiley InterScience. www.interscience.wiley.com, https://doi.org/10.1002/aqc.998

Bugenyi FWB, Magumba KM (1996) The present physicochemical ecology of Lake Victoria, Uganda. In: Johnson TC, Odada E (eds) The limnology, climatology and paleoclimatology of the East African Lakes. Gordon and Breach, Toronto, pp 141–154

Christensen JH, Krishna KK, Aldrian E, An SI, Cavalcanti IFA, de Castro M, Dong W, Goswami P, Hall A, Kanyanga JK, Kitoh A, Kossin J, Lau NC, Renwick J, Stephenson DB, Xie SP, Zhou T (2013) Climate phenomena and their relevance for future regional climate change. In: Climate change 2013: the physical science basis. Contribution of working group I to the fifth assessment report of the intergovernmental panel on climate change [Stocker TF, Qin D, Plattner GK, Tignor M, Allen SK, Boschung J, Nauels A, Xia Y, Bex V and Midgley PM (eds).]. Cambridge University Press, Cambridge, United Kingdom and New York, NY, USA

Codd GA, Metcalf JS, Morrison LF, Krienitz L, Ballot A, Pflugmacher S, Wiegand C, Kotut K (2003) Susceptibility of flamingos to cyanobacterial toxins via feeding. Vet Rec 152:722–723

Conway D (2002) Extreme rainfall events and lake level changes in East Africa: recent events and historical precedents. In: Odada EO, Olago DO (eds), The East African great lakes: limnology,

palaeolimnology and biodiversity. Advances in Global Change Research, Kluwer Academic Publishers, pp 63–92
Emma Jayne Tebbs EJ (2014) Remote sensing for the study of ecohydrology in East African Rift lakes. PhD Thesis, Department of Physics & Astronomy, University of Leicester, UK
Endris HS, Omondi P, Jain S, Lennard C, Hewitson B, Chang'a L, Awange JL, Dosio A, Ketiem P, Nikulin G, Panitz HJ, Büchner M, Stordal F, Tazalika L (2013) Assessment of the performance of CORDEX regional climate models in simulating East African rainfall. J Clim 26:8453–8475. https://doi.org/10.1175/jcli-d-12-00708.1
ENSO Project (2003) Summary of the ENSO project. http://members.tripod.com/tanganyika/id39_m.htm. Accessed 18 March 2003
Finch JW, Hall R (2006) Evaporation from lakes. In: Encyclopedia of Hydrological Sciences. https://doi.org/10.1002/0470848944.hsa047
Gownaris NJ, Pikitch EK, Aller JY, Kaufman LS, Kolding J, Lwiza KMM, Obiero KO, Ojwang WO, Malala JO, Rountos KJ (2017) Fisheries and water level fluctuations in the world's largest desert lake. Ecohydrology. https://doi.org/10.1002/eco.1769
Hann N, Farmer G, Wheeler R (2003) vulnerability to food insecurity in Kenya 2001. In: A WFP pilot study for improving vulnerability analysis
Harper DM, Tebbs E, Bell O, Robinson VJ (2016) Conservation and management of East Africa's soda lakes. In Schagerl M (ed). Soda lakes of East Africa. Springer International Publishing Switzerland, pp 345–364
Hecky RE, Bugenyi FWB, Ochumba P, Talling JF, Mugidde R, Gophen M, Kaufman L (1994) Deoxygenation of the deep water of Lake Victoria, East Africa. Limnol Oceanogr 39:1476–1481
Hodbod J, Stevenson EJG, Akall G, Akuja T, Angelei I, Bedasso EA, Buffavand L, Derbyshire S, Eulenberger I, Gownaris N, Kamski B, Kurewa A, Lokuruka M, Mulugeta MF, Okenwa D, Rodgers C, Tebbs E (2018) Social-ecological change in the Omo-Turkana basin: a synthesis of current developments. Ambio (2019). https://doi.org/10.1007/s13280-018-1139-3
Hughes RH, Hughes JS (1992) A directory of African Wetlands. IUCN/UNEP/WCMC
ICRAF (2000) Improved land management in the Lake Victoria Basin—Final Technical Report Startup Phase July 1999 to June 2000. ICRAF Working Paper Series, Nairobi, Kenya
ILEC (2007) Integrated lake basin management: an introduction. International Lake Environment Committee Foundation, Kusatsu, Japan
ILEC and RCSE Shiga University (2012) Primer: development of ILBM platform process, evolving guidelines through participatory improvement, 26 p. http://www.ilec.or.jp
IPCC (2014) Summary for policymakers. In: Climate change 2014: impacts, adaptation, and vulnerability. Part A: global and sectoral aspects. contribution of working group II to the fifth assessment report of the Intergovernmental Panel on Climate Change [Field CB, Barros VR, Dokken DJ, Mach KJ, Mastrandrea MD, Bilir TE, Chatterjee M, Ebi KL, Estrada YO, Genova RC, Girma B, Kissel ES, Levy AN, MacCracken S, Mastrandrea PR, White LL (eds).]. Cambridge University Press, Cambridge, United Kingdom and New York, NY, USA, pp 1–32
Jenkin PM (1957) The filter feeding and food of flamingoes ((Phoeniconaias minor). Philos Trans R Soc B 240:401–493
Johnson TC, Scholz CA, Talbot RM, Kelts K, Ricketts D, Ngobi G, Beuning RM, Ssemanda I, McGill JW (1996) Late pleistocene desiccation of Lake Victoria and rapid evolution of cichlid fishes. Science (New York, N.Y) 273:1091–1013. https://doi.org/10.1126/science.273.5278.1091
Kaggwa MN, Gruber M, Oduor SO, Schagerl M (2013) A detailed time series assessment of the diet of Lesser Flamingos: further explanation for their itinerant behaviour. Hydrobiologia 710:83–93
Kallqvist T, Lien L, Liti D (1988) Lake Turkana limnological study 1985–1988. Norwegian Institute for Water Research (NIVA) and Kenya Marine Fisheries Research Institute (KEMFRI)
Kavembe GD, Meyer A, Wood CM (2016) Fish population in East African saline lakes, In: Schagerl M (ed). Soda lakes of East Africa, pp 227–257. Springer International Publishing Switzerland
Kiema JBK (2013) Remote sensing application supporting IWRM in Kenya. In: Paron P, Olago D, Omuto CT (eds) Kenya: a natural outlook geoenvironmental resources and hazards. Publ by Elsevier B.V, Amsterdam, The Netherlands, pp 51–66

Kolding J, van Zwitten PAM (2012) Relative lake level fluctuations and their influence on productivity and resilience in tropical lakes and reservoirs. Fish Res 115–116:99–109

Krienitz L, Ballot A, Kotut K, Wiegand C, Pütz S, Metcalf JS, Codd GA, Pflugmacher S (2003) Contribution of hot spring cyanobacteria to the mysterious deaths of Lesser Flamingos at Lake Bogoria, Kenya. FEMS Microbiol Ecol 1437:141–148

Kuya EK (2016) Precipitation and temperatures extremes in East Africa in past and future climate. Master of Science Thesis, University of Oslo, Oslo, Norway

Lwenya CA, Lwenya KR, Abila R, Omwega R (2006) Gender participation in fisheries management of Lake Victoria, Kenya. In: Odada EO, Olago DO, Ochola W, Ntiba M, Wandiga S, Gichuki N, Oyieke H (eds). 11th world lakes conference Nairobi, Kenya, 31 October to 4th November 2005, proceedings volume I: Socioeconomic aspects of lakes and lake management. Ministry of Water and Irrigation, Kenya and International Lake Environment Committee (ILEC), Japan, pp 266–272

Masson-Delmotte V, Schulz M, Abe-Ouchi A, Beer J, Ganopolski A, González Rouco JF, Jansen E, Lambeck K, Luterbacher J, Naish T, Osborn T, Otto-Bliesner B, Quinn T, Ramesh R, Rojas M, Shao X, Timmermann A (2013) Information from paleoclimate archives. In: Climate change 2013: The physical science basis. Contribution of working group I to the fifth assessment report of the Intergovernmental Panel on Climate Change [Stocker TF, Qin D, Plattner GK, Tignor M, Allen SK, Boschung J, Nauels A, Xia Y, Bex V, Midgley PM (eds).]. Cambridge University Press, Cambridge, United Kingdom and New York, NY, USA

Melack JM, Kilham P (1974) Photosynthetic rates of phytoplankton in East African alkaline, saline lakes. Limnol Oceanogr 19:743–755

Metcalf JS, Morrison LF, Krienitz L, Ballot A , Krause E, Kotut K, Pütz S, Weigand C, Pflugmacher S, Codd GA (2006) Analysis of cyanotoxins anatoxin-a and microcystins in Lesser Flamingo feathers. Toxicol Environ Chem 88:159–167

MEWNR & JICA (2013) The project on the development of the National Water Master Plan 2030. The Ministry of Environment, Water and Natural Resources MEWNR) and the Water resource management Authority (WRMA), Republic of Kenya. Japan International Cooperation Agency (JICA) and Nippon Koei Co. Ltd. October 2013, GED/JR/13–202

Misiani HO (2013) Assessment of temperature and precipitation extremes over Kenya using the coordinated regional downscaling experiment model outputs. Master of Science Thesis, University of Nairobi, Nairobi, Kenya

Mogaka H, Gichere S, Davis R, Hirji R (2005) Climate variability and water resource degradation in Kenya improving water resources development and management (English). World Bank working paper series; No. 69. World Bank, Washington, DC. http://documents.worldbank.org/curated/en/990211468283546915/Climate-variability-and-water-resource-degradation-in-Kenya-improving-water-resources-development-and-management

Mogaka H, Gichere S, Davis R, Hirji R (2005) Climate variability and water resource degradation in Kenya improving water resources development and management. Working Paper No. 69. The World Bank, Washington DC

Moinde-Fockler NN, Oguge NO, Karere GM, Otina D, Suleman MA (2006) Human and natural impacts on forests along lower Tana river, Kenya: implications towards conservation and management of endemic primate species and their habitat. Biodivers Conserv 16:1161–1173. https://doi.org/10.1007/s10531-006-9096-8

Mumma A, Lane M, Kairu E, Tuinhof A, Hirji R (2011) Kenya: groundwater governance case study. In: Water Papers. Water Partnership Programme. www.worldbank.org/water

Negishi JN, Nakamura M (2006) Current status and challenges in managing the upper rural areas of world lake basins: preliminary implications from the GEF-project Lake Briefs. In: Odada EO, Olago DO, Ochola W, Ntiba M, Wandiga S, Gichuki N, Oyieke H (eds). 11th World lakes Conference Nairobi, Kenya, 31 October to 4th November 2005, proceedings volume I: socioeconomic aspects of lakes and lake management. Ministry of Water and Irrigation, Kenya and International Lake Environment Committee (ILEC), Japan, pp 40–47

Nelson YM, Thampy RJ, Motelin GK, Raini AJ, DisSante CJ, Lion LW (1998) Model for Trace metal exposure in filter feeding flamingos at alkaline Rift Valley lakes Kenya. Environ Toxico Chem 17:2302–2309

Njiru J, Waithaka E, Aloo PA (2017) An overview of the current status of Lake Naivasha fishery: challenges and management strategies. Open Fish Sci J 10:1–11

Nyingi DW, Gichuki N, Ogada MO (2013) Freshwater ecology of Kenyan highlands and lowlands. In: Paron P, Olago D, Omuto CT (eds) Kenya: a natural outlook geoenvironmental resources and hazards. Publ by Elsevier B.V, Amsterdam, The Netherlands, pp 199–218

Odada EO, Olago DO (2005) Holocene climatic, hydrological and environmental oscillations in the tropics with special reference to Africa. In: Low Pak Sum (ed) Climate change and Africa. Cambridge University Press, Cambridge

Odada EO, Olago DO (2007) Challenges of an ecosystem approach to water monitoring and management of the African Great Lakes. J Aqu Ecosyst Health Manag 9(4):433–446

Odada EO, Olago DO, Kulindwa K, Ntiba M, Wandiga S (2004) Mitigation of environmental problems in Lake Victoria, East Africa: Causal chain and policy option analyses. Ambio 33(1–2):13–23. UNEP (2006) Africa environment outlook (AEO) 2: our environment, our wealth. ISBN: 92-807-2691-9

Oduor SO, Kotut K (2016) Soda lakes of the East African Rift System: the past, the present and the future, In: Schagerl M (ed). Soda lakes of East Africa, pp 365–374. Springer International Publishing Switzerland

Ogallo LJ (1989) The spatial and temporal patterns of the East African seasonal rainfall derived from principal components analysis. Int J Clim 9:145–167

Ojwang WO, Obiero KO, Donde OO, Gownaris N, Pikitch EK, Omondi R, Agembe S, Malala J, Avery ST (2016) Lake Turkana: World's Largest Permanent Desert Lake (Kenya). In: Finlayson C, Milton G, Prentice R, Davidson N (eds) The wetland book. Springer, Dordrecht

Okia TO, Kipkore WK, Mngube FM (2006) Management of a shared lake ecosystem: a case study of Lake Victoria. In: Odada EO, Olago DO, Ochola W, Ntiba M, Wandiga S, Gichuki N, Oyieke H (eds). Proceedings of the 11th world Lakes conference Nairobi, Kenya, 31 October to 4th November 2005, vol I. Socioeconomic aspects of lakes and lake management. Ministry of Water and Irrigation, Kenya and International Lake Environment Committee (ILEC), Japan, pp 75–81

Olago DO, Umer M, Ringrose S, Huntsman-Mapila P, Sow EH, Damnati B (2007) Palaeoclimate of Africa: An overview since the last glacial maximum. In: Otter L, Olago DO, Niang I (eds). Global change processes and impacts in Africa: a synthesis. East African Educational Publishers Ltd., Nairobi

Olago DO (2013) Quaternary Evolution. In: Paron P, Olago D, Omuto CT (eds) Kenya: a natural outlook geoenvironmental resources and hazards. Publ by Elsevier B.V, Amsterdam, The Netherlands, pp 31–38

Olago DO, Opere A, Barongo J (2009) Holocene palaeohydrology, groundwater and climate change in the lake basins of the Central Kenya Rift. Hydrol Sci J 54(4):765–780

Olaka LA, Odada EO, Trauth MH, Olago DO (2010) The sensitivity of East African rift lakes to climate fluctuations. J Paleolimnol 44:629–644. https://doi.org/10.1007/s10933-010-9442-4

Oliver N (2019) Waterbird count results in the Rift Valley, Nairobi and central, Kenya for July 2018. National Museums of Kenya, PO Box 40658-00100, Nairobi, Kenya. NMK Centre for Biodiversity Research Reports: Ornithology No. 84, February 2019

Omwega RN (2006) Community involvement in fish harvesting around Lake Victoria (Kenya) In: Odada EO, Olago DO, Ochola W, Ntiba M, Wandiga S, Gichuki N, Oyieke H (eds). Proceedings of the 11th world lakes conference Nairobi, Kenya, 31 October to 4th November 2005, vol. I. Socioeconomic Aspects of Lakes and Lake Management. Ministry of Water and Irrigation, Kenya and International Lake Environment Committee (ILEC), Japan, pp 245–251

Ong'ang'a O (2010) Lake victoria (Kenya) and its environs: resources, opportunities and challenges. OSIENALA, Kisumu, Kenya

Onywere S, Shisanya C, Obando J, Ndubi A, Masiga D, Irura Z, Maragia H (2013). Geospatial Extent of 2011–2013 flooding from the Eastern African rift valley lakes in Kenya and its implication on the ecosystems

Onywere SM, Getenga ZM, Mwakilila SS, Twesigye CK, Nikaranda JK (2011) Assessing the challenge of settlement in Budalangi and Yala swamp area in Western Kenya using landsat satellite imagery. Open Environ Eng J 4:97–104. https://doi.org/10.2174/1874829501104010097

O'Reilly CM, Alin SR, Plisnier P-D, Cohen AS, McKee BA (2003) Climate change decreases aquatic ecosystem productivity of Lake Tanganyika, Africa. Nature 424:766–768

Owen R, Crossley R, Johnson T, Tweddle D, Kornfield I, Davison S, Eccles D, Engstrom D (1990) Major low levels of Lake Malawi and their implications for speciation rates in cichlid fisheries. Proc R Soc London B 240:519–553

Raini JA (2006) The eastern Africa flamingo Lakes: Building partnerships for sustainable resource management. In: Odada EO, Olago DO, Ochola W, Ntiba M, Wandiga S, Gichuki N, Oyieke H (eds). Proceedings of the 11th world lakes conference Nairobi, Kenya, 31 October to 4th November 2005, vol I. Socioeconomic Aspects of Lakes and Lake Management. Ministry of Water and Irrigation, Kenya and International Lake Environment Committee (ILEC), Japan, pp 276–281

Raini JA (2009) Impact of land use changes on water resources and biodiversity of Lake Nakuru catchment basin, Kenya. Afr J Ecol 47 (1):39–45. Blackwell Publishing Ltd

Ramsar (2001) Information Sheet on Ramsar Wetlands (RIS) Kenya, Lake Baringo 2002. http://www.ramsar.org/ris_kenya_baringo1.htm

Ramsar Regional Center–East Asia (2017) The designation and management of Ramsar Sites–a practitioner's guide. www.ramsar.org and www.rrcea.org

Republic of Kenya (2015) Kenya green economy strategy and implementation plan (GESIP), Maanzoni-1 Draft, May 2015

Richardson JL, Dussinger RA (1986) Paleolimnology of mid-elevation lakes in the Kenya Rift Valley. Hydrobiologia 143:167–174

Ropelewski CF, Halpert MS (1987) Global and regional scale precipitation patterns associated with the El Niño/southern oscillation. Mon Wea Rev 115:1606–1626

Rowell DP, Ben BB, Booth BBB (2015) Reconciling past and future rainfall trends over East Africa. J Clim 28:9768–9788. https://doi.org/10.1175/JCLI-D-15-0140.1

Schagerl M, Burian A (2016) The ecology of African soda lakes: Driven by variable and extreme conditions, In: Schagerl M (ed). Soda lakes of East Africa. Springer International Publishing Switzerland, pp 295–320

Schagerl M, Renaut RW (2016) Dipping into the soda lakes of East Africa. In: Schagerl M (ed). Soda lakes of East Africa. Springer International Publishing Switzerland, pp 3–24

Schagerl M, Burian A, Gruber-Dorninger M, Oduor SO, Kaggwa MN (2016) Algal communities of Kenyan soda lakes with a special focus on Arthrospira fusiformis. Fottea 15:245–257

Schneider P, Hook SJ (2010) Space observations of inland water bodies show rapid surface warming since 1985. Geophys Res Lett 37:L22405. https://doi.org/10.1029/2010GL045059

Seehausen O (1996) Lake Victoria rock cichlids: taxonomy, ecology, and distribution. Verduijn Press, Zevenhuizen, The Netherlands

SEI (2009) Economics of Climate Change Kenya, Stockholm Environment Institute, Project Report–2009, Stockholm Environment Institute, Oxford Office, Oxford, UK. http://kenya.cceconomics.org/

Shepherd K, Walsh M, Mugo F, Ong C, Hansen TS, Swallow B, Awiti A, Hai M, Nyantika D, Ombao D, Grunder M, Mbote F, Mungai D (2000) Improved land management in the Lake Victoria Basin: linking land and lake, research and extension, catchment and lake basin. Final Technical Report, Startup Phase, July 1999 to June 2000, working paper series, working paper 2000–2. International Centre for Research in Agroforestry and Kenya Ministry of Agriculture and Rural Development, Soil and Water Conservation Branch, National Soil and Water Conservation Programme, Nairobi, Kenya

Shongwe ME, van Oldenborgh GJ, van den Hurk B, van Aalst M (2011) Projected changes in mean and extreme precipitation in Africa under global warming part ii: East Africa. J Clim 24:3718–3733

Sitoki L, Gichuki J, Ezekiel C, Wanda F, Mkumbo CO, Marshall BE (2010) The environment of Lake Victoria (East Africa): current status and historical changes. Int Rev Hydrobiol 95(3):209–223. https://doi.org/10.1002/iroh.201011226

Spigel RH, Coulter GW (1996) Comparison of hydrology and physical limnology of the East African Great Lakes: Tanganyika, Malawi, Victoria, Kivu and Turkana (with reference to some North American Great Lakes). In: Johnson TC, Odada E (eds). The Limnology, Climatology and Paleoclimatology of East African Lakes: Gordon and Breach, Toronto, pp 103–139

Stocker TF, Qin D, Plattner GK, Alexander LV, Allen SK, Bindoff NL, Bréon FM, Church JA, Cubasch U, Emori S, Forster P, Friedlingstein P, Gillett N, Gregory JM, Hartmann DL, Jansen E, Kirtman B, Knutti R, Krishna Kumar K, Lemke P, Marotzke J, Masson-Delmotte V, Meehl GA, Mokhov II, Piao S, Ramaswamy V, Randall D, Rhein M, Rojas M, Sabine C, Shindell D, Talley LD, Vaughan DG, Xie SP (2013) Technical summary. In: Climate change (2013) The physical science basis. Contribution of Working Group I to the Fifth Assessment Report of the Intergovernmental Panel on Climate Change [Stocker TF, Qin D, Plattner GK, Tignor M, Allen SK, Boschung J, Nauels A, Xia Y, Bex V and Midgley PM (eds).]. Cambridge University Press, Cambridge, United Kingdom and New York, NY, USA

Stoof-Leichsenring KR, Junginger A, Olaka LA, Tiedemann R, Trauth MH (2011) Environmental variability in Lake Naivasha, Kenya, over the last two centuries. J Paleolimnol. https://doi.org/10.1007/s10933-011-9502-4

Street-Perrott FA (1982) The response of lake levels to climatic change implications for the future. In: Rosenweig C, Dickinson R (eds) Climate-vegetation interactions. OIES and UCAR, Boulder, Colorado, pp 77–80

Street-Perrott FA, Harrison SP (1985) In: Hecht AD (ed) Paleoclimate analysis and modeling. Wiley, New York, pp 291–340

Swallow B, Walsh M, Nyantika D, Muriithi S, Noordin Q, Ong C, Shepherd K, Place F, Awiti A, Hai M, Ochieng O, Cohen M, Mugo F, Oyasi S, Omuto C, Cohen L, Okono A (2002) Improved land management in the Lake Victoria Basin. Annual Technical Report July 2001 to June 2002. World Agroforestry Centre (ICRAF), Natural Resource Problems, Priorities and Policies Programme, Working Paper Series, Working Paper 2002-2. World Agroforestry Centre and National Agricultural and Livestock Extension Programme of the Kenya Ministry of Agriculture and Rural Development, Nairobi, Kenya

Tamatamah RA, Hecky RE, Duthie HC (2005) The atmospheric deposition of phosphorus in Lake Victoria (East Africa). Biogeochemistry 73:325–344. https://doi.org/10.1007/s10533-004-0196-9

Thenya T (2001) Challenges of conservation of dryland shallow waters, Ewaso Narok swamp, Laikipia District, Kenya. Hydrobiologia 458:107–119. https://doi.org/10.1023/A:1013196500456

Thiery W, Davin EL, Seneviratn SI, Bedka K, Lhermitte S, van Lipzig NPM (2016) Hazardous thunderstorm intensification over Lake Victoria. Nat Commun 7. https://doi.org/10.1038/ncomms1278

Tiercelin JJ, Lezzar KE (2002) A 300 Million years history of rift lakes in central and east Africa: an updated broad view. In: Odada EO, Olago DO (eds) The East African great lakes: limnology, paleolimnology and biodiversity. Kluwer, Dordrecht, pp 3–62

Tierney JE, Smerdon JE, Anchukaitis KJ, Seager R (2013) Multidecadal variability in East African hydroclimate controlled by the Indian Ocean. Nature 493:389–392

Uhe P, Philip S, Kew S, Shah K, Kimutai J, Mwangi E, van Oldenborgh GJ, Singh R, Arrighi J, Jjemba E, Cullen H, Otto F (2018) Attributing drivers of the 2016 Kenyan drought. Int J Climatol 38:e554–e568. https://doi.org/10.1002/joc.5389

UNEP (2004) Odada EO, Olago D, Kulindwa KAA, Bugenyi F, West K, Ntiba M, Wandiga S, Karimumuryango, J East African Rift Valley Lakes, GIWA Regional assessment 47. University of Kalmar, Kalmar, Sweden

UNEP (2006) Lake Victoria Basin Environment Outlook: Environment and Development. UNEP, Nairobi

UNEP (2009) Kenya: atlas of our changing environment. United Nations Environmental Programme, Nairobi, Kenya, pp 1–39

Vareschi E (1978) The ecology of Lake Nakuru (Kenya) I. Abundance and feeding of the Lesser Flamingo. Oecologia 32:11–35

Verburg P, Hecky RE, Kling H (2003) Ecological consequences of a century of warming in Lake Tanganyika. Science 301(5632):505

Waktola A (1999) Exploratory study of two regions in Ethiopia to identify target areas and partners for intervention. Drylands Coordination Group, Report No. 6

Woldegabriel G, Olago D, Dindi E, Owor M (2016) Genesis of the East African Rift System. In: Schagerl M (ed). Soda Lakes of East Africa. Springer International Publishing Switzerland, pp 25–59

Xu Y (2008) Editor's message: building capacity and partnerships towards sustainable utilization of groundwater in Africa. Hydrogeol J 16:405–406

Chapter 10
Hydrology and Climate Impacts on Streamflow and Sediment Yield in the Nyando River Basin, Kenya

J. U. Kitheka, S. Okoyo, and N. Mboya

Abstract This study seeks to establish the extent to which climate change and/or variation is affecting streamflow variability and sediment transport in the Nyando River Basin in Kenya. The basin is one of the country's most important, because it is a source of water for domestic use and irrigation. The river also discharges water and sediment into Lake Victoria, one of the largest freshwater lakes in the world. Data used in this study were obtained from the Water Resources Management Authority (WARMA) and Kenya Meteorological Department (KMD), and include river discharge, rainfall, and total suspended sediment for stations located in the basin. Statistical methods and application of various hydrological data analyses of time-series, flow duration, and Gumbel Distribution for flood flow analysis were used. Land-use change was determined through analysis of Landsat satellite images and use of overlay GIS techniques. Results show that there is significant variability in rainfall and river discharge of the Nyando River, which can be attributed to climate variability. There has also been major change in land use in the basin, which is evidenced by heavy deforestation in the catchment areas for agriculture and settlement of the rapidly increasing population. This increase in degraded areas, coupled with an increase in runoff, has led to increased peak flood flows. The increase in frequency of flood flows is attributed to rapid flow of surface runoff in degraded areas with low water infiltration and percolation rates. However, the relationship between rainfall and river discharge is strongly positive, reflecting the rapid hydrologic response to rainfall input in the basin. There is also a strong positive relationship between sediment discharge and river discharge. The increase or decrease of river discharge leads to a corresponding change in sediment yield. As a result of land degradation and high hydrologic response, the basin experiences high rates of soil erosion and sediment yield. High hydrologic response also leads to flooding in the low-lying Kano plains. Extreme streamflow variability in the plains has both direct and indirect impacts that are ultimately increasing poverty levels in the basin. This study shows that the effects of climate change and/or variability on the hydrologic systems in tropical Africa are being compounded by population growth that has driven land-use change.

J. U. Kitheka (✉) · S. Okoyo · N. Mboya
School of Environment, Water and Natural Resources, Department of Hydrology and Aquatic Sciences, South Eastern Kenya University, Kitui County, Kenya
e-mail: jkitheka@seku.ac.ke; kolbio_kolbio@yahoo.com

S. Diop et al. (eds.), *Climate Change and Water Resources in Africa*,
https://doi.org/10.1007/978-3-030-61225-2_10

Therefore, it is unlikely that efforts to reduce sediment load and control floods in the basin can be successful without a deliberate effort to mitigate the root causes of land degradation in the basin: rapid population growth and land-use changes.

Keywords Streamflow · Climate variability · Land use · Sediment loads · Floods · Nyando river basin · Kenya

1 Introduction

Climate change is affecting many river basins in Africa through increased occurrence of extreme hydrologic events, such as floods and droughts. In Kenya, like in other countries in Africa, the manifestation of climate change is evident in increased variability of streamflows. There is evidence that the frequency of occurrence of both extremely low and extremely high streamflows has increased in the recent past. The Nyando River Basin in Western Kenya is typical in experiencing increased streamflow variability with increased risk of floods and droughts due to climate change (see also Opere and Ogallo 2006; Opere and Okello 2011a, b). Extreme hydrological events in the basin have major socio-economic impacts because extreme variability of streamflow is causing changes to ecosystems on which local communities depend for survival. This has increased the vulnerability of communities to the impacts of climate change, and is expected to increase more in the near future due to the compounding impacts of climate change that can be attributed to land-use change due to the rapidly increasing population and urbanization in the basin.

The status of the Nyando Basin is similar to other basins in Africa where occurrence of extreme hydrologic events has increased noticeably in the recent past, affecting millions of people and significantly decelerating economic development. As a result, considerable effort has been invested in the mitigation of drought- and flood-induced damage over the last decade. In Kenya, floods are second to droughts in terms of frequency and severity of impacts on land use, as droughts continue to cause physical damage in major river basins, including the Nyando Basin. Within the Nyando River Basin, interventions to address the impacts of extreme hydrologic events have not yielded the desired results. This is partly because most mitigation measures have not been based on hydrologic research and have also largely ignored local perceptions and experiences. There has also been relatively limited research to characterize the hydrologic response of tropical river basins in this era of climate change. Similarly, there have been few hydrologic studies targeting degraded tropical river basins such as Nyando that are already experiencing the impacts of climate change. The lack of such studies limits the effectiveness of drought and flood management intervention. There is therefore a need for studies to determine the extent to which tropical river basins in Africa are being impacted by climate change in terms of streamflow variability, soil erosion, and sediment yield. This need is amplified by the fact that extreme hydrologic events have cumulative impacts that can in turn have major socio-economic impacts on communities. However, these are generally undervalued by both national and international disaster management and development policies and represent a challenge for disaster risk reduction (Mutua 2001).

The problem of sediment production and transport subject to climate change within the basin must be examined, because it has major socio-economic implications for communities and the country. Past studies have shown that land degradation occurring on fragile slopes in the Nyando River Basin is responsible for high rates of soil erosion (ICRAF 2000), which leads to high sediment load, and is thought to be influenced by factors such as land use, vegetation cover, runoff dynamics, and rainfall variability. There is, however, limited data and information on the extent to which extreme hydrologic events and land-use change have altered the sediment load delivered to Lake Victoria by the Nyando River. Because of this, assessment of the influence of rainfall, river discharge, and land-use change is important because these factors are highly variable as a result of climate and land-use change, with big implications for the Nyando River Basin. The object of this study is therefore to establish the extent to which climate change is affecting streamflow variability and sediment transport in the Nyando River Basin in Kenya, an important aspect of the current debate on the impacts of climate on water resources in Africa.

2 Description of the Study Area

2.1 Location

The Nyando River Basin is located in the Western region of Kenya between longitudes 0° 50'S and 0° 10'S and between latitudes 33° 05'E and 34°25'E (Fig. 1). The river originates from the Nandi Hills and Nyabondo Escarpment and drains into the Winam Gulf in Lake Victoria (LBDA 1992; Republic of Kenya 2001a). It has a total length of 170 km and a surface area of 3,517 km^2. The main source of the rivers is in the forested highlands at an altitude of 3,000 m above sea level (Kamau 2016).

2.2 Climate

The Nyando River Basin experiences two rainfall seasons. The long rainy season is between March–April–May (MAM), and the short rainy season is between September–October–November (SON). The average annual total rainfall ranges from about 1100–1600 mm. The highland areas of the basin experience higher amounts of rainfall (1800 mm) compared to the middle and lower areas that receive between 600 and 1100 mm of rainfall annually (ICRAF 2000 and 2002). The mean minimum annual temperatures ranging from 14 °C to 18 °C are recorded from June to July. The

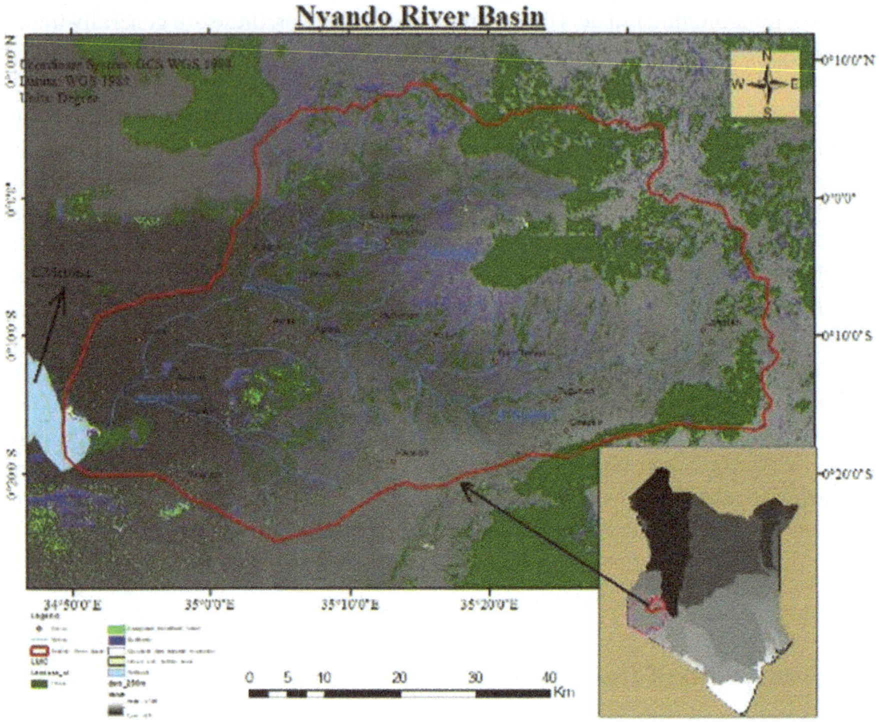

Fig. 1 The Nyando River Basin in Kenya

highest annual temperatures, ranging from 27 °C to 32 °C, are recorded in August and September. The monthly mean evaporation ranges from 1300 to 2200 mm.

2.3 Vegetation Cover

Vegetation cover in the Nyando River Basin also shows significant spatial variation; the influence of altitude is predominant in determining vegetation types. In the highlands, vegetation consists of evergreen broadleaf forest, and there are also many tea plantations, on both large- and small-scale farms. Large areas are covered with plantation forests, especially *Pinus patula* and *Cupressus lustanica.* Eucalyptus trees have also been planted, especially on tea estates. The lowland areas are covered with grasslands and scattered acacia trees.

2.4 Land Use

Land use and land cover types in the Nyando Basin are diverse (Van Der Kwast and Zomer 2002; Onyango 2003). The main land-use types are forestry and agriculture. Forests are found on government-designated land in Timboroa, Tinderet, Londiani, Western Mau, and parts of South Nandi. Some parts of these forests are planted with exotic tree species (*Pinus patula, Pinus radiate* and *Cupressus* spp), which are used for commercial purposes. The lower parts of the basin are composed of a flood-prone lakeshore area which is mostly used for subsistence farming of maize, beans, and sorghum, and commercial production of sugar cane and irrigated rice. The vegetation in the basin has changed considerably from the woods of the past to the present shrubs. The dominant land use in the highlands is for tea, both on large estates and small-scale farms. Other crops found in these areas are maize, potatoes, pyrethrum, wheat, and cabbage. In the last 40 years, the basin has experienced dramatic land-use changes, as the land has been converted from small-scale farming to intense smallholder cultivation (Verchot 2008; Swallow et al. 2002; Omuto 2003; ICRAF 2000 and 2002).

3 Methodology

3.1 Data Types and Sources

Data used in this study were gathered from both primary and secondary sources. These include unpublished reports of the National Environment Management Authority (NEMA), Directorate of Water Development (DWD), Water Resource Management Authority (WRMA), and LVEMP. Reports from other NGOs and the private sector also provide valuable information on land-use activity in the basin. The main data used in this study are rainfall, river discharge, suspended sediment concentrations, sediment loads, and land-use change. These data were obtained from the Water Resources Management Authority (WRMA) and Kenya Meteorological Department (KMD).

3.2 Secondary Data

Data on daily river discharges and Total Suspended Sediment Concentrations (TSSCs) for the periods between 2000 and 2015 were collected at River Gauging Station (RGS) 1GD07 (Nyando: Long. 35 10‘E and Lat. 0 10‘S) and RGS 1GD04 (Ainomotua: Long. 335 02‘E and Lat. 0 07‘S). River discharge data were obtained from the WRMA Regional Offices in Kisumu. These two river gauging stations were considered suitable for this study because the area upstream of the basin is the most affected by changes in rainfall and river discharge levels. The mean monthly rainfall data were obtained from the Kenya Meteorological Department (KMD) rainfall stations located at Ahero, Kibos, Koru, Lumbwa, and Nyando. Rainfall data were used to determine the relationship between rainfall and streamflow.[1] There is, however, a huge gap in the sediment data because the only available data on TSSC for RGS 1GD04 were for the period between 2000 and 2015.

3.3 Analysis of Land-Use/Land Cover Changes

Cloud-free Landsat images were downloaded from the United States Geological Survey (USGS) website for the years 2000 and 2015. To obtain complete coverage of the study area, two scenes, of paths 169 and 170, and raw satellite images were subjected to atmospheric correction and then projected to UTM 37 using WGS 84 data. Bands 4, 3, 2 for these images, and 5, 3, 2 for OLI images, were used to obtain a false color composite. Three subsets were created using the shape file of the study area. Images were enhanced to facilitate image interpretation. To obtain a rough idea of the number of classes in the study area, unsupervised clustering of ISODATA was used. Training sites were developed from Google maps. Supervised classification using maximum likelihood algorithm was used to classify the images. The classified images were then subjected to post-classification smoothing using a majority filter to remove the "salt and pepper" appearance and enhance cartographic display. Post-classification change detection was used.

3.4 Methods of Data Analysis

Data were analyzed using both qualitative and quantitative methods. The frequency distribution model (Gumbel Extreme Value) was used to determine the return periods of various categories of streamflow of the Nyando River. In this analysis, the magnitude of floods was related to frequency of occurrence. The frequency and magnitude of flood flows were determined using the Gumbel Extreme Value Probability Model that also compares different return periods to predict future flows. Return periods of rainfall and river discharges were also determined in this study. The inverse of probability is generally expressed in percentage, given the estimated time interval between events of a similar size or intensity. Trend analysis was undertaken to determine seasonal and annual variations of both rainfall and river discharges. The study also employs the use of measures of central tendency in the analysis of data (Lane 2008). Various quantitative methods used in data analysis include regression and correlation analyses and Analysis of Variance (ANOVA).

[1]Data collected were coded, processed, and analyzed using Microsoft Excel.

4 Results

4.1 Variability of Rainfall

Streams contributing to the streamflow of the Nyando River originate from high altitude (1700–3000 m asl). These are high rainfall (1200–2400 mm) areas located in Kericho and Nandi counties (Fig. 1). The river flows through V-shaped valleys in the upper mountainous region with a channel width of 20 m (NEMA 2004). At the middle region, the width of the river is about 40 m, but this increases to a channel width of 50 m further downstream in the low-lying region (WARMA 2004; Kamau 2016). The peak rainfall experienced in the month of April during the long rainy season ranges between 150 and 270 mm (Fig. 2). There is also a peak rainfall in August, when it ranges from 125 to 170 mm, as well as in November during the short rains when the peak ranges between 70 and 140 mm. Peak river discharge is also experienced in the same periods. The lowland area lies in a region of low rainfall, receiving a mean of between 800 and 1,200 mm annually, while sections of the basin in the north and south experience rainfall >1,600 mm per annum. The highest rainfall occurs in the Nandi Hills toward the northeastern region, which gradually decreases in the southeastern direction. The mean annual rainfall in the basin has been estimated using Thiessen's Polygon method to be 1431 mm (Kamau 2016).

The results of the analysis of hydrological data show that there was an increase in rainfall in the period between 2003 and 2012 (Fig. 3). Due to lack of data, the trend for earlier years is not established. It is important to note that the increasing trend in the period 2003–2012 could be attributed to normal inter-annual rainfall variability.

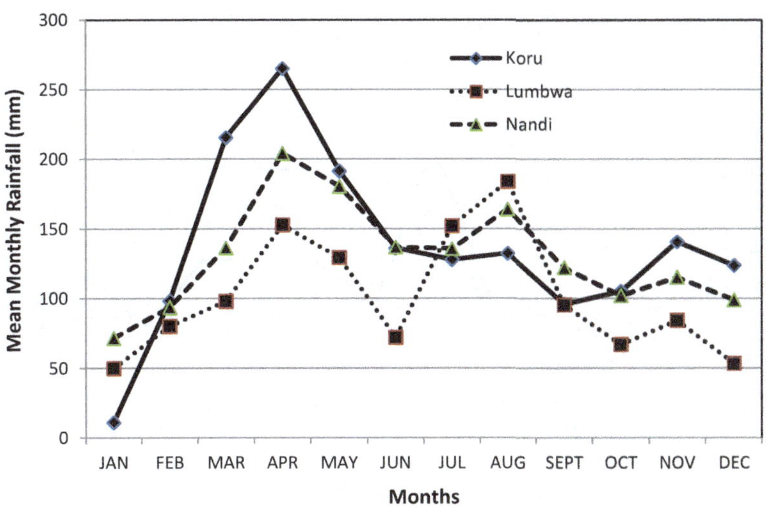

Fig. 2 The mean monthly rainfall in the Nyando River Basin. The plots are based on averages of 11 years for Koru and Lumbwa rainfall stations and 41 years for Nandi Rainfall Station

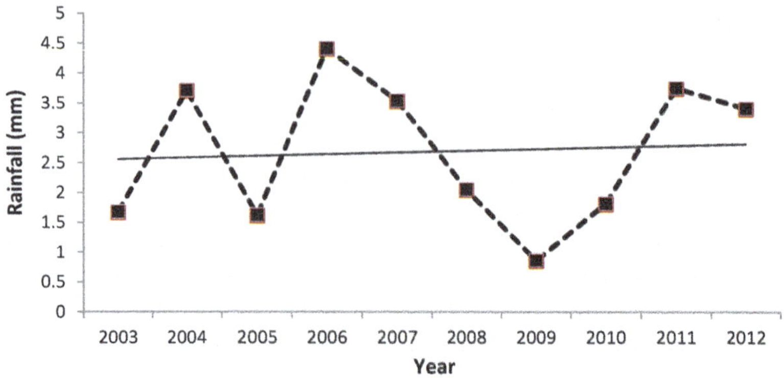

Fig. 3 Inter-annual variation of rainfall in the Nyando River Basin, 2003–2012, based on records obtained at RGS IGD03

As with rainfall, river discharges also show normal year-to-year variations that are related to fluctuation of rainfall (Fig. 3). There is also indication of an increase in the degree of variability, with an increase in the frequency and magnitude of extreme rainfall observed in the basin. It is possible that the declining trend is either a result of climate variability and/or land-use change. It is, however, difficult at this stage to determine the main driver. While rainfall shows a declining trend, the degree of variability seems to have increased since 2013 (Fig. 4).

The basin experiences distinctive seasonal variation throughout the year, with two maximum rainy seasons from April to May and also from October to November (Figs. 5, 6, and 7). The lowland area lies in a region of low rainfall, with a mean between 800 and 1200 mm annually, while sections of the basin in the north and south experience >1600 mm of rainfall. The greatest rainfall of >2000 mm occurs

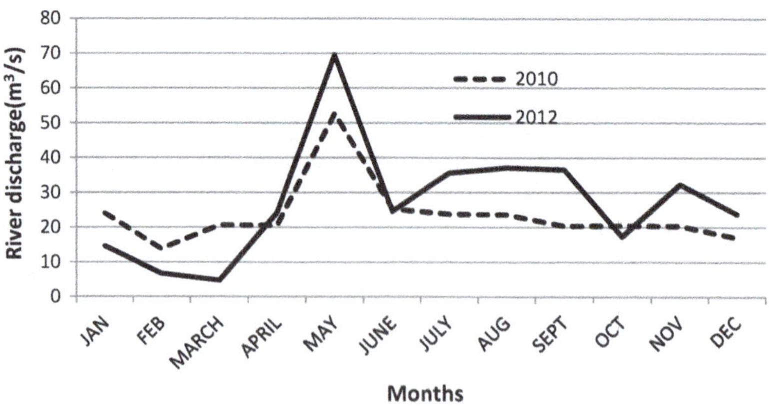

Fig. 4 Seasonal variations of Nyando River discharge between 2010 and 2012 based on records at RGS IGD03

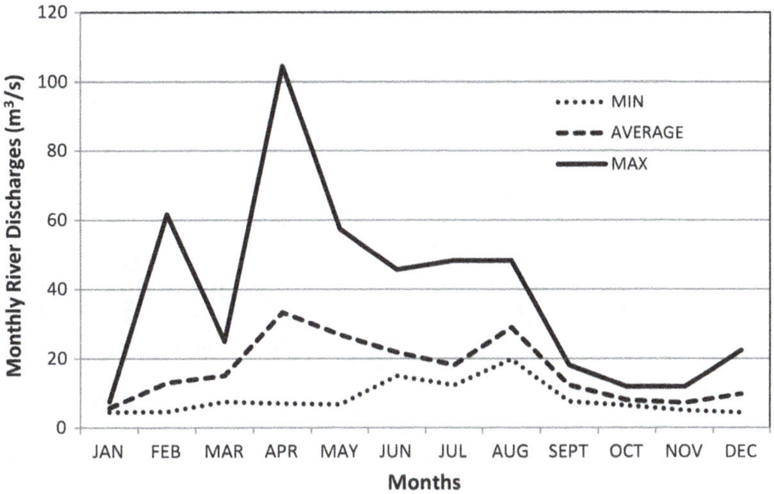

Fig. 5 Mean monthly river discharge in the Nyando River Basin at RGS 1GD4

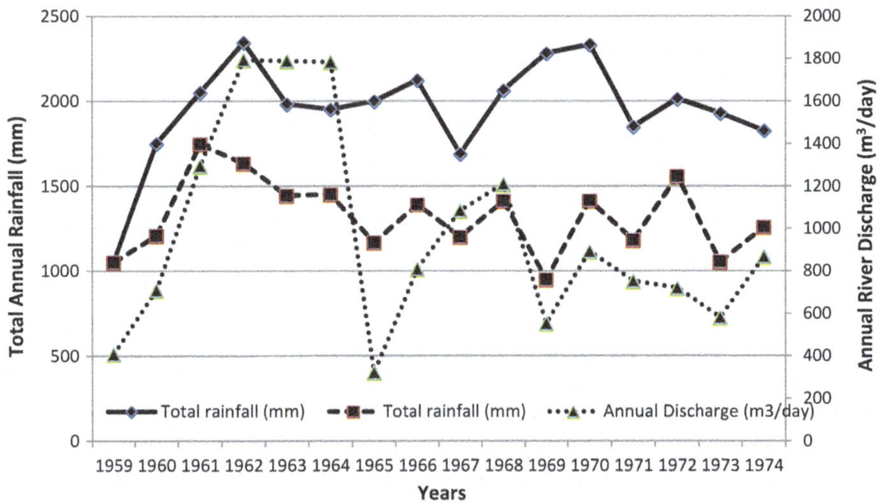

Fig. 6 Annual river discharge and total annual rainfall in the Nyando River Basin at Koru and Lumbwa Rainfall stations

in the Nandi Hills, and gradually decreases in the southeastern direction, bringing variations in discharge.

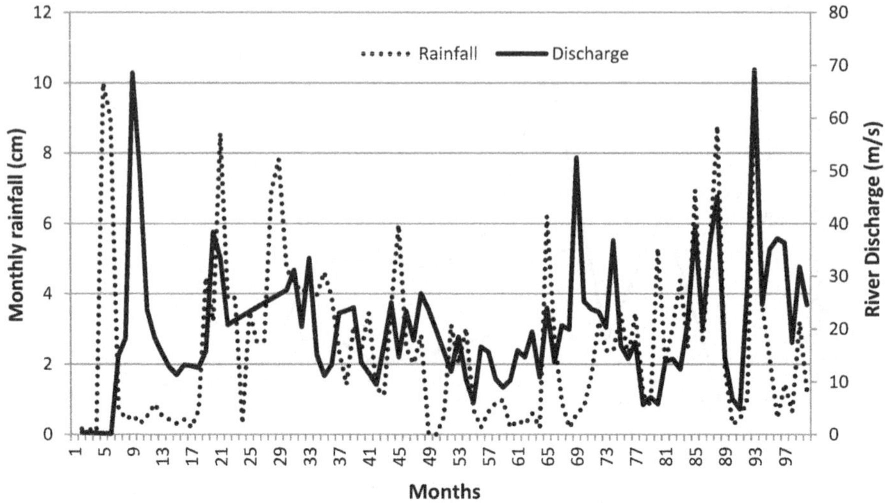

Fig. 7 Variations of rainfall and river discharge from 2003 to 2012 based on records obtained at RGS IGD03

4.2 Variability of River Discharges

River discharge in the Nyando River shows significant seasonal and inter-annual variability (Fig. 5, 6, and 7). High streamflows are experienced during the long rainy season (March–May) and also during the two short rainy seasons (August–September and October–December), as shown in Fig. 6. The highest streamflows are experienced during the long rainy season (April–May) and are of the order 969 m^3/s. Low streamflows in the dry period between December and February are on the order 7 m^3/s. Peak river discharges in the tributaries of the river are, however, much lower. The high inter-annual variation in river discharge is also discerned from records. For instance, the peak river discharge in May 2013 was 89 m^3/s, while in 2015, the river discharge was well below 50 m^3/s. However, in 2013, discharge was relatively higher, and greater than that experienced in other years.

Although there is an indication of an increase of rainfall, the corresponding increase in streamflow is not clear in the data, although previous studies have reported an increase (GOK 2009; Opere and Okello 2011a, b). High seasonal and inter-annual variability of streamflow is, however, evident in the data. This could be linked to climate change or to normal inter-annual rainfall variability (Figs. 7 and 9). Previous studies have indicated that the increasing variability of rainfall in the basin has made it difficult to predict high flood flows with certainty (ICRAF 2000 and 2002).

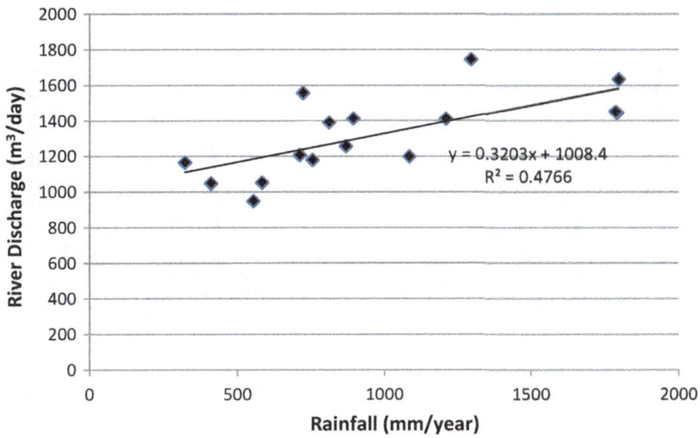

Fig. 8 Relationship between total annual rainfall and annual river discharge in the Nyando River, 1956–1974

4.3 Relationship Between Rainfall and River Discharge

The relationship between rainfall and river discharge in the Nyando River Basin is complex. Data for the period 1956–1974 yield a correlation coefficient r of 0.71 with a coefficient of determination R^2 of 0.51. The relationship between rainfall and streamflow in the basin established using 2003–2012 data yield a R^2 of 0.63 and correlation coefficient r of 0.80. These results could be interpreted to indicate that rainfall in the Nyando River Basin explains 51%–63% of the variation in river discharge. Therefore, in addition to rainfall, catchment characteristics could also be contributing to variation in river discharge. These catchment characteristics include land use, vegetation cover, and topography. The results also show that there exists a strong relationship between rainfall and river discharge in the Nyando Basin (Fig. 8). An increase in rainfall in the basin therefore leads to a corresponding increasing in river discharge.

4.4 River Discharge and Rainfall Deficits

Rainfall and river discharge deficits were computed as the difference between monthly values and long-term mean values. Figure 9 shows rainfall and river discharge deficit from 2003 to 2012 at RGS IGD03, which indicate that river discharge was above the long-term mean value 16 times. River discharge was also below the long-term mean value 16 times. However, the magnitude of above-average flows seems to be much higher than the below-average river discharge. Above-average flows normally last four to eight months, and below-average flows last about the same.

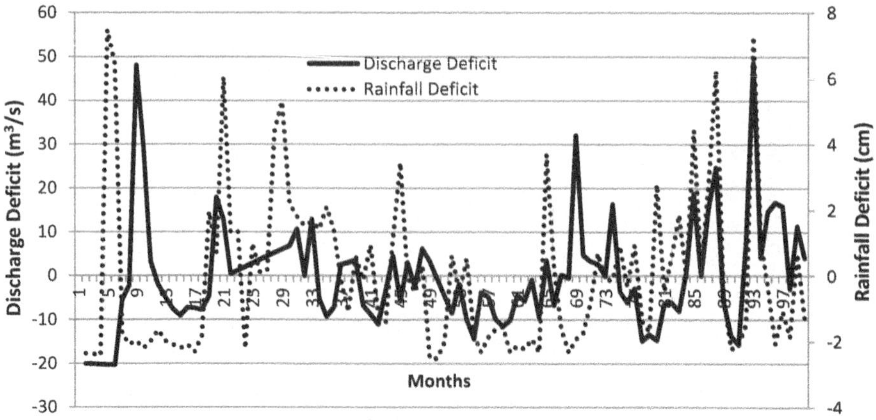

Fig. 9 Rainfall and river discharge deficit in the period 2003–2012 at RGS IGD03. The rainfall deficit was computed as the difference between monthly discharge and long-term mean discharge

The results also show that rainfall deficits are more dominant than the shortages associated with droughts. Positive deficits are associated with high rainfall and flooding. The duration of excess and shortages of rainfall are similar (Fig. 9). Results also show that periods of river discharge deficit correspond to those of rainfall deficits. Above-average river discharge occurs in periods of above-average rainfall. Results generally show that the occurrence of above-average rainfall in the Nyando River Basin is more dominant than the occurrence of below-average rainfall over the same period (2003–2012).

4.5 *Catchment Characteristics and Flood Flows*

Floods in the Nyando River Basin are normally associated with peak streamflows. Flooding is predominant in the lower parts of the basin, which are generally low-lying Kano plains. Figure 10 shows the flow duration curve of the Nyando River, which provides an important indication of the nature of the basin, including the frequency of occurrence of flow of a given magnitude. The curve indicates that the basin is highly degraded and experiences rapid hydrologic response. Also, most of the flows in the basin are low magnitude, with 95% of river discharges being on the order 3 m^3/s. In contrast, 50% of river discharges are on the order of 10 m^3/s. Streamflows that are greater than 100 m^3/s are relatively infrequent as they occur less than 1.0% of the time. Therefore, high flood flows have a lower probability of occurrence in the Nyando Basin compared to other river basins in Africa. The magnitudes of 50- and 100-year floods were computed as 1135 m^3/s and 1200 m^3/s, respectively. River discharge data for 1956–1974 for RGS 1GD4 shows that the highest flood flows between 1961 and 1964 were 605 m^3/s and 823 m^3/s, respectively. After 1964, there has continued to be

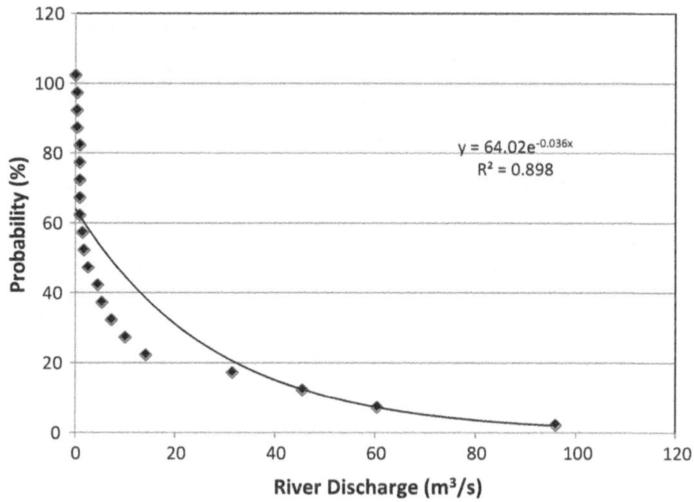

Fig. 10 Flow duration curve of the Nyando River

much lower annual peak river discharges, which in most cases are less than 300 m^3/s. In fact, river discharge data showed a declining trend after 1964. In view of the lack of long-term river discharge data, it is not certain if this decrease can be attributed to climate change. It is important to note that in Kenya, rainfall amounts have generally been declining, as reflected by the growing frequency of droughts in most part of the country (ICRAF 2000 and 2002). Excess rainfall occurrences have been cyclical, as evidenced by El Niño episodes. It can therefore be argued that an increase in rainfall since the late 1970s, together with changes in catchment characteristics, has led to changes in the magnitude and duration of peak river discharges in the Nyando Basin.

4.6 River Discharge and Suspended Sediment Load

The Total Suspended Sediment Concentration (TSSC) in the Nyando River varies from 0.08 g/l to 0.66 g/l, with a mean TSSC of 0.44 g/l. There is a significant relationship between TSSC and river discharge, with a correlation coefficient r of 0.90. Therefore, as river discharge increases, the suspended sediment concentration in river water also increases. Variations in the concentration of suspended sediments in the Nyando River can be attributed largely to variations in river discharge. Low flow river discharge events are characterized by relatively low TSSC (< 0.2 g/l), while high river discharges are characterized by TSSC > 0.40 g/l. The relatively high TSSC during high flows can be attributed to increased capacity of the river to transport a higher volume of sediment load capacity under high streamflow conditions.

The influence of rainfall on TSSC in the basin is complex, with a correlation coefficient r of 0.40. This means that in addition to river discharge, rainfall variability influences variations in TSSC in the Nyando River. The TSSC, and therefore water turbidity, increases as rainfall increases. The relationship between rainfall and TSSC is complicated by the availability of sediment within a given stage of the river during the rainy season. During early stages of the rainy season, low vegetation cover means that more loose sediment is available for transport by runoff into the river channel. In later stages, improvement of vegetation cover reduces sediment availability, leading to relatively low TSSC (and low turbidity), even when river discharge may be increasing. TSSC usually increases with river discharge because the increased flow generally results from direct surface runoff, which is associated with soil erosion. The supply of suspended sediments usually limits the variation of TSSC in the river.

The relationship between sediment load of the river and river discharge is strong, with a correlation coefficient r value of 0.99 and coefficient of determination R^2 of 0.98. This indicates a strong correlation between sediment load and river discharge. Increase in river discharge results in an increase in the total suspended sediment load. The computation of sediment load of the Nyando River is based on the mean and maximum river discharge of 20 m^3/s and 89 m^3/s, respectively. We used the mean TSSC of 0.4 g/l, which seems to be the long-term mean for the river. The results show that the mean sediment load is 691 tons/day, which translates to 252,288 tons/year. But the maximum sediment load is 3.07×10^3 tons/day, which translates to 1.12×10^6 tons/year. The sediment load of the Nyando River can therefore be assumed to vary from 0.25×10^6 to 1.12×10^6 tons/year. It should, however, be noted that during periods of extreme flood flows (> 400 m^3/s), the sediment load of the river can be 100 times greater than average. During flood flows, the river discharges 22,809 tons of sediment per day, which for a period of one month is equivalent to 684,288 tons. Therefore, during periods of flood flows, the river can discharge 50% of the total annual sediment load in just one month. Based on the above data, we compute the sediment production rate in the basin to be in the range of 72–318 tons/km^2/year. This is comparable to those reported for the Upper Tana Basin in the Central Kenya Highlands (Njogu et al. 2018; Mwendwa et al. 2019; Kitheka et al. 2019).

4.7 *Analysis of Land-Use Change*

There has been significant change in land use in the Nyando River Basin from the pre-colonial to post-colonial periods (Table 1). In the past 100 years, there have been major changes in land use and vegetation cover in the basin (see Swallow et al. 2002; Onyango 2003). The analysis of Landsat satellite images and aerial photographs for the period between 2003 and 2012 show that forest cover increased by 8.32% in this period due to reforestation activities in the basin. There was, however, in the same period a 100% increase in the area under human settlement. The wetlands area decreased by 82% in the lower parts of the basin in the same period due to conversion

Table 1 Land-use changes in the Nyando River Basin for the years 2000 and 2015

Land Cover	2000	2015	Change (Area)	Change (%)
Cropland (km^2)	2,760	2,727	−34	−1.2
Forestland (km^2)	669	725	56	8
Grassland (km^2)	156	162	6	3.8
Settlements (km^2)	2	4	2	100
Wetlands (km^2)	36	7	−29	−81.7

of wetlands for agricultural use (Fig. 11). There was also a slight decrease in the area under cultivation, which is attributed to the impacts of flooding in the lower parts of the basin.

Soil erosion is a serious problem in the Nyando River Basin. It is estimated that 60% of the basin is experiencing accelerated soil erosion, making it one of the major sources of sediment discharged into Lake Victoria (Swallow et al. 2002; Onyango 2003; Walsh et al. 2004). The high rates of soil erosion in the basin can be attributed to inappropriate land-use practices. Most cultivation is done on steep slopes on highly erodible soils without the application of appropriate soil erosion protection measures.

Most of the land-use/land cover changes observed in the basin can be attributed mainly to rapid population growth and expansion in the basin. For instance, the population in the basin has increased from fewer than 20,000 people in the 1960s to the 2009 population of more than 750,000 people (KNBS 2009). Although the population growth rate appears to be slowing, the relatively high population growth

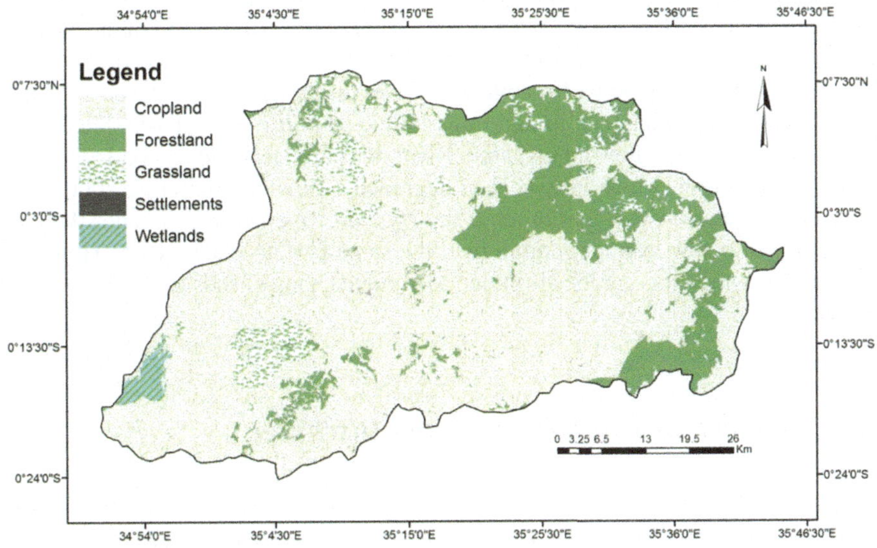

Fig. 11 Current land use in the Nyando River Basin. The proportion of land under cultivation in the basin is relatively high compared to other river basins in the Lake Victoria Basin

rate (3.7% per annum) in the last three to four decades led to an increase in the clearance of land for agriculture and settlements. More than 80% of the population in the basin depends on agriculture for its livelihood. Poverty in the basin is among the highest in Kenya, with 65% of the population in the basin living below the poverty line (KNBS 2009; ICRAF 2000). The majority of the people living in the basin are highly dependent on land for their daily sustenance. Therefore, without addressing poverty in the basin, it is unlikely that the current high rates of soil erosion and land degradation in the basin can be adequately addressed.

5 Discussion

5.1 Relationship Between Rainfall and Discharge

The influence of rainfall on Nyando River discharge is depicted in flow duration curves and also in hydrograph patterns. Data analyzed in this study reveal a rainfall trend with normal inter-annual variability, with no clear long-term trend that can be attributed to climate change. River discharges show normal year-to-year variation that is related to rainfall variability in the basin. Although the relationship between rainfall and river discharge is generally weakly correlated, rainfall variation explains only 50%–60% of the variation in river discharge. Therefore, there are parameters other than rainfall in the basin that contribute to streamflow variations. The imperfect correlation could also be due to the lag between rainfall and river discharge. Peak flow at the outlet of a river usually occurs some hours after peak rainfall in the highlands located 170 m away. When storms occur at the extreme end of the basin, it takes up to a day before this is reflected as peak flow in the lower parts of the basin. The strong relationship between rainfall and river discharge can also be attributed to land-use changes, which lead to rapid hydrologic response of the basin due to land degradation. The data show that Nyando River discharge tends to respond rapidly to rainfall in the basin, which is typical of degraded basins (cf. Hai et al. 2000). The occurrence of above-average rainfall in the Nyando River Basin is greater than the occurrence of below-average rainfall over the same period (2003–2012). However, the extent to which this can be attributed to climate change needs to be investigated further.

5.2 Sediment Yield and Rainfall and Discharge

There has been significant change in land use/land cover in the Nyando River Basin in the last 100 years. While there are year-to-year variations in land use in the basin, the overall trend is toward increased land degradation which can be attributed to rapid population growth and expansion (Oruma et al. 2017). There has also been a massive increase in human settlement and cultivation on steep slopes without application of appropriate soil conservation measures, which accounts for the high rates of soil erosion in the basin (see also Walsh et al. 2004). Soil erosion in the basin is influenced by several factors, including topography, rainfall pattern, land-use type, degree of vegetation cover, and soil erodibility (cf. ICRAF 2000 and 2002; Swallow et al. 2002; Walsh et al. 2004). Soil erosion and resulting land degradation have cumulatively altered the hydrologic response of the basin such that streamflow responds quite rapidly to rainfall storms. The basin is characterized by relatively high rates of sediment load which are estimated to range between 252,288 tons/year and 1.12×10^6 tons/year (see also ICRAF (2000 and 2002). These results are consistent with those reported from other degraded river basins in the Kenya highlands (Oruma et al. 2017; Njogu et al. 2018; Mwendwa et al. 2019; Kitheka et al. 2019).

During flood periods, the sediment load of the Nyando River increases more than 100-fold, although this occurs for a relatively short period of less than one month. This study shows that during flood flows, the river discharges 684,288 tons of sediment into Lake Victoria, which means that during flood conditions, the river discharges almost 50% of its total annual sediment load in just one month. This indicates that the Nyando River discharges a high volume of sediment to Lake Victoria on an annual basis. There is a need to implement a long-term sediment load monitoring program in order to establish the significance of discharge contributions from other rivers to total river sediment load discharged into Lake Victoria (cf. Oruma et al. 2017). The entire basin is characterized by severe soil erosion and land degradation (ICRAF 2002).

There is a relationship between sediment transport and river discharge. An increase in river discharge results in an increase in TSSC and consequently an increased suspended sediment load. The TSSC usually increases with river discharge because the increased flow generally originates from direct surface runoff on degraded lands (cf. Hai et al. 2000). Because of significant inter-annual variations in rainfall that subsequently induce inter-annual variations in streamflow, it is expected that sediment yield will also exhibit significant inter-annual variation. There appears to be an increase in the degree of variability, including an increase in frequency and magnitude of extreme rainfall events and high streamflow in the basin. It is therefore expected that the variability of sediment yield will also increase. These variations may be a result of climate change and/or of normal climatic variation. Previous studies have shown that within the basin, most of the surface runoff is generated from degraded lands which are also the source areas of sediment transported by the river. Therefore, any serious soil conservation program must focus on reducing erosion of soil from degraded areas (cf. Swallow et al. 2002).

5.3 Flooding in the Lower Nyando Basin

Flooding in the Nyando River Basin is mainly limited to the lower parts of the Kano plains. Most floods occur during the long rainy season between March and May.

Three major flood events have occurred in the recent past: 1961/1964, 1997/1998, and 2003/2004. During floods, the entirety of the Kano Plains is usually inundated with water for a period of up to one month. Data show that since the 1961/1964 floods, subsequent floods in the plains have been of relatively lower magnitudes. There is, however, an indication that the frequency of occurrence of these low-magnitude floods has increased.

The data do not indicate clearly whether there is a long-term increase in streamflow that could be related to increasing rainfall (see also Opere and Okello 2011a, b). Although data for the period 2003–2012 show an increasing trend in streamflow, it declined in the period 1964–1974. This suggests that in addition to inter-annual variation in river discharges, there may also be inter-decadal variations in streamflow that could be explained by inter-decadal climatic variability. This requires further hydrological and climatological research.Regardless, it is significant that land-use/cover change has substantial effects on streamflow in the basin. While forest cover has been increasing slowly due to the implementation of reforestation programs in the basin, more must be done, as suggested by previous studies that have shown that increased forest cover in the basin results in decreased peak flood flows there (Opere and Okello 2011a, b).

Flooding in the Kano plains has both direct and indirect impacts, which can have positive and negative effects. Impact analysis based on various reports (see ICRAF 2002) shows that direct negative impacts of floods include water quality deterioration, habitat degradation or alteration, river channel modification, siltation of water bodies (dams, water pans, canals), destruction of fishing grounds, reduced fish catches, destruction of infrastructure (water supplies, roads, bridges), loss of human life, loss of livestock, creation of habitat for disease vectors, damage to farms, loss of crops, reduced crop production, interruption of education/school programs, destruction of property (e.g., houses, buildings and pit latrines), contamination of water supplies, lack of clean water and disruption of social and economic activity.

Indirect negative impacts include loss of income, unemployment, economic stagnation, waterborne disease, and other health impacts (diarrhea, cholera, malaria, typhoid, skin diseases, stress, and anxiety), increased incidence of livestock diseases, a decline in education performance, lack of economic diversification, lack of diversification of income, reduced human productivity, conflicts, increased rural-urban migration, increased dependency on donor assistance and grants, increased government expenditure (repairs, evacuation, health, relief), and degradation of recreation opportunities.

Positive impacts of flooding include recharge of groundwater aquifers, supply of fertile silt to the flood plains (including paddy fields), recharge of wetlands and riparian zones, increased nutrient supply to Lake Victoria, and regeneration of wetlands.

The overall adverse impact of flooding in the basin is increased poverty levels among the majority of the people living in the basin. More than 65% of the population there live below the poverty line (see also Swallow 2005).

6 Conclusions

The Nyando River Basin experiences significant seasonal and inter-annual variability in both rainfall and river discharge. There has been an increase in the variability of both rainfall and streamflow in the basin since 1970. However, it is difficult to fully attribute this variability to climate change due to lack of long-term data. It is possible that increased variability in streamflow is also due to increased degradation of the catchment areas and land-use change as a result of population increase. Because of land degradation, the basin is characterized by a relatively high sediment load. The lower parts of the basin also experience periodic flooding that impacts the socio-economy of local communities, including ecosystems in the basin. The overall impact of flooding has been increased poverty among the communities living in the lower parts of the basin. Without sustainable land management in the basin, climate change and land degradation will continue to reduce streamflow significantly, leading to major impacts on the livelihoods of the local communities. A comprehensive soil and water conservation program for the entire river basin is necessary in order to control soil erosion and reduce the high sediment load of the river.

References

Hai M, Ong C, Mungai D (2000) The impact of land use and rainfall on runoff and sedimentation in the Lake Victoria Basin. In: Paper presented at the KL2000 UNESCO/IUFRO symposium: forest–water–people in the humid tropics: past, present and future hydrological research for integrated land and water management. University of Kebangsaan, Malaysia, 30 July 2000, 4 p

ICRAF (2000) Improved land management in the Lake Victoria Basin: linking land and lake, research & extension, catchment & Lake Basin. First Report of Progress and Results, July 1999 to March 2000. Nairobi, ICRAF and Ministry of Agriculture and Rural Development

ICRAF (2002) Improved land management in the Lake Victoria Basin, Annual Technical Report, working paper-2, World Agroforestry Centre, Nairobi, Kenya

Kamau BK (2016) Hydrological study of Nyando river basin. Unpublished project report, Department of Civil Engineering, University of Nairobi, 87 p

Kitheka JU, Mwangi S, Mwendwa PK (2019) The effect of rainfall variability and land use/land cover change in a small tropical river basin in Kenya. Int J Hydrol 3(1) 58–64. https://doi.org/10.15406/ijh.2019.03.00163

KNBS (2009) Kenya population and housing census. 'counting our people for the implementation of vision 2030, vol 1 A

Mutua FM (2001) Report on the impacts of climate change on water resources in the Kenya

Mwendwa PK, Kitheka JU, Mwangi M, Otieno H (2019) The drivers of water abstraction and river water diversion in tropical rivers: a case of south west upper tana basin, Kenya. Int J Hydro 3(2):149–157

Njogu IN, Kitheka JU, Otieno H (2018) Streamflow variability and sediment yield in North-West Upper Tana basin, Kenya. Hydrol Curr Res 9(4). https://doi.org/10.4172/issn2157-7587.1000305

Omuto CT (2003) Rapid mapping of field-saturated hydraulic conductivity in tropical watersheds. MSc Thesis, University of Nairobi, Kenya

Onyango DO (2003) The influence of land use patterns on runoff and soil erosion in the River NyandoBasin at Kobong'o area in Kenya. MSc thesis, Moi University, Eldoret, Kenya

Opere AO, Ogallo L (2006) A natural disaster in lake Victoria Basin (Kenya): causes and impacts on environment and livelihoods, In Odada EO, Olago DO, Ochola W (eds) Environment for development: an ecosystems assessment of Lake Victoria Basin. Nairobi, UNEP)/ Pan African START Secretariat (PASS)

Opere AO, Okello BN (2011a) Hydrologic analysis for river Nyando using SWAT–hessd. University of Nairobi, Departments of Meteorology and Civil Engineering

Opere AO, Okello BN (2011b) Hydrologic analysis for river Nyando using SWAT. Hydrol Earth Syst Sci Discuss 8:1765–1797. https://doi.org/10.5194/hessd-8-1765-2011

Oruma SK, Kitheka JU, Mwangi M (2017) The study of the effects of Mau catchment degradation on the flow of the Mara River, Kenya. J Environ Earth Sci 7(2). www.iiste.org ISSN 2224-3216 (Paper) ISSN 2225-0948 (Online)

Swallow B (2005) Poverty reduction strategies for addressing multi-level poverty traps: reflections from Kenya. In: Ellis F, Freeman HA (eds) Rural livelihoods and poverty reduction policies. Routledge, London, pp 311–331

Swallow B, Walsh M, Nyantika D, Muriithi S, Noordin Q, Ong C, Shepherd K, Place F, Awiti A, Hai M, Ochieng O, Cohen M, Mugo F, Oyasi S, Omuto C, Cohen L, Okono A (2002) Improved land management in the Lake Victoria basin: Annual technical report, July 2001 to June 2002. International Centre for Research in Agroforestry Natural Resource Problems, Priorities and Policies Programme. Working Paper Series. Working Paper 2002–2

Van der Kwast J, Zomer R (2002) Land cover classification and DEM generation: using SPOT and landsat images in the Lake Victoria Basin, Western Kenya, Walsh, Markus, 35 p

Verchot (2008) Baseline report: Nyando River Basin, Western Kenya integrated ecosystem management project findings from the baseline surveys, International Centre for Research in Agroforestry, Nairobi, Kenya

Walsh M, Shepherd K, Verchot L (2004) Identification of sediment sources and sinks in the Nyando river basin. In Mungai D, Swallow B, Mburu J, Onyango L, Njui A (eds) Proceedings of a workshop on reversing environmental and agricultural decline in the Nyando River Basin, ICRAF: Nairobi, Kenya, pp 27–32

Chapter 11
Saharan Agriculture in the Algerian Oasis: Limited Adaptation to Environmental, Social and Economic Changes

M. Hadeid, T. Ghodbani, O. Dari, and S. A. Bellal

Abstract Climate change and its management through public policy have had a profound impact on the natural balances of oases in the Sahara of Maghreb. Modern irrigation techniques are practiced increasingly on the margins of traditional oases with appropriate adaptation methods. The pressure on water resources has led to intense power dynamics between water owners and new agricultural investors. The juxtaposition of local traditional knowledge in the use of resources to current approaches to technical and administrative management is the basis of a new territorial dynamic that is currently shaping the oasis space of the Maghreb. In this chapter, we propose a multiple-criteria analysis of the dynamics of transformation at the human-nature interface. The subject of this study was derived from an analysis of several research projects conducted over the past two decades (1999–2019).

Keywords Climate change · Public policies · Oases · Irrigation · Adaptation · Water resources

1 Introduction

For more than a century, the Algerian region of the Sahara has been the scene of significant social and spatial changes. The water management baseline considered in this chapter was established during the territorial and economic integration and appropriation of the colonial period, after which the State took action, which accelerated and consolidated water policy and incentives. As they entered the urban era, the Algerian Saharan territories were profoundly reconfigured, which gave rise to multifaceted tensions. These environmental, economic and social transformations have challenged the traditional balances and functioning of these territories. The gradual regrouping of populations and the many fold increase in urban populations have led to a weakening of ecosystems, an increase in economic inequality and, more generally, the emergence of structural territorial disparities. Spatial planning

M. Hadeid (✉) · T. Ghodbani · O. Dari · S. A. Bellal
EGEAT Research Laboratory, University of Oran 2 Mohamed Ben Ahmed, Bir El Djir, Algeria
e-mail: hadeid009@yahoo.fr

S. Diop et al. (eds.), *Climate Change and Water Resources in Africa*,
https://doi.org/10.1007/978-3-030-61225-2_11

and management policies designed to develop these areas economically have not prevented the creation of clear spatial inequalities, despite huge investments to open up the social facilities and economic infrastructure of this vast and constrained area by the creation of a relatively dense transportation network.

The regions of Touat, Gourara and Western Tidikelt are administratively part of the Wilaya[1] of Adrar, a vast Saharan area occupying nearly 18% of national territory but containing less than 1.2% of the national population and more than 12% of the total Saharan population, according to a 2008 survey (Kouzmine 2007; Hadeid et al. 2018). Having experienced the same development policies (administrative divisions, equipment program, agricultural development), this region of southwestern Algeria has suffered from the same policy exclusions, leading to the marginalization of a number of sectors and, consequently, a part of the population. Faced with this marginalization, traditional practices (sometimes informal), are maintained or even developed in the face of insufficient state support, leading to the disruption of local ecosystems. In some cases, these practices reflect the failures of civil society to address existing spatial disparities and their environmental and health effects in this area. This has resulted in the pauperization of the oases and a consequent mass departure of the most vulnerable populations, from the least coveted ksour to the most dynamic cities. The new rural policy launched in the 1980s through access to agricultural land ownership and the PNDA[2] in 2000 came respectively with the aim of correcting the disparities induced by this urbanization policy, but the application of this vast program has also been subject to criticism insofar as users have been much more attracted by the generous subsidies of this program than by a genuine interest in reviving agricultural activity in these arid and semi-arid spaces. The balance sheet, particularly after the State gradually withdrew its commitment to fund these agricultural development operations, has been mixed. Faced with the disengagement of the State, the saturation of the tertiary sector and the instability of oasis agriculture, a part of oasis society has sought mechanisms to revive their activity and improve their standard of living. Although interesting as well as ambitious, its solutions do not seriously address the issue of sustainability and hence their effects on non-renewable water resources in this vast, arid region. Thus, social strategies, or "social technologies" are really only a form of temporary adaptation of a society to an immediate crisis situation in the face of the State's failure to create a public policy for effective management of these arid, hostile spaces.

In addition to this territorial complexity, there are new challenges related to climate change. The local expertise in water mobilization and tillage of people in the far south has become a model for adaptation to climate change—not only to increasingly extreme natural conditions but also to frequent social and economic changes. How

---

[1]The Wilaya is an administrative entity, the equivalent of a "department" in France.

[2]PNDA (National Agricultural Development Plan). This program represents, in a way, a return of the State, particularly to boost agricultural activity after a period of disengagement that lasted more than a decade (Hadeid 2011). The orientations contained in this agricultural program are intended to ensure socio-economic and technical obligations (improving Algeria's agricultural competitiveness, increasing production and yields), but they also and mainly aim at rebuilding the agricultural territory, protecting fragile ecosystems and developing land for agricultural use (Bessaoud 2002).

do old techniques for the exploitation of natural resources still survive? What are the weaknesses of the new state-led development model? Can lessons be learned to better mitigate the effects of human activity on sustainable adaptation?

2 Method

This study is a synthesis of the research carried out in the Adrar region in order to determine the basis of the oasis system and its evolution over time. It also examines many case studies identified through continuous investigation of oases over more than two decades.

Two approaches were used:

- An evolutionary approach in order to identify the transformation of the oasis system through different historical phases, with an emphasis on the postcolonial period. This is the period during which our study reveals the most profound socioeconomic changes.
- A multi-factorial approach that accounts for an increasingly arid environmental framework, dynamic social capital and the actions of the State in different contexts. Direct surveys were conducted at the household level to determine the basic activity and challenges faced by individuals and families. What has driven and informed people as they develop strategies to protect their livelihoods? Who are the actors involved? What is the role of the State (passive and/or active)? How do oasis adaptation systems fit into global mitigation of climate change?

Areas for the survey were selected based on field observations, the dynamics of agricultural land use, the use of new irrigation techniques, the improvement of traditional means of access to water, and other factors. The twenty oases studied represent the nexus of traditional and new techniques in water use.

3 Theoretical and Conceptual Framework

Based on the different solutions adopted by the oasis societies studied, this research seeks to learn how "social technologies" can be applied to water use adaptation in the face of climate change. This is a concept that bridges the social values of solidarity and modern technological solutions for sustainable adaptation. It was first characterized at the University of Chicago by Albion Woodbury Small in the late 19th century, who described social technology as using knowledge and laws of social life to apply rational social objectives (Small 1898). Today, "social technology" has come to mean the use of technology in service to social, societal and citizen action, such as mitigating inequalities, building social solidarity and catalyzing social innovation. Investment in technological capabilities promises the ability to act in the public interest. This concept, with the similar concept of "social innovation", is used in Europe as a lever for territorial development. At a time when economic and social models are being undermined by the shocks of climate change, addressing unmet social needs is a particular challenge. According to the CSESS[3], "*there is a tremendous inventiveness on the part of citizens, civil society actors and companies that only needs a little favourable ground to grow, develop and provide significant responses to the main societal challenges: increased impoverishment of the population, limitation of energy resources, digital divide, suppression of public services in rural areas, ageing population, isolation of the elderly, increased demand for organic food... These challenges are reflected in the need for innovative solutions in terms of energy, accommodation, mobility,* etc. *On all these subjects and many others, the collective imagination is essential to bring out new responses in a context of scarce public funding.*" The CSESS has proposed a fulsome definition of "social technologies" that can be applied to any country in the world: the development of new responses to new or insufficiently- met social needs under current market and social policy conditions, involving the participation and cooperation of all actors concerned, in particular the users. These innovations concern both the product or service, as well as the mode of organization and distribution, in broad areas such as aging, early childhood, housing, health, the fight against poverty, exclusion, discrimination, etc. The stages of social process development consist of emergence, experimentation, dissemination and evaluation. This desire to integrate social technologies into the development process of the territories is not exclusive to France—the entire European Community supports it. In a speech to the European Parliament, the President of the European Commission, José Manuel Barroso, made it clear that innovation in general and social innovation in particular should be recognized as factors for sustainable growth, job creation and strengthening the competitiveness of the territories.

Several doctoral theses have also applied the concept of social needs, for example, in France (Patureau 2010) and on the water in Brazil (Coutinho 2015). Patureau (2010) uses social technology as a means to address major social issues: reduce poverty and fight exclusion, protect the environment, promote the integration of people with disabilities, enable young people to integrate professional and civic life, combat unemployment, manage an aging population, support the education and development of children, etc. In other words, the social technology movement is the recognition that neither government policy nor the market alone can provide complete solutions to big societal challenges. Coutinho, for example, studies the semi-arid region of Paraïba, Brazil. After the failure of efforts by the Brazilian government to mitigate inequalities in access to the water supply through the construction of dams and large capacity reservoirs, various civil society networks implemented a new development model based on the introduction of social technologies for access to water using tanks, stone reservoirs (reservoirs built on impermeable outcrops) and underground dams. The deployment of these techniques has resulted in a social network that has transformed the spaces in rural Paraíba. Another example is the use of social technologies to enable the local population to manage their new

[3]CSESS: Higher Council of Social Economy and Solidarity in France.

city. The philosophy on which Isle d'Abeau[4] was developed is to enable people to organize themselves instead of responding to external requests to organize. This approach, while not unique, was well-suited to L'Isle d'Abeau where the population is committed to building and managing its own development projects (Hominal 1982).

These social technologies are not only useful in their direct role in addressing social needs, but also because they facilitate new relationships among different stakeholders. This frames the current study on the arid southwestern Algerian region, with emphasis on identifying the actors who are effective social "translators" between state planners and the social networks that take up these innovations.

In the Sahara of North Africa, the development of the oasis areas is increasingly complex. Despite their small surface area compared to the size of the desert, oases play a fundamental role in maintaining the population in the great south of the Maghreb (Bisson 1996). Extensive research on this topic has been contributed by French geographers (Despois 1967; Bisson 1996 and 2003) as well as German (Suter 1959) scientists, who considered natural constraints, including water scarcity, in Saharan environments. Richter (1995) proposed a typology of oases based on four oasis resources: springs at the foot of a mountain, on rivers, in the foggara and with drainage tunnels. Water mobilization techniques are different in each setting, but have common requirements for effective water management. Sociological studies of the oases of the Maghreb note a stratification based on race (Touat, Gourara, Ouarzazat, Tafilalet) and/or other ethnic criteria (Ghardaïa). Marouf (1980, 2010a, b) and Moussaoui (2011) highlight inequalities in access to water and agricultural land, with land and foggara in these oasis societies considered socially inferior because it is traditionally the domain of a class of black descendants of slaves from Sub-Saharan Africa. In contrast, land and water ownership is often reserved for Chorfa (nobility). This distinction, extant since the Middle Ages, has diminished markedly since the independence of the Maghreb countries but still arises, for example, in marriage between men and women belonging to different groups, sharing of natural resources or in local power politics. Studies of transformations that have impacted society and the environment in a profound way have served as interesting laboratories to observe the evolution of the relationship between society and its environment. For example, Kassah (2010) looked at Southern Tunisia, Côte (2002) at the Souf, Schmit (2008) at the Mzab and Bencherifa and Popp (1992) at the Todra valleys in Algeria and the Drâa in Morocco. They all highlight the role of centralized public policies in the creation of a new socio-spatial order, as well as the failures of efforts to implement sustainable development in these oases.

[4]Old French New Town located in Nord-Isere, south-east of Lyon (France), whose creation was decided in 1968.

4 Forms of Social Adaptation to Government Disengagement

Oasis society naturally tries to address its own unsatisfied needs. The cases studied predominantly concern water and agriculture. For example, oasis agriculture in southwest Algeria has developed indigenous responses to overcoming an environment hostile to settlement, supplying water to both the oasis populations and the caravanners who cross the Sahara (Côte 2002). Traditional economic frameworks change over time as non-agricultural activities are introduced, leading to a diversification in employment and a corresponding decrease in dependence on oasis agriculture, which is based on a marked social stratification. Indeed, by introducing other types of employment, particularly in the tertiary and even secondary sectors (south-eastern Algeria and, recently, southwestern Algeria), the independent state has severely disrupted this traditional agriculture. As a result of the disruption of the social stratification established over centuries, agriculture has lost significant ground to other introduced activities. State policy based on agricultural development has been in place since 1983 in the arid and semi-arid regions (Steppe and Sahara) in order to replace traditional agriculture that operates in difficult (scarce labor, dried foggaras, etc.) and less profitable environments. This directly affects cereal production, negating previous social and feudal organization. Despite generous government subsidies, government policy has not achieved its objectives. Recent years have been marked by a gradual disengagement of the State, leaving new agricultural dynamics to emerge in both the modern and traditional agricultural sectors.

5 Relaunch Traditional Oasis Agriculture and Revitalize Modern Agriculture

The withdrawal of the State has left a void filled predominantly by farmers from the Sétif[5] region, who have started growing market gardening products, especially cucumbers, used in the manufacture of cosmetic products in the center of the country (Algiers, Blida). This activity has filled a market gap, particularly in the south-eastern Ziban region, taking advantage of the relatively conducive growing conditions there during the cold winter period in the north (Table 1). Moreover, these cucumbers transport and weigh well. By renting land from indigenous farmers, who themselves have been less enthusiastic about farming after the decline in State subsidies, the Sétif people have been able to institutionalize greenhouse cultivation in the Aougrout[6] region, known for its soil and water potential, in few years.

[5]Located 270 km east of Algiers, this important city in eastern Algeria is home to more than 300.000 inhabitants.

[6]Large commune of the Gourara region between Adrar and Timimoun.

Table 1 Greenhouse cucumber compared to other vegetable products in Aougrout (2015/2016)

	Area (ha)	(%)	Production (Qx)	%
Cucumber	27,50	**63,0**	23.165	**77,8**
Other market garden products	16,15	37,0	6.594	22,2
Total	43,65	100,0	29.759	100,0

Source Reconstitution of the data of the subdivision of agriculture, Aougrout (Adrar)

People from Sétif had already been in direct contact with the traders in the region for the transport of food products. Despite the unavailability of reliable data from the Wilaya of Adrar, our study estimates the greenhouse agricultural area cultivated at more than 70 ha in the community of Aougrout alone (Hadeid et al. 2018), and it continues to increase; currently, more than 2200 greenhouses are identified in Aougrout by satellite imagery. However, there is a danger that mining, use of fossil water and other unsustainable practices that support irrigation are significant and without reliable evaluation by authorities (Directorate of Hydraulics, ANRH). The number of wells drilled in the entire Adrar region exceeds 900, with 75% used for agriculture. Monoculture and the absence of fallow periods and mineral salt leaching cycles lead to a displacement of greenhouses and causes soil degradation cycles every 2–3 years; uncontrolled use of fertilizers also represents an enormous risk of contamination of aquifers.

The artisanal pivot (Photos 1 and 2) technique is another interesting application of social technology for irrigation. It has already been used, for about 20 years, in the Souf region (South-East Algeria), allowing the emergence of market garden production basins by mobilizing deep groundwater on the margins of oases. Based on the principle proposed by Akrich et al. (1988) that "adopting an innovation means adapting it; it needs to be transformed, modified according to the site where it is implemented", Souf craftsmen and farmers have been able to design a small irrigation system (less than 1 ha) that meets the needs of farmers who do not have financial resources or whose land is not cultivable for large pivots (between 30 and 50 ha), which in turn demand more expensive and technical resources for operation (Fig. 1). Having seen the success of this irrigation technique in the Souf region, farmers in the Aougrout region have begun to use it where plasticulture is the most widespread.

Photo 1 and 2 Handcrafted pivot used in Aougrout (Adrar). (Hadeid M. 2018)

Fig. 1 Size of the pivots used in the irrigated perimeters of Aougrout. *Source* Image Google Earth, 2016

Indeed, about ten farms have equipped themselves with this tool, as demonstrated by small pivots in the irrigated areas north of Aougrout. It is a very low price—a one-ha pivot costs 50 times less than a professional 30 ha pivot—lowers a major barrier to adoption. An area of about one hectare in the Souf region can be irrigated by a single artisanal pivot, the purchase price of which is €1000 (about US$1140) (Ould Rebai et al. 2017).

Drip equipment for one hectare costs about €2000 (US$2280) and a sprinkler kit to fully irrigate a hectare costs between €2000 (US$2280) and €5000 (US$5700), depending on the quality of the equipment. The more expensive systems require removal during tillage and harvesting to allow access, unlike the artisanal pivot that can be left in place.

The sufficiency of the artisanal pivot irrigation system for potato production, with its low cost, has resulted in the rapid expansion of cultivated areas.[7] But the question remains about its impact on natural resources (soil depletion and lower yields, nematode[8] invasion and spread of phytopathogens,[9] overexploitation and groundwater pollution), as experienced in the Souf region, where its use has been intensive (Ould Rebai et al. 2017). It is understandable that smallholder farmers consider short term revenue a priority over long term environmental impact, including water resilience and saving the long term sustainability of agricultural development. In fact, the artisanal pivot irrigation system is inconsistent with the National Climate

[7]In the Souf region, 35% of potato production is provided by this method of irrigation.

[8]Nematodes are very small worms, most of whose species are invisible to the naked eye. Many species live freely in the soil, and parasitize many plant species, without causing them great harm, as long as the plants are not exposed to too much stress.

[9]A phytopathogenic agent is a living or almost living organism (bacterium, virus, fungus, etc.). that can infect plants and trigger diseases. Phytopathogenic agents belong to all pests of crops alongside pests and weeds.

Plan (NCP 2018), National Land Use Plan (SNAT 2010) and the local-scale Wilaya d'Adrar Development Plan (PAW 2015).

Inevitably oasis agriculture will thrive as modern agriculture thrives generally. For example, a good part of the existing palm groves in the Adrar region is maintained by members within the same oasis community, where individuals contribute according to their experience (education, administration, trade, etc.) and work the grove in addition to their primary livelihood, in the evening and on weekends, including children and women.

Apart from the broadly-disseminated practice of market gardening, oasis people dedicated themselves to tomato cultivation after the reopening of the Reggane factory. The field sales contracts with farmers throughout the Wilaya allows the company to guarantee the marketing of their products. The government of the Wilaya of Adrar has even implemented an awareness campaign targeting farmers to drive massive production of industrial tomatoes, with a guarantee of purchase by the factory. In Tilouline (a large ksar located in the commune of In-Zeghmir), farmers report that the tomato harvest lasts almost 4 months, a significant contribution to the income of the oasis population there.

Tobacco cultivation is beginning to gain momentum in several palm groves in Touat, a speculative crop in a region where the populations of Sub-Saharan African origin have increased. Tobacco is in high demand for local use, and can be exported to the Sahel countries. Tobacco growing is popular because it is profitable: a kilo of semi-processed tobacco can sell for 600 to 1000 DA (Hadeid et al. 2018). Our field missions over the last 5 years have enabled us to see the progressive extent of this crop in terms of land use. "Dried tobacco leaves are marketed in their raw state. In most cases, they are used to make snuff using traditional formulas. The contribution of tobacco cultivation to local agricultural income is important, although harvesting and processing remain traditional and artisanal" (Touzi1 and Merzaia-Blama 2008). The revival of traditional oasis agriculture has contributed to the stabilization of a significant number of families in the oasis population, but questions remain about the sustainability of their crops and methods, which are more demanding of irrigation water than the foggara can provide.

6 Solutions to Ensure Irrigation Water: New Challenges, New Effects

Oases are one of the driest environments in the world, where rainfall is less than 40 mm/year. Moreover, heat waves have increased in frequency in the last 50 years (1970–2019). However, this vast, arid space contains huge underground, though non-renewable, water reserves. The underground reserves are created by nappe, known as "Continental Intercalaire" or "Albian nappe", a fossil reservoir that covers an area of 1,100,000 km^2 over Algeria, Tunisia and Libya. This aquifer covers the entire northern Algerian Sahara, where access to water between 120 and 150 m has enabled

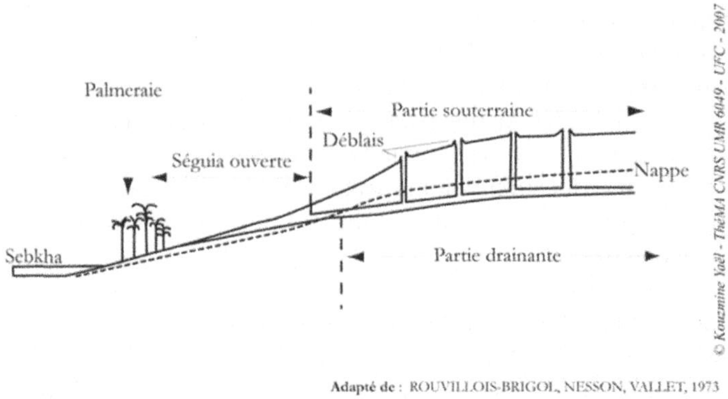

Fig. 2 Sketch explaining how a foggara works

irrigation in the Adrar region. The exploitation of this aquifer has been continuous over nearly 11 centuries by inhabitants of the oases, using an ingenious technique called "foggara": the collecting of groundwater that flows by gravity from below the plateau (Fig. 2).

The water is transported underground to the land to be irrigated. The tunnels that transport the water are perforated by vertical shafts for access and ventilation and have four major elements: the kasria (distributor), seguia (canal), madjen (collection basin) and guemoun (garden) (Fig. 3). However, maintenance of the foggara suffered in the post-independence period with insufficient financing by the central government, complicated by social traditions of servility in the ksour in addition to disregard for the foggara as a valid traditional irrigation technique. This explains much of the distrust between landowners and the government (Bendjelid et al. 1999).

The people of Tilouline (a large ksar located in Touat) offer another example of a local community collectively developing a strategy to provide as much palm grove irrigation as possible despite a diminishing water supply from the foggara (Fig. 4).

Ksourian population diverted drinking water intended for households in order to extend the plots[10] under cultivation. Ksourian households pay a flat quarterly charge of 1500 DA (US$13) for drinking water, so Oasis farmers take advantage of the fixed-rate by installing a long pipe[11] from their houses to their gardens (Photos 3, 4, 5, 6, 7 and 8).

It is striking that although the network of pipes required for the diversion is apparent, indicating that local authorities are aware of the "illegal piracy" of drinking

[10]Since it's difficult to define in detail the plots irrigated by drinking water and the houses affected by this operation, it can still be estimated that more than 40% (96 hectares) of the palm grove of Tilouline is partially irrigated by drinking water.

[11]These pipes can reach 300 m in some cases.

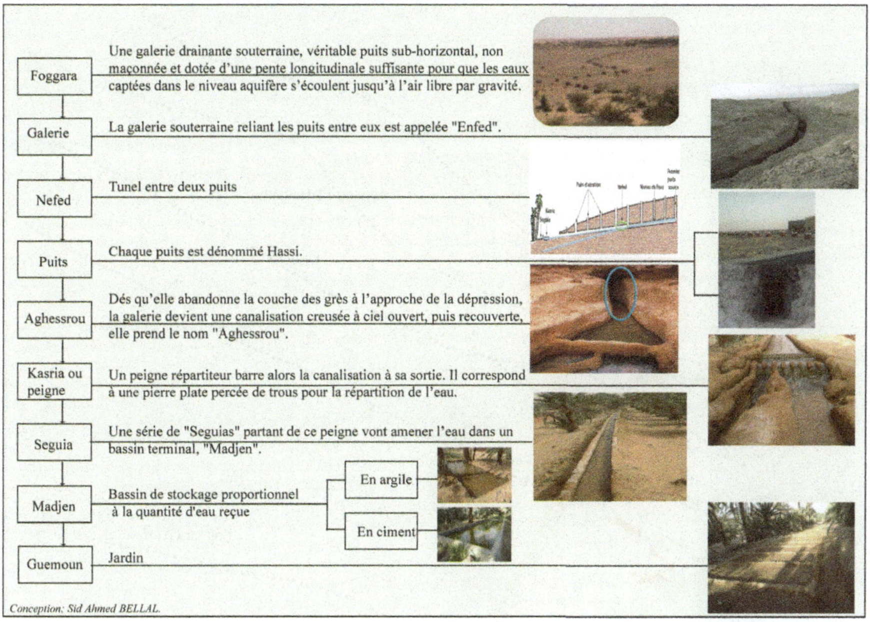

Fig. 3 Components of the hydraulic system of the foggara

water, they turn a blind eye to the practice as long as it strengthens agricultural activity. The question is whether this can be considered a "social technology" or simply a form of adaptation to the shortage of irrigation water because neither the foggara nor the wells drilled to reinforce them supply sufficient water to the oasis? (Fig. 5).

7 Conclusion

In an environment as naturally arid as the Algerian Sahara, it is difficult to attribute its transformation to climate change. Clearly, government policy in the past decade (2007–2019) has had the most immediate effect on water and its management in the oases. Faced with a gradual disengagement of the government, local people have developed strategies to revive their agricultural activity in response to their immediate challenges: poverty, driven by the decrease of state subsidies. These strategies are collectively known as "social technologies" because they respond to specific needs for which the government has declined to take responsibility. Defined as the use of technology for social action to mitigate inequalities and create social cohesion, Oasis society has readily embraced this approach. While often ingenious, these strategies tend to ignore the long term issues of sustainability in favor of short-term survival. Thus they result in excessive pumping of non-renewable water from the Albian groundwater, health risks and soil degradation. It is therefore imperative that local

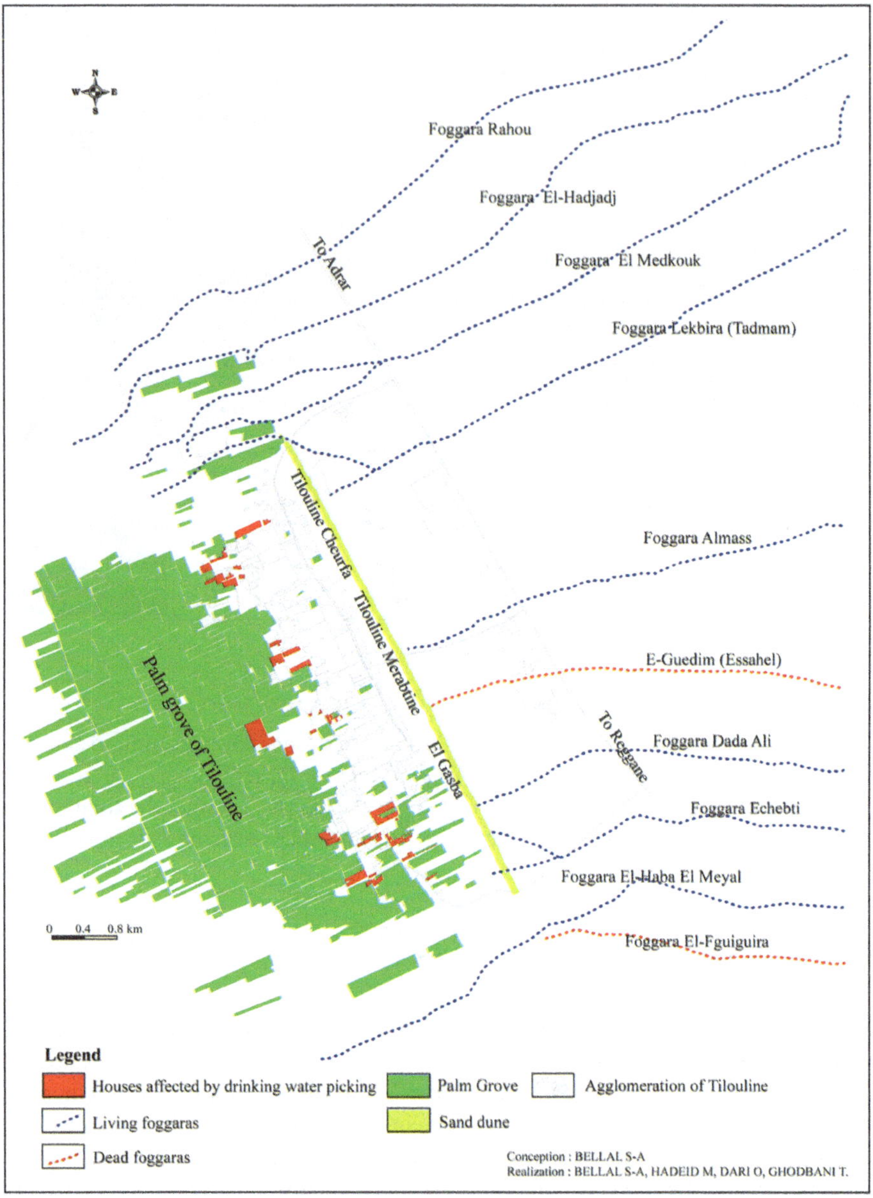

Fig. 4 Location of houses affected by the illegal connection to the drinking water network

Photos 3 to 8 Some landscape aspects of the diversion of drinking water for irrigation in the Tittaf and Tilouline palm groves. (Hadeid M. 2018)

strategies be integrated with the National Climate Plan, which will enable the use of renewable energy in water pumping, cultivation of genetically adapted local seeds, and adoption of complementary approaches to integrate traditional oasis agriculture and modern development.

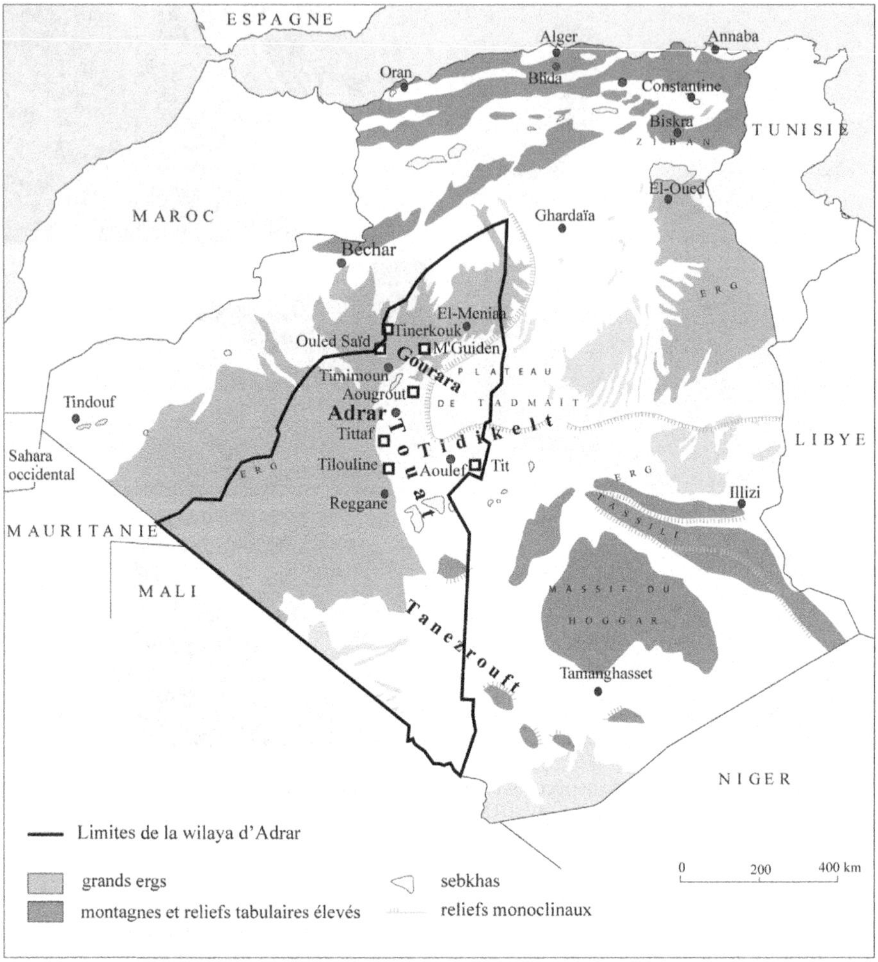

Fig. 5 Geographic location of the study area

References

Akrich M, Callon M, Latour B (1988) À quoi tient le succès des innovations ? 1: l'art de l'intéressement; 2: le choix des porte-parole, Gérer et Comprendre. Annales des Mines, pp 11–12

Bendjelid A, Hadeid M, Bellal SA, Gacem F, Belmahi MN, Hani S (1999) Mutations sociales et adaptation d'une paysannerie ksourienne du Touat: Ouled-Hadj-Mamoun (wilaya d'Adrar, Algérie), Insaniyat, n°7 (vol III, 1), CRASC d'Oran, pp 39–53

Bensaâd A (2011) *L'eau et ses enjeux au Sahara*. Iremam, Karthala, 242 p

Bessaoud O (2002) L'agriculture algérienne : des révolutions agraires aux réformes libérales (1963–2002). In Pierre B (dir.), Du Maghreb au Proche-Orient : les défis de l'agriculture. Paris, L'Harmattan: pp 73–99

Bisson J (1996) Paysanneries du Sahara maghrébin: dynamiques locales et politiques de développement. In Fontaine J (eds). Sociétés sahariennes, entre mythes et développement, Cahiers d'URBAMA, Cahiers de l'Université Ouverte, Tours Besançon, pp 63–80
Bisson J (2003) Le Sahara: mythes et réalités d'un désert convoité. Paris, L'Harmattan, 479 p
Côte M (2002) In Le Sahara, cette « autre Méditerranée » Méditerranée, tome 99 ed by Fernand Braudel, Des oasis aux zones de mise en valeur: l'étonnant renouveau de l'agriculture saharienne, pp 5–14
Coutinho A (2015) Organisation territoriale et technologies sociales d'alimentation en eau: politiques publiques et pratiques pour l'accès à l'eau dans le semi-aride de la Paraeïba (Brésil). Thèse de doctorat en Géographie. Université du Maine, 295 p
Despois J, Raynal R (1967) Géographie de l'Afrique du Nord-Ouest. Payot, Paris, p 570p
Ghodbani T, Dari O, Bellal SA, Hadeid M (2017) Entre perte de savoirs locaux et changement social: les défis et les enjeux de la réhabilitation des foggaras dans le Touat, Sahara algérien, Autrepart 81, 91–114. https://doi:10.3917/autr.081.0091
Hadeid M (2011) La politique de mise en valeur agricole en milieu steppique algérien: un essai de bilan dans les Hautes Plaines sud-oranaises (Algérie). Insaniyat 51–52:99–118
Hadeid M, Bellal SA, Dari O, Ghodbani T (2018) L'agriculture au Sahara du sud-ouest algérien : entre développement agricole moderne et permanences de l'agriculture oasienne traditionnelle. Cah Agric 27. https://doi.org/10.1051/cagri/2017060
Hominal J (1982) Technologies sociales en quartiers nouveaux, in Les Annales de la recherche urbaine, D'un quartier l'autre, 17:70–78
Kassah A (1980) In Actes du colloque Gestion des ressources naturelles et développement durable des systèmes oasiens du Nefzaoua, 25–27 février 2009, Douz, Tunisie, Oasis et aménagement en zones arides, Enjeux, défis et stratégies, pp 1–6
Kouzmine Y (2007) Dynamiques et mutation territoriales du Sahara algérien, vers de nouvelles approches fondées sur l'observation, Thèse de doctorat, Université de Franche-Comté, Besançon, France, 423 p
Marouf N (ed). (2010a) L'eau, la terre, les hommes: Passé et présent des Oasis Occidentales (Algérie). Paris, Balzac, 198 p
Marouf N (ed) (2010b) Lecture de l'espace oasien. Paris, Sindbad, 281 p
Ould Rebai A, Hartani T, Chabaca MN, Kuper M (2017) Une innovation incrémentielle: la conception et la diffusion d'un pivot d'irrigation artisanal dans le Souf (Sahara algérien). Cah. Agricul https://doi.org/10.1051/cagri/2017024
Patureau C (2010) Innover pour répondre aux besoins sociaux, Thèse professionnelle, Mastère Spécialisé HEC-Entrepreneurs, Paris, 97 p
Popp H, Benchrifa A (1992) L'oasis de Figureuig, persistance et changement. Passau, 109 p
Richter M (1995) In Les Oasis au Maghreb Mise en Valeur et Développement. Les oasis du Maghreb. Typologie et problèmes agro-écologiques, pp 29–56
Schmit T (2008) in Les pays du Maghreb, Contributions de la géographie humaine allemande, Société allemande de Géographie, Protection du patrimoine culturel et transformation socioculturelle: La vallée du M'zab, pp 38–49
Small AW (1998) Notes du séminaire: la méthodologie du problème social. Division I, Les sources et les utilisations du matériel. Amer J Sociol 4 (1):113–144
Stewart J (2007) in the domestication of information and communication technologies. Inform Commun Soc Local Exp:547–69
Suter K (1959) In Mitteilungen der Österreichischen Geographischen Gesellschaft, Seite, Die Palmgärten der Mzab. Zur Anthropogeographie einer Region der nordalgerischen Sahara, pp 63–71
Touzi A, Merzaia-blama A (2008) In Revue des Energies Renouvelables SMSTS'08, Alger La conservation des denrées agro-alimentaires par séchage dans les régions sahariennes, pp 267–272

Chapter 12
Water Management Policy for Freshwater Security in the Context of Climate Change in Senegal

Cheikh Faye, Amadou Abdoul Sow, and Sidy Dieye

Abstract The impact of climate change on freshwater systems and its management have been ongoing for many years and is projected to intensify as temperature, sea level and rainfall variability increases. Senegal's water resources are now threatened by human activity of various origins and by the harmful effects of climate change. This chapter addresses the extent to which water management policy ensures freshwater security in the context of climate change. Given the scale of water scarcity, this study is limited to characterizing the relationship between water management policy and freshwater security in the climate change context, focusing on data and information from institutions with authority for water policy and minimization of water scarcity. The discussion includes a case study on the impact of climate change on water resources in the Gambia River Basin, which indicates that while there is a great deal of attention paid to freshwater security policy, there is no consensus on how to mitigate the effects of climate change to ensure water security. Water and climate change are managed by different ministries, resulting in predictable conflicts of authority and responsibility. Thus, it is necessary for the government to establish an effective water policy, with clear, sustainable strategies to mitigate the impacts of climate change for the sustainable freshwater security of the country.

Keywords Climate change · Water security · Water policy · Water management · Water scarcity

C. Faye (✉)
Département de Géographie, UFR Sciences et Technologies, UASZ, Laboratoire de Géomatique et D'Environnement, BP 523 Ziguinchor, Senegal
e-mail: cheikh.faye@univ-zig.sn

A. A. Sow · S. Dieye
Département de Géographie, Faculté Des Lettres et Sciences Humaines, UCAD, BP 5005, Dakar-Fann, Senegal

S. Diop et al. (eds.), *Climate Change and Water Resources in Africa*,
https://doi.org/10.1007/978-3-030-61225-2_12

1 Introduction

Freshwater is essential for the survival of natural ecosystems and for human activity. Until the mid-twentieth century, water was considered inexhaustible and accessible for anyone to use, own and exploit according to their needs (Honegger and Tabarly 2011). These needs vary by economic sector, but all economic development and industrial activity depend on water. In fact, water withdrawals have increased six-fold since the 1900s, twice the rate of population growth (UNESCO 2006). This has led to competitive use of water resources among and within sectors, which is a root cause of many challenges such as political and territorial conflicts, deterioration of water quality, degradation of the environment, disparities in water distribution and a decline of access to water resources (Batcho 2008).

Senegal's total renewable water resources are about 38.97 km^3 per year, of which about 25.8 km^3 come from within the country. Per capita renewable water volume is estimated at about 4747 m^3 per year (CONGAD 2009), well below the average of sub-Saharan Africa (7000 m^3 per person per year) and the world average (8210 m^3 per person per year). Senegal's internal renewable surface water resources are estimated at 23.8 km^3/year and renewable groundwater resources are on the order of 3.5 km^3/year. The overlap between surface water and groundwater is estimated at 1.5 km^3/year and internal renewable water resources are estimated at 25.8 km^3/year. The average annual rainfall is estimated at 686 mm (FAO 2016). The relative abundance of water and the fact that available freshwater supplies are 24.5 times greater than water consumption (1591 million m^3 in 2000) (FAO 2016) hide the real potential for water scarcity in Senegal.

The Senegal, Gambia and Kayanga rivers, all of which originate in Guinea, together irrigate a large part of Senegal. Most surface water reserves are located in the Senegal and Gambia river basins, whose waters come from the Fouta Djallon Massif in the Republic of Guinea (Sané 2015). Smaller rivers, the Casamance and the Kayanga, with intermittent flows, and their main tributaries, the Anambé, Sine and Saloum and coastal creeks (Fig. 1), complement the water supplied by the large rivers. A number of other lakes and basins complete the hydrographic network, the most important of which are Guiers Lake, the bolongs of estuaries and the bounds of the Niayes region of the northern coast and Ferlo. Available water in any area depends on the amount of surface and groundwater available from rivers, lakes, reservoirs and aquifers (Chitambi 2017).

Water scarcity is defined as insufficient water resources to meet long-term average needs. Variation from the norm results in long-term water imbalances when water availability from natural recharge is lower than the demand level (EU-Commission 2007). This, in turn, results in a shortage of rapid and reliable access to water resources for different economic and social groups. The risk of water scarcity and the potential to pre-empt or effectively address it depends not only on the physical endowment of water but also on the institutional, technical, political, economic and infrastructure capacity of a country to manage water to the satisfaction of all stakeholders (Chitambi 2017).

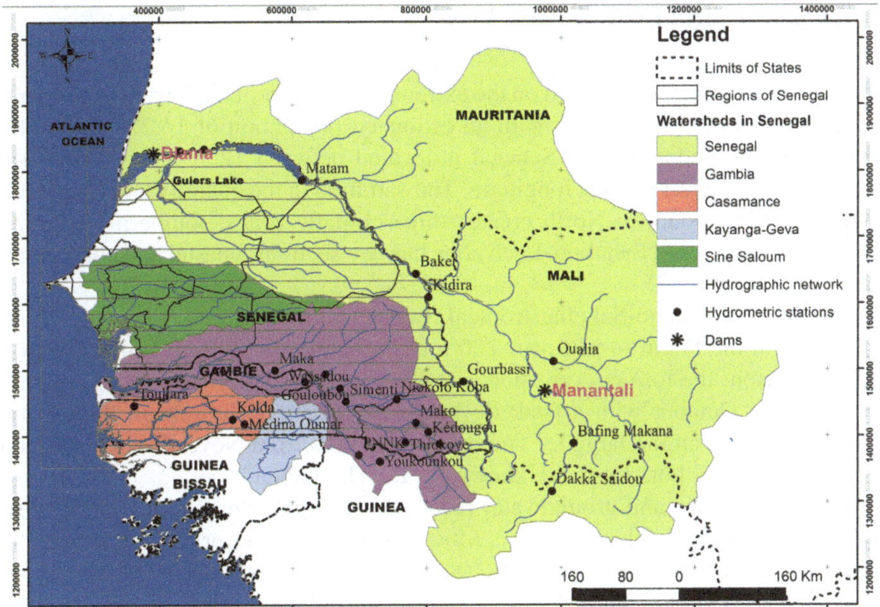

Fig. 1 Watersheds draining Senegalese territory (*Source* DGPRE)

The International Water Management Institution (IWMI 2007), classifies Senegal as a country with economic water scarcity. Although water resources are abundant, they are not effectively applied to the country's development goals because of limited institutional, technical and financial capacity. Water scarcity is driven by variation in rainfall, the vulnerability of water resources, disparity in spatial distribution, current and/or threatened conflicts caused by water exploitation and over-exploitation, degradation of the volume and quality of available water, inequity of access among and within zones and social groups, and various socioeconomic activities (MH/DGPRE 2007).

Good governance and accountability, with a view of the interests of all stakeholders within an appropriate legal and institutional framework, must be properly planned, exploited and maintained for effective infrastructure and capacity development (UN-Water 2011). Every aspect of security worldwide—for example, food, economic and health—depends on water (Boge 2006).

To improve water management, Senegal is working to comply with the recommendations of the World Summit (Rio-Dublin, January 1992 and Johannesburg, August 2002). Senegal's water policy as it applies to achieve the Sustainable Development Goals (SDGs) is managed by the Directorate of Water Resources Management and Planning (DGPRE).

2 Description of the Study Area

The Republic of Senegal, located on the extreme western tip of the African continent, covers an area of 196,722 km^2 with an estimated population of 16,200,000. With 700 km of Atlantic coastline, Senegal is located between 12.5° and 16.5° North latitude and 12° and 17° West longitude. The variation in the quantity and frequency of rainfall from South to North creates three climatic zones: southern Sudanian, northern Sudanian and Sahelian. Each zone has two sections: coastal and continental zone (Faye et al. 2017).

Senegal has a Sudano-Sahelian tropical climate with annual rainfall between about 1,250 mm in the South to just over 200 mm in the North. There is a rainy season and dry season: the former corresponds to the monsoon period, extending roughly from June to October with a peak in August-September. The rain in this period varies depending on the latitude. The dry season lasts from November to June and is characterized by hot, dry wind, known as "the Harmattan."

Senegal's surface and groundwater resources are rich, with heterogeneous soils and a dense hydrographic network. Most surface water reserves are located in the Senegal and Gambia river basins (Sané 2015). Groundwater consists of four major aquifer systems: shallow, intermediate, deep and basement (CONGAD 2009). Senegal currently has sufficient water resources to meet demand, but it is threatened by drought and climate change.

3 Methodology

3.1 Characterizing Climate Change Impacts on Water Resources

3.1.1 Context

Climate change is defined by trends in the global mean sea and atmospheric temperatures, variation and quantity of rainfall, and increase of the magnitude and frequency of natural disasters and extreme events (drought, floods, etc.) (IPCC 2007). Global warming, combined with increased variability of rainfall, is driving an upsurge in extreme events, including floods and low flows, which in turn will increase in frequency and in intensity across the African continent. Various studies reveal the evolution of flows in rivers and their impacts on natural and human ecosystems, subject to the hydro-climatic conditions in various regions (Faye et al. 2015a).

The African continent is among the most impacted by climate fluctuations on water resources (Kanohin et al. 2009). Several studies carried out in West and Central Africa have shown a decrease in surface and groundwater flows since the 1970s as a result of a decrease in rainfall (Dione 1996; Sow 2007; Faye 2013; Faye et al. 2015b), but an increase in flows (Niang et al. 2008; Ali and Lebel 2009; Ozer et al. 2009;

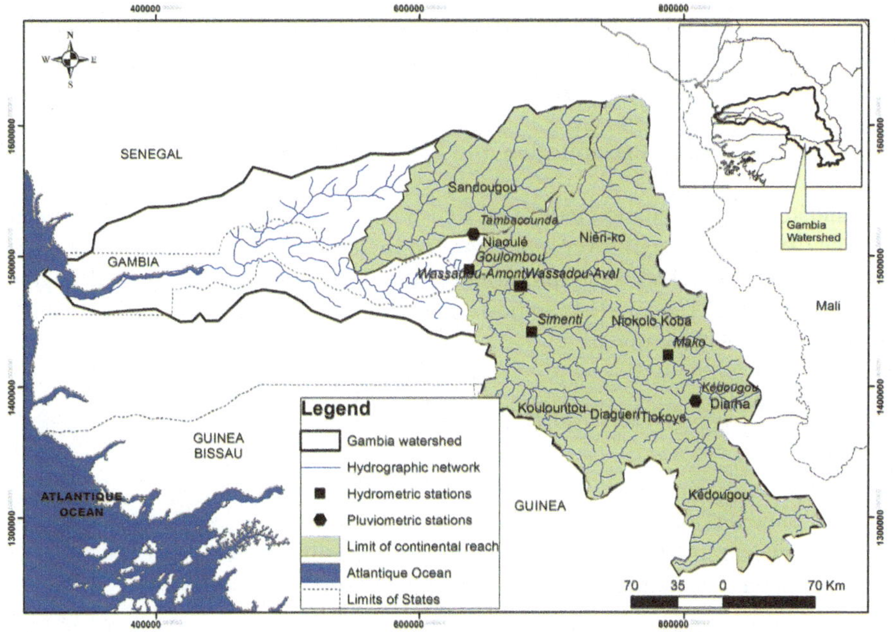

Fig. 2 Gambia River Basin situation

Ouoba 2013; Bodian 2014) since the 1990s, suggesting an improvement of water balance in this period.

The Gambia Basin covers an area of nearly 77,100 km, shared between three states (Lamagat 1989): Guinea (where it originates at 1125 m. altitude around Labé); Senegal (from which it drains almost all the Tambacounda region and part of Upper Casamance and southern Sine-Saloum); Gambia (of which it is the backbone). It extends, in latitude, from 11°22 North (in Fouta-Djalon) to 14°40 North (in the south-eastern Ferlo) and, in longitude, from 11°13 West (Fouta-Djalon) to 16 "42 West (in Banjul at the mouth). It is of interest to the territories of Guinea, Senegal and Gambia. It is divided into two distinct zones: the continental basin and the estuary basin, respectively, east and west of Gouloumbo, the last station where the flow of freshwater inputs is measured (Dione 1996; Sow 2007) (Fig. 2).

3.1.2 Data

Rainfall and high- and low-temperature data were measured at Tambacounda and Kedougou stations from 1970 to 2012 to represent climate change in the Gambia River Basin. Data from the hydrological stations at Gouloumbou, Mako, Simenti, Wassadou Upstream and Wassadou Downstream from 1970 to 2012 indicate changes in surface water in the Gambia River Basin. Groundwater measurements for the basin were taken from Mako station over the same years. The selected stations followed consistent criteria for each parameter measured to ensure the duration of available information and data quality.

3.1.3 Standardized Precipitation and Flow Rates Indices

A standardized precipitation index (McKee et al. 1993; Hayes 1996) was developed to quantify rainfall deficit over time, adopted in 2009 by the World Meteorological Organization (WMO) as a global instrument for measuring meteorological droughts. According to the "Lincoln Declaration on Drought Indices" (Jouilil et al. 2013), it is expressed mathematically as:

$$SPI(SFI) = \frac{(X_i - X_m)}{S}$$

where

X_i = annual rainfall (or flow) i

X_m = mean rainfall of series over the considered time scale

S = standard deviation of series over the considered time scale

This standardized flow rate index (SFI) is similar to the index used for hydrology and has been developed to quantify the water deficit over time to reflect the impact of drought on the availability of different types of water resources for a given period (Sharma and Panu 2010).

3.1.4 Depletion Coefficient and Water Volume Mobilized by Aquifers

The depletion of water from all the aquifers of the basin is an important factor in the tropical hydrological regime (Briquet et al. 1995). The calculation of the depletion coefficient is based on the Maillet model (Sow 2007; Faye 2013; Faye et al. 2015a), which demonstrates the degree to which rivers are drying up. Maillet's model shows that, by factoring out precipitation, depletion corresponds to the exponential decay of flow as a function of time. This corresponds to the emptying of groundwater as the only contribution to the water flow of a basin. The Maillet model is represented by:

$$\mathbf{Q_t = Q_0 e^{-\alpha(t-t_0)}}$$

where

Q_0 is the initial flow at time t_0, $(t-t_0)$: the time expressed in days between the observation of the flow rate Q_0 and that of the flow rate Q_t (flow at the end of the dry period)

α is the dry-out coefficient

The volume of water mobilized by aquifers is represented by:

$$V_{\text{mobilized}} = \frac{Q_0}{k}$$

The Maillet model makes it possible to determine the temporal evolution of depletion coefficients and water volumes mobilized by aquifers in the Gambia watershed, and to assess the duration of the drying up of rivers under the effect of climate change, enabling the use of methods based on trend detection and break in series.

3.1.5 Pettitt and Mann–Kendall Tests

The Pettitt test (1979) represents a break at an unknown moment in series from a formulation derived from that of the Mann–Whitney test. This test represents differences in values that make up the sample, corresponding to a resulting time series. The calculation of the p-value indicates whether or not the threshold break is statistically significant.

The Mann–Kendall test represents the linear trend (up or down) within a time series. This tendency test was first studied by Mann (1945), then Kendall (1975), then improved by Hirsch and Slack (1984). The test has been validated by several comparison studies by Yue and Wang (2004).

3.2 Characterization of Water Policy in a Climate Change Context

Data and information from various institutions were selected based on the institution's mandate for water policy, security/water management and/or climate change. Among the institutions and ministries from which data were derived are the Ministry of Hydraulics and Sanitation, the Ministry of Environment and Sustainable Development, the Directorate of Management and Planning of Water Resources, the National Agency of Civil Aviation and Meteorology, the National Agency of Statistics and Demography, the Organization for the Development of the Senegal River and the National Water Company of Senegal. Some semi-structured interviews of key players were conducted to complement policy analysis. Data were sourced mainly from the documentary/policy reviews of selected institutions. Institutional documents, progress reports and institutional water resource management libraries were used to gain a deeper perspective on water policy and water security in the context of climate change. The analysis focuses on water security, water resource management and climate change.

4 Results and Discussion

4.1 *Characterization of Climate Change Impacts: The Gambia Watershed*

4.1.1 Temperature and Precipitation Trend

Trend shifts and/or breaks in annual temperature and precipitation are used to characterize climate change in the Gambia River Basin. Table 1 shows the results of the Pettitt and Mann–Kendall tests using data for minimum, maximum and average temperatures from Kédougou and Tambacounda stations for 1970–2012. Both tests indicate the presence of a break and/or trend: for example, in 1987 in Kédougou and in 1990 in Tambacounda for T_X and T_N. These breaks are confirmed by the Mann–Kendall test, which shows positive and significant Kendall τ of 0.511 °C for T_X and 0.3321 °C for T_N at Tambacounda. At Kedougou station, Kendall τ, although not significant, is positive: 0.0171 °C for T_X and 0.7261 °C for T_N.

To quantify the variation of temperatures through the break date, we cut the time series into two subperiods, before and after the break dates. A comparison of the two sub-periods shows that the latter one reveals a surplus of 2.61% for T_X and 1.35% compared to the earlier one for T_N at Tambacounda.

Neither the Pettitt nor the Mann–Kendall tests show a significant break or trend in precipitation, with the p-values for both tests >0.01. The trend in precipitation rises slightly after 1994 and breaks are noted in 1994 and 2003 in Tambacounda and Kedougou, respectively. Thus, indices are generally negative from 1970 to 1994 and positive between 1994 and 2012, as illustrated by positive Kendall τ with −0.074 mm at Tambacounda and 0.098 mm at Kédougou. These suggest that, in addition to the

Table 1 Results of annual temperature and precipitation tests (1970–2012)

			Mann–Kendall test			Pettitt test	
			p-value	τ of Kendall	Slope of Sen	Date of break	% surplus or deficit
Tambacounda	Temperature	TX	<0.0001	0.511*	0.035	1990	2.61
		TN	0.958	−0.006	0	1979	−0.65
		TM	0.002	0.332*	0.014	1990	1.35
	Annual precipitation		0.492	0.074	2.158	1994	13.3
Kedougou	Temperature	TX	0.017	0.259	0.015	1987	−2.61
		TN	0.141	−0.167	−0.019	1987	−4.27
		TM	0.726	0.041	0.002	1987	−0.76
	Annual precipitation		0.359	0.098	2.257	2003	21.0

(−): negative trend; (+): positive trend; (*): significant trend; TX = maximum temperatures; TN = minimum temperatures; TM = average temperatures drought of the 1970s as confirmed by multiple studies (Sow 2007; Faye 2013; Faye et al. 2015b), another important change in rainfall pattern is observed at the turn of the twenty-first century as indicated by studies (Ali and Lebel 2009; Ozer et al. 2009; Ouoba 2013; Bodian 2014) that suggest the end of Sahelian drought during the 1990s. Thus, in the year before and year after the break, rainfall increased by 13.3% in Tambacounda and 21% in Kédougou.

4.1.2 Impacts of Climate Change on Surface Waters in the Gambia Basin

The impact of climate change on surface waters in the Gambia River Basin have been analyzed using standardized flow indices. Deriving these indices from different stations allows better comparison of station data as basins of different sizes drain. The indices subjected to Mann–Kendall and Pettitt tests reveal significant fluctuations with multiple consequences for the environment, making them worthy of study. The analysis shows that there is an upward trend and a break in 1994 (Table 2) at the five stations. However, neither the trend nor the break was significant at the 1% level. Thus, distinct periods are established before and after 1994. The first one, from 1970 to 1994, shows a negative outflow trend relative to the drought of the 1970s (Faye et al. 2015b). The second, from 1994 to 2012, although not significant, shows a rising trend and corresponds to an improvement in rainfall conditions starting in the 1990 s (Ozer et al. 2009). This upward trend in a runoff in the Gambia River Basin is consistent with the findings of Ali and Lebel (2009) in the Sahelian zone, Ouoba (2013) in Burkina Faso, Ozer et al. (2009) in Niger, Niang (2008) in Mauritania and Bodian (2014) in Senegal.

The trend analysis indicates that this variability is synchronized with two hydro-climatic periods: a dry period between 1970 and 1994, and a wet period between 1994 and 2012. Thus, beyond the hydrological drought of the 1970s, a new hydrological

Table 2 Results of Pettitt and Mann–Kendall runoff tests analyzed in the Gambia River Basin (1970–2012)

	Mann–Kendall test			Pettitt test		
	p-value	τ of Kendall	Slope of Sen	p-value	Date of break	% surplus or deficit
Gouloumbou	0.611	−0.062	−0.010	0.651	1981	23.6
Mako	0.479	0.094	0.020	0.025	1994	48.8
Simenti	0.241	0.154	0.030	0.011	1994	64.7
Wassadou upstream	0.710	−0.052	−0.008	0.564	1979	51.5
Wassadou Downstream	0.065	0.227	0.026	0.002	1994	44.6

(−): negative trend; (+): positive trend; (*): significant trend change occurred again in the 1990 s, with river flows rising. We believe that, on an annual scale, rainfall measured at the stations is quite similar to the date of the break in 1994. However, the magnitude of the surplus is smaller (23.6%) in Gouloumbou than at other stations where it exceeds 50% (64.7% in Simenti, 51.5% in Wassadou and 48.8% in Mako).

Flow changes in the Gambia River Basin are caused by various upheavals, including climate change. One of its impacts was the severe drought of the 1970s (Sow 2007; Ozer et al. 2009; Faye et al. 2015b). On the other hand, during the 1990s, the improvement in rainfall conditions (increase in rainfall) led to an increase in runoff observed at the various stations.

4.1.3 Climate Change Impacts on Groundwater in the Gambia Watershed

While accumulated rainfall deficits during the 1970s resulted in diminished underground reserves of river basins (Briquet et al. 1995), the increase in flow in the Gambia River Basin has increased the volumes drained by groundwater. The natural low water level in the Sudano-Sahelian rivers are directly affected by changes in climatic conditions. This depletion is typical of all the aquifers in the basin and is an important characteristic of the Gambia River Basin. Depletion data are provided in Table 3 and show that variation at Mako station is particularly dramatic.

Depletion coefficients show that before 1994, the year of break, according to the Pettitt test (Table 3), values are fairly steady. Drying coefficients are generally low and range from 0.024 (in 2011–2012) to 0.076 (in 1977–1978) in Mako. In contrast, the 1994–2012 coefficients average 0.044, and correspond to the highest water table

Table 3 Results of the Pettitt and Mann–Kendall tests on drying variables in the Gambia River Basin at Mako station (1970–2012)

Mako	Mann–Kendall test				Pettitt test			
	p-value	τ of Kendall	Significance Threshold (between 10% and 1%)	Sensitivity of trend	p-value	Date of break	Significance Threshold (between 10% and 1%)	% deficit or surplus
Q0	<0,0001	0,44	Presence of trend	Rise	0,0001	1986	Presence of break	102
Qt	0,07	0,19		Rise	0,0105	1994		285
t	0,06	0,20		Rise	0,004	1985		9,1
K	0,29	−0,11		Decline	0,10	1994		−15
V m^3/an	<0,0001	0,50		Rise	<0,0001	1993		115

Q_0: flow at the beginning of the dry period; and Qt: flow at the end of the dry period; t: the number of days; α: the drying coefficient; V m^3 / year: the support volume of the slicks volumes. Over this period, the average support volume for aquifers is 6531 m^3/year, with a maximum of 16,454 m^3/year (in 2012–2013) in Mako.

Table 3 indicates that starting from the 1994 break, the rise in flows corresponds to a real drop in the depletion coefficient, as shown by Kendall's τ on the order of -0.11 in Mako. This decrease in the coefficient resulted in an increase in the volume of groundwater contribution to the general flow of the Gambia River Basin, in turn resulting in a positive net Kendall τ of 0.50 (in Mako), part of a trend of the increase in groundwater contribution in recent période (1994–2014) in the basin (115% in Mako).

The decrease in the coefficient of dryness in the current dry period corresponds essentially to an increase in the extension and width of the groundwater tables in the basin. The 1994–2014 period, compared to the 1970–1993 period, is in surplus for the duration of the dry period and for support volumes, and in deficit for the coefficient of dryness.

4.2 Water Policy During Climate Change

4.2.1 Water Management Policy and Freshwater Security

Water policy as it relates to freshwater security is addressed in the sectoral water policy of 2005. In response, programs and projects (National Water Partnership of Senegal in 2002; PAGIRE in 2007) were implemented in close collaboration with the responsible ministries to contribute to and protect the environment, including water resources. We believe that these efforts represent a strong commitment of the Government of Senegal to promote Integrated and Sustainable Water resource management through:

- improving knowledge and means of water resource management;
- creating an environment conducive to the application of Integrated Water Resources Management (IWRM) through legal, organizational and political reforms;
- improving communication, information, education and awareness about water resources, etc.

Table 4 lists laws regulating water use, pollution control and conservation as part of natural resources management. They complement water policy objectives for water resource management and development to enhance collaboration and cooperation among key stakeholders in Senegal. Water resource management and water use in Senegal are based on a combination of sectoral policy papers (urban and rural water policies), laws, decrees, and ministerial and inter-ministerial orders, circulars and codes on water, hygiene, environment, etc.).

Sectoral policy on water and sanitation was the dominant instrument for the implementation of the Millennium Drinking Water and Sanitation Program (PEPAM). After 2015, extending the progress made by PEPAM, a coordination and monitoring

Table 4 Water policy laws relating to freshwater security

Enabling legislation	Purpose of the water sector policy
Law No. 81-13 of March 4, 1981 on the Water Code	– Establish the protection and safeguarding of water resources, good water management (especially in the sanitary field); – Assure good planning of resources, their good management and equitable distribution between different uses and each according to their needs within the framework of strict respect of general interest
Law No. 2008-59 of September 24, 2008 on the organization of public service for drinking water and collective sanitation of domestic wastewater	– Organize modernization and rationalization, in the longer term, of public water and sanitation service, in order to face the challenges of the future and meet the needs of Senegal; – Delegate the public water and/or sanitation service to maintain the water and/or sanitation installations in good working order
Law No. 83-71 of July 5, 1983 on the Hygiene Code	– Ensure the protection of water quality; – Establish a protection perimeter to be respected around water intake points intended for human consumption; – Protect the works of the catchment, treatment, storage and elevation of water
Law No. 90-07 of June 26, 1990 establishing the SONES–PEPAM	– Realize conditions for a better division of roles and the profitability of sub-sector of urban hydraulics through a more adapted private management mode and a more assertive commercial dimension
Law No. 2001-01 of January 15, 2001, on the Environment Code	– Establish good management and protection of the environment, which is one of the concerns of the public authorities in Senegal; – Put in place a national policy for the protection of environmental resources
Law No. 2008-43 of August 20, 2008 on the Town Planning Code	– Manage the projects of drinking water supply and sewerage (rainwater and black water) of the districts; – Manage urban green spaces, including areas of wet depressions, urban planes and waterways
Law No 2009-24 of July 8, 2009 bearing the Sanitation Code	– Enable people to have adequate sanitation and access for all to the rule of law regarding sanitation in Senegal

(continued)

Table 4 (continued)

Enabling legislation	Purpose of the water sector policy
Decree No. 2013-1270 of September 23, 2013, on the powers of the Minister of Environment and Sustainable Development	– Ensure the protection of the environment (protection of nature, fauna and flora); – Fight against pollution; – Participate in the implementation of water and soil conservation policy through the construction of retention basins and artificial lakes
Decree No 2012-654 of July 4, 2012 relating to the attributions of the Minister of Hydraulics and Sanitation	– Ensure the supply of drinking water to populations in rural, urban and peri-urban areas; – Ensure the quality of water supplied to households and businesses; – Ensure the availability of water for the satisfaction of the needs of agriculture and livestock over the national territory
Law No. 2016-32 of November 8, 2016 on Mining Code on Mining Code	– Ensure the protection of the environment during exploration, extraction and abandonment of mines
Law No. 96-06 of March 22, 1996, on the Local Authorities Code	– Provide and maintain water supplies; – Characterize the regime and the methods of access and use of water points of all kinds; – Manage inland waters (excluding national and international rivers); – Ensure the protection of groundwater and surface water
Constitutional Law No. 2016-10 of April 5, 2016, revising the Constitution	– The guiding principles of state policy set a framework within which regulation and allocation of water can take place

unit for water and sanitation programs was created (DGPRE 2016; ANSD 2017). Senegal is committed to the IWRM approach as a strategic option consistent with the Africa Water Vision 2025 and other international water policies. However, a lack of adequate information on water availability limits the strategic planning of water resources and constrains efforts to regulate the development of surface water and groundwater.

The objective of water resource management is to conserve and maintain acceptable quality standards (Chitambi 2017). To achieve this level of water resource management, the following goals must be realized by the appropriate ministries to:

- reduce the impact of water-related disasters such as drought and floods by the implementation of early warning systems;
- create institutions for the management of shared watercourses within Senegal in collaboration with national institutions to ensure the protection of Senegal's interests;

- prevent water shortages by the collaborative implementation of early warning systems;
- coordinate regional and international organizations to respond effectively to emergencies.

Water resource knowledge and monitoring are essential to the development and implementation of good policy to preserve resources from all forms of pollution and degradation as population growth inevitably increases pressure on water resources. The population of Senegal is projected to increase to over 19,390,000 in 2025. A reliable water monitoring system that allows for input by all stakeholder constituencies is essential to build an understanding of water availability, abstraction and consumption in various contexts. Senegal has improved access to drinking water, but at the same time faces weak access to basic social services and reliable databases to support harmonized, comprehensive and sound policymaking.

The proportion of the population with access to drinking water in Senegal is estimated to have been 87.2% in 2015, compared to 84.1% in 2014. Even if Senegal had reached the water target set by the Millennium Development Goals (MDGs), geographic coverage of 58.3% (in 2016) combined with the new more ambitious objectives of SDG 6 mean that much greater progress must be made.

4.2.2 Water Management Policy and Climate Change

Although Senegal has sufficient water for its population, the water sector faces many challenges related to climate change which threaten access to sufficient and clean water and, therefore, the socioeconomic development of the country. Poor distribution of surface water in many parts of the country, particularly in the north, impedes wealth creation, poverty reduction and disease prevention, resulting in major challenges to environmental regulators, and ultimately to Senegal's economy, health, food and security.

According to the National Agency of Civil Aviation and Meteorology (ANACIM), Senegal is among the African countries most exposed to the negative effects of climate change, which is causing an increase in temperatures and frequency of extreme events (floods and droughts) and late-onset and/or early cessation of rainfall (resulting in shorter seasons and more intense rainfall). Adverse effects of climate change have impacted water reserves in marine waters in Casamance and Sine-Saloum, and have driven the drying up of the Ferlo and associated valleys, a general decline of groundwater levels, a decrease in water flow in continental rivers and other flood plains and the salinization of fresh water and cultivated land in coastal areas.

The climatic deterioration of recent years (from the years 2000 to now), combined with overexploitation of the resource, has led in places (West of the country) to a drop-in water table (sometimes with withdrawals exceeding the renewal capacity) and marked saline intrusions in the low valleys of the Sine Saloum, at the level of the deltas of the Casamance and Senegal rivers, as well as the Grande Côte (Niayes area). At the same time, in some areas, especially the Dakar region, surface water tables are polluted by discharges linked to sanitation deficits (bacteria, chemicals, heavy metals, nitrates). In these areas, the quantity of water available is not the problem, but rather water quality and the cost of its mobilization are the major concerns (Senegal National Blue Book Committee 2010).

Solutions have been proposed to reduce vulnerability and strengthen the adaptive capacity of populations and ecosystems. Ensuring freshwater security through more efficient water management will contribute to development objectives, adaptation to climate change and disaster risk reduction (AMCOW 2013). Climate disruption introduces new constraints and exacerbates those already faced by the Senegalese government and its water partners. Variables include water availability, variations in weather and extreme weather phenomena, and increasing uncertainty (World Water Council 2016).

Water is inextricably linked to climate, so global climate change and how it is managed worldwide has serious implications for water resources and regional development. Indeed, climate change impacts on water resources are already visible in the Sahel countries, including Senegal, in the forms of variation in the average and geographical distribution of rainfall, increase of evapotranspiration, recrudescence of periods of drought and heavy rainfall (World Water Council 2016) and recurrent droughts and floods. These resurgent effects have socioeconomic impacts which are weighing on the country financially, necessitating that the government implements urgent action to mitigate their effects and build resilience. These additional pressures, compounding the existing over-exploitation in many parts of the country, intensify the challenges ahead.

Senegal has implemented measures to mitigate the impact of climate change (Table 5), including awareness campaigns to ensure that the public is informed about climate change and related issues, such as mitigation and adaptation measures (Chitambi 2017), and adoption of new strategic planning instruments such as the National Action Plan for Environment (NAPE), the National Action Plan to Combat Desertification (NAP/TCD), the Senegal Forest Action Plan (SFAP), the National Strategy for the Implementation of the Framework Convention on Climate Change, the Program of Action on Biological Diversity, the Action Plan for the Protection of the Ozone Layer and the Hazardous Waste Management Plan (Environment Code). Senegal has also increased water supply strategically through a hydraulic mobilization policy, using dams and wells. Since 2015, Senegal has been actively implementing its National Adaptation Plan (NAP) and benefitted in 2016–2019 from the Scientific Support Project to the National Adaptation Plans processes (PAS-PNA).

Laws and codes are often essential for managing freshwater security in Senegal. The policy is developed through a decentralized process that increases local capacity that can be otherwise limited by the lack of financial and technical resources (Assetto et al. 2003). For water management to be effective and politically feasible, it must balance policy, economics, the environment and national security (Allen 2002). For example, Law No. 2001-01 (Environment Code), Law No. 81-13 (Water Code) and Law No. 2009-24 (Sanitation Code) remain to a large extent adequate to ensure the safety of the water. However, as circumstances evolve and weaknesses are identified, some codes are being revised in Senegal to adapt them to the current conditions.

Table 5 Water policy laws relating to climate change

Enabling legislation	Objective of climate change policy
Decree No. 2013–1270 of September 23, 2013 on the attributions of the Minister of Environment and Sustainable Development	– To establish the development of environmental education; – Manage a mechanism for monitoring trends in climate change and changes in the state of the environment; – Participate in international technical meetings dedicated to the protection of the environment, sustainable development, climate and biodiversity
Law No. 2001-01 of January 15, 2001 on the Environment Code	– Adopt new strategic planning instruments; – Put in place a national strategy for implementing Framework Convention on Climate Change; – Addressing desertification (land degradation in arid, semi-arid and dry sub-humid areas) as a result of various factors, including climatic variations and human activities; – Establish sustainable development that meets the needs of the present without compromising the ability of future generations to meet theirs; – Conduct environmental impact studies, such as effects on climate and atmosphere.
Law No. 81-13 of March 4, 1981 on the Water Code	– Deal with harmful situations related to water problems such as floods (and some floods) and droughts that are increasing with climate change
Law No. 2009-24 of July 8, 2009 bearing on the Sanitation Code	– Provide pumping stations to transport rainwater from their source to a treatment plant or receiving natural environment; – Develop a rainwater collection system that allows, after a rain, the effective evacuation of runoff water without causing the inundation of other public or private places, near or far
Law No. 96-06 of March 22, 1996 on the Local Authorities Code	– To ensure prevention, by suitable precautions, and intervention, by the distribution of necessary help, in the event of calamitous scourges, such as floods are accentuated with Climatic Changes

Some laws, such as Law No. 96-06 (Code of Local Authorities) have proven to be ineffective from the beginning because they were formulated without sufficient accounting for interdependencies and supported by limited municipal budgets.

Projects described in Table 6 allow sustainable development of water resources, including the improvement of knowledge of hydro-climatological data, water

Table 6 Some achievements in adaptation to climate change in Senegal (*Source* Ministry of Environment and Sustainable Development)

Sectors	Actions/infrastructure	Actors
Coastal zone	– Coastal protection works Direction – Dams and protective wall; – Reforestation at littoral level; – Coastal Act	Direction of Environment and Classified Establishments; Network of Parliamentarians for the Protection of Environment in Senegal; Third World Environment and Development
Agriculture/ Water Resources / Livestock	– Control of soil salinization; – Control of land degradation; – Landscaping and watersheds; – Anti-salt dikes; – Hydro-agricultural developments; – Sustainable land management practices; – Agroforestry; – Basins of retention, pastoral ponds	Direction of Environment and Classified Establishments; Retention Basins and Artificial Lakes Branch; Agency of Great Green Wall; Sustainable and Participatory Energy Management Project; Water and Forest Service; National Institute of Pedology; Senegalese Institute of Agricultural Research
Local governance	– Development of integrated territorial climate plans mapping of zones' vulnerability, carbon footprint; – Setting up and training of Regional Committees on Climate Change	*Texas Advanced Computing Center*; *United Nations Development Programme;* Direction of Environment and Classified Establishments; Regional Committees on Climate Change
Risk and disasters Management	– Establishment of an interdepartmental Crisis Management Operational Center conditioned by an integrated early warning system; – Floods management; – Installation of undersized structures against floods	Continuous Professional Development; National Agency for Civil Aviation and Meteorology; Direction of Environment and Classified Establishments; Direction of Planning Restructuring of Flood Areas; National Office of Sanitation of Senegal
Fisheries and biodiversity	– Creation of Protected Mayors Areas; – Aquaculture development; – Community Nature Reserves	Municipal Development Agency; Direction of Marine Protected Areas; National Parks Direction; International Union for Conservation of Nature

conservation, development of hydraulic infrastructures, management of risk, and environmental preservation, particularly in sensitive and fragile areas.

Senegal has also developed new guidelines, commonly known as National Adaptation Plans (NAPs), in coordination with the 16th Conference of the Parties on Climate Change (COP16), held in 2010 in Mexico. Thus, adaptation to climate change has been integrated into water resource planning in the medium and long term. Good water policy in Senegal is also based on promoting the active contributions of stakeholders in the design, implementation and management of water resource programs and projects.

Although Senegal has a sufficient supply of water during the rainy season, poor geographical distribution and high rainfall variability as well as inadequate management, lead to water scarcity in some areas. Thus, in spite of significant advances, water availability remains uncertain for areas facing quality problems (fluoride, water pollution), supply challenges (overexploitation of aquifers) and access, because of the very high cost of resource mobilization. Compounding these difficulties are insufficient budgets for the water sector. Regulation and enforcement are not consistently coordinated among ministries, and there has been inadequate investment in storage infrastructure due to recurring droughts and floods. These needs have been shortchanged because water resource management tends not to prioritize freshwater security in the climate change context.

In Senegal, water monitoring is mainly the responsibility of the Directorate of Water Resources Management and Planning (DGPRE) in collaboration with other national entities. However, water management in Senegal is implemented at the sector level, which limits the development of a comprehensive national water resource management strategy. Many different national authorities oversee various aspects of water management, such as policy, law creation and enforcement, delivery of services and consumption (Chitambi 2017). Thus institutional arrangements can be unclear and/or conflicting.

The number and cross-cutting nature of various authoritative entities creates dysfunction and therefore serves as a major constraint for freshwater safety. A big issue is the fragmentation of skills and lack of coordination among organizations responsible for implementing policies in Senegal and the other countries that share its basins, as well as questions of authority within and between countries and duplication of effort. Jointly, this fragmentation has undermined sustainable livelihoods (Murenga 2003) and driven economic, social and ecological costs to human societies and to the environment (Kouam-Kenmogne et al. 2006). At the same time, a certain level of redundancy has advantages in assessing natural land surface waters and integrating water resource management at the national level. In addition, Senegal must strengthen regional cooperation on water resource management as well as improve research, data collection and information-sharing on water resource assessment and management.

4.2.3 National Initiatives and Strategies for Responding to Climate Change in the Water Sector

The laws and codes described are a key mechanism for the adaptation of water resource management to climate change, but their implementation must balance mitigation and adaptation policies, and claim sufficient legal scope in relation to other land-use planning documents. Adaptation strategies emphasize water stress and drought management, which addresses adaptation for water demand and development of freshwater supply and flood risk management, which in Senegal has favored infrastructure protection (retention basins, dikes, etc.) over prevention and infrastructure adaptation. Water stress and flood risk management, as part of the National Flood Prevention Plan, must be developed and adapted in all vulnerable areas.

Responding to a "water crisis" is in itself insufficient. The real challenge is to adjust strategies to local contexts in order to overcome obstacles inherent to climate change. This requires the adoption of the right laws, the right policies, the right institutional guidelines and clarification of the roles and responsibilities of all stakeholders. As water governance is also a global concern complicated by cross-border water-sharing along the Senegal river (represented by the OMVS and OMVG frameworks), it is critical to pool resources and efforts among all partners and neighboring countries (predominantly Mali, Guinea, Mauritania and Gambia) in order to deal with the consequences of climate change, to strengthen the resilience of populations.

Governments must also coordinate their respective laws and policies that touch on a broad range of societal issues that impact, directly or indirectly, water resources and align management, service delivery and demand. Appropriate measures are likely to support a shared vision of the water sector and the pooling of means and ideas in order to ensure sustainable development of water resources in Senegal.

Achieving the Sustainable Development Goals–particularly Goal 6–requires effective and sustainable resource management. A reliable supply of water is a prerequisite for the economic and social development of any country. This, in turn, requires protection and equitable sharing of water resources (Guesnier 2010) through fair and reasonable governance and the management of water resources as a common, local and also global heritage whose vital value is recognized by all (Baudru and Maris 2002). Given the role of water resource management for climate change mitigation and adaptation, freshwater safety is clearly a development priority for Senegal, which has integrated various strategies for water protection, provision, saving, distribution, etc. Access to drinking water and sanitation in Senegal must be enabled by integrated management (Senegal National Blue Book Committee 2010), which is why Senegal has invested in reconciling economic development with environmental protection across the country for the benefit of present and future generations to align with the SDGs (including SDG 6).

Because of the urgency of water resource management to comply with the recommendations of various world water summits, Senegal has been dedicated to coordinating and enabling national and regional actors and programs that mobilize and secure water resources, such as the Office of Lakes and Waterways and the Agency for Promotion of the National Hydrographic Network (APRHN) (formerly the Office of the Lake de Guiers) to enhance adequate and sustainable management of resources to meet demand and supply for freshwater.

5 Conclusion

Water resources in Senegal are fragile and subject to the many pressures driven by human activity and compounded by climate change. The effects of climate change have become real as the frequency and intensity of droughts and floods increase. Water pollution in its various forms is increasing and contributes to the degradation of water quality.

Freshwater security strategies must account for demand, ensuring that any calculation and approach reflect "real" needs as expressed by the population. Solutions must first address demand management before increasing net water supply, where possible. Mobilization of additional water resources to support the overall demand must include conventional as well as unconventional water resources (e.g., rivers, lakes and water re-use).

Programs must focus on water resource preservation, including from wetlands and other aquatic environments. Particular attention must be paid to the preservation of groundwater resources, including appropriate measures for avoiding over-exploitation and pollution. Effective water management is based on relevant, comprehensive and reliable information and decision support tools. Water quality and quantity measurement networks must be maintained and modernized and information must be accessible to stakeholders involved in water management as stipulated by principles of good governance. To monitor progress and disseminate good practices, observatories and water information systems to monitor climate change effects and impacts on the water resources of the country must be established.

References

Agence Nationale de la Statistique et de la Démographie (2013) *Recensement Général de la Population et de l'Habitat, de l'Agriculture et de l'Elevage (RGPHAE)*. Rapport provisoire, mars 2014, 36 p

Ali A, Lebel T (2009) The Sahelian standardized rainfall index revisited. Int J Climatol 29(12):1705–1714

Allen JA (2002) The middle east water question, hydro politics, and the global economy. London and New York

AMCOW (2013) Sécurité en eau et développement résilient au changement climatique. Document d'information technique, 149 p

ANSD (2017) Cadre institutionnel, État de la mise en œuvre & Défis : Cas du Sénégal. Atelier sur l'environnement institutionnel robuste, la coopération, le dialogue et les partenariats pour la production et l'utilisation des indicateurs des ODD. Libreville, Gabon, 3–5 Juillet 2017, 15 p

Assetto VJ, Hajba E, Mumme SP (2003) Democratization, decentralisation, and local environmental policy capacity: hungary and Mexico. Social Sci J 40:249–268

Baudru D, Maris B (2002) Quatre modèles de gestion, de l'eau patrimoine de l'humanité, in *le Monde Diplomatique,* Manière de voir n°65, septembre-octobre

Bodian A (2014) Caractérisation de la variabilité temporelle récente des précipitations annuelles au Sénégal (Afrique de l'Ouest) », *Physio-Géo* [En ligne], Vol 8 l, mis en ligne le 12 janvier 2014, Consulté le 17 février 2015. http://physio-geo.revues.org/4243. https://doi.org/10.4000/physio-geo.4243

Boge V (2006) Water governance in Southern Africa co-operation and conflict prevention in trans boundary river basins. Boom International Centre for Conversion. Brief 33, Bonn, Germany

Briquet JP, Mahé G, Bamba F (1995) Changements climatiques et modification du régime hydrologique du fleuve Niger à Koulikoro (Mali). In: Actes de la conférence de Paris. IAHS Publication, Paris, pp 113–124

Chitambi M (2017) An evaluation of adequacy of water policy stipulation in ensuring water security in the context of climate change. Ethiopian J Environ Stud Manag 10(1):55–63

Comité National du Livre Bleu Sénégal (2010) L'eau, la vie, le développement humain. CONGAD, Le secrétariat international de l'eau, 12 p

Conférence internationale de l'eau et du climat (2016) Sécurité hydrique pour une justice climatique. Rabat, 11–12 juillet 2016, 25 p

Conseil Mondial de l'Eau (2016) Eau et climat, Livre bleu. Ministère Délégué chargé de l'Eau, Royaume du Maroc, 48 p

DGPRE (2016) Rapport phase pilote du suivi des indicateurs de l'ODD6 de l'initiative GEMI au Sénégal. Rapport final, Ministère de l'Hydraulique et de l'Assainissement, DGPRE, décembre 2016, 69 p

Dione O (1996) Evolution climatique récente et dynamique fluviale dans les hauts bassins des fleuves Sénégal et Gambie. Thèse de doctorat, Université Lyon 3 Jean Moulin, 477 p

European Union Commission (2007) Water scarcity and droughts, second interim reports

FAO (2016) Site web AQUASTAT. Organisation des Nations Unies pour l'alimentation et l'agriculture. Site consulté le 21/05/2017. Disponible sur. http://www.fao.org/statistics/fr/

Faye C (2013) Evaluation et gestion intégrée des ressources en eau dans un contexte de variabilité hydroclimatique : cas du bassin versant de la Falémé. Thèse de Doctorat, Université Cheikh Anta Diop de Dakar, 309 p

Faye C, Diop EHS, Mbaye I (2015a) Impacts des changements de climat et des aménagements sur les ressources en eau du fleuve Sénégal : caractérisation et évolution des régimes hydrologiques de sous-bassins versants naturels et aménagés. Belgeo 4:1–22

Faye C, Sow AA, Ndong JB (2015b) Étude des sècheresses pluviométriques et hydrologiques en Afrique tropicale : caractérisation et cartographie de la sècheresse par indices dans le haut bassin du fleuve Sénégal. Physio-Géo Géographie Physique et Environment 9:17–35

Faye C, Ndiaye A, Mbaye I (2017) Une évaluation comparative des séquences de sècheresse météorologique par indices, par échelles de temps et par domaines climatiques au Sénégal. J Wat Environ Sci 1(1):11-28

Guesnier B (2010) L'eau et le dévelopement durable : un couple en break sans gouvernance sociétale et coopération décentralisée. Développement durable et territoires 1(1):1–15

Hirsch RM, Slack JR (1984) A nonparametric trend test for seasonal data with serial dependence. Water Resour Res 20:727–732

Honegger A, Tabarly S (2011) La "gestion durable des ressources en eau" dans le bassin du Rhône, de la théorie à la pratique. Université de Lyon, UMR/CNRS 5600 EVS) pour Géoconfluences, 12 p

International Water Management Institute (2007) Water for life water for food. A comprehensive assessment of water management in Agriculture. A summary report

IPCC (2007) Résumé à l'intention des décideurs. In : Changements climatiques 2007 : Les éléments scientifiques, Contribution du Groupe de travail I au quatrième Rapport d'évaluation du Groupe d'experts intergouvernemental sur l'évolution du climat, [Solomon S., A. Quin, M. Manning, Z. Chen, M. Marquis, K.B. Averyt, M. Tignor et H.L. Miller (éds).], Cambridge University Press, Cambridge, UK et New York, NY, USA, 18 p

Jouilil I, Bitar K, Salama H, Amraou I, Mokssit A, Tahiri M (2013) Sécheresse météorologique au bassin hydraulique OUM ER RBIA durant les dernières décennies. Larhyss J 12:109–127

Kanohin F, Saley MB, Savané I (2009) Impacts de la variabilité climatique sur les ressources en eau et les activités humaines en zones tropicale humide: Cas de la région de Daoukro en Ivory Coast. Eur J Sci Res 26(2):209–222

Kendall M (1975) Multivariate analysis. Charles Griffin & Company, London, p 202

Kouam-Kenmogne GR, Mpakam HG, Ndonwy SA, Bopda SLD, Ekodeck GE (2006) Gestion intégrée des ressources en eau et objectifs du millénaire pour le développement en Afrique : Cas du Cameroun, VertigO-la revue électronique en sciences de l'environnement, 7 (2). Disponible sur. http://vertigo.revues.org/2319

Lamagat JP (1989) Monographie hydrologique du fleuve Gambie Collection M&m. ORSTOM-OMVG, 250 p

Mann HB (1945) Nonparametric Tests against Trend. Econometrica 13(3):245–259

Ministère de l'Hydraulique/DGPRE (2007) Plan d'Action de Gestion intégrée des ressources en eau du Sénégal. Global Water Partnership West Africa, 61 p

Murenga K (2003) The Impact of Integrated Water Resources Management (IWRM) on Poverty Alleviation. In: 4th WaterNet/Warfsa symposium: water, science, technology & policy convergence and action by All, 15–17 Oct 2003, 7 p

Niang AJ (2008) Les processus morphodynamiques, indicateurs de l'état de la désertification dans le Sud-Ouest de la Mauritanie. Approche par analyse multisource. Thèse de Doctorat, Université de Liège (Belgique), 286 p

Ouoba AP (2013) Changements climatiques, dynamique de la végétation et perception paysanne dans le Sahel burkinabè. Thèse de Doctorat Unique, Université de Ouagadougou (Burkina Faso), 305 p

Ozer P, Hountondji YC, Laminou MO (2009) Évolution des caractéristiques pluviométriques dans l'Est du Niger de 1940 à 2007. Geo-Eco-Trop, Revue de Géographie et d'Écologie Tropicale 33:11–30

Pettitt AN (1979) A non-parametric approach to the change-point problem. Appl Statist 28(2):126–135

Sane M (2015) Note sur les ressources en eaux du Sénégal : zones potentielles pour le transfert d'eau. Directeur de l'hydraulique, République du Sénégal, 8 p

Sharma TC, Panu US (2010) Analytical procedures for weekly hydrological droughts: a case of Canadian rivers. Hydrol Sci J 55(1):79–92

Sow AA (2007) L'hydrologie du Sud-est du Sénégal et de ses Confins guinéo-maliens : les bassins de la Gambie et de la Falémé. Thèse (PhD). Université Cheikh Anta Diop de Dakar, 1232 p

UNESCO-WWAP (2006) Water development report 2

UN-Water (2011) Analytical brief. Water security UNESCO-WWAP (2006). Water development report 2

Yue S, Wang CY (2004) The Mann-Kendall test modified by effective sample size to detect trend in serially correlated hydrological series. Water Resour Manag J 18:201–218

Chapter 13
Assessment of Hydrological Impacts of Climate Change on the Diarha Watershed

Ibrahima Thiaw, Bakary Faty, Honoré Dacosta, Anastasie Mendy, Abel Vincent Manga, and Amadou Abdoul Sow

Abstract This article assesses the impacts of climate change on surface water resources in the Diarha watershed, a tributary of the Gambia River, in West Africa. The analysis is based on the parallel application of two hydrological models (GR4J and SAC-SMA) with RS Minerve software and includes an automatic optimization based on a multi-criteria function that aggregates the Nash, Nash-ln, KGE' and R criteria according to the Shuffled Complex Evolution-University of Arizona (SCE-UA) method. The hydrological model that best replicates the rainfall-runoff relationship during the cross-validation phase (1975–1992 and 1998–2003) was fed with daily climate data projected for the 2050 horizon, while evaluation of future impacts of climate change on basin hydrology was based on the use of three Regional Climate Models (IPSL-CM5A-LR, INM-CM4, and GFDL-ESM2G) from CORDEX climate projections. CDF-t bias correction was applied to these scenarios before application to the hydrological model. Flow rates of the Diarha river basin were subsequently calculated in order to characterize their contribution to the projected climate scenarios at the 2050 horizon. Results indicate that the GR4J model is better able to reproduce the observed hydrographs of the Diarha river basin than the SAC-SMA model: values of the Nash-ln, KGE' and R criteria were greater than 80% in both calibration and validation, while the Nash model's performance varied between 91% (in calibration) and 62% (in validation). The models predict that climate change will have a significant impact on the Diarha's future flows, characterized by a decrease in annual streamflows of 5% under the RCP4.5 scenario and 25% under RCP8.5. The models furthermore project a decrease in the characteristic high flows (DMAX, DCC_10j, DCC_20j, DC1, DC2 and DC3) in both emission scenarios.

Keywords Climate change · *CORDEX* · Diarha watershed · Hydrological impacts · RS minerve

I. Thiaw (✉) · B. Faty · H. Dacosta · A. Mendy · A. V. Manga · A. A. Sow
Laboratory of Morphology and Hydrology, Cheikh Anta Diop University, P.O.B 5005, Dakar, Senegal
e-mail: ibrahima4.thiaw@ucad.edu.sn

S. Diop et al. (eds.), *Climate Change and Water Resources in Africa*,
https://doi.org/10.1007/978-3-030-61225-2_13

1 Introduction

The 1990s and 2000s are marked by a slight resumption of rainfall in the Sahelian zone (Lebel and Ali 2009; Panthou et al. 2014; Fall 2014; Marega 2016) after a long period of drought that not only weakened the functioning of hydrosystems but also affected human activities such as agriculture, fisheries and livestock (Dacosta 1989; Servat et al. 1998). This return of rainfall intensity has been observed across several West African watersheds, including the Senegal (Cissé et al. 2014), Gambia (Faye 2018), and Diarha (Thiaw 2017) rivers, putting their hydrological functioning at risk. It has been suggested that the return to humid conditions observed in some West African watersheds over the past few decades could be caused by anthropogenic contributions to climate change (Paeth and Hence 2004). However, the monitoring and assessment of these hydrosystems are hampered by a low density of monitoring networks associated with a lack of financial resources for national hydrological services (Bodian et al. 2015), so the attribution of hydrological impacts of climate change to rainfall or any other cause or causes would not be based on robust data.

Spot data for temperature, precipitation and potential evapotranspiration available at several monitoring stations are key inputs to existing water balance models, compensating for a lack of flow data (Bop et al. 2014); in our study, a rainflow model transforms time series describing the climatic conditions of a watershed into a series of flows. This indirect assessment is an essential tool to understand the future of rainwater and its dynamics in a watershed, the preferred scale for hydrological studies. Given the complexity of this process, often several mathematical models are used in parallel: global conceptual models, semi-distributed and distributed models. The accuracy of these models to simulate flows is improved by calibrating parameters (Gutpa et al. 2003; Vrugt et al. 2006)—modifying them until the output of the model corresponds to an observed data set (Liu 2009)—which has enabled automatic optimization techniques for hydroclimatic models in recent decades (Duan et al. 1992; Vrugt et al. 2003).

Hydrological impacts of climate change include an increase in evaporation coupled with rainfall variability, which can significantly affect surface runoff, frequency and intensity of floods and droughts, soil moisture and water available for irrigation and hydroelectric production (Setegn et al. 2011). In West Africa, the work of Roudier et al. (2014) shows that previous studies focused mainly on individual watersheds and were based on Global- (GCM) or Regional Climate Model (RCM) projections. More recently, the CORDEX (Regional Climate Reduction Experience Coordinated, Giorgi et al. 2009; Famien et al. 2018) initiative indicates an overall decrease in rainfall of 20%-30% in Africa, resulting in a decrease in river flow ranging from 40 to 60% (Biao 2017).

In Senegal, studies of the Gambia and Senegal river basins (Ardoin-Bardin et al. 2009; Bodian et al. 2013; Roudier et al. 2014; Mbaye et al. 2015; Bodian et al. 2018; Stanzel et al. 2018) suggest a decrease in the flow of its rivers by the end of the twenty-first century, correlating with a decrease in precipitation. Future impacts of climate change have not been projected for the Diarha sub-basin. The few studies of the large basins of the Gambia, at Gouloumbou (Ardoin-Bardin et al. 2009; Stanzel et al. 2018) and Mako (Bodian et al. 2018), project a reduction in water resources in the Gambia of between −4 and 26%, but these studies do not provide disaggregate projections at the level of each sub-basin of the Gambia. It is essential to assess the sensitivity of water resources to future climate change at finer spatial scales in order to understand fluctuations, given the significant uncertainties of hydroclimatic models (Brulebois 2016; Stanzel et al. 2018). In addition, the local analysis provides a basis for integrating climate into sustainable water resource management and exploitation policies (e.g., water management design, hydroelectric development, the definition of appropriate adaptation strategies to reduce the vulnerability of ecological systems to irregular rainfall and flow variability).

Two models of RS Minerve (GR4J and SAC-SMA) were compared to evaluate the hydrological impacts of climate change on the Diarha river basin. The CORDEX approach among three regional climate models (IPSL -CM5A-LR, INM-CM4 and GFDL-ESM2G) that best reflects the rainfall-runoff relationship during the cross-validation phase (1975–1992 and 1998–2003) will be applied to project flows of the Diarha in the near future (2020–2050).

2 Materials and Methods

2.1 Study Area

The Diarha catchment is a tributary of the left bank of the Gambia River. It extends between the parallels 12°26' and 12°63' and the meridians −12°86' and −12°57'. The physical environment of the basin has been described in earlier work (Thiaw et al. (2017)). With an area of 759.3 km^2 and a perimeter of 145.1 km, the Diarha extends 45 km upstream-to-downstream. Geologically, the basin lies entirely in the Paleozoic and Proterozoic formations of the Birrimian basement, which has given rise, as a result of the physicochemical alteration of the rocks, to low permeability and very fragile soils. Its Sudano-Guinean climate is marked by the spatial and temporal variability of precipitation since the 1967 climate break. From that time to the year 2014, precipitation in the catchment has decreased by 10.5%. This decrease in precipitation is attributable to the drought of the 1970s, which has been continuous since 1968 for all the rainfall stations studied, and the magnitude of which, analyzed by the Standardized Precipitation Index (SPI), is greater than that of the offsetting wet period (1921–1967) (Thiaw 2017). In addition to changing its hydrological regime and increased water erosion problems (Thiaw and Dacosta, 2017), this drought has resulted in diminished rainfed and dry season farming systems, now threatening food self-sufficiency. However, a slight return of rainfall conditions has been noted over the last two decades (1991–2010) (Thiaw 2017) which also implies potential changes in the hydrological behavior of the watershed.

2.2 *Data and Tools*

2.2.1 Observed Data

The observed data (temperatures and daily rains) used in this study are from the national meteorological agencies of Senegal and Guinea Conakry. Figure 1 shows that the rainfall stations in the basin are very poorly distributed, with only three stations in the basin itself, with another located in the east, far from the basin.

Table 1 provides the inventory of available climatic data and the percentage of days that are unmeasured for each station during the reference period 1960–2012. Although not common to all stations, this period is consistent with observed temperature and rainfall data and has relatively few gaps.

Hydrological data from the Diarha station at the road bridge are from the database of the Water Resources Management and Planning Directory (DGPRE). The Diarha was followed regularly between 1972 and 1992, with only a two-year gap between 1973 and 1974. Its average annual flow is 7 m^3/s, which equates to a volume of about 212.10^6 m^3/year (Thiaw 2017).

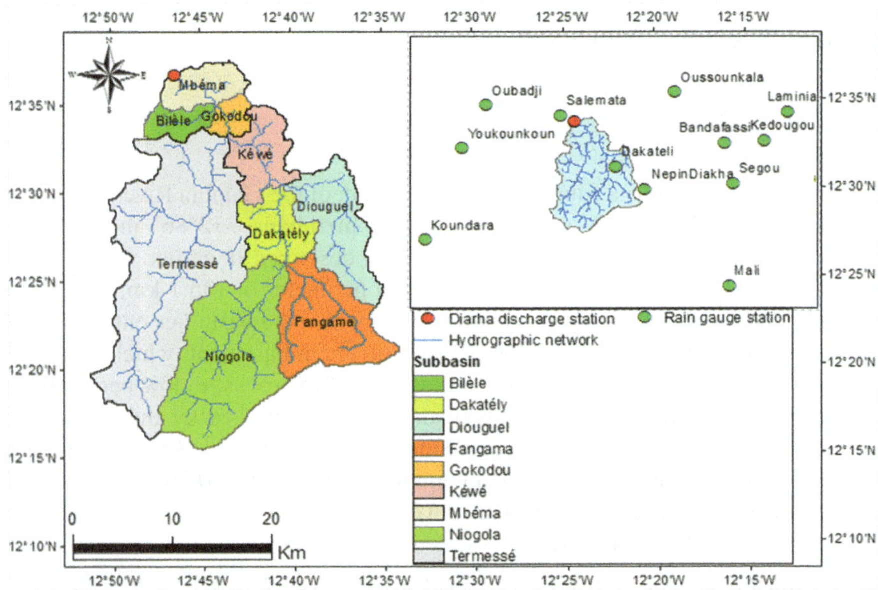

Fig. 1 Diarha watershed—Location map of monitoring networks

Table 1 Inventory of rainfall and daily temperature data observed

Longitude	Latitude	Station Id.	Data Description	Start of record	End of record	Gap %
−12.15	12.25	Bandafassi	Daily Rainfall	1/1/1975	1/1/2014	0.05
−12.64	12.46	Dakately	Daily Rainfall	1/1/1980	1/1/2015	26.51
−12.01	12.25	Fongolimbi	Daily Rainfall	1/1/1963	1/1/2015	0.02
−12.13	12.34	Kédougou	Daily Rainfall	1/5/1918	12/31/2016	0.02
			Daily Temperature	1/1/1961	12/31/2013	0.01
−13.11	12.47	Koundara	Daily Rainfall	1/1/1950	12/31/1994	0.004
−12.18	11.19	Labé	Daily Rainfall	1/1/1923	1/1/2013	1.2
			Daily Temperature	1/1/1950	12/31/2012	1.1
−12.11	12.64	Laminia	Daily Rainfall	1/1/1989	12/31/2002	0.5
−12.56	12.39	Nepin-Diakha	Daily Rainfall	1/1/1980	12/31/2001	2.0
−12.69	10.81	Pita	Daily Rainfall	1/1/1950	12/31/1982	0.10
−12.81	12.63	Salemata	Daily Rainfall	1/5/1973	1/1/2015	4.05
−11.47	12.47	Saraya	Daily Rainfall	1/1/1948	1/1/2015	0.01
−11.40	11.26	Tougué	Daily Rainfall	1/1/1968	11/9/2014	1.26
−13.30	12.50	Youkounkoun	Daily Rainfall	1/1/1950	1/1/1995	5.89
−11.23	12.85	Kéniéba	Daily Temperature	1/1/1950	31/12/2012	1.1
−12.08	10.36	Mamou	Daily Temperature	1/1/1960	31/12/2006	0.01

2.2.2 RS Minerve Model

RS Minerve is a hydrological and hydraulic modeling software program developed by the Center for Research on the Alpine Environment (CREALP) and Hydro Cosmos SA, in collaboration with the Polytechnic University of Valencia, Hydro10 and the Federal Polytechnic School of Lausanne. It simulates the production and transfer of flows over a complex watershed according to a semi-distributed conceptual scheme and an object-oriented approach. It incorporates several rain-flow models, such as GSM, SOCONT, SAC-SMA, GR4J and HBV.

The RS Expert module of the software allows a thorough evaluation of simulation results. Automatic calibration is applied using different algorithms, such as Shuffled Complex Evolution, developed by the University of Arizona (SCE-UA) (Duan et al. 1994) and the Uniform Adaptive Monte Carlo (Gilks et al. 1998; Liu 2001). The coupled method, Latin Hypercube (Rosenbrock 1960), calculates the best hydrological parameters according to a user-defined Objective Function (OF). For fluvial propagation, Lag-Time, Kinematic-Wave, Muskingum-Cunge and St-Venant are applied. In addition, the scenario-simulation module simulates weather scenarios or other parameters to study the variability and sensitivity of model results.

The two RS Minerve models GR4J and SAC-SMA tested in this study are calibrated in a semi-distributed conceptual mode to distinguish the respective contributions of each sub-basin to the natural flow of the large basin (Figs. 2 and 3). They require (i) a spatial meteorological database (virtual stations) which provide daily rainfall and temperature for each climate station, and (ii) the vector layers of the watersheds for the automatic calculation of mean rainfall and potential evapotranspiration (ETP). The Diarha basin has been described accordingly in the ArcMap environment to create Hydrological Response Units (HRU), which include area, the center of gravity (x, y and z) and the longest flow-path of each HRU (Table 2). Using the RS Minerve GIS module, the vector layers of the HRUs were imported into the

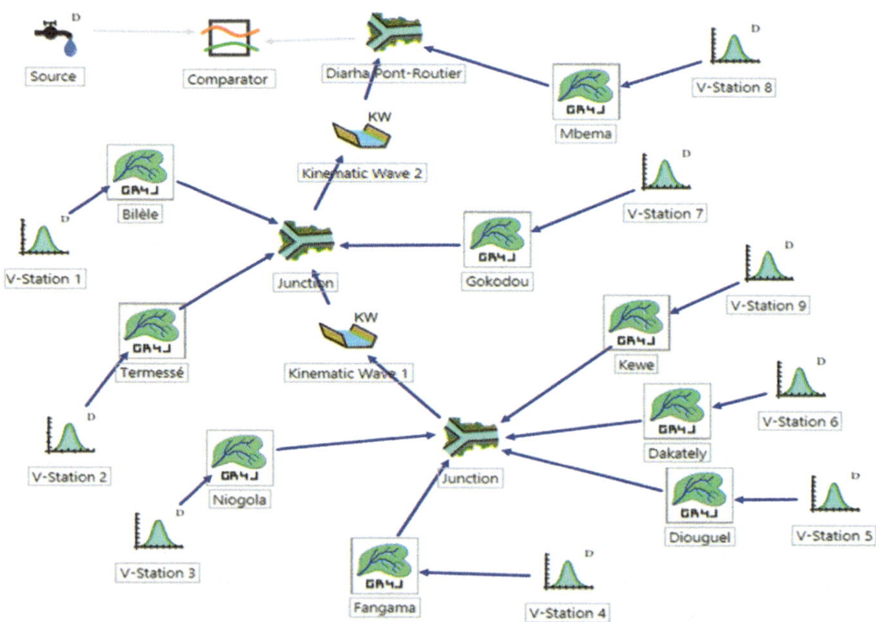

Fig. 2 Topology of the full model GR4J: production, transfer, then transport

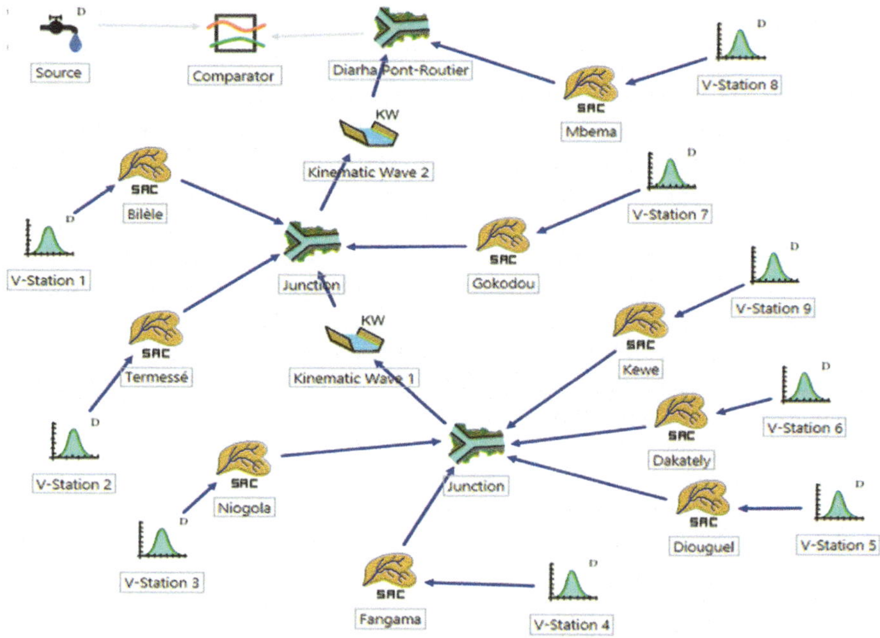

Fig. 3 Topology of the full model SAC-SMA: production, transfer, then transport

Table 2 Parameters of hydrological response units

HRU	Area (km^2)	Longest Flow path (m)	X (m)	Y (m)	Z (m)
Termessé	274.8	40,300	741,466	1,375,370	132
Niogola	159.3	28,500	747,162	1,365,570	174
Fangama	93.7	20,600	756,923	1,369,230	160
Bilèle	19.7	10,080	742,252	1,390,280	66
Diouguel	73.93	24,180	758,190	1,379,220	93
Dakately	44.03	12,470	752,031	1,379,030	107
Gokodou	14.43	8,212	747,620	1,390,750	74
Kéwé	44.42	12,380	750,577	1,386,850	92
Mbéma	31.57	10,724	744,745	1,394,010	62

program; because each sub-basin is associated with a virtual weather station (Figs. 2 and 3), the coordinates for the centers of gravity (x, y and z) have been used for spatial interpolation using the Thiessen polygon method (Roche 1963). The ETP was calculated using Oudin's methodology (2006), which is based on average daily temperatures and the latitude of the study area.

2.3 Methods

In this study, two hydrological models of RS Minerve (GR4J and SAC-SMA) were compared. The model that best reproduced the hydrographs observed during the calibration (1975–1992) and validation (1998–2003) phases were used to reconstruct flow data that was missing for the Diarha over three periods (1961–1974; 1993–1997 and 2004–2012). This enabled consideration of a continuous period of discharges from 1961 to 2012. In addition, the same model was used to project future flows (horizon 2050) to anticipate the impact of climate change on surface water resources of the Diarha catchment. CORDEX simulations produced climate outputs that were subject to three RCM (IPSL-CM5A-LR, INM-CM4, and GFDL-ESM2G). In addition, characteristic high (DCC) and low (DCE) flow rates were applied to projected climate scenarios through 2050.

2.3.1 Structure of Hydrological Models

GR4J Hydrological Model

The GR4J model (Perrin et al. 2003) is based on two reservoirs (production and routing) and two unitary hydrographs (UH1 and UH2). The model simulates flows through the production and transfer functions (Fig. 4). It first neutralizes equivalent daily rainfall (Peq) by potential daily evapotranspiration to determine net rain (Pn) and net evapotranspiration (En) (Garcia Hernandez et al. 2018):

$$\begin{aligned} Pn &= Peq - ETP \\ En &= 0 \quad \text{Si} \quad P \geq ETP \\ Pn &= 0 \quad \text{Si} \quad P > ETP \\ En &= ETP - Peq \end{aligned} \tag{1}$$

When Pn deviates from zero, a part of Pn denoted (Ps) feeds the production reservoir (S), as presented in Eq. 2. Similarly, if En is not zero, the evapotranspiration of the production reservoir (Es) is calculated as a function of the level of water contained in Ps as described in Eq. 3.

$$Ps = \frac{d}{dt}\frac{X1.\left(1 - \left(\frac{S}{X1}\right)^2\right).\tanh\left(\frac{Pn.dt}{X1}\right)}{1 + \left(\frac{S}{X1}\right).\tanh\left(\frac{Pn.dt}{X1}\right)} \tag{2}$$

$$Es = \frac{d}{dt}\frac{S.\left(2 - \left(\frac{S}{X1}\right)\right).\tanh\left(\frac{En.dt}{X1}\right)}{1 + \left(1 - \frac{S}{X1}\right).\tanh\left(\frac{En.dt}{X1}\right)} \tag{3}$$

where

Ps intensity of the rain feeding the production reservoir

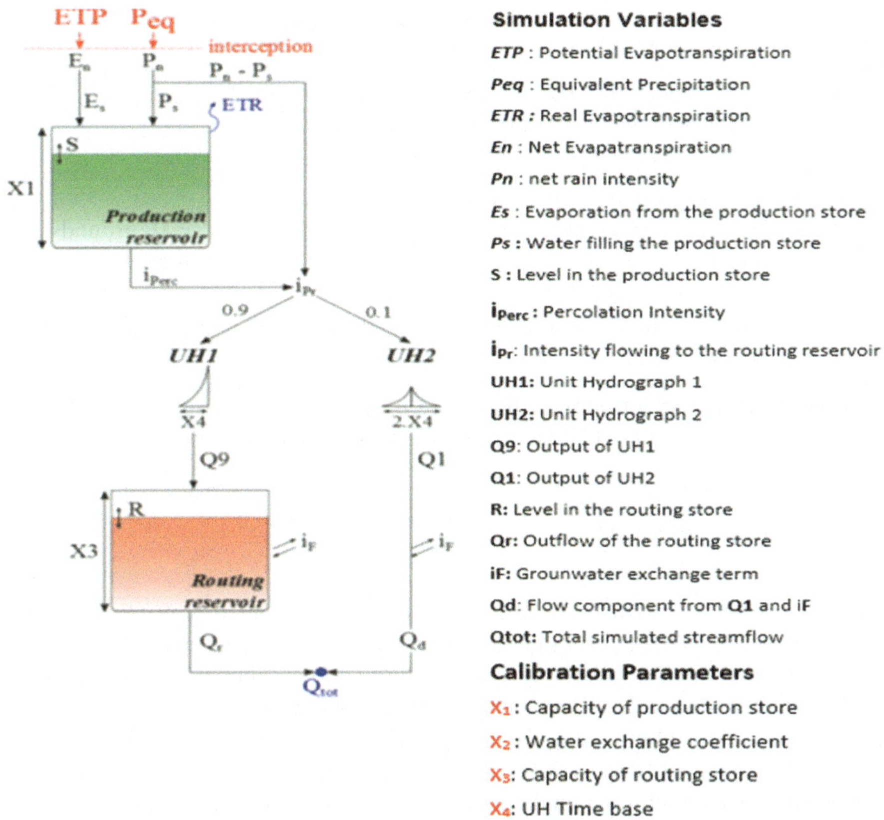

Fig. 4 Conceptual scheme of the GR4J model (Garcia Hernandez et al. 2018)

X1 maximum capacity of the production reservoir;
S water content in the production reservoir
Es evapotranspiration of the production reservoir

The percolation (iPerc) from the production reservoir is therefore calculated as:

$$iPerc = \frac{d}{dt}(S + (Ps - Es).dt).\left(1 - \left(1 + \left(\frac{4}{9}.\frac{S + (Ps - Es).dt)}{X1}\right)4\right)^{-\frac{1}{4}}\right) \tag{4}$$

The variation of the production reservoir (S) is:

$$\frac{dS}{dt} = Ps - Es - iPercS \geq 0 \tag{5}$$

The amount of water routed to the routing reservoir (iPr) is determined by:

$$iPr = iPerc + (Pn - Ps) \quad (5)$$

where 90% of iPr is transported by unit hydrograph (UH1) and by the routing reservoir and the remaining 10% by a second unit hydrograph (UH2).

Four parameters are used to calibrate the model:

- X1 the size of the production reservoir
- X2 exchanges between surface water and groundwater (if X2 < 0, groundwater feeds runoff and vice versa if X2 > 0, runoff feeds groundwater)
- X3 maximum capacity of the transfer reservoir
- X4 the base time of the flood (Fig. 4) on which the two-unit hydrographs UH1 and UH2 depend

A more detailed description of the GR4J model is available in the RS Minerve software technical manual (Garcia Hernandez et al. 2018) and in Perrin et al. (2003).

SAC-SMA Hydrological Model

SAC-SMA or SACRAMENTO (Fig. 5) is a semi-distributed conceptual model, developed in the 1970s (Burnash et al. 1973, Burnash 1995) to facilitate the simulation of river flows by optimizing soil moisture and percolation. It is widely used for flood forecasting in river forecast centers across the United States. Satisfactory results in hydrological modeling are demonstrated by Ajami et al. (2004); Anderson et al. (2006); Gan and Burges (2006); Laura et al. (2016). The model runs on a daily basis with rain and ETP as input data. The ability of the model to simulate flow rates depends primarily on the initial parameters and conditions presented in Table 3.

2.3.2 Optimization of Hydrological Models

The automatic optimization of the models is based on the use of a multi-criteria function, which aggregates the Nash, Nash-ln, KGE' and R (Table 4) criteria, to avoid the biases that could result from the use of just one criterion (Nash).

The Shuffled Complex Evolution-University of Arizona (SCE-UA) method was used for the automatic calibration of models. SCE-UA is a global optimization method (Duan et al. 1992) based on the best characteristics of several algorithms (Kan et al. 2016) and on the Nelder and Mead (1965) simplex method for linear operations. This approach was developed to improve the calibration of hydroclimatic models (Ajami et al. 2004, Muttil and Liong 2004, Blasone et al. 2007)—in this case, rainflow conceptual models (CRR models). It has recently been applied successfully to watershed resource management (Zhu et al. 2006; Liu et al. 1998; Jeon et al. 2014). SCE-UA synthesizes several techniques: (i) combining probabilistic and deterministic approaches, (ii) systematic evolution of a complex of points covering the parameter space, (iii) competitive evolution and, (iv) the Shuffling complex.

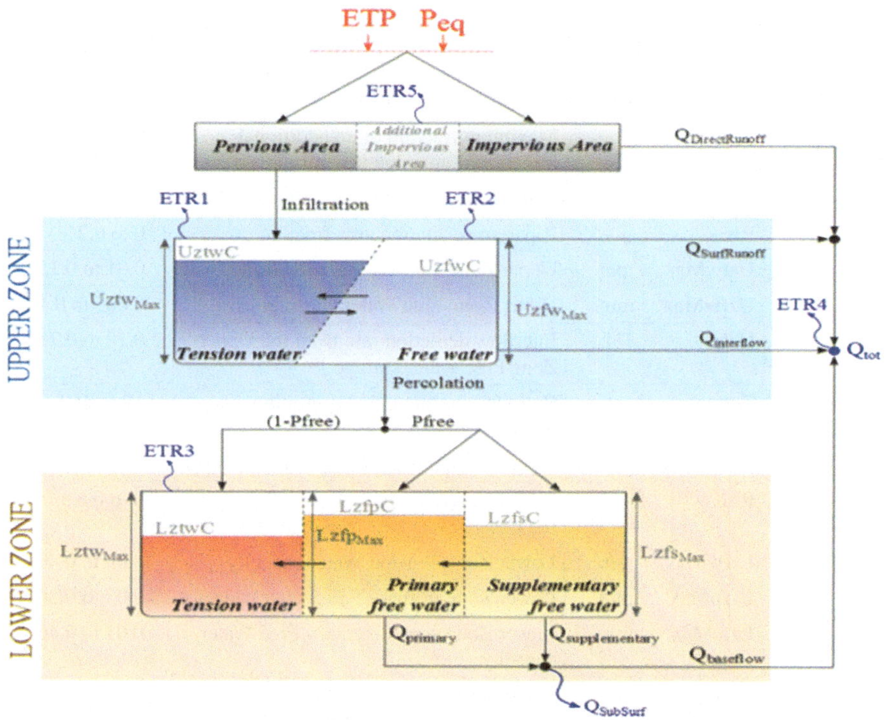

Fig. 5 Conceptual scheme of the SAC-SMA model (Garcia Hernandez et al. 2018)

Numerous studies using different CRR models, including hydrological (HYMOD), six parameters (SIX-PAR) and SAC-SMA (Duan et al 1992, Sorooshian et al. 1993, Eckhardt and Arnold 2001) have shown that SCE-UA is an efficient, consistent and flexible algorithm for the automatic calibration of environmental models.

SCE-UA optimization is based on the number of objective criteria selected and the maximum number of iterations (MAXN) until the optimal value is reached, based on a defined Objective Function (OF). A random draw of 7,000 combinations of parameters for each model is then run to identify optimal values. Finally, 295 iterations of the functions of production and transfer by the models were run, to which local adjustment of the parameters were applied. This process allows a choice of model parameters to best reflect the relationship of rain to flow.

2.3.3 Downscaling of Regional Climate Models

In this study, the CDF-t method developed by Michelangeli et al. (2009) is used to adjust the regional climate models IPSL-CM5A-LR, INM-CM4, and GFDL-ESM2G produced by CORDEX climate projections. According to Famien et al (2018), this method consists of matching the CDF of a climate variable simulated by a model

Table 3 Initial parameters and conditions of the SAC-SMA model

Object	Name	Units	Description	Regular range
SAC-SMA	S	km^2	Area of the watershed	>0
	Adimp	–	Maximum fraction of an additional impervious area due to saturation	0 to 0.2
	Pctim	–	Permanent impervious area fraction	0 to 0.05
	Riva	–	Riparian vegetation area fraction	0 to 0.2
	UztwMax	mm	Upper Zone Tension Water capacity	0.01 to 0.15
	UzfwMax	mm	Upper Zone Free Water capacity	0.005 to 0.10
	Uzk	1/d	Interflow depletion rate from the Upper Zone Free Water storage	0.10 to 0.75
	Zperc	–	Ratio of maximum and minimum percolation rates	10 to 350
	Rexp	–	Shape parameter of the percolation curve	1 to 4
	Pfree	–	Percolation fraction that goes directly to the Lower Zone Free Water storage	0 to 0.6
	LztwMax	mm	Lower Zone Tension Water capacity	0.05 to 0.40
	LzfpMax	mm	Lower Zone primary Free Water capacity	0.03 to 0.80
	LzfsMax	mm	Lower Zone supplementary Free Water capacity	0.01 to 0.40
	Rserv	–	Fraction of Lower Zone Free Water not transferable to Lower Zone Tension Water	0 to 1
	Lzpk	1/d	Depletion rate of the Lower Zone primary Free Water storage	0.001 to 0.03
	Lzsk	1/d	Depletion rate of the Lower Zone supplemental Free Water storage	0.02 to 0.3
	Side	–	Ratio of deep percolation from Lower Zone Free Water storage	0 to 0.5
	Adimlni	mm	Initial Tension Water content of the Adimp area	–
	Uztwlni	mm	Initial Upper Zone Tension Water content	–
	Uzfwlni	mm	Initial Upper Zone Free Water content	–
	Lztwlni	mm	Initial Lower Zone Tension Water content	–
	Lzfplni	mm	Initial Lower Zone Free supplementary content	–
	Lzfslni	mm	Initial Lower Zone Free primary content	–

to the CDF of the same variable in observations through a mathematical function (the transform). The CDF-t approach can be considered an extension of Q-matching, because it directly provides CDFs. Instead of applying the quantile–quantile correction (Déqué 2007), it takes into account only the probability distribution. The CDF-t adjusts model outputs based on the historical data and gauge observations of transformations from the CDFs. The method assumes a transformation (T) that allows

Table 4 Performance indicators

Criterion	Scale	Idea value
Nash (NSE)	$-\infty$ to 1	1
Nash-ln	$-\infty$ to 1	1
Pearson (R)	-1 to 1	1
KGE' (modified KGE-statistic)	0 to 1	1

Nash is the Nash-Sutcliffe coefficient; $X_{sim,t}$ is the simulated flow at time step t; $X_{ref,t}$ is observed flow at time step t; KGE' is modified KGE-statistic; R is correlation coefficient between simulated and reference flow; β is the ratio between the mean of the simulated flows and the mean of reference ones; γ is variability ratio (ratio between the coefficient of variation of the simulated flows and the coefficient of variation of the reference ones)

Table 5 Regional Climate Models (RCM) used in this study

Modeling Center or Group	Model Name	Resolution	Scenarios RCP	Present Period	Historical	Future Period
NOAA Geophysical Fluid Dynamics Laboratory	GFDL-ESM2G	0.5° × 0.5°	4.5 and 8.5	1961–2012	1950–2005	2020–2050
Institute for Numerical Mathematics	INM-CM4	0.5° × 0.5°	4.5 and 8.5	1961–2012	1950–2005	2020–2050
Institute Pierre-Simon Laplace	IPSL-CM5A-LR	0.5° × 0.5°	4.5 and 8.5	1961–2012	1950–2005	2020–2050

translation of the CDF of a GCM/RCM variable to be the predictor of the CDF representing the local climate at a given weather station (Michelangeli et al. 2009). F_{Sh} stands for the CDF of observed and historical daily temperature/precipitation data, and F_{Gh} stands for the CDF of model outputs at a given grid cell for the same period. F_{Gf} and F_{Sf} are the CDF equivalent to F_{Gh} and F_{Sh} but for a future period.

$$T\,(F_{Gh}(x)) = F_{Sh}(x) \tag{7}$$

Define u = FGh (x) and thus x = F − 1Gh (u) with u ∈ [0, 1]. Replacing (x) in Eq. 7 allows the following definition for transform T:

$$T\,(u) = F_{Sh}\left(F_{Gh}^{-1}(u)\right) \tag{8}$$

In which *T* denotes the functional relationship between the CDF of observation and simulation results over a historical period. Thus, assuming that the relationship in Eq. 8 will remain valid in the future, and the corrected CDF is given by:

$$F_{Sf}(x) = F_{Sh}\left(F_{Gh}^{-1}(F_{Gf}(x))\right) \quad (9)$$

As with all methods, CDF-t results and technical constraints have their strengths and weaknesses. An important advantage of CDF-t is that it has been extensively tested and referenced in various publications since 2009 (Michelangeli et al. 2009; Vrac et al. 2012, Vrac and Friederichs 2015, 2016; Vigau et al. 2013; Grouillet et al. 2016; Famien et al. 2018). The Regional Climate Models used in this study are presented in Table 5. They are bias-corrected at the scale of West Africa within CORDEX climate projections (Famien et al. 2018).

3 Results

3.1 *Efficiency of the SAC-SMA Model in the Cross-Validation Phase*

The comparison of observed and simulated hydrographs shows that the SAC-SMA model reproduces average flows of the Diarha better than it does low flows and flood peaks. Parameters calibrated between 1975 and 1992 give the best Nash criteria (0.89 for calibration and 0.62 for validation; the Nash calculated from the logarithm (Nash-ln) ranges from 0.52 (calibration) to 0.37 (validation)) (Figs. 6, 7, 8 and 9).

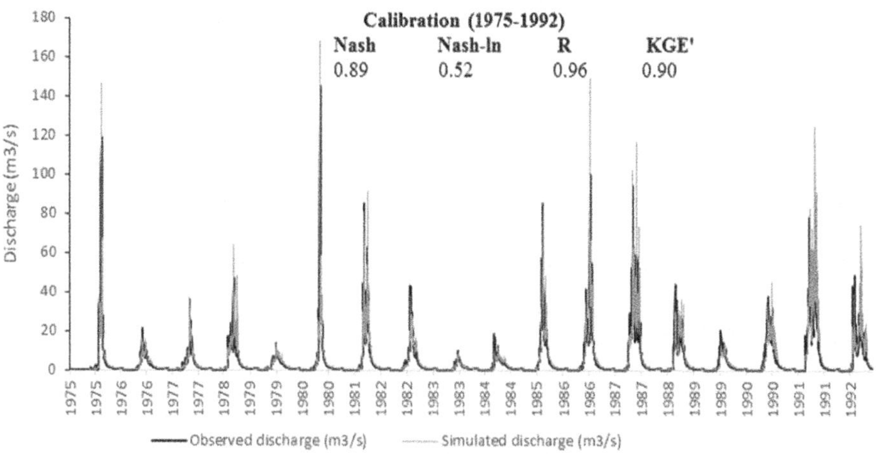

Fig. 6 Calibration of the SAC-SMA model over the period 1975–1992

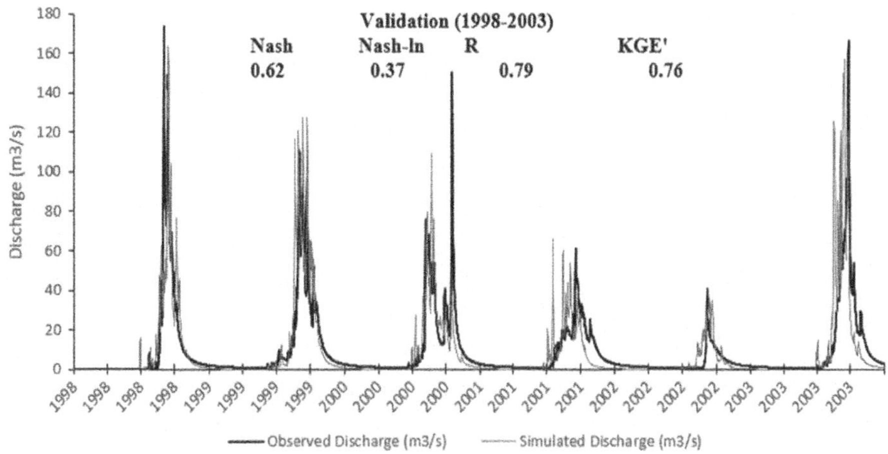

Fig. 7 Validation of the SAC-SMA model over the period 1998-2003

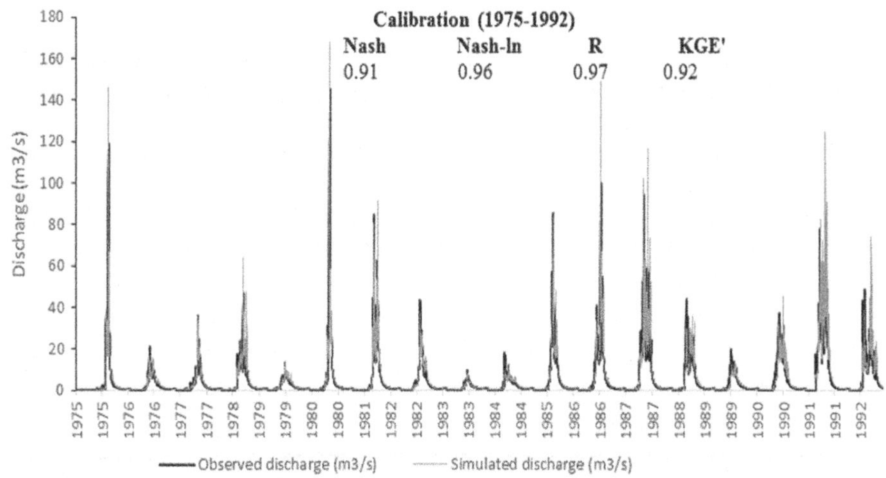

Fig. 8 Calibration of the GR4J model over the 1975–1992 period

3.2 *Efficiency of the GR4J Model in the Cross-Validation Phase*

Performance criteria calibrated with the GR4J model over the calibration (1975–1992) and validation (1998–2003) periods show good superimposition of the observed and simulated daily hydrographs. The calculated Nash logarithm (Nash-ln) gives the best results (0.96 for calibration and 0.90 for validation). Pearson (R) and Kling Gutpa Efficiency (KGE') values are greater than 80% in both calibration and validation. These criteria (R and KGE'), although very sensitive to extreme values

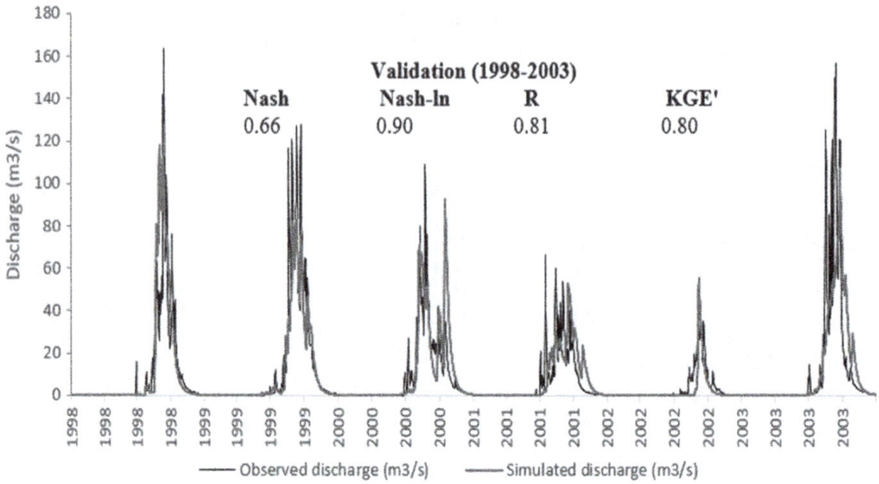

Fig. 9 Validation of the GR4J model over the period 1998–2003

(Legates and McCabe 1999), remain good indicators for the simulation of Diarha streamflows. The output for calibration (1975–1992) and validation (1998–2003) shows that the GR4J model has been well-calibrated to simulate future flows.

Results indicate that the GR4J model repruces the observed Diarha flow more accurately than does the SAC-SMA model. For this reason, parameters (X1, X2, X3 and X4) of its 1975–1992 calibration period were used to reconstruct the missing flows over the periods 1961–1974; 1993–1997 and 2004–2012 (Fig. 10).

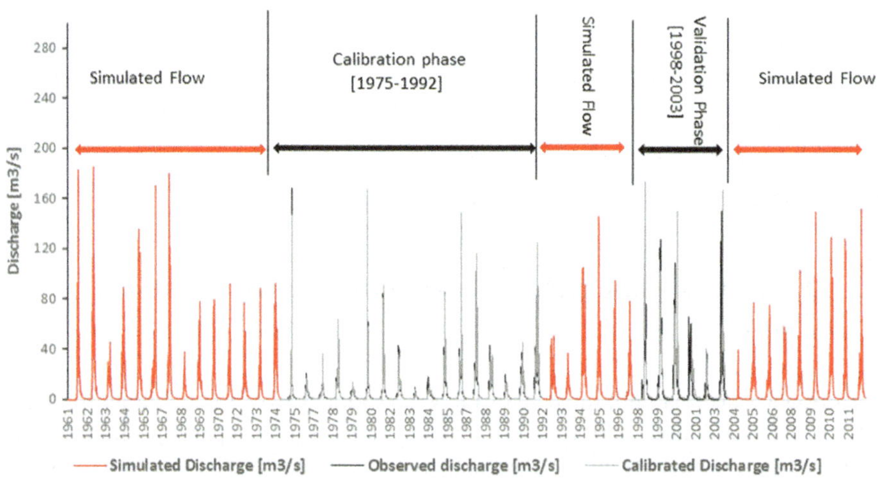

Fig. 10 Observed and simulated daily hydrograph of Diarha

Table 6 Segmentation of the annual flow series of the Diarha Basin

Beginning	End	Mean discharge (m^3/s)
1961	1967	9.82
1968	1993	6.01
1994	2012	8.2

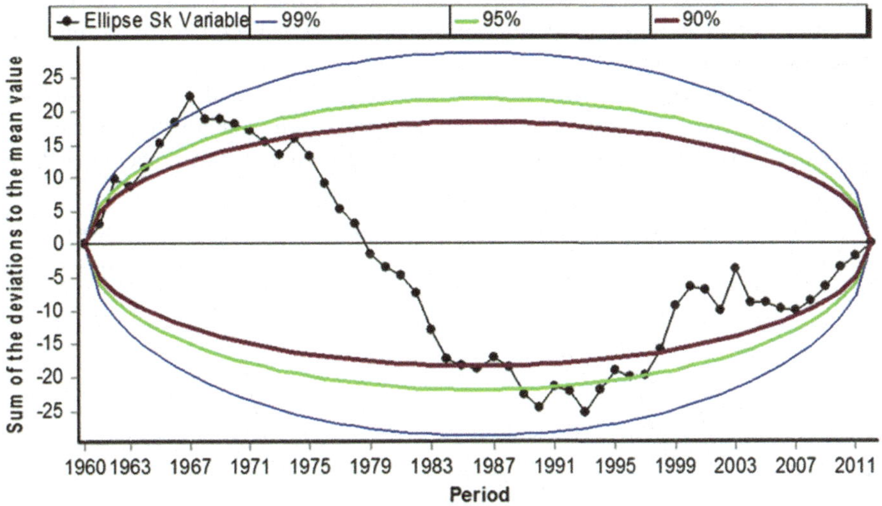

Fig. 11 Demonstration of the presence of rupture at all thresholds of confidence by the Bois' Ellipse on the annual flows of Diarha

Buishand's test (Buishand 1982) and Hubert's segmentation method (Hubert et al. 1998) were applied to streamflow chronicles to analyze their stability, using Khrono-Stat software (IRD 1998). Results indicate that the streamflow series of the Diarha is not stationary, exhibiting several breaks as revealed by the segmentation of Hubert (Table 6) and the Bois ellipse (Fig. 11).

Annual mean flows of the Diarha show breaks in 1967, 1990 and 2012, characterized as:

- a wet phase (1961–1967) with an average annual flow of 9.82 m^3/s;
- a dry phase (1968–1993) with an average flow of 6.01 $m^3/s/year$;
- a wet phase (1994–2012), with an average flow of 8.2 $m^3/s/year$.

3.3 Future Temperature and Precipitation Changes

The Cumulative Distribution Function transformation (CDFt) was applied to daily temperature and precipitation data to assess the capacity of regional climate models (INM-CM4, GFDL-ESM2G and IPSL-CM5A-LR) to forecast precipitation and temperature of the Diarha catchment. This method has been applied in several studies to correct the extreme signals of climate models (Guo et al. 2018; Mbaye et al. 2018).

Results show that the CDFt method effectively corrects the variation of future temperature (Fig. 12) and rainfall (Fig. 13) of the Diarha catchment.

The IPSL-CM5-A-LR, INM-CM4 and GFDL-ESM2G models predict, under the RCP4.5 scenario, an increase in mean temperatures of 1.4 °C; 1.7 °C and 0.9 °C, respectively in 2050 horizon. This increase is more pronounced under the RCP8.5 scenario, in which the same models forecast an increase of 1.99 °C; 2.04 °C and 1.09 °C, respectively, in 2050.

The CDFt projection shows that in both the RCP4.5 and RCP8.5 emission scenarios, there is a downward trend in rainfall in the Diarha catchment. Compared to the period 1961–2012, precipitation is expected to decrease in 2050 by between 2.1%

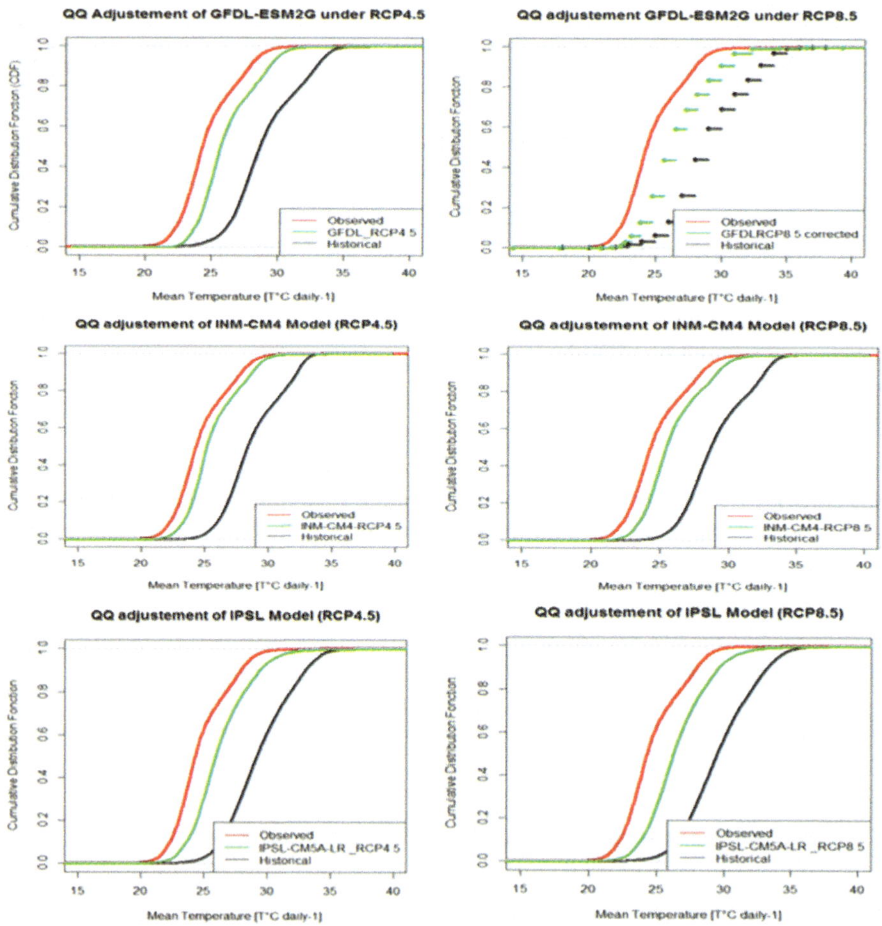

Fig. 12 Temperature biases corrected under RCP4.5 and RCP8.5

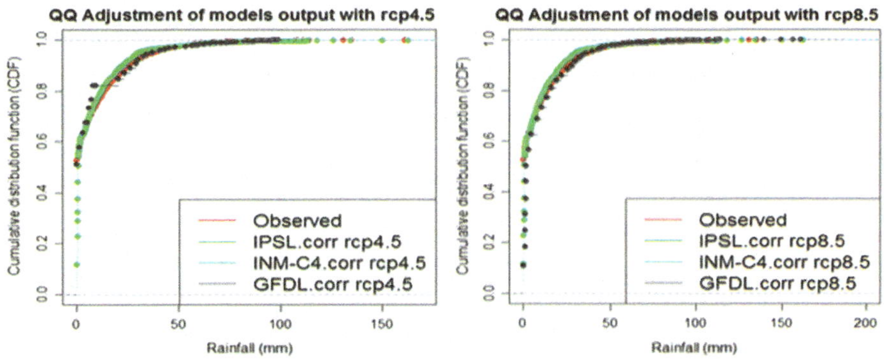

Fig. 13 Precipitation biases corrected under RCP4.5 and RCP8.5

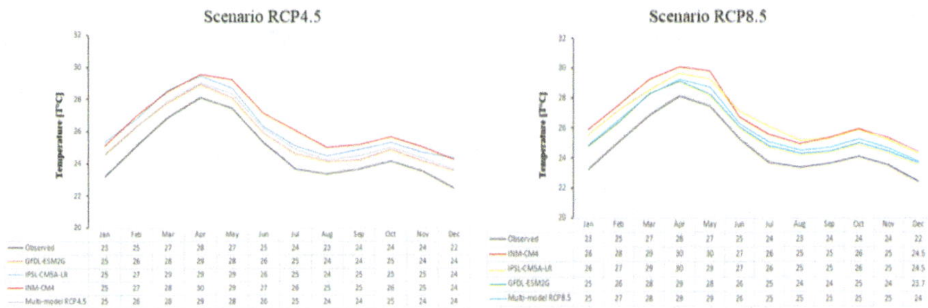

Fig. 14 Projected average temperatures of the Diarha basin under RCP4.5 and RCP8.5

(RCP4.5) and 12% (RCP8.5), signaling that the warming and decreased precipitation that results from climate change will probably affect the hydrology of the Diarha watershed (Figs. 14, 15 and 16).

3.4 Evolution of Annual Streamflow and Characteristic Flow Rates by 2050

The corrected outputs of Regional Climate Models (RCM) were submitted to the GR4J model to simulate the evolution of Diarha streamflows by 2050. Compared to the 1961–2012 reference period, the GFDL-ESM2G, IPSL-CM5A-LR and INM-CM4 models predict a decrease in streamflows of 0.5%, 1.5% and 12.4%, respectively.

For the RCP8.5 scenario, the decrease in precipitation, as well as that of flows, is more severe. Models predict a decrease in precipitation of 7.6% (GFDL-ESM2G) to 17.2% (INM-CM4). For flows, the same models predict a decrease of 16% to 35%. The overall average RCP8.5 model predicts a decrease in rainfall of 12%, which

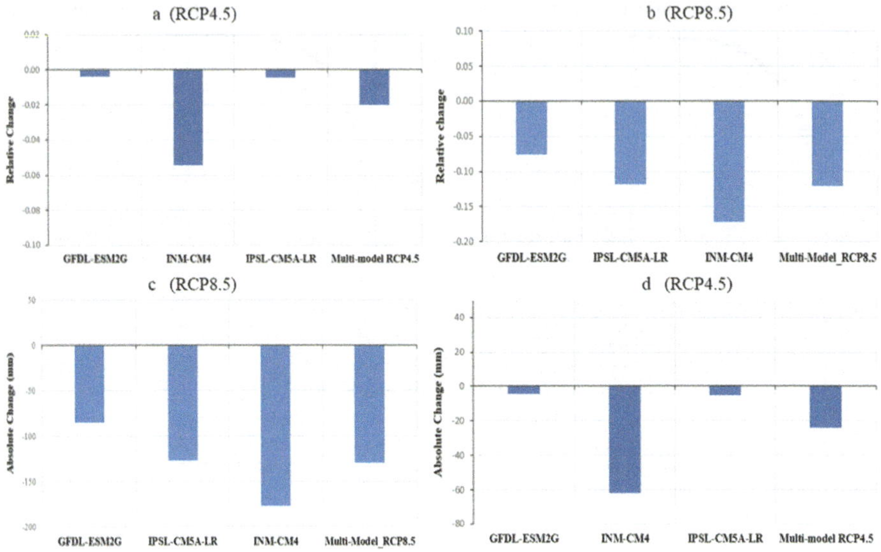

Fig. 15 Relative (**a**, **b**) and Absolute (**c**, **d**) changes of annual rainfall for the Diarha River under RCP4.5 and RCP8.5

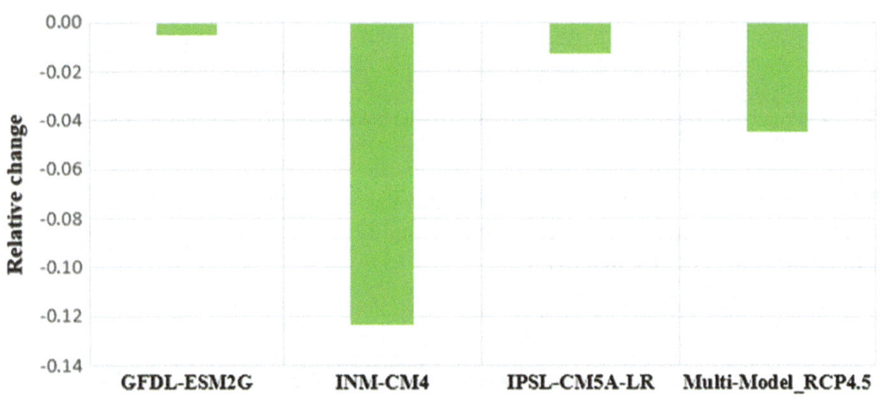

Fig. 16 Relative change of mean annual flow for the Diarha River under RCP4.5

translates as a reduction in the flow of the Diarha of 25% (Fig. 17). These results show how streamflow forecasting is highly sensitive to the quality of RCM output data.

In addition, Diarha's high (DCC) and low (DCE) characteristic flow rates were forecast for 2050 climate scenarios, as compared to the 1961–2012 reference period. Table 7 presents the characteristic flow rates used; Appendices A (RCP4.5) and B (RCP8.5) show the relative projected changes in DCC and DCE under their respective scenarios. Figures 18 and 19 show the variation of the projected DCC and DCE (in

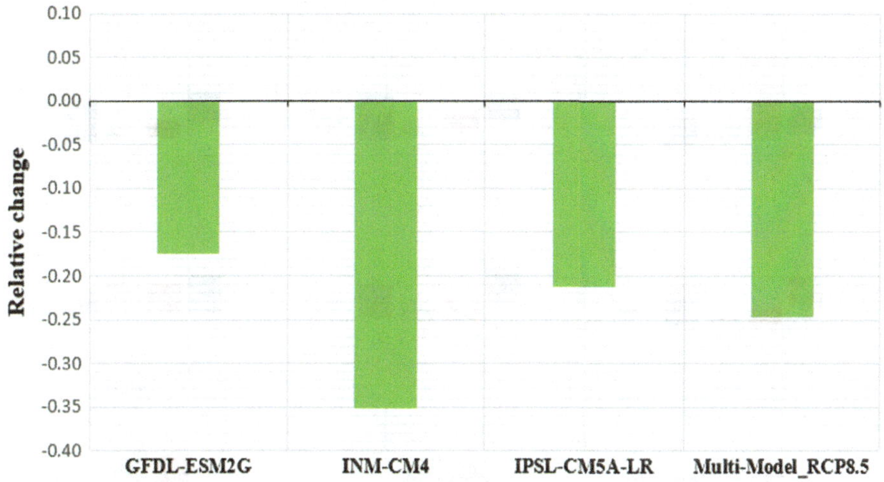

Fig. 17 Relative change of mean annual flow for the Diarha River under RCP8.5

Table 7 Characteristic flow rates used

Characteristics	Acronym	Description
High flow	DMAX	Maximum Daily flow
	DCC_10j	Characteristic flood flow exceeded 10 days per year
	DCC_20j	Characteristic flood flow exceeded 20 days per year
	DC1	Characteristic flow exceeded 1 month in the year
	DC2	Characteristic flow exceeded 2 months in the year
	DC3	Characteristic flow exceeded 3 months in the year
	DC4	Characteristic flow exceeded 4 months in the year
	DC5	Characteristic flow exceeded 5 months in the year
	Median	Flows exceeded six months in the year
Low flow	DC7	Characteristic flow exceeded 8 months in the year
	DC8	Characteristic flow exceeded 8 months in the year
	DC9	Characteristic flow exceeded 9 months in the year
	DC10	Characteristic flow exceeded 10 months in the year
	DC11	Characteristic flow exceeded 11 months in the year
	DCE_20j	Characteristic low flow on a 20-day period
	DCE_10j	Characteristic low flow on a 10-day period
	DMIN	Minimum Daily flow

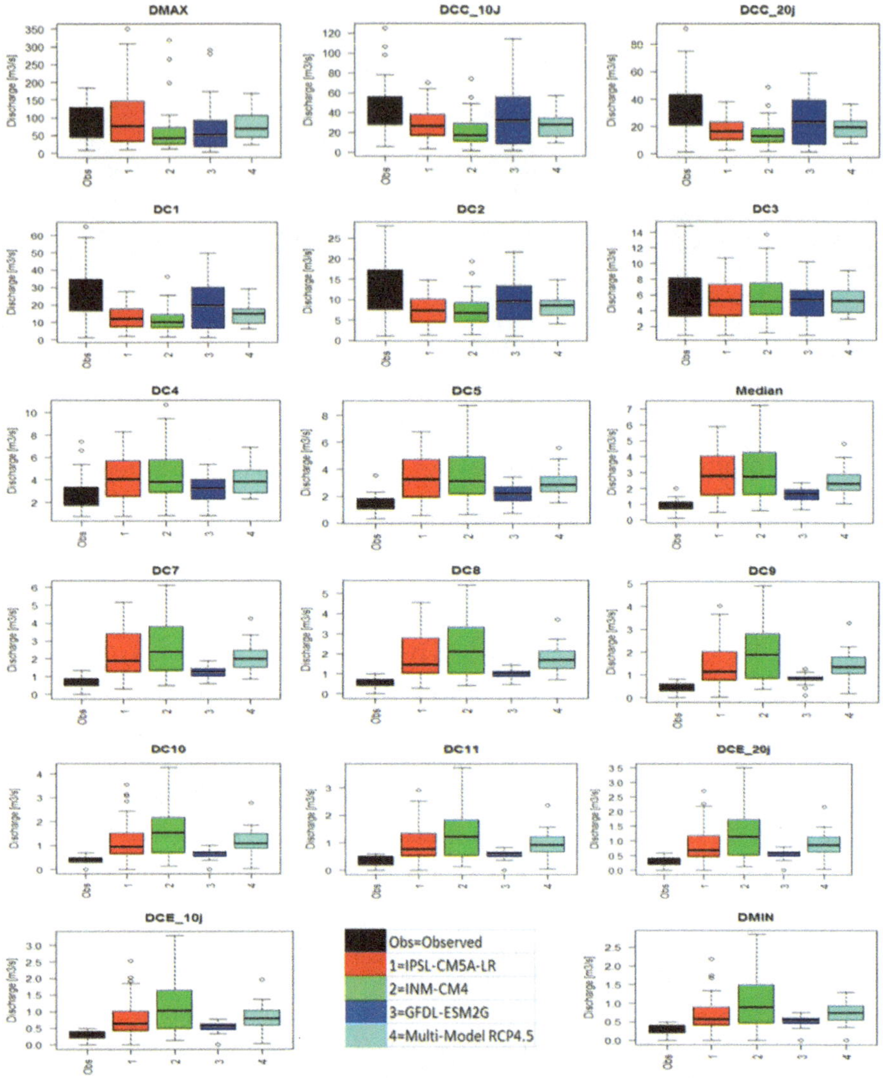

Fig. 18 Evolution of characteristic flow rates (DCC et DCE) by 2050 under the RCP4.5 scenario

RCP4.5 and RCP8.5 scenarios) for the different RCM, compared to the reference period 1961–2012.

The simulation results project a reduction of 9% (GFDL-ESM2G model) to 59% (INM-CM4 model) of the maximum daily flow rates (DMAX) under the RCP4.5 scenario. Only the IPSL-CM5A-LR model forecasts a slight increase of 7%. The RCP4.5 multi-model set (average of the three RCM) predicts a 15% and 14% decrease in peak flow achieved or exceeded for 10 days (DCC_10j), 14% for 20 days (DCC_20j), 16% for one month (DC1), 20% for two months (DC2) and 7% for three

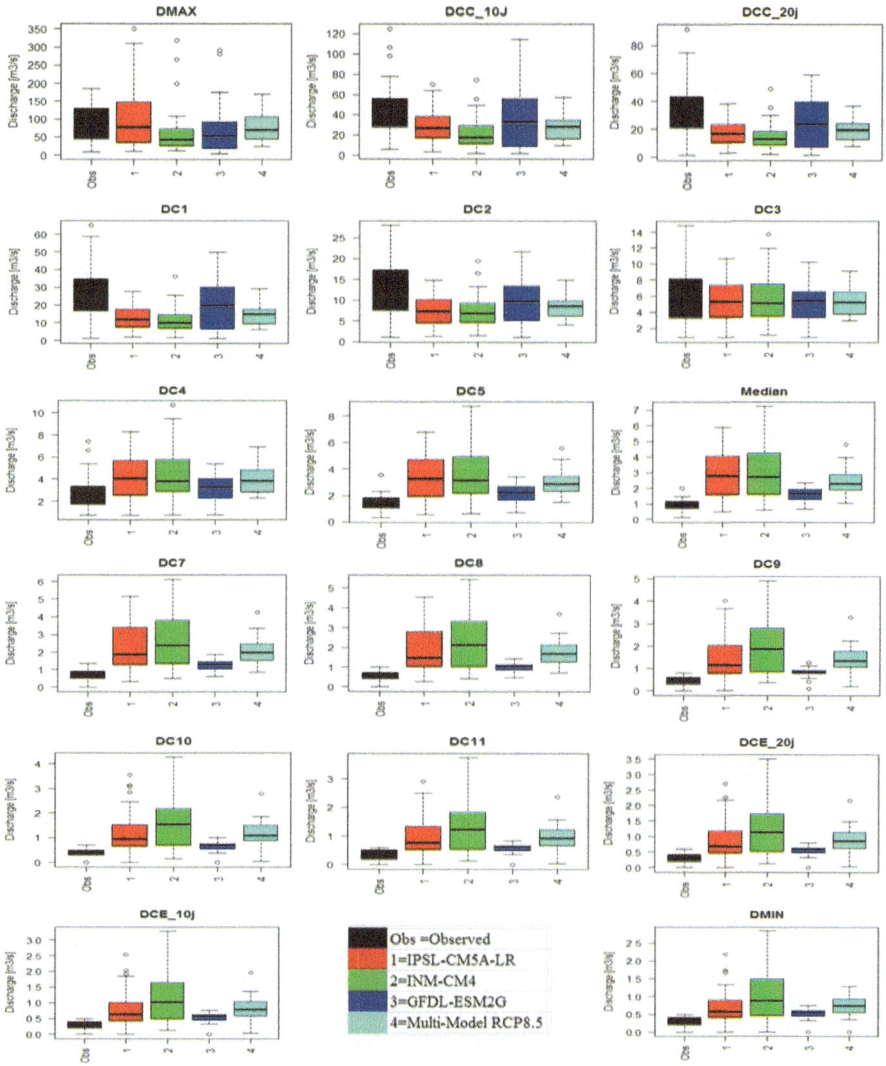

Fig. 19 Evolution of characteristic flow rates (DCC et DCE) by 2050 under the RCP8.5 scenario

months (DC3). However, all models forecast an increase of 23%, 42% and 50% of the characteristic flows reached or exceeded for four months (DC4), five months (DC5) and six months (median flow), respectively, for the period 2020–2050. Low flow (DC7, DC8, DC9, DC10, DC11, DCE_20j, DCE_10j, and DMIN) models project an average increase of 59% under the scenario RCP4.5.

The decrease in characteristic rates is more pronounced under the RCP8.5 scenario. The GFDL-ESM2G and INM-CM4 models predict a decrease of 35% and 47% in DMAX; 24% and 97% (DCC_10j); 30% and 93% (DCC_20j); 27% and

98% (DC1); 27% and 63% (DC2); and 4% and 11% (DC3), respectively. Model IPSL-CM5A-LR, predicts a 20% increase in DMAX, and a chronic decrease in DCC_10j (57%), DCC_20j (86%), DC1 (96%) and DC2 (67%), by 2050. Overall averages with RCP8.5 (multi-model-RCP8.5) predicts a decrease of 41% (DMAX), 66% (DCC_10j), 80% (DCC_20j and DC1), 51% (DC2) and 7% (DC3). The models also predict by 2050 an increase of 23%, 32% and 33% of the DC4, DC5 and median flows, respectively.

For low flow characteristic rates (DC7, DC8, DC9, DC10, DC11, DCE_20j, DCE_10j, and DMIN), models predict an average increase of 63% by 2050, under RCP8.5. In sum, the hydrological model submitted to the outputs of regional climate models projects a decrease in DMAX, DCC_10j, DCC_20j, DC1, DC2 and DC3, and an increase in Diarha low flow characteristic rates under both emission scenarios (RCP4.5 and RCP8.5). The hydrological model only predicts an increase in DMAX with the outputs of the IPSL-CM5A-LR model under both RCP scenarios.

4 Discussion and Conclusions

Predicting the variation of water volumes in a watershed is essential because it enables planners to assess the correlation between water demand and water availability in order to anticipate and manage potential conflicts among users and sectors. In this study, two hydrological models, GR4J and SAC-SMA, of the RS Minerve were calibrated based on existing rainfall data for the periods 1975–1992 and 1998–2003. Results show that the GR4J model is better correlated with actual observed flow rates of the Diarha over the calibration (1975–1992) and validation (1998–2003) phases. The parameters X1, X2, X3 and X4 were determined based on the calibration period and used to simulate flow rates over the missing periods. This allowed for the extension of data to represent discharges continuously from 1961 to 2012. In addition, randomness verification tests show the absence of stability on Diarha flow chronicles, confirming the sensitivity of climate variables to climate change. (Stanzel et al. 2018). The work of Thiaw (2017) confirms that rainfall series in the Diarha catchment are marked by high interannual variability and long-term fluctuations, from a very wet period between 1921 and 1967 to a very dry period between 1968 and 1990, followed by a slight return of rainfall starting in the 2000s.

At the same time, it should be noted that flow rates do not depend solely on climate. Several factors related to the hydrological structure of the watershed (geology, pedology, land use, etc.) may also impact model outputs. Unfortunately, the GR4J model, though calibrated in semi-distributed conceptual mode, does not integrate land use dynamics in the flow simulation. In addition, several studies have shown that this model does not fully account for the relationship between groundwater and surface water (Fabre et al. 2014, Bodian et al. 2018). Therefore, depending on the physical characteristics of the basins modeled, the GR4J model overestimates or underestimates low flows and peak flows. There are other hydrological models that better take into account the interaction between groundwater and surface water

(Pulido-Velazquez et al. 2007); however, their application at the scale of the Diarha catchment is difficult because of the scarcity of hydrogeological data. Overall, despite its inability to integrate the atmosphere-water-vegetation cycle, the semi-distributed conceptual approach of the GR4J model correlates simulated and observed Diarha hydrographs well.

After calibrating the GR4J model and taking into account the interannual variability of the Diarha flows, the outputs of the RCM (GFDL-ESM2G, INM-CM4 and IPSL-CM5A-LR) of the CMIP5 (Coupled Model Intercomparison Project 5) experiment under two Representative Concentration Pathways (RCP4.5 and RCP8.5) were used to simulate the trend of flows by 2050, compared to the 1961–2012 reference period. The models predict that by 2050 there will be a decrease in annual mean flows of the Diarha compared to the 1961–2012 reference period.

Under the RCP4.5 scenario, the models predict a decrease of 2.1% in precipitation and an average temperature increase of 1.4 °C, resulting in a decrease of 5% in Diarha flows. The decrease of precipitation and, therefore, of streamflow is more pronounced under the RCP8.5 scenario, which estimates a rainfall decrease of 12% and an increase in temperatures of 1.8 °C, causing a net decrease of 25% in rainflow. The models project, beyond the decline in annual flows, a decrease in the characteristic flow rates (DMAX, DCC_10j, DCC_20j, DC1, DC2 and DC3) and an increase in low flows in the two emissions scenarios (RCP4.5 and RCP8.5). This increase in characteristic low flows (DCE) may be due to the limitation of the hydrological model (GR4J) to model an extremely dry climate.

These results are consistent with those of Bodian et al. (2018) in the Gambia River at Mako station and Senegal River at Bakel station. In that study, the authors used six GCMs (CanESM2, CNRM, CSIRO, HadGEM2-CC, HadGEM2-ES and MIROC5) from CMIP5, and the GR4J model to evaluate the hydrological impact of climate change. They predicted a decline in Senegal's streamflow of 8% (RCP4.5) to 16% (RCP8.5), and a decline in the Gambia' from 22% (RCP4.5) to 26% (RCP8. 5) by 2050. In addition, Mbaye et al. (2015), using a CORDEX simulation of the regional REMO climate model as input to the MPI-HM hydrological model, found a decline in flows, runoff, ETR, and soil moisture in the upper basin of the Senegal River by 2071–2100, under the scenarios RCP4.5 and RCP8.5. The authors predicted a decline of more than 50% in the water resources of the Senegal River, especially in the northern part.

Furthermore, Setegn et al. (2011) used the outputs of nine GCMs under the SRES-A2 emission scenario and the SWAT model to assess the impact of climate change on the water resources of Lake Tana, in Ethiopia. They found a significant decrease in average annual flows by 2080–2100, consistent with the work of Biao, (2017) in the sub-basins Betou and Bonou of the Oueme River in Benin. In that study, the HyMoLAP hydrological model was submitted to the outputs of the HIRHAM5 and RCA4 climate models under the RCP4.5 and RCP8.5 scenarios to assess the hydrological impact of climate change. It noted, under the RCP4.5 scenario, a drop in river flows with a magnitude ranging respectively from 25% to -39% and from 20% to 37% by 2020 (2011–2040), 2050 (2041–2070) and 2080 (2071–2100), and a drop between 15% and 34%; and 18% to 36% over the same time horizons under the RCP8.5 scenario.

On the other hand, these results contradict the work of Azari et al. (2015) in the Gorganroud watershed of Iran. To simulate the hydrological impact of climate change on this basin, the authors submitted the outputs of the A1F1, A2 and B1 emission scenarios to the SWAT model. They found an increase in the mean annual flow of the Gorganroud of 5.8% (A1F1), 2.8% (A2) and 9.5% (B1). In addition, Ardoin-Bardin et al. (2009), using a hydrological model at monthly intervals (GR2M), using GCM outputs (CSIRO-Mk2, ECHAM4, HadCM3 and NCAR-PCM) from the IPCC third report, underscored the failure of these climate models to reproduce actual precipitation volumes in the Sahelian zone as well as their failure to simulate the seasonal dynamics of rainfall in the Guinean zone. For this reason, these studies used two climate scenarios based on the variations predicted by the HadCM3-A2 model to generate precipitation and ETP data to the end of the twenty-first century. These data then became inputs to the GR2M model to assess the future impacts of climate change on the flows of the Senegal, Gambia, Sassandra and Chari rivers. The results indicate a drop-in flows from the Gambia and Senegal rivers and an increase in the Sassandra and Chari rivers. Variables such as the differences between rainfall-flow models, the quality of Global/Regional Climate Models used, and the geography under study (difference in climate, soil, relief, land use, etc.) can be decisive factors. Future studies should use rainfall-flow models that integrate, in addition to the climate, other variables that impact the hydrological dynamics of a watershed (hydrogeology, topography, pedology, land use, etc.), for example, the HEC-HMS and SWAT models.

This study demonstrates the negative impact of climate change, especially for characteristic high flows (DCC), and suggests the importance of integrating analysis into adaptive management programs. These results will be useful to decision-makers of the Gambia River Basin Development Organization (OMVG) and the Support Program for Agricultural Development and Rural Entrepreneurship (PADAER) to build and improve adaptive water resource management practices in the context of climate change. However, assumptions based on these results should be cautious due to weak monitoring networks in the Diarha river basin. Future studies would benefit from the implementation of a robust monitoring network to provide more reliable data on water resources in order to improve approaches for long-term exploitation and sustainable management. Also, this research has not specifically studied the impact of climate change-related changes in streamflow and characteristics in relation to the water demand-availability balance. Future studies should examine how climate change will impact different users of Diarha's water resources and study the correlation between demand and available water resources in different time horizons and scenarios. This study will serve as a foundation for this future work.

Author Contributions Ibrahima Thiaw designed the study, developed the methodology and wrote the manuscript. Ibrahima Thiaw and Bakary Faty collected and processed the data, while Abel Vincent Manga, Anastasie Mendy, Honoré Dacosta and Amadou Abdoul Sow read and corrected the manuscript.

Table 8 Relative change in characteristic high flows for different used RCM under RCP4.5

River Basin	Flow rates	GFDL-ESM2G	INM-CM4	IPSL-CM5A-LR	Multi-Model_RCP4.5
Diarha river RCP4.5	DMAX	– 0.09	– 0.59	0.07	– 0.20
	DCC_10j	– 0.07	– 0.26	– 0.12	– 0.15
	DCC_20j	– 0.07	– 0.18	– 0.18	– 0.14
	DC1	– 0.06	– 0.16	– 0.27	– 0.16
	DC2	– 0.08	– 0.19	– 0.32	– 0.20
	DC3	– 0.04	– 0.05	– 0.12	– 0.07
	DC4	0.21	0.24	0.24	0.23
	DC5	0.39	0.41	0.46	0.42
	Median	0.46	0.48	0.57	0.50

Table 9 Relative change in flows below the median for different chosen RCM under RCP4.5

River basin	Flow rates	GFDL-ESM2G	INM-CM4	IPSL-CM5A-LR	Multi-Model_RCP4.5
Diarha river RCP4.5	DC7	0.49	0.52	0.64	0.55
	DC8	0.51	0.54	0.68	0.57
	DC9	0.53	0.53	0.71	0.59
	DC10	0.39	0.52	0.73	0.55
	DC11	0.53	0.54	0.76	0.61
	DCE_20j	0.52	0.54	0.76	0.61
	DCE_10j	0.53	0.54	0.77	0.62
	DMIN	0.54	0.56	0.78	0.63

Conflicts of Interest The authors declare no conflict of interest.

Appendix A

See Tables 8 and 9.

Appendix B

See Tables 10 and 11.

Table 10 Relative change in characteristic high flows for different used RCM under RCP8.5

River Basin	Flow rates	GFDL-ESM2G	INM-CM4	IPSL-CM5A-LR	Multi-Model_RCP8.5
Diarha river RCP8.5	DMAX	– 0.35	– 0.47	0.20	−0.41
	DCC_10j	– 0.24	– 0.97	– 0.57	−0.59
	DCC_20j	– 0.30	– 0.93	– 0.86	– 0.70
	DC1	– 0.27	– 0.98	– 0.96	– 0.70
	DC2	– 0.27	– 0.63	– 0.62	– 0.51
	DC3	– 0.11	– 0.04	– 0.07	– 0.07
	DC4	0.13	0.29	0.26	0.23
	DC5	0.22	0.38	0.36	0.32
	Median	0.23	0.39	0.36	0.33

Table 11 Relative change in flows below the median for different chosen RCM under RCP8.5

River Basin	Flow rates	GFDL-ESM2G	INM-CM4	IPSL-CM5A-LR	Multi-Model_RCP8.5
Diarha river RCP8.5	DC7	0.44	0.74	0.70	0.63
	DC8	0.46	0.76	0.71	0.65
	DC9	0.48	0.78	0.72	0.66
	DC10	0.45	0.77	0.71	0.64
	DC11	0.47	0.77	0.68	0.64
	DCE_20j	0.46	0.75	0.66	0.63
	DCE_10j	0.48	0.74	0.64	0.62
	DMIN	0.49	0.73	0.61	0.61

References

Ajami K, Gupta N, Wagener HT, Sorooshian S (2004) Calibration of a semi-distributed hydrologic model for streamflow estimation along a river system. J Hydrol 298(1–4):112–135. https://doi.org/10.1016/j.jhydrol.2004.03.033

Anderson RM, Koren VI, Reed SM (2006) Using SSURGO data to improve Sacramento Model a priori parameter estimates. J Hydrol 320(1–2):103–116. https://doi.org/10.1016/j.jhydrol.2005.07.020

Ardoin-Bardin S, Dezetter A, Servat E, Paturel JE, Mahé G, Niel H, Dieulin C (2009) Using general circulation model outputs to assess impacts of climate change on runoff for large hydrological catchments in West Africa. Hydrol Sci J 54(1):77–89. https://doi.org/10.1623/hysj.54.1.77

Azari M, Moradi HR, Saghafian B, Faramarzi M (2015) Climate change impacts on streamflow and sediment yield in the North of Iran. Hydrol Sci J 61(1):123–133. https://doi.org/10.1080/02626667.2014.967695

Biao E (2017) Assessing the Impacts of Climate Change on River Discharge Dynamics in Oueme River Basin (Benin, West Africa). Hydrology 4(4):47. https://doi.org/10.3390/hydrology4040047

Bodian A, Dezetter A, Dacosta H, Ardoin-Bardin S (2013) Impact du changement climatique sur les ressources en eau du haut bassin du fleuve Sénégal. Rev. Géogr. Lab. Leïd. 2515:236–251

Bodian A, Dezetter A, Dacosta H (2015) Rainfall-runoff modeling of water resources in the upper Senegal River basin. Int J Water Resour Dev 32(1):89–101. https://doi.org/10.1080/07900627.2015.1026435

Bodian A, Dezetter A, Diop L, Deme A, Djaman K, Diop A (2018) Future Climate Change Impacts on Streamflows of Two Main West Africa River Basins: Senegal and Gambia. Hydrology 5(1):21. https://doi.org/10.3390/hydrology5010021

Bop, M.; Amadou, A.; Seidou, O.; Kébé, C.M.F.; Ndione, J.A.; Sambou, S.; Sanda, I.S. Modeling the Hydrological Dynamic of the Breeding Water Bodies in Barkedji's Zone. J. Water Resour. Prot. 2014, 6, 741–755. [Google Scholar]

Blasone R-S, Madsen H, Rosbjerg D (2007) Parameter estimation in distributed hydrological modeling: comparison of global and local optimisation techniques. Nord Hydrol 38(4–5):451. https://doi.org/10.2166/nh.2007.024

Brulebois E. (2016). Impacts du changement climatique sur la disponibilité de la ressource en eau en Bourgogne : aspects quantitatifs et qualitatifs. Sciences de la Terre. Université de Bourgogne. Français. < NNT:2016DIJOS027 > .<tel-01488290>

Buishand TA (1982) Some methods for testing the homogeneity of rainfall records. J Hydrol 58(1–2):11–27. https://doi.org/10.1016/0022-1694(82)90066-x

Burnash, R. J., Ferral, R. L., & McGuire, R. A. (1973). A generalized streamflow simulation system, conceptual modeling for digital computers

Burnash RJC (1995) The NWS river forecast system-catchment modeling. Computer models of watershed hydrology 188:311–366

Cisse MT, Sambou S, Dieme Y, Diatta C, Bop M (2014) Analyse des écoulements dans le bassin du fleuve Sénégal de 1960 à 2008. Rev Sci Eau 27(2):167. https://doi.org/10.7202/1025566ar

DACOSTA H. (1989) - *Précipitation et écoulements sur le Bassin de la Casamance.* Thèse de 3ème cycle, Faculté des Lettres et Sciences Humaines, Université Cheikh Anta Diop de Dakar, 278 p

Déqué M (2007) Frequency of precipitation and temperature extremes over France in an anthropogenic scenario: model results and statistical correction according to observed values. Global Planet Change 57:16–26

Duan Q, Sorooshian S, Gupta V (1992) Effective and efficient global optimization for conceptual rainfall-runoff models. Water Resour Res 28(4):1015–1031. https://doi.org/10.1029/91wr02985

Duan Q, Sorooshian S, Gupta VK (1994) Optimal use of the SCE-UA global optimization method for calibrating watershed models. J Hydrol 158(3–4):265–284. https://doi.org/10.1016/0022-1694(94)90057-4

Eckhardt K, Arnold JG (2001) Automatic calibration of a distributed catchment model. J Hydrol 251(1–2):103–109. https://doi.org/10.1016/s0022-1694(01)00429-2

Fabre J, Ruelland D, Dezetter A, Grouillet B (2014) Simulating long-term past changes in the balance between water demand and availability and assessing their main drivers at the river basin management scale. Hydrol Earth Syst Sci Dis 11(11):12315–12364. https://doi.org/10.5194/hessd-11-12315-2014

FALL A. (2014), Le Ferlo sénégalais. Approche géographique de la vulnérabilité des anthroposystèmes sahéliens, thèse de doctorat, Université Paris 13 Sorbonne Paris-cité, 380 p

Famien AM, Janicot S, Ochou AD, Vrac M, Defrance D, Sultan B, Noël T (2018) A bias-corrected CMIP5 dataset for Africa using CDF-t method. Earth Syst. Dynam. Discuss., under review, A contribution to agricultural impact studies. https://doi.org/10.5194/esd-2017-111

Faye C (2018) Climatic Variability and Hydrological Impacts in West Africa: Case of the Gambia Watershed (Senegal). Environmental and Water Sciences, public Health and Territorial Intelligence Journal 2(1):54–66

Gan TY, Burges SJ (2006) Assessment of soil-based and calibrated parameters of the Sacramento model and parameter transferability. J Hydrol 320(1–2):117–131. https://doi.org/10.1016/j.jhydrol.2005.07.008

García Hernández, J., Paredes Arquiola, J., Foehn, A. and Roquier, B. (2018). RS MINERVE—Technical manual v2.13. RS MINERVE Group, Switzerland

Gilks WR, Roberts GO, Sahu SK (1998) Adaptive Markov Chain Monte Carlo through Regeneration. Journal of the American Statistical Association 93(443):1045. https://doi.org/10.2307/2669848

Giorgi, F. Jones, C. & Asrar G.R (2009). Addressing climate information needs at the regional level: the CORDEX framework. WMO Bull., 58 (3) (2009), pp. 175–183

Guo L-Y, Gao Q, Jiang Z-H, Li L (2018) Bias correction and projection of surface air temperature in LMDZ multiple simulation over central and eastern China. Advances in Climate Change Research 9(1):81–92. https://doi.org/10.1016/j.accre.2018.02.003

Gupta, H. V., Bastidas, L. A., Vrugt, J. A., & Sorooshian, S. (2003). Multiple criteria global optimization for watershed model calibration. Calibration of Watershed Models, 125–132

Grouillet B, Ruelland D, Ayar PV, Vrac M (2016) Sensitivity analysis of runoff modeling to statistical downscaling models in the western Mediterranean. Hydrol Earth Syst Sci 20:1031–1047

Hubert P., Servat E., Paturel J-E., Kouamé B., Bendjoudi H., Carbonnel J-P., et Lubes-Niel H., 1998. La procédure de Segmentation, 10 ans après. http://horizon.documentation.ird.fr/exl-doc/pleins_textes/divers09-04/010017937.pdf

IRD (1998)-KhronoStat version 1.0. Logiciel, IRD, Montpellier. IRD, (1998)-KhronoStat version 1.0. logiciel, IRD, Montpellier. URL : http://www.hydrosciences.org/spip.php?article239

Jeon J-H, Park C-G, Engel B (2014) Comparison of Performance between Genetic Algorithm and SCE-UA for Calibration of SCS-CN Surface Runoff Simulation. Water 6(11):3433–3456. https://doi.org/10.3390/w6113433

Kan G, Liang K, Li J, Ding L, He X, Hu Y, Amo-Boateng M (2016) Accelerating the SCE-UA Global Optimization Method Based on Multi-Core CPU and Many-Core GPU. Advances in Meteorology 2016:1–10. https://doi.org/10.1155/2016/8483728

Legates DR, McCabe GJ (1999) Evaluating the use of "goodness-of-fit" Measures in hydrologic and hydroclimatic model validation. Water Resour Res 35(1):233–241. https://doi.org/10.1029/1998wr900018

Lebel T, Ali A (2009) Recent trends in the Central and Western Sahel rainfall regime (1990–2007). J Hydrol 375(1–2):52–64. https://doi.org/10.1016/j.jhydrol.2008.11.030

Liu JS (2001) *Monte Carlo Strategies in Scientic Computing*. Springer Verlag, New-York, p 343p

Liu Y (2009) Automatic calibration of a rainfall–runoff model using a fast and elitist multi-objective particle swarm algorithm. Expert Syst Appl 36(5):9533–9538. https://doi.org/10.1016/j.eswa.2008.10.086

Lujano Laura E, Sosa Sarmiento JD, Lujano Laura A, Lujano Laura R (2016) Modelación hidrológica semidistribuida en la región hidrográfica del Titicaca: caso de estudio cuenca del río Ramis, Perú. Revista de Investigaciones Altoandinas - Journal of High Andean Research 18(4):431. https://doi.org/10.18271/ria.2016.235

MAREGA O. (2016), Changements socio-environnementaux et évolution des pratiques agropastorales en Afrique sahélienne : étude comparée entre le Ferlo (Sénégal), le Gourma (Mali) et le Fakara (Niger) » , thèse de doctorat, Université Diderot-Paris 7, 700 p

Mbaye ML, Haensler A, Hagemann S, Gaye AT, Moseley C, Afouda A (2015) Impact of statistical bias correction on the projected climate change signals of the regional climate model REMO over the Senegal River Basin. Int J Climatol 36(4):2035–2049. https://doi.org/10.1002/joc.4478

Mbaye, M. L., Diatta, S., & Gaye, A. T. (2018). Climate Change Signals Over Senegal River Basin Using Regional Climate Models of the CORDEX Africa Simulations. Lecture Notes of the Institute for Computer Sciences, Social Informatics and Telecommunications Engineering, 123–132

Michelangeli P-A, Vrac M, Loukos H (2009) Probabilistic downscaling approaches: application to wind cumulative distribution functions. Geophys Res Lett 36:L11708. https://doi.org/10.1029/2009GL038401

MUTTIL, N., & LIONG, S.-Y. (2004). ARE COMPLEXES IN SCE-UA NECESSARY? Hydroinformatics, 874–881

Nelder JA, Mead R (1965) A Simplex Method for Function Minimization. Comput J 7(4):308–313. https://doi.org/10.1093/comjnl/7.4.308

Oudin L (2006) Une formule simple d'évapotranspiration potentielle pour la modélisation pluie-débit à l'échelle du bassin versant. La Houille Blanche 6:113–120. https://doi.org/10.1051/lhb:2006109

Paeth H, Hense A (2004) SST versus Climate Change Signals in West African Rainfall: 20th-Century Variations and Future Projections. Clim Change 65(1/2):179–208. https://doi.org/10.1023/b:clim.0000037508.88115.8a

Panthou G, Vischel T, Lebel T (2014) Recent trends in the regime of extreme rainfall in the Central Sahel. Int. J, Clim

Perrin C, Michel C, Andréassian V (2003) Improvement of a parsimonious model for streamflow simulation. J Hydrol 279(1–4):275–289. https://doi.org/10.1016/s0022-1694(03)00225-7

Pulido-Velazquez, D., Sahuquillo, A., Andrieu, J., & Pulido-Velazquez, M. (2007). An efficient conceptual model to simulate surface water body-aquifer interaction in conjunctive use management models. Water Resources Research, 43(7). https://doi.org/10.1029/2006wr005064

Roudier P, Ducharne A, Feyen L (2014) Climate change impacts on runoff in West Africa: a review. Hydrol Earth Syst Sci 18(7):2789–2801. https://doi.org/10.5194/hess-18-2789-2014

Roche M. (1963).Hydrologie de surface

Rosenbrock HH (1960) An Automatic Method for Finding the Greatest or Least Value of a Function. Comput J 3(3):175–184. https://doi.org/10.1093/comjnl/3.3.175

SERVAT É., PATUREL J.E., KOUAME B., TRAVAGLIO M., OUEDRAOGO M., BOYER J.F., LUBES-NIEL H., FRITSCH J.M., MASSON J.M., MARIEU B. (1998) - Identification, caractérisation et conséquences d'une variabilité hydrologique en Afrique de l'Ouest et Centrale. *In*: *Water Resources Variability in Africa during the XXth Century* (Abidjan'98 Conference), É. SERVAT, D. HUGUES, J.M. FRITSCH, M. HULME édit., Publication AISH, n° 252, p. 323–337

Setegn, S. G., Rayner, D., Melesse, A. M., Dargahi, B., & Srinivasan, R. (2011). Impact of climate change on the hydroclimatology of Lake Tana Basin, Ethiopia. Water Resources Research, 47(4). https://doi.org/10.1029/2010wr009248

Sorooshian S, Duan Q, Gupta VK (1993) Calibration of rainfall-runoff models: Application of global optimization to the Sacramento Soil Moisture Accounting Model. Water Resour Res 29(4):1185–1194. https://doi.org/10.1029/92wr02617

Stanzel P, Kling H, Bauer H (2018) Climate change impact on West African rivers under an ensemble of CORDEX climate projections. Climate Services 11:36–48. https://doi.org/10.1016/j.cliser.2018.05.003

Thiaw I (2017) Rainfall Variability and Water Supplies in the Diarha Watershed (Tributary of Gambia River). Hydrology 5(4):41. https://doi.org/10.11648/j.hyd.20170504.11

Thiaw I, Dacosta H (2017) Mapping of Soil Erosion Risk in the Diarha Watershed Using Rusle, RS and GIS. American Journal of Remote Sensing 5(4):30. https://doi.org/10.11648/j.ajrs.20170504.11

Thiaw I, Dacosta H, Mendy A, Sow AA (2017) Topographic characterization and mapping of soil erosion in the Diarha watershed using RUSLE, RS and GIS. International Journal of Advanced Research 5(12):657–675. https://doi.org/10.21474/ijar01/6005

Vigau N, Vrac M, Caballero Y (2013) Probabilistic downscaling of GCM scenarios over southern India. Int J Climatol 33:1248–1263

Vrac M, Friederichs P (2015) Multivariate-Intervariable, Spatial, and Temporal-Bias Correction. J. Climate 28:218–237. https://doi.org/10.1175/JCLI-D-14-00059.1

Vrac M, Noël T, Vautard R (2016) Bias correction of precipitation through Singularity Stochastic Removal: Because occurrences matter. J. Geophys. Res. Atmos. 121:5237–5258. https://doi.org/10.1002/2015JD024511

Vrac M, Drobinski P, Merlo A, Herrmann M, Lavaysse C, Li L, Somot S (2012) Dynamical and statistical downscaling of the French Mediterranean climate: uncertainty assessment. Nat Hazards Earth Syst Sci 12:2769–2784. https://doi.org/10.5194/nhess-12-2769-2012

Vrugt, J. A., Gupta, H. V., Bastidas, L. A., Bouten, W., & Sorooshian, S. (2003). Effective and efficient algorithm for multiobjective optimization of hydrologic models. Water Resources Research, 39(8). https://doi.org/10.1029/2002wr001746

Vrugt, J. A., Clark, M. P., Diks, C. G. H., Duan, Q., & Robinson, B. A. (2006). Multi-objective calibration of forecast ensembles using Bayesian model averaging. Geophysical Research Letters, 33(19). https://doi.org/10.1029/2006gl027126

Zhu, X., Wu, J., & Wu, J. (2006). Application of SCE-UA to Optimize the Management Model of Groundwater Resources in Deep Aquifers of the Yangtze Delta. First International Multi-Symposiums on Computer and Computational Sciences (IMSCCS'06). https://doi.org/10.1109/imsccs.2006.192

Chapter 14
Water Resources in the Sahel and Adaptation of Agriculture to Climate Change: Burkina Faso

Babacar Lèye, Cheick Oumar Zouré, Roland Yonaba, and Harouna Karambiri

Abstract In West Africa, a shift in climate conditions since the 1970s is evidenced by a decrease in average annual rainfall and an increase in temperatures between 1931–1960 and 1968–1990. The latter period was followed by a rainfall recovery since the late 1990s in the western Sahel. The prolonged rainfall deficits of the twentieth century have, however, strongly influenced the hydrological environment of the major river basins. As a result, nowadays, there is a reduction in the vegetation cover as compared to the early 1950s, thus an increase in runoff, along with the gradual disappearance of permanent streams and the appearance of large floodplains. Climate change has also had negative consequences on agricultural activity, which is still mainly rainfed. The current state has motivated local stakeholders to undertake programs and strategies to address the consequences of climate change. In this chapter, an inventory of water resources and management practices in Burkina Faso is presented and the impacts of climate change on the river hydrology and the agricultural sector, as well as adaptation measures to these changes by the affected communities, are investigated. These strategies include adaptation, mitigation and strategic planning. The impacts on all stakeholders, including the rural communities, government authorities and international agencies, are discussed.

Keywords Agriculture · Burkina faso · Climate change

1 Introduction

In West Africa, particularly in Burkina Faso, climate variability has increased since the early 1970s. Between 1931–1960 and 1968–1990, there was a decrease in average annual rainfall of 20–40% (Intergovernmental Panel on Climate Change (IPCC 2014)). According to the IPCC, average temperatures in Africa increased by 0.6 °C just between 1900 and 1996, and is anticipated to increase by another 3.5 °C during

B. Lèye (✉) · C. O. Zouré · R. Yonaba · H. Karambiri
Laboratory of Water, HydroSystems and Agriculture (LEHSA), International Institute for Water and Environmental Engineering (2iE), Rue de La Science, 01, PO Box 594, Ouagadougou 01, Burkina Faso
e-mail: babacar.leye@2ie-edu.org

S. Diop et al. (eds.), *Climate Change and Water Resources in Africa*,
https://doi.org/10.1007/978-3-030-61225-2_14

the twenty-first century in the Sahel zone (Dia and Duponnois 2013). In contrast, Burkina Faso is experiencing an increase in rainfall since the 1990s (Nicholson 2005; Hountondji et al. 2009), although a trend towards a significant decrease in the number of rainy days still persists (Hountondji et al. 2009; Lebel and Ali 2009).

Rainfall deficit in Burkina Faso has strongly influenced the hydrological regimes of the major river basins. The average annual flows of the major rivers in the region have decreased by 30 to 60% between the 1960s and the early 2000s (Ardoin Bardin 2004). In addition, significant population growth (about 3% per year) since the beginning of the 1990s has led to a profound change in management methods and the use of natural resources (Dia and Duponnois 2013). Farmers have had to expand the areas under cultivation to compensate for lower yields and to meet the growing needs of an increasing population. As a result, there is a reduction in the vegetation cover, thus promoting runoff, the disappearance of permanent river streams and the appearance of large floodplains.

Climate change has other negative consequences for agricultural activity, which is mainly rainfed. The agricultural sector employs 86% of the rural population and contributes up to 30% of the gross domestic product. Because irrigation is practiced on less than 2% of cultivated areas (FAO 2005), large numbers of people remain vulnerable to dry spells that can reduce agricultural production.

This situation has motivated stakeholders in the field (including rural populations, non-governmental organizations and government technical services) to undertake programs and strategies to address the negative consequences and effects of climate change. Thus, farmers and researchers have used techniques including water conservation and harvesting, soil restoration, varietal adaptation and the use of climate information (Sawadogo 2011; Zougmoré et al. 2014; Zongo et al. 2016). These initiatives have been supported by the government through the development and implementation of national climate change adaptation programs.

The general objective of these programs is to contribute to sustainable development by providing appropriate solutions to water-related problems so that limitations of water do not limit socioeconomic development. To wit, meeting demands for water must respect ecosystems, protect against erosion, flooding and pollution, and reduce the burden to the public by reallocating costs to the beneficiaries of investments. Strategies must also seek to prevent conflicts with neighboring countries. At the regional level, many new projects facilitate agricultural planning in response to climate hazards (Redelsperger et al. 2006; Roudier et al. 2014). These techniques make it possible to recover large areas of degraded land and improve agricultural yields. Irrigation is also essential to increase agricultural production. For this reason, the government and its partners have constructed dams on many rivers to ensure the availability of water resources. Small reservoirs are now an essential component of rural landscapes and an undeniable element of water resource management (Cecchi et al. 2009). These reservoirs mitigate the negative effects of droughts through temporal regulation of available water and provide flood protection. One of the additional benefits of these hydraulic infrastructures is their contribution to the recharge of groundwater (Koussoubé 1996). The management of water resources, therefore, deserves the highest attention.

The main objective of this study is to summarize adaptation strategies implemented in Burkina Faso in response to the impact of climate change on water resources. It does this by first presenting an inventory of water resources and management practices, second by presenting the impacts of climate change on the agricultural sector, hydrology of the catchment areas and the perceptions of the population regarding climate change, and third by presenting some of the climate change adaptation policies and strategies adopted. These policies and strategies are discussed at the levels of the rural community, government authorities and international agencies through empirical research.

2 Water Resources

The demand for domestic water is estimated at 104 million m^3 per year, of which 40 million m^3 serves urban and semi-urban demand, and 64 million m^3 serves rural demand. The supply of drinking water to urban centers is the responsibility of the National Office for Water and Sanitation. The Office currently manages 56 centers, 37 of which are supplied from groundwater, 12 from surface water and seven from a combination of the two. In rural areas, the main source of modern drinking water is provided through wells equipped with human-powered pumps. According to the official database of the National Inventory of Structures, Burkina Faso had 58,003 Modern Water Points as of 31 December 2015 (MEA/BF 2015), including 48,808 boreholes (i.e., small-diameter wells) and 8,258 drillings (i.e., larger diameter and deeper wells).

The three sources of water in Burkina Faso are rainwater, surface water and groundwater; each is discussed in the following sections.

2.1 *Rainwater*

Burkina Faso has three climatic zones: the Sahelian in the North (600 mm of rain per year), the Sudano-Sahelian (between 600 and 900 mm) and the Sudano-Guinean (between 900 and 1200 mm). These climatic zones are illustrated in Fig. 1.

The country has two seasons: the rainy season, from June to September, and the dry season, from October to May. Rainfall is limited to the rainy season and all water resources depend on it, as do natural and cultivated plants. At the annual or the inter-annual scale, precipitation and evapotranspiration control the dynamics of the country's hydrosystems (Jalota and Arora 2002). For example, Fig. 2 represents climate data from 1961 to 2015 at the Ouahigouya station, located in the Sahelian climatic zone (Mounirou et al. 2012; Zouré et al. 2019), which shows a significant annual water deficit except in July and August.

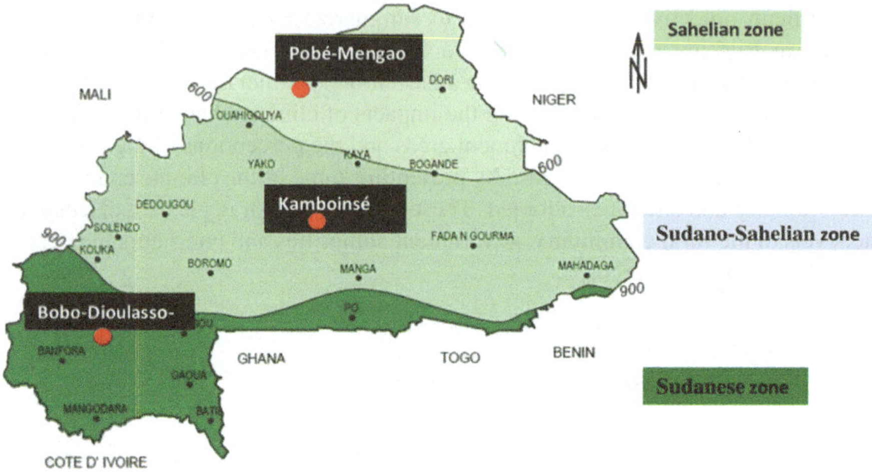

Fig. 1 Climatic zones of Burkina Faso (1981–2010). *Source* Agence Nationale de la Météorologie du Burkina Faso

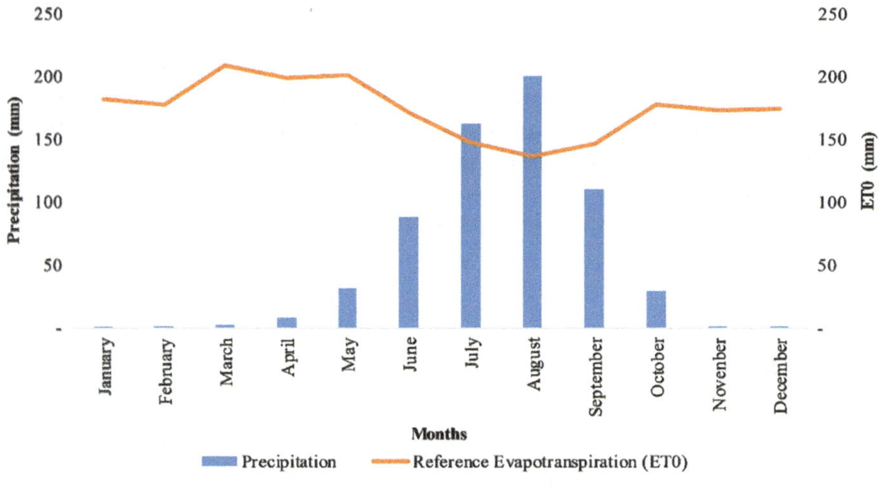

Fig. 2 Interannual evolution of the standardized rainfall index in Ouahigouya during the period 1961–2015

Figure 3 shows the evolution of the standardized precipitation indices at the Ouahigouya station. There is great interannual variability in precipitation and drought between 1969 and 1990, and between 1970 and 1994.

In the Sahel region, since the 1990s, there has been an average of four dry spells per rainy season (Fig. 4a), which is higher than the previous period (1978–1990). A dry spell is defined as a consecutive number of days without effective rainfall (i.e., a rainfall amount which is likely to trigger runoff) during the rainy season (Zouré

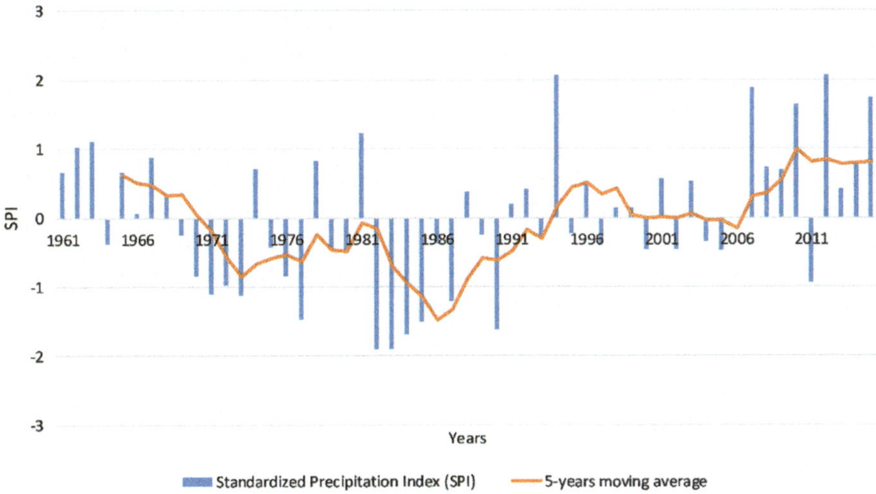

Fig. 3 Climate diagram based on climate data from the Ouahigouya station during the period 1961–2015 (*Source* Agence Nationale de la Météorologie du Burkina Faso)

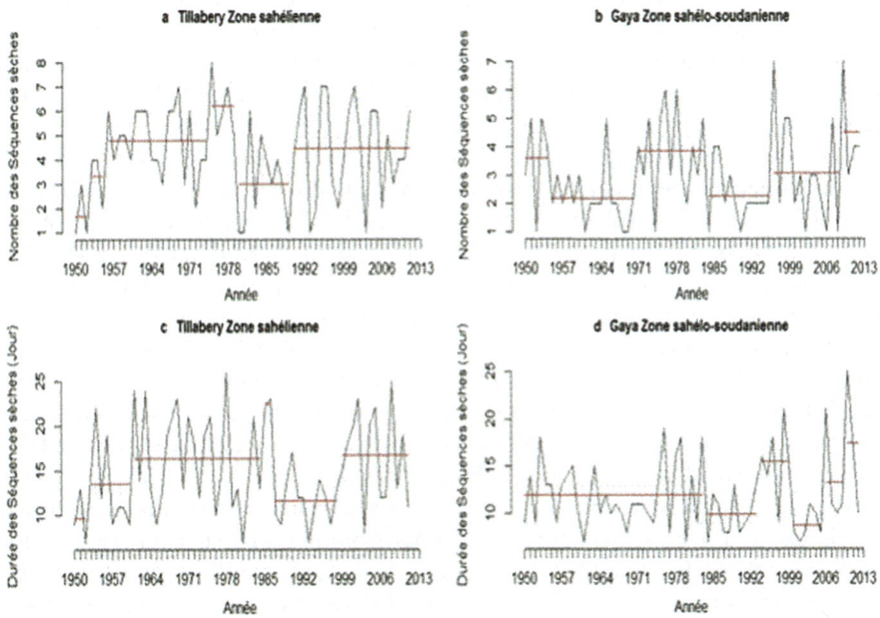

Fig. 4 Number of dry spells (**a**) and (**b**) and dry spell lengths (**c**) and (**d**) at the Tillabery and Gaya stations in Niger (Nassourou et al. 2018)

et al. 2019). In the same area, starting from the year 1998, dry spells of more than 16 days have been observed (Fig. 4c). In the Sahelo-Sudanian zone, the length of dry spells are somewhat stable and generally do not exceed 15 days (Fig. 4d), with a number of dry spells between three and four (Fig. 4b). Nevertheless, from 1984 to 2013, there was more variability (8–17 days) in the duration of dry spells.

As compared to the 1986–2005 IPCC baseline, climate projections predict a decrease in rainfall of 3.4% in 2025 and 7.3% in 2050. The decrease in rainfall will be coupled with high interannual and seasonal variability. Rainfall in July, August and September will decrease by 20–30%, while rainfall in November will experience an increase from 60 to 80% (MECV/BF 2007).

2.2 *Surface Water*

Surface waters consist of streams, lakes, ponds and artificial mobilization of surface water, such as is created by dams. The storable potential of these water reservoirs in Burkina Faso is estimated at about 5 billion m^3, of which 0.8 billion m^3 represents the storage capacity of small dams (Leemhuis et al. 2009). In the Sahelian zone, Class A pan measurements have been used to estimate daily evaporation at 7–8 mm during the rainy season and can reach 10 mm during the dry season, with an average of 2,920 mm per year, far greater than the cumulative annual rainfall (Niang 2006). Daily temperatures vary from 16 to 45 °C over the year.

Reservoirs are made up of 56% dams, 31% artificial lakes, 12% ponds and 1% natural lakes (Naturama 2015). The country has a hydrometric network of about 95 functional stations that realize varying amounts of the water height, volume and flow. According to the Ministry of Agriculture and Hydraulics, Burkina Faso had 1,806 surface water reservoirs in 2015 (MEA/BF 2015). Furthermore, the National Program for Environmental Information Management, which is responsible for improving the relevance, quality and availability of environmental information, developed a database in 1999 that lists 1,457 dams throughout the country (MEA/BF 2015). Nearly half of these dams were built between 1974 and 1987, a period that coincides with the severe droughts of the 1970s and 1980s, rapid population growth and the promotion of small-scale irrigation in West Africa.

The use of small basins has focused on domestic water use, manufacture of dried clay bricks and livestock watering. The water from these ponds is rarely used for supplemental irrigation, but they do help mitigate the negative effects of droughts and act as flood protection structures. They also contribute to the recharge of groundwater (Koussoubé 1996). Figure 5 shows the map of water reservoirs and watersheds across the country.

Several recent studies have recorded climate parameters (rainfall, temperature, evapotranspiration) and surface water resource in Burkina Faso (Op de Hipt et al. 2018). The interannual variation of the mean temperature is illustrated in Fig. 6. Climate models projections indicate an increase in average temperatures of 0.8 °C by 2025 and 1.7 °C by 2050 (MECV/BF 2007) as compared to the 1986–2005

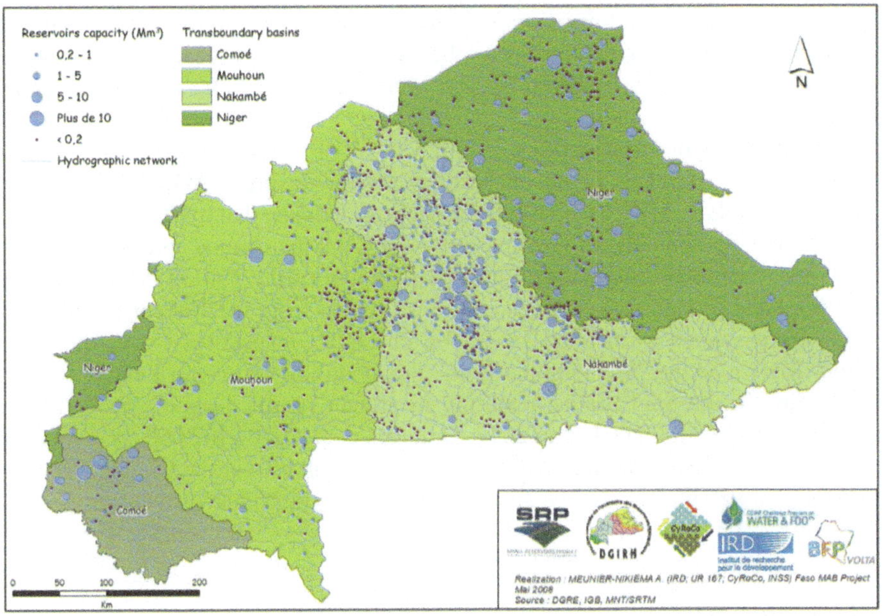

Fig. 5 Map of the water reservoirs in Burkina Faso (Cecchi et al. 2009)

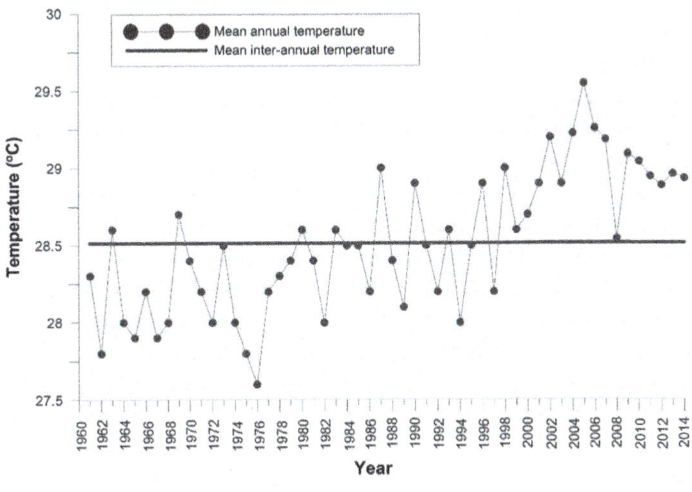

Fig. 6 Interannual variation of the annual mean temperature at the Ouaga-Aero station (*Source* Agence Nationale de la Météorologie du Burkina Faso)

IPCC baseline. However, this increase is subject to seasonal variation: the months of December, January, August and September will become significantly warmer than usual, while the months of November and March will experience smaller increases in heat.

The annual volume of water flow by 2025 and 2050 for each of the country's four watersheds were forecast based on these projections using MAGICC/SCENGEN climate model data (MECV/BF 2007). Projections indicate that by 2025, the volume of water flowing annually will decrease by 45.6% on the Comoe and 54.7% on the Mouhoun watersheds compared to the mean annual of the 1961–1990 period. On the other hand, the annual volumes of water flowing from the Nakanbe and Niger basins are projected to increase by 35.9% and 47%, respectively, over the same period. Known as the "Sahelian paradox," this contradiction is attributable to land degradation in the Said basins, resulting in greater runoff despite a reduction in rainfall (Albergel 1987; Gal et al. 2017; Descroix et al. 2018). In 2050, as compared to the 1961–1990 period, there will be a decrease of 68.9% for the Comoe, 73% for the Mouhoun, 29.9% for the Nakanbe and 41.4% for the Niger.

2.3 Groundwater

Two main types of aquifer formation are found within the crystalline basement, which constitutes 82% of the territory, and the sedimentary zones. Groundwater is related to cracking, fracking or alteration of rocks. Sedimentary zones range from southwest to north and south-east. Flow rates in the crystalline basement are generally between 0.5 and 20 m^3/h, whereas sedimentary flows can reach 100 m^3/h or more. A map illustrating aquifer types and productivity is in Fig. 7.

The volume of groundwater is estimated at 32.43 billion m^3. Groundwater is replenished annually by rainwater, especially through runoff. Therefore, when rainfall decreases, groundwater recharge is highly reduced.

Groundwater fluctuation data show that after a steady decline around the 1980s, there was an increase in the 1990s, followed by another decline and then some stabilization. The various sedimentary aquifers discharge their renewable waters into the Mouhoun (127 million m^3), the Comoe (123 million m^3) and finally into the Bafing (0.438 million m^3), with a total discharge of 250,438 million m^3/year (MEE/DGH 2001). Because of the incidence of waterborne disease caused by the consumption of raw water from reservoir dams, it is generally recommended to use water from drillings and boreholes, which have become the main sources of water supply for the rural population.

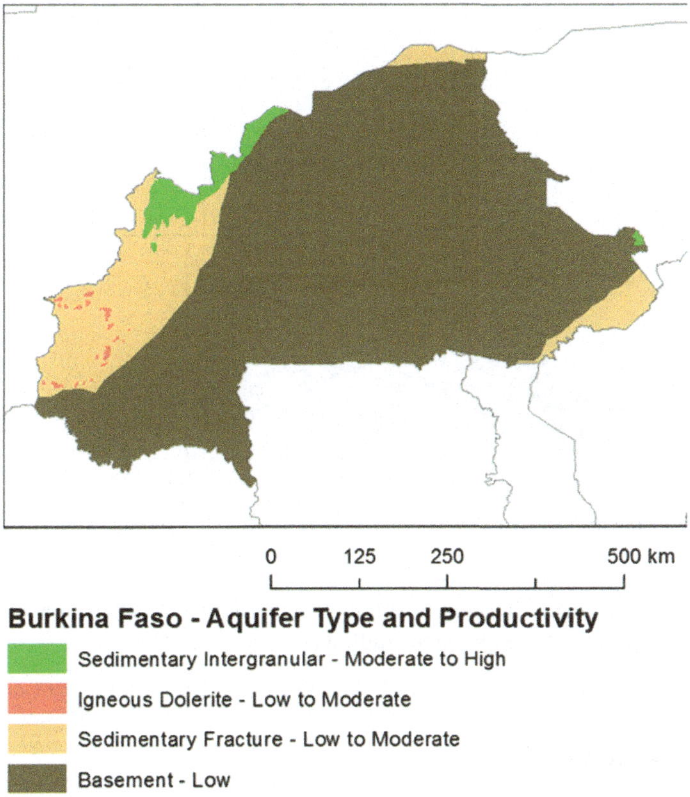

Fig. 7 Burkina Faso aquifer type and productivity (BGS Earthwise 2013)

3 Impacts of Climate Change

3.1 The Agricultural Sector

Irrigation accounts for 64% of total water consumption. In Burkina Faso, as in most Sahelian countries, climate variability and change pose a threat to agriculture and, by extension, to food security. The major rainfall disruptions experienced during the 2007–2008 agricultural season, characterized by an early end of the rainy season at the beginning of September, caused a 16% drop in cereal production compared to the 2006–2007 season and an 11% drop compared to the average of the previous five seasons (Ibrahim et al. 2014). . Agricultural water conservation techniques adopted by farmers have limited effect in stabilizing agricultural production when drought is prolonged over two to three weeks (Barbier et al. 2009). This results in the cumulative and irreversible decrease in the availability of fodder, which in turn pushes people and herds southward. In addition, there is an increased frequency of drought and

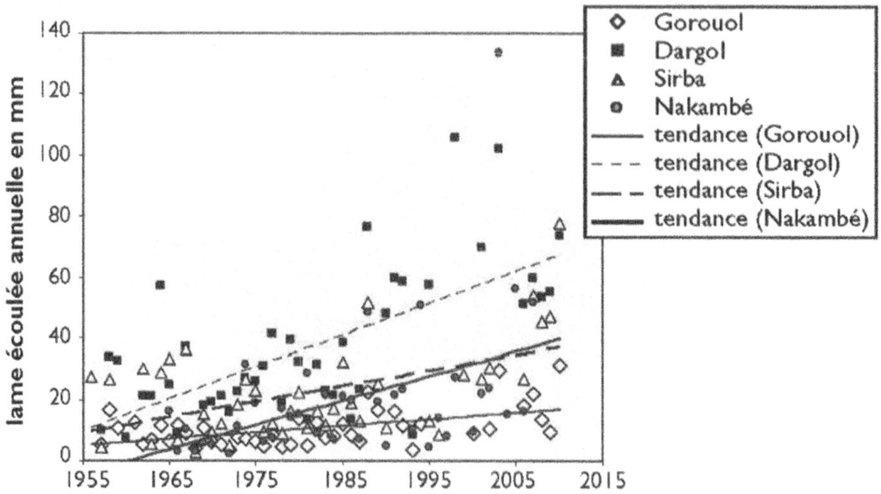

Fig. 8 Increase in Sahelian river flows during drought conditions (Dia and Duponnois 2013)

floods, which can make the land unsuitable for agriculture. The impact of climate change on irrigation has rarely been studied (Fischer et al. 2007).

3.2 *The Watersheds*

Studies conducted over the past three decades (since 1990) have shown that runoff coefficients have increased significantly, despite a marked decrease in regional rainfall (Albergel 1987; Diello 2007; Gal et al. 2017; Descroix et al. 2018). Rainfall simulations associated with a pedological study have shown that the infiltration capacity of soils is dependent on the properties of the soil surface, in particular, it is crusting. A decrease in hydraulic conductivity has been observed on degraded soils compared to cultivated soils and fallows (Mounirou et al. 2012), showing that soil crusting and sealing result in a significant increase in runoff (between 50 and 80%), due to the significant decrease in hydraulic conductivity (between 2 and 20 mm/h). The crusting process associated with increased runoff has resulted in increased flows at outlets of the Sahelian basins (Fig. 8). This phenomenon has led to a decrease in the water retention capacity of the basins as well.

3.3 *The Population Perspective*

Several studies have shown that the increase in cultivated and bare land, compared to natural vegetation, has an impact on precipitation and runoff (Legesse et al. 2003).

While some regions are facing severe drought due to the decrease in total rainfall, others are flooding. The drought periods between 1973–1974 and 1983–1984 greatly affected eco- and production systems. A decrease in rainfall is also accompanied by an extension of the dry season and a shortening of the duration of the rainy season. As articulated by a farmer, *There is a climate change. I was away for 11 years. When I came back, I saw the difference. The rains have become messy, less regular and more intense. Years before, it could rain continuously for three days; nowadays it rains for 30 min. Similarly, when I was a child, it would not rain during the month of February, while nowadays it could happen* (Benoît 2008).

4 Adaptation Policies and Strategies

4.1 The Rural World

Historically, populations cope with climatic disturbances based on their own adaptation techniques and knowledge. Local and intergenerational knowledge of agriculture, including seed selection, agricultural practices and scheduling agricultural interventions, has been perpetuated over generations. However, development policies implemented by the authorities often emphasize the use of equipment, fertilizer extension and seeds, and do not always take into account the know-how of farmers in developing adaptive agricultural practices. Not surprisingly, the population is, therefore, often reluctant to adopt new measures (Oualbéogo 2007). As a result, poor agricultural productivity has, over time, led to an increase in cultivation to compensate. As cultivatable land becomes scarcer, lowlands have been developed, and the perimeters downstream of water reservoirs have been irrigated more intensely. A case in point is the development of lowlands for rice and market gardening production, supported by the government, which provides mineral fertilizer. At the same time, rural populations have erected cropping fields on secure high land (mounds, hills) to prevent the excess runoff overflow during heavy rains. Other examples of adaptation are the building of stone rows and gullies, digging zaï (pits), half-moons and rock dams, and the application of mulching and grass strips.

4.1.1 Agricultural Practices

Stone rows (Fig. 9) reduce runoff and increase water infiltration, which reduces crop water stress during dry spells. This technique has proven effective in reducing soil erosion, and 20–25% of areas in the north and north-central regions can be subjected to the technique with corresponding gains in yield per hectare of 30–68% for sorghum and 21–48% for millet (Ouédraogo 2005).

The zaï technique (Fig. 10) involves the introduction of organic matter into pits containing the crop. The plants are sowed under favorable conditions for growth. This

Fig. 9 Stone rows (Zouré et al. 2019)

Fig. 10 Zaï technique (Zouré et al. 2019)

technique contributes to reducing the effects of dry spells by improving water filtration into the soil through the retention of runoff water. However, it is not adaptable to the humid southern Sudanese area. The practice of zaï has increased the income from sorghum cultivation by 30 GBP/ha (38 USD/ha) (Zongo 2016).

Half-moons (Fig. 11) are also used to recreate soil moisture and fertile conditions that are favorable to crop development on previously degraded soils. Their half-moon

Fig. 11 Half-moon technique (Zouré et al. 2019)

shape rises downstream to retain runoff water. This technique is especially suitable for Sahelian and Sudano-Sahelian areas. It is less useful in the Sudanian zone because the half-moons prevent the mechanization of agricultural work and require significant maintenance. These structures are generally found only in dry areas and are suitable only for soils that are relatively heavy and have high-water retention capacity: silty and clay-like soils. Similar to zaï, the effectiveness of half-moons is usually limited to the needs of rainfed crops over just two to 3 weeks of a dry spell (Roose et al. 1999).

Permeable rock dams (Fig. 12) are made of free stones built in a gully. They improve water infiltration and therefore contribute to adaptation to rainfall variability. They also are effective in restoring land degraded by gullying and contribute to groundwater recharge. They are well suited to the Sudano-Sahelian zone, given their

Fig. 12 Permeable rock dams (WOCAT 2017)

capacity to reduce flow velocity. On degraded soils, they can create new fields of crops through sedimentation. However, these permeable rock dams do not maintain soil moisture during long dry spells.

Mulching (Fig. 13) consists of covering the ground with a layer of grass that has been cut to ensure soil cover against wind erosion and to stimulate termite activity. Mulching contributes to the improvement of soil moisture retention, water infiltration and the reduction of water erosion. However, its impact on yields is small: data from surveys in the Sudano-Sahelian zone of Burkina Faso reveal an increase of cereal yield of just 2–5% using this technique (Ouédraogo 2005).

Grass strips (Fig. 14) are barriers made of perennial grass. They are planted perpendicular to the direction of the runoff and according to corresponding characteristics

Fig. 13 Mulching (GIZ 2012)

Fig. 14 Grass strips (GIZ 2012)

of the soils, slopes and land use. They contribute to reducing the effects of drought by promoting water infiltration and slowing runoff due to their rough surface. This technique enables farmers to plant crops with shorter cycles than traditional crops. For example, sorghum varieties have decreased from 120–150 days to 70–90 days during the 2000–2015 period in some environments. Over 40 years (1970–2010), the average adoption rate for stone rows is 69.3%, 49.1% for zaï, 39.1% for improved seeds and 26.2% for grass strips. The rate of adoption of other innovations (half-moons, bunds, mulching) is less than 10% in total (Zongo 2016).

4.1.2 Seasonal Forecasts

Seasonal forecasts are also among farmers' traditional strategies to reduce climate risk for rainfed crop production. The agricultural calendar is established by farmers according to pedoclimatic conditions: increase in temperatures, strong winds followed by dust, and the clear visibility of the stars and the moon reflect a regular distribution of rainfall for the upcoming season. Early and significant flowering of *Erythrina abyssinica* and *Brachystegia speciformis* from July to November has also been identified as one of the signals of a good rainy season. In contrast, abundant production of fruit by the *Boscia albitrunca* and *Adansonia digitata* and the increase of elephants appearing near watering points indicate an upcoming dry season. The flowering of the plant species *Brachystegia spiciforms* and *Julbernardia globiora* indicates the beginning of the rainy season, while an abundance of fruit from *Uapaca kirkiana* and *Parinari curatellifolia* indicate drought for the upcoming season. In addition to observing the behavior of animals and plants, ritual specialists use visions, dreams and divine revelations to predict the weather. As climate change has increased, these ritual forecasts have become less reliable (Mogotsi et al. 2011).

4.2 Public Authorities

Burkina Faso implemented a national climate change adaptation program in 2007. The plan seeks to develop adaptive capacity and resilience and to integrate the anticipated consequences of climate change into policies and programs within relevant sectors. Its purpose is to produce a reference to serve as a basis to reduce the vulnerability of natural, social and economic systems. The plan was developed rigorously at the institutional, technical and financial levels. Researchers from the University of Ouagadougou have prepared climate projections for Burkina Faso up to the year 2100 and have assessed the vulnerability of the various development sectors. The Ministry of Environment and Sustainable Development has appointed a Technical Committee to prepare and monitor the adaptation plan. Funding will be provided by a combination of Federal funds, partners, international foundations, the private sector and Non-Governmental Organizations. Other measures that are being developed by the Government in Burkina Faso to help mitigate the adverse effects of climate change include:

- the reorganization of the agrarian land system (through law No 034-2012/AN);
- the implementation of the environmental code (law No 006/97/ADP) of 8 February 2001, which modifies the Water Management Policy Act No. 002-2001/AN of 31 January 1997;
- the creation of the Integrated Water Resources Management Action Plan;
- the introduction of the National Drinking Water Supply and Sanitation Program, and
- building water resources management structures (water agencies, the National Water Council, Local Water Committees, etc.).

This system, established by the Government with the support of development partners, seeks to improve the environmental management of water resources. At the reservoir level, management is often entrusted to water user associations that often lack the management tools and scientific and technical capacity to serve their mission effectively.

A range of development partners have worked with the Government to implement the commitments of the Government of Denmark to improve the drinking water supply (the governments of Burkino Faso and Denmark have been in a development cooperative agreement since 1973). Denmark has mobilized technical assistance to strengthen the capacities of stakeholders to program, invest and manage integrated water resource management. In addition to policy-level interventions, efforts target conservation, extraction and mobilization of water resources. Water resource challenges in Burkina Faso have also been the subject of empirical research, the results of which are expected to enable the Government to improve its technological, legal, institutional and economic strategies to support access to water.

4.3 Scientific Research

4.3.1 The Anthropogenics of Hydrological Modeling

Changes in the rainfall-runoff relationship depend on an anthropogenic dimension, in addition to climate variability. It is, therefore, necessary to take into account the dynamics of land use in the hydrological modeling of watersheds. The work of Diello (2007) simulates human activity in and around watersheds in the Sahel, accounting for land use dynamics and how they affect soil water retention capacity. The study shows changes in Nakanbe River Basin natural vegetation, cultivation, bare soil and water bodies in 1972, 1986, 1992 and 2002. Cultivated lands experienced a two-stage dynamic between 1972 and 2002. From 1972 to 1992, the proportion of land that was cultivated increased significantly (from 0.87 million ha in 1972 to 1.37 million ha in 1992). Between 1992 and 2002, the proportion of land that was

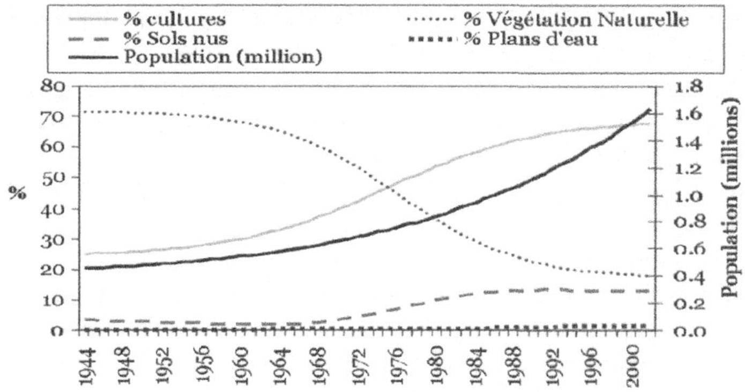

Fig. 15 Temporal evolution of land use indicators in the Nakanbe Basin (Diello 2007)

cultivated stabilized and even underwent a downward trend. This indicates that at the beginning of the 1990s, the potential of arable land was exploited to the detriment of natural vegetation (which shrank from 10,475 km^2 in 1972 to 3,439 km^2 in 2002). The bare soil surfaces increased in the basin between 1972 and 2002 (from 4.5% in 1972 to 13.8% in 2002). Figure 15 provides a summary of the dynamics of population changes and their impact on four indicators in the Nakanbe Basin.

The simulations carried out by Diello (2007) demonstrate that changes in land use magnified changes in the water retention capacity of the soil. Hence, it is critical to account for these dynamics to better represent, through modeling, hydrologic response and balance at the watershed scale.

4.3.2 Analysis of the African Monsoon

One cause of climate variability in West Africa is the West African Monsoon. Since 2001, more than 60 European, African and American laboratories have jointly operated a vast research program, the Multidisciplinary Analysis of the African Monsoon (AMMA), to advance knowledge on the African Monsoon. It seeks to improve the scientific basis for addressing water resources and food security using a large observatory, the AMMA-CATCH observing system (Lebel and Ali, 2009), with locations throughout West Africa. Its main objectives are to (i) analyze long-term evolution of the eco- and hydro-systems at the regional scale; (ii) better understand critical zone processes and their variability, and (iii) coordinate with decision-makers to address socioeconomic and development needs (Galle et al. 2018).

4.3.3 Optimization of Small Dams

Researchers are trying to optimize the management of water resources for small dams. The work of Fowe (2015) examines the Boura reservoir dam, located in southern

Burkina Faso, as a pilot site representing a large number of small reservoirs in the Volta basin subject to the same constraints. Once the water balance was determined, it became possible to characterize the functioning of the Boura hydrosystem at different time steps in the filling and destocking cycle. Findings are that of the water mobilized by the reservoir (9.23 million m^3), 60% evaporates, 24% infiltrates and 19% is levied to meet user demands, especially irrigation. The load factor went from 2.3 for the 2012–2013 season to only 0.7 for the following (2013–2014) season. This indicates that even over a short period, the water regime of the Boura dam is greatly variable, mainly driven by variations in input controlled by rainfall. Calibrated hydrologic models based on these data are then used to make projections of inputs to the Boura reservoir, of agricultural water demand and of the functioning of the Boura reservoir under different scenarios of climate change and socioeconomic development. The conclusion is that it would be difficult to satisfy irrigation demands by considering only the surface reservoir as a source of supply, and that the aquifer must also be considered a water source to meet demand.

4.3.4 Supplemental Irrigation

In the Sahelian countries, farmers, the scientific community, governments and their partners seek innovative agricultural practices to respond to periods of drought. The thesis work of Zongo (2016) considers the adoption of agricultural innovations in the Sahelian and Sudano-Sahelian areas of Burkina Faso, including the practice of supplementing irrigation from individual basins on farms. Complementary irrigation delivers water to crops during the long periods of drought during the rainy season. This is runoff water stored in small basins built near crop fields (Fig. 16). The practice increases maize yield by 1.08 T/ha (88.3%) compared to a lack of supplemental irrigation. About 78.4% of households believe that supplemental irrigation is an interesting option to mitigate the effects of drought periods on agricultural production. The study, however, notes the need to establish a system of financial credit to enable such a scheme.

4.3.5 Agriculture and Climate Information

Several models have been developed to generate and communicate climate information to different users according to their needs. The source of information is seasonal climate forecasts based on meteorological data. The most relevant variables used to create forecasts are the start of heavy rains, the beginning of sowing, the end of the rainy season and the distribution of rainfall within the season (intra-seasonal distribution). The thesis of Zongo (2016) evaluates the impact of climate information on maize and sorghum production in the Bam province in the Sahel region during the 2013–2014 and 2014–2015 agricultural seasons. It compares the dates of seedlings, yields and contribution to cereal requirement from experimental to control plots and shows that producers use the sowing date information in the experimental group to

(a) Excavation of a basin from family labor (b) Basin filled with runoff water

(c) Observer plot (without irrigation) (d) Experimental plot

Fig. 16 Complementary irrigation (Zongo 2016)

greater effect than in the control group. Farmers believe that the planting dates of the experimental plots have been more accurate. As a result, 93% of households indicate that they need climate information to guide their decision-making.

5 Conclusion

In Burkina Faso, the stakes of climate change are considerable. Several recent studies have analyzed historic climate parameters (rainfall, temperature, evapotranspiration) and surface water resources in the country. In the Nakanbe basin, results show an increasing trend in surface runoff, suggesting that it could face a highly variable climate and the hydrological regime in the coming decades, ranging from heavy rainfall that leads to flooding to severe drought that leads to water scarcity and a reduction in crop yields.

Agricultural innovation is a critical strategy to reduce household vulnerability to climate change. According to a 2009 report by CILSS and the Centre for International

Cooperation, most villages that practice natural management techniques have experienced poverty reduction. The biggest constraints to achieving the most effective practices are poor organization and the lack of access of farmers to extension agents. The majority of agricultural households believe that the practice of supplemental irrigation and the provision of climate information is a viable alternative to mitigate the drastic effects of dry spells on agricultural production. However, it is premature to draw conclusions about the validity of information and the science on which it is based. Efforts must be made to further improve seasonal forecasts by integrating indigenous and scientific knowledge.

The role of the Government has been better appreciated over time, particularly at the decentralized level, where minimal resources have been allocated to regional water departments, limiting their ability to fulfill their missions. Local authorities, civil society and the private sector are being coordinated better in their common goal of improving water resource management. Since the early 2000s, there has been a return to large hydraulic projects (e.g., the Samandeni dam in Burkina Faso, the Bui dam in Ghana, the Fomi dam in Guinea and the Kandadji dam in Niger). It is imperative to strengthen the capacity of technical services and skills, and human resources, to sustainably improve resilience.

Despite harsh and unfavorable natural conditions in Burkino Faso, solutions exist to help the country move toward sustainable development. To this end, the national climate change adaptation program provides a reference framework for coordinating efforts across stakeholders to reduce structural vulnerability, increase resilience and sustain development.

References

Albergel J (1987) Sécheresse, désertification et ressources en eau de surface : application aux petits basins du Burkina Faso. In: The influence of climate change and climatic variability on the hydrologic regime and water resources, pp 355–365

Ardoin Bardin S (2004) Variabilité hydroclimatique et impacts sur les ressources en eau de grands bassins hydrographiques en zone soudano-sahélienne. PhD Thesis, Université de Montpellier 2, IRD

Barbier B, Yacouba H, Maïga AH, Mahé G, Paturel JE (2009) Le retour des grands investissements hydrauliques en Afrique de l'Ouest : les perspectives et les enjeux. Géocarrefour, pp 31–41. https://doi.org/10.4000/geocarrefour.7205

Benoît É (2008) Les changements climatiques : vulnérabilité, impacts et adaptation dans le monde de la médecine traditionnelle au Burkina Faso. VertigO. https://doi.org/10.4000/vertigo.1467

BGS Earthwise (2013) Hydrogeology of Burkina Faso [WWW Document]. http://earthwise.bgs.ac.uk/index.php/Hydrogeology_of_Burkina_Faso. Accessed 25 June 2019

Cecchi P, Meunier Nikiema A, Moiroux N, Sanou B (2009) Towards an atlas of lakes and reservoirs in Burkina Faso. IRD, 23 p

Descroix L, Guichard F, Grippa M, Lambert L, Panthou G, Mahé G, Gal L, Dardel C, Quantin G, Kergoat L, Bouaïta Y, Hiernaux P, Vischel T, Pellarin T, Faty B, Wilcox C, Malam Abdou M, Mamadou I, Vandervaere JP, Diongue-Niang A, Ndiaye O, Sané Y, Dacosta H, Gosset M, Cassé C, Sultan B, Barry A, Amogu O, Nka Nnomo B, Barry A, Paturel J-E (2018) Evolution of

surface hydrology in the sahelo-sudanian strip: an updated review. Water 10:748. https://doi.org/10.3390/w10060748
Dia A, Duponnois R (eds) (2013) Le projet majeur africain de la Grande Muraille Verte: Concepts et mise en œuvre. IRD Éditions. https://doi.org/10.4000/books.irdeditions.2106
Diello P (2007) Interrelation climat-homme-environnement dans le Sahel Burkinabé: impacts sur les états de surface et la modélisation hydrologique (Thèse de Doctorat). Université de Montpellier II, Sciences et Techniques du Languedoc
FAO (2005) La Situation Mondiale de l'Alimentation et de l'Agriculture. Le commerce agricole et la pauvreté: le commerce peut-il être au service des pauvres? Collection FAO, Agriculture N° 36. FAO, Rome (Italy)
Fischer G, Tubiello FN, van Velthuizen H, Wiberg DA (2007) Climate change impacts on irrigation water requirements: Effects of mitigation, 1990–2080. Technol Forecast Soc Chang 74:1083–1107. https://doi.org/10.1016/j.techfore.2006.05.021
Fowe T (2015) Simulation et optimisation du fonctionnement du barrage de Boura en zone soudanienne du Burkina Faso. PhD Thesis, 2iE, Ouagadougou
Gal L, Grippa M, Hiernaux P, Pons L, Kergoat L (2017) The paradoxical evolution of runoff in the pastoral Sahel: analysis of the hydrological changes over the Agoufou watershed (Mali) using the KINEROS-2 model. Hydrol Earth Syst Sci 21:4591–4613. https://doi.org/10.5194/hess-21-4591-2017
Galle S, Grippa M, Peugeot C, Moussa IB, Cappelaere B, Demarty J, Mougin E, Panthou G, Adjomayi P, Agbossou EK, Ba A, Boucher M, Cohard JM, Descloitres M, Descroix L, Diawara M, Dossou M, Favreau G, Gangneron F, Gosset M, Hector B, Hiernaux P, Issoufou B-A, Kergoat L, Lawin E, Lebel T, Legchenko A, Abdou MM, Malam-Issa O, Mamadou O, Nazoumou Y, Pellarin T, Quantin G, Sambou B, Seghieri J, Séguis L, Vandervaere JP, Vischel T, Vouillamoz JM, Zannou A, Afouda S, Alhassane A, Arjounin M, Barral H, Biron R, Cazenave F, Chaffard V, Chazarin J-P, Guyard H, Koné A, Mainassara I, Mamane A, Oi M, Ouani T, Soumaguel N, Wubda M, Ago EE, Alle IC, Allies A, Arpin-Pont F, Awessou B, Cassé C, Charvet G, Dardel C, Depeyre A, Diallo FB, Do T, Fatras C, Frappart F, Gal L, Gascon T, Gibon F, Guiro I, Ingatan A, Kempf J, Kotchoni DOV, Lawson FMA, Leauthaud C, Louvet S, Mason E, Nguyen CC, Perrimond B, Pierre C, Richard A, Robert E, Román-Cascón C, Velluet C, Wilcox C (2018) AMMA-CATCH, a Critical Zone Observatory in West Africa Monitoring a Region in Transition. Vadose Zone J 17:1–24. https://doi.org/10.2136/vzj2018.03.0062
GIZ (2012) Bonnes pratiques de conservation des eaux et des sols. Contribution à l'Adaption au changement climatique et à la résilience des producteurs au Sahel. Bonn (Allemagne)
Hountondji Y, Sokpon N, Nicolas J, Ozer P (2009) Ongoing desertification processes in the sahelian belt of West Africa: an evidence from the rain-use efficiency. Taylor & Francis Group
Ibrahim B, Karambiri H, Polcher J, Yacouba H, Ribstein P (2014) Changes in rainfall regime over Burkina Faso under the climate change conditions simulated by 5 regional climate models. Clim Dyn 42:1363–1381. https://doi.org/10.1007/s00382-013-1837-2
IPCC (2014) The physical science basis: working group contribution to the fifth assessment report of the intergovernment panel on climate change. Cambridge, United Kingdom, Cambridge University Press
Jalota SK, Arora VK (2002) Model-based assessment of water balance components under different cropping systems in north-west India. Agric Water Manag 57:75–87
Koussoubé Y (1996) Hydrogéologie en milieu de socle cristallin du Burkina Faso. Cas du bassin versant du bas-fond de Bidi (province du Yatenga). PhD Thesis, Université Cheikh Anta Diop de Dakar, Dakar
Lebel T, Ali A (2009) Recent trends in the Central and Western Sahel rainfall regime (1990–2007). J Hydrol 375:52–64. https://doi.org/10.1016/j.jhydrol.2008.11.030
Leemhuis C, Jung G, Kasei R, Liebe J (2009) The Volta Basin water allocation system: assessing the impact of small-scale reservoir development on the water resources of the Volta basin, West Africa. Adv Geosci 21:57–62. https://doi.org/10.5194/adgeo-21-57-2009

Legesse D, Vallet-Coulomb C, Gasse F (2003) Hydrological response of a catchment to climate and land use changes in Tropical Africa: case study South Central Ethiopia. J Hydrol 275:67–85. https://doi.org/10.1016/S0022-1694(03)00019-2

MEA/BF (2015) Programme national d'Approvisionnement en Eau Potable et Assainissemnt (PN-AEPA): rapport bilan annuel au 31 décembre 2015 (Inventaire). Ministère de l'Eau et de l'Assainissement, Ouagadougou

MECV/BF (2007) Programme National d'Adaptation à la Variabilité et aux Changements Climatiques (PANA) du Burkina Faso. Ministère de l'Envrionnement et du Cadre de Vie (MECV), Ouagadougou

MEE/DGH (2001) Etat des lieux des ressources en eau du Burkina et de leur cadre de gestion. Ministère de l'Environnement et de l'Eau | Direction Générale de l'Hydraulique, Ouagadougou

Mogotsi K, Nyangito MM, Nyariki DM (2011) The perfect drought? constraints limiting Kalahari agro-pastoral communities from coping and adapting. Afri J Environ Sci Technol 5:168–177

Mounirou LA, Yacouba H, Karambiri H, Paturel JE, Mahé G (2012) Measuring runoff by plots at different scales: understanding and analysing the sources of variation. CR Geosci 344:441–448. https://doi.org/10.1016/j.crte.2012.08.004

Naturama (2015) Les Ressources en eau au Burkina [WWW Document]. http://www.naturama.bf/web/index.php/actualites/item/84-les-ressources-en-eau-au-burkina. Accessed 20 June 2019

Niang D (2006) Fonctionnement hydrique de différents types de placages sableux dans le sahel burkinabè [WWW Document]. Infoscience. https://doi.org/10.5075/epfl-thesis-3667

Nicholson S (2005) On the question of the "recovery" of the rains in the West African Sahel. J Arid Environ 63:615–641. https://doi.org/10.1016/j.jaridenv.2005.03.004

Op de Hipt F, Diekkrüger B, Steup G, Yira Y, Hoffmann T, Rode M (2018) Modeling the impact of climate change on water resources and soil erosion in a tropical catchment in Burkina Faso, West Africa. CATENA 163:63–77. https://doi.org/10.1016/j.catena.2017.11.023

Oualbéogo TA (2007) Intégration des savoirs locaux traditionnels et des nouvelles méthodes de production des cultures de rente dans la protection de l'environnement : cas des villages de Kamandena et de Sambouali

Ouédraogo (2005) Intensification de l'agriculture dans le Plateau Central du Burkina Faso: Une analyse des possibilités à partir des nouvelles technologies. s.n

Redelsperger JL, Thorncroft CD, Diedhiou A, Lebel T, Parker DJ, Polcher J (2006) African monsoon multidisciplinary analysis: an international research project and field campaign. Bull Am Meteor Soc 87:1739–1746. https://doi.org/10.1175/BAMS-87-12-1739

Roose E, Khamsouk B, Lassoudière A, Dorel M (1999) Origine du ruissellement et de l'érosion sur sols bruns à halloysite de Martinique : premières observations sous bananiers. Bulletin Réseau Erosion 139–147 multigr

Roudier P, Muller B, d'Aquino P, Roncoli C, Soumaré MA, Batté L, Sultan B (2014) The role of climate forecasts in smallholder agriculture: lessons from participatory research in two communities in Senegal. Clim Risk Manag 2:42–55. https://doi.org/10.1016/j.crm.2014.02.001

Sawadogo H (2011) Using soil and water conservation techniques to rehabilitate degraded lands in northwestern Burkina Faso. Int J Agricu Sustain 9:120–128

WOCAT (2017) WOCAT SLM Technologies [WWW Document]. wocat.net. https://qcat.wocat.net/en/wocat/technologies/view/technologies_1617/ Accessed 20 June 2019

Zongo B (2016) Stratégies innovantes d'adaptation à la variabilité et au changement climatiques au Sahel : cas de l'irrigation de complément et de l'information climatique dans les exploitations agricoles du Burkina Faso. Université de Liège, Liège, Belgique

Zongo B, Diarra A, Barbier B, Zorom M, Yacouba H, Dogot T (2016) Farmers' perception and willingness to pay for climate information in Burkina Faso. J Agric Sci 8:175–187. https://doi.org/10.5539/jas.v8n1p175

Zougmoré R, Jalloh A, Tioro A (2014) Climate-smart soil water and nutrient management options in semi-arid West Africa: a review of evidence and analysis of stone bunds and zaï techniques. Agric Food Secur 3:16. https://doi.org/10.1186/2048-7010-3-16

Zouré C, Queloz P, Koïta M, Niang D, Fowé T, Yonaba R, Consuegra D, Yacouba H, Karambiri H (2019) Modeling the water balance on farming practices at plot scale: case study of Tougou watershed in Northern Burkina Faso. CATENA 173:59–70. https://doi.org/10.1016/j.catena.2018.10.002

Chapter 15
Impacts of Climate Change on Water Resources in the Volta River Basin: Reducing Vulnerability and Enhancing Livelihoods and Sustainable Development

George A. Manful and Yaw Opoku-Ankomah

Abstract The Volta River Basin is a transboundary basin in West Africa with a population of about 24 million people in six riparian countries (Benin, Burkina Faso, Cote d'Ivoire, Ghana, Mali and Togo) with low gross domestic products. Water resources are needed for domestic and industrial use, agriculture, hydropower generation, environmental health and sustainability and ecosystem functioning. Assessment of the impact of climate change on water resources is therefore crucial for the socioeconomic development of all six countries within the basin, including measures designed to alleviate poverty. A number of such studies assess the impacts of climate change and variability on water resources, agriculture and human health. Climate change adaptive measures and management practices to increase resilience and contribute to the Sustainable Development Goals have been analyzed using historical temperature and precipitation data to determine climate variability. Climate forcing from anthropogenic emissions of greenhouse gases is, in most cases, evaluated using Regional Circulation Models downscaled from General Circulation Models. Mid-level Representation Concentration Pathways and high-level forcing scenarios indicate that before the 1980s, the basin was cooler than the average of the reference period 1976–2005. Between 2006–2100, as GHG forcing increases, warming will increase faster than the period 1976–2005, with increases in temperature between 1.5 and 6 °C projected by 2100. Precipitation anomalies show complex patterns in the basin with high inter-annual variability. There is no clear trend for the magnitude of variation for precipitation. A shift and shortening of rainy seasons and an increase in the frequency of dry spells occurring within the rainy seasons is forecast under climate change. These changes in rainfall characteristics will have a serious impact on agricultural production, which is largely rain-fed. Agricultural productivity is a principal pathway out of the vicious cycle of poverty within the Volta river basin; negative impacts on agriculture can be averted if large scale irrigation practices are implemented as part of a comprehensive water management strategy.

G. A. Manful (✉) · Y. Opoku-Ankomah
Climate Change Consultant, Post Office Box CT 4602, Cantonments, Accra, Ghana
e-mail: gmanful@gmail.com

S. Diop et al. (eds.), *Climate Change and Water Resources in Africa*,
https://doi.org/10.1007/978-3-030-61225-2_15

Keywords Volta river basin · Climate change · Reducing vulnerabilities · Enhancing resilience · Sustainable development

1 Introduction

The Volta River Basin is a transboundary basin that spans parts of six riparian countries in West Africa: Benin, Burkina Faso, Cote d'Ivoire, Ghana, Mali and Togo. The basin is saddled with socioeconomic challenges associated with inadequate water management practices and national development efforts. These countries depend, to a large extent, on the exploitation of their natural resources, including water, for their survival and to make progress toward their development aspirations. Throughout the basin, agriculture, a major economic activity, is predominantly rain-fed, including for cotton, millet, rice and sorghum, which are cultivated in the northern part of the basin, and yam, cassava, plantain, cocoyam and cocoa, which are predominantly grown in the south. Limited irrigation is used to grow crops like tomatoes and onions throughout the basin in the dry season. Animals are also raised alongside crop production. About 77% of water extracted from the basin is used for crop irrigation and livestock farming; the remaining 23% is used for urban and rural water supply for drinking (UNEP-GEF Volta Project 2013). Large-scale irrigation in the basin is not used and has a little socioeconomic impact. The total amount of water extracted from the basin is 928 million cubic meters per annum. However, in order to meet the agriculture demands of an increasing population coupled with variable and uncertain rainfall intensity and distribution, serious irrigation practices must be introduced across the basin. Water for irrigation alone is anticipated to increase from 69% of the total volume of water used in 2000 to about 82% in 2020 (UNEP-GEF Volta Project 2013). This proportion is subject to variation when climate change is factored into planning for water management. Complicating planning, the effects of climate change on rainfall distribution in the Volta River basin are not well understood; more research is needed in this area.

Water supply for domestic and industrial use is also important for socioeconomic development in the basin. Domestic water use there is projected to increase sharply from 360 million cubic meters in 2000 to about 1,058 million cubic meters in 2025 (UNEP-GEF Volta Project 2013). In urban and rural areas with large populations, surface water is treated and sufficient to meet human needs, and rural areas and villages with smaller populations are served by water from springs and streams, as well water. In general, urban populations have better access to safe drinking water than rural populations. In Burkina Faso, for example, only 11.5% of the urban population does not have access to safe water, compared to 35% of the rural population (Lemoalle and de Condappa 2009). The potential for climate change to reduce surface and groundwater resources can threaten the sufficient supply of water if proper management plans are not implemented. In the downstream basin, water is used for hydropower generation; reduction in flows as a result of climate change and large scale atmospheric and global oceanic phenomena like El Niño and Southern Oscillation (ENSO) would pose further serious challenges to riparian countries which jointly depend on hydropower as a renewable energy resource. The Akosombo and Kpong reservoirs downstream produce 4,800 GWh/year of hydroenergy; Bui in mid-Volta produces 980 GWh/year. Upstream in Burkina Faso, hydroenergy production is 65, 19 and 17 GWh/year at Bagre, Kompienga and Samandeni reservoirs, respectively (Sonabel 2009; VRA 2009; Lemoalle and de Condappa 2009).

The basin is underdeveloped, and economies are based predominantly on agriculture and hydropower generation. The availability of water is subject to the climate, the environment and the infrastructure available for water storage and use. The primary source of water in the basin is rainfall modulated by general circulation, evaporation and transpiration from vegetation. Rainfall becomes a surface and groundwater resource through hydrological processes. Temperature can cause water to be lost to the atmosphere without having been used. Climate change, caused to a large extent by anthropogenic activity, is impacting global climates and thus the availability of water for human survival and socioeconomic activities, including agricultural development, hydropower generation and water supply for domestic and industrial use. This literature review looks at the evolution of climate change and its impacts on water resources in the basin in the past, and projects impacts in the future. Temperature and rainfall projections for the basin developed by the IPCC and used in simulations in the Regional Climate Models (RCMs) and Global Circulation Model (GCMs) are discussed.

2 The Volta Basin

2.1 Physical and Geopolitical Settings

The Volta Basin is the ninth largest in West Africa. It lies between latitudes 5°30'N and 14°30'N and longitudes 2°00'E and 5°30'W (UNEP-GEF Volta Project 2013). It covers an area of about 400,000 km^2 and spreads over Benin, Burkina Faso, Cote d'Ivoire, Ghana, Mali and Togo. Lemoalle and de Condappa (2009) noted that 85% of the basin area is in Burkina Faso and Ghana. The Volta Basin is delimited at its eastern and western borders by a series of hills and mountain ranges. At the eastern border, there are the Akwapim ranges, Togo Mountains, Fazao Mountains and the Atakora Mountains in Benin. The Kwahu Plateau and Banfora Plateau constitute the western border. Ninety-five percent of the basin lies at an altitude below 400 m, with an average elevation of 257 m. The basin slope index is 0.25 m/km.

2.2 The River System

The Volta River has three main tributaries: the Black Volta, which originates from Burkina Faso as Mouhoun; the White Volta, also from Burkina Faso where it is called the Nakambe, and the Oti River, originating in Benin, where it is known as the Pendjari (Lemoalle and de Condappa 2009). The Oti passes through Togo before it joins the main Volta in Ghana. All these tributaries drain into Lake Volta, which, together with the Lower Volta downstream, also receives flows from other small tributaries, including the Pru, Sene and Afram. Therefore, the basin is considered to have four sub-basins: the lower Volta, the black Volta, the white Volta and the Oti (see Fig. 1).

2.3 Climate Conditions

The climate of the basin is determined by the north-south migration of the Inter-Tropical Convergence Zone (ITCZ), an interface between the hot, dry and dusty North-East Trade Winds blowing from the Sahara Desert, and the warm, moist tropical maritime climate from the Atlantic Ocean known as the Monsoon. In addition, squall Lines (SLs) originate over West Africa and move in a westerly direction. They are mesoscale disturbances and are the most important convective rainfall system of the West Africa Sahel region, yielding precipitation in the summer months from June to September (Peters and Tetzlaff 1988). SLs, which join larger-scale features such as the African Easterly Jet, the Easterly Waves and the Monsoon Front, produce fields of flow and moisture (Peters and Tetzlaff 1988) and generate about 80% of annual precipitation of the Sahel (Dhonneur 1981). In southern West Africa, SLs contribute to annual rainfall, but its contribution to total rainfall decreases from 71 to 56% in the coastal zone (Maranan et al. 2018).

In general, there are three types of climatic zones: the humid south, with two distinct rainy seasons (March/April-May/June, September-October), the tropical transition zone, also with two rainy seasons which (May/June, September-October) are narrowly separated, and the semi-arid north with only one season of rainfall (May-September), with a peak around August. The south-north gradient of rainfall is used to identify the main agro-climatic zones of the basin according to the FAO classification for West Africa (FAO/GIEWS 1998). They are the Sudano-Sahelian zone, the Sudanian zone and the Guinean zone. The Sudano-Sahelian zone receives rainfall between 500 and 900 mm, the Sudanian zone 900 to 1100 mm and the Guinean zone more than 1100 mm (Lemoalle and Condappa 2009).

The mean annual temperatures range from 24 °C in the South to about 36 °C in the North (Oguntunde 2004; Hayward and Oguntoyinbo 1987). The high temperatures in the basin amplify the impact of evapotranspiration, which is mainly driven by heat

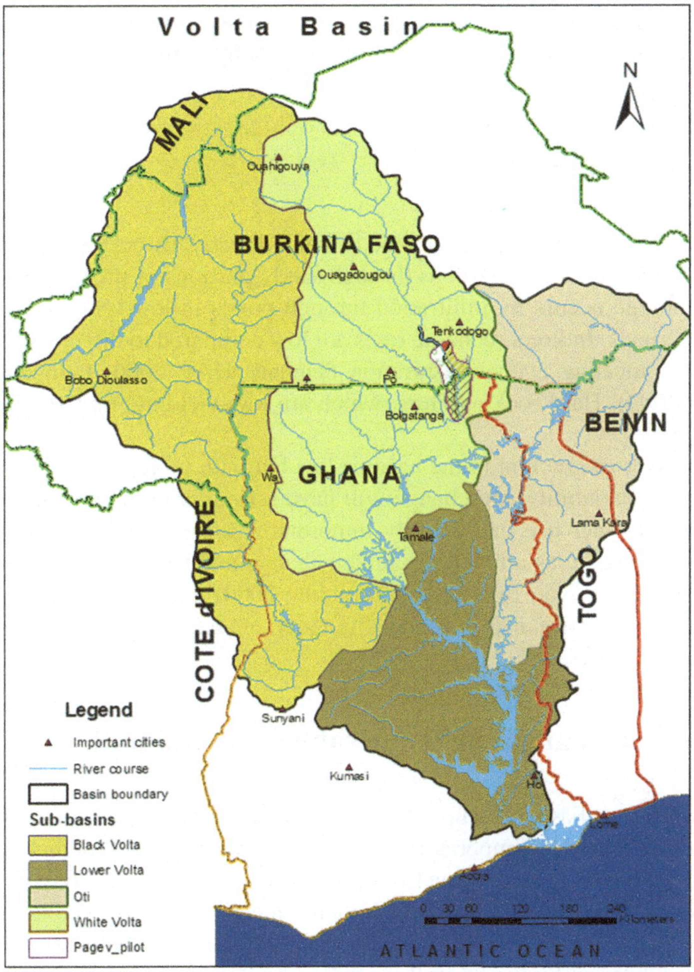

Fig. 1 The network of major rivers in the Volta Basin (*Source* Water Resources Commission, Ghana)

fluxes, wind and relative humidity. Increasing temperatures as a result of anthropogenic greenhouse gas emissions or climate change may increase water deficits via evapotranspiration.

Precipitation is the primary source of water to the hydrological cycle. Studies in the Volta basin examine the repartition of precipitation into various components: evapotranspiration, the outflow of the basin, seepage to groundwater and other losses (de Condappa et al. 2008; Lemoalle and de Condappa 2009). Lemoalle and de Condappa (2009) compute a total volume of the annual rainfall of about 395 km^3 and ET of 346 km^3 (88%), which is notably high. Outflow and groundwater recharge are only

33 km^3 (8%) and 13 km^3 (3%), respectively; various losses are 1%. Fresh water availability in the basin is thus very limited, especially in the drier northern part.

2.4 Groundwater

Aquifers in the basin are recharged directly and indirectly. Direct recharge is through fractured and fault zones as well as through sandy portions of the weathered zones. Indirect recharge occurs by infiltration through ponds in low-lying areas and also through riverbeds. Indirect recharge rates are low compared to direct recharge from rainfall after meeting evapotranspiration demand (HAP 2006; Lemoalle and de Condappa 2009). Hard rock aquifers, which are low-yielding, underlie about 90% of the basin.

Groundwater abstraction in the basin is low but with high potential for development. Climate variability and change will have serious impacts on surface waters, making groundwater use increasingly important. Currently, multiple streams and springs are drying up as a result of climate change and variability, as well as poor land use practices (Gyau-Boakye and Tumbulto 2000). Knowledge of aquifer yield and other characteristics in the basin are limited; more research in this area is needed.

3 Observed Changes in Temperature and Rainfall

Ghana and Burkina Faso make up 85% of the total area of the basin (Lemoalle and de Condappa 2009); hence temperatures and rainfall in the agro-climatic zones of the basin used for this study were taken from national statistics of these two countries sent to the UNFCCC through their national communication reports. The agro-climatic zone of the Burkina Faso part of the basin is largely Sudano-Sahelian (see Sect. 2.3), and that of Ghana is Sudanian in the north and Guinean in the midsection of the country.

3.1 Changes in Historical Temperatures

Over the period 1961–2008, the average annual temperatures in Burkina Faso increased by at least 0.5 °C at all the synoptic stations (UNFCCC 2015a). The minimum and maximum temperatures also increased in magnitude. Average minimum temperatures varied between 20 and 22 °C over 1961–1990 and 1971–2000, respectively, with the variability of about 2 °C. Maximum temperatures varied between 0.4 and 3.6 °C, and minimum temperatures for the period 1960–2000 in Sudan and Guinea Savannah zones in Ghana (in the lower part of the basin) showed a 3.7% increment. For maximum temperatures, rates of change of 6.1% and 2.7% were observed for Sudan and Guinea Savannah, respectively (UNFCCC 2015b).

3.2 *Changes in Historical Rainfall Amounts*

Rainfall has been variable across the basin and over time (Oguntunde et al. 2006). From 1901 to 2002, the 1970s and 1980s was the driest period, while the 1930s and 1950s were the wettest (Oguntunde et al. 2006). Inter-annual variability was more pronounced in the northern part than the southern part of the basin (Kasei 2009, van de Giesen et al. 2010). The northern part has only one rainfall season from about May/June to September/October, and rainfall totals are lower in the north than in the south, with about 500 mm or less in the Sahelian zone in the far north and about 1100 mm or more in the Guinean zone in the south (Lemoalle and de Condappa 2009). Hence, rainfall variability produces stress on human lives and socioeconomic activities, especially in the upper reaches of the basin. Droughts there have become very frequent since the 1970s (van de Giesen et al. 2010). Mean annual rainfall for 1901–1969 was 1100 mm; for 1970–2002 the mean was 987 mm. Furthermore, if the year 1970 alone is eliminated from rainfall trend analysis, no rainfall trend is evident, meaning that rainfall amounts varied from 1970 to 2002 without any particular trend.

Analysis of historical data from Burkina Faso's National Communication (UNFCCC 2015a) showed a movement of isohyets for the periods 1931–1960, 1961–1990 and 1971–2000 as shown in Fig. 2. The wettest 30-year period was 1931–1960, followed by 1961–1990. 1971–2000 was the driest thirty-year period. From 1931 to 1960, a 1200 mm isohyet was discernible in the southwestern corner of the country. whereas in the latter periods it was completely absent. This observation is consistent with findings that the isohyets in the Sahel and Sudano-Sahelian zone in the Volta basin moved about 150 km south during the climate transition period of the early 1970s (L'Hote and Mahe 1996). Furthermore, even though 1971–2000 was the driest period, there was no significant change in the number of rainy days compared to 1961–1990. Because the two 30-year periods 1961–90 and 1971–2000 overlap, further research is required to validate the migration of isohyets southwards in the basin. In the southern part of the basin, the Sudan and Guinean agro-climatic zones in Ghana also show variable rates of change in rainfall without any clear trend (UNFCCC 2015b).

4 Climate Scenarios and Projection of Climate Change in the Basin

Methods for assessing climate change in the basin have been varied, and range from simple to robust. Some national climate change reports used old emission

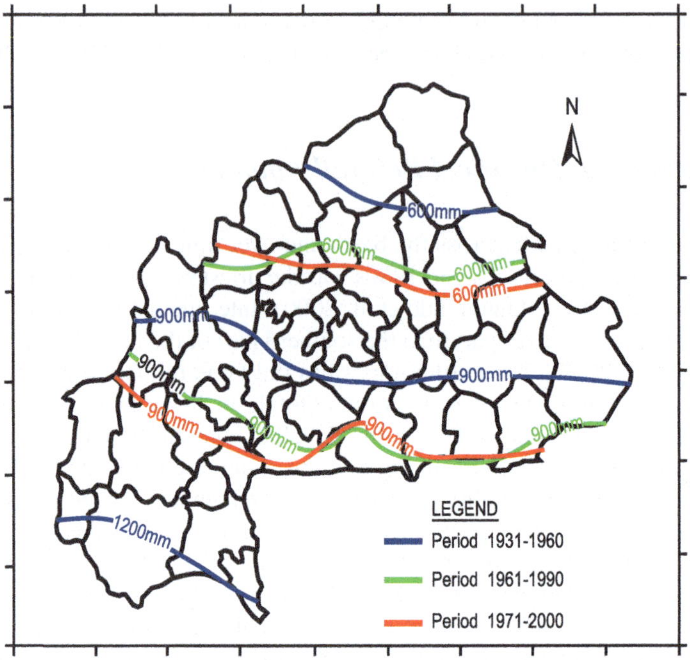

Fig. 2 Migration of isohyets from 1931 to 2000 (*Source* Third National Communication of Burkina Faso)

scenarios produced by the IPCC Special Report on Emission Scenarios (SRES) (IPCC 2000). These projections integrate emissions, climate change and climate impact data, including an assessment of their inherent uncertainties.

An advanced method is used in the IPCC Representative Concentration Pathways (RCPs). The RCPs comprise a set of four GHG concentration trajectories (RCP2.6, RCP4.5, RCP6.0, RCP8.5) spanning a large range of plausible human-caused climate forcing (van Vuuren et al. 2011). RCP2.6 is the best-case scenario, with much effective reduction of Greenhouse Gases (GHG); RCP8.5 is the worst case, used as a reference point for "business as usual" (i.e., taking no action to mitigate the effects of climate change). Regional Climate Models are applied to simulations to downscale GCMs to project future climate of the basin under different levels of anthropogenic GHG forcing. This method was used to develop climate projection in the 5th Coupled Inter-comparison Project (Taylor et al. 2012). Sylla et al. (2016) used CORDEX (Coordinated Regional Climate Downscaling Experiment; Giorgi et al. 2009) simulation to estimate climate change in the Volta Basin. CORDEX uses a number of regional climate models in downscaling GCMs at a spatial resolution of 50 km. CORDEX has also been validated over West Africa (Gbobaniyi et al. 2014 and Klutse et al. 2015). Two IPCC RCPs (RCP4.5 and RCP8.5) were used by Sylla et al. (2016) for temperature and rainfall studies under climate change for a historical period of 1951–2005 using 1976–2005 as a reference period. Plots of anomalies of temperatures from the mean of the reference period show that the basin was cooler than the average temperature before the 1980s. This was confirmed by multi-model ensembles of CORDEX RCMs and observations from Climate Research Unit (Harris et al. 2014). Also, in recent decades, warming has grown at a sharp and unprecedented rate, and is projected to increase. Warming estimates based on the two scenarios diverge in the 2040s, with RCP8.5 warming at higher rate than RCP4.5. RCP4.5 is projected to have temperature anomalies in the range of 1.5 to 4 °C and RCP8.5 in the range of 4 to 6 °C by 2100. Overall, the basin is likely to increase in temperature from 1.5 to 6 °C by 2100 (Sylla et al. 2016). These findings confirm the projection for increased temperatures in the 3rd National Communication of Ghana (UNFCCC 2015b) and 2nd National Communication of Burkina Faso (UNFCCC 2015a) submitted to the UNFCCC. Temperatures in Ghana are projected to increase between 1 and 7 °C by 2080 in all the agro-ecological zones and in Burkina Faso by 0.9 °C by 2025 and 1.5 °C by 2050 in the whole country. In addition to the different time horizons of these projections, the extent of increase in the temperatures also vary among the studies. Standardization and harmonization of studies in the basin and West Africa sub-region, in general, will be necessary.

Relating to precipitation, Sylla et al. (2016) showed great inter-annual variability in the basin, with rainfall measured between −0.5 and 0.5 mm/decade, making detection of negative or positive trends in rainfall very difficult. However, positive anomalies dominated, even with some negative anomalies projected. This suggests that the occurrence of wetter conditions is highly probable in the basin in the future. Mean annual rainfall in the Guinea Savannah of Ghana (in the Volta Basin) is projected to decrease by 3.5% in 2021–2040, 0.9% in 2041–2060 and 3.1% in 2061–2080. However, for the Sudan Savannah, the mean annual rainfall is projected to decrease by 3.2% by 2021–2040, increase by 0.8% in 2041–2060 and decrease by 2.3% by 2061–1080. The results also show no clear trend in the decadal analysis. Available projection of rainfall in Burkina Faso's 2nd National Communication (UNFCCC 2015a) did not use GCM-RCM modeling. Moreover, the study was for the whole country and did not focus on agro-ecological zones. The results of that study indicate that rainfall could decrease by 6.4% by 2025 and by 11% by 2050 in the worst-case scenario. This result does not support evidence of a negative or downward trend in projected rainfall.

The variability of inter-annual rainfall magnitudes bears on the availability of water for socioeconomic development. Variability can impact on surface and groundwater resources needed to supply water for domestic, industrial and irrigation use, hydro-power generation and hydro-ecology for the sustenance of flora and fauna. Inadequate rainfall in the basin as experienced in 1970 and 1983–1984 can have a significant negative impact on rain-fed agriculture, a major economic activity in the basin. While inter-annual rainfall variability is important, rainfall distribution – onset, duration and intensity – and its variability within each year is also of great importance to agricultural productivity. These studies have examined the impacts of climate change on the variability of rainfall distribution, and have shown that the onset of the rainy season has been shifting forward from April to May (van de Giesen et al. 2010 and Laux et al. 2007). Knowing when the rainy season will begin is essential to farmers so that they can prepare the land and sow in order to use the early rains for crop development, especially in an environment where rainfall is not very reliable and can fail. While onset is variable, studies show that the end of the rainy season remains fixed, resulting in a shortening of rainy seasons. Furthermore, the average annual rainfall amount has not changed as the rainy season duration has decreased, meaning that there is an increase in the frequency of extreme rainfall events. Studies carried out by Sylla et al. (2016) show that extreme precipitation intensity will tend to increase across the basin in the near future (2036–65) to the late twenty-first century when greater GHG forcing will produce greater precipitation intensity. The southern part of the basin is estimated to experience an increase of more than 30%, and the northern part an increase of 10% to 20% compared to the reference period 1976–2005.

Dry spell duration during the rainy season or lack of continuity of the rainy season can disrupt rain-fed agriculture. Dry spell length is defined as the maximum number of consecutive dry days that rainfall is less than 1 mm/day. Projection of dry spell length is very small, from −5 to 5%, for mid-level GHG forcing in the near future. In contrast, dry spell length for high-level GHG forcing for the same period is about 10%, and by the late 21st century, projections for high-level GHG forcing could be up to 30% in the south and middle part of the basin. Under these conditions, farmers in the basin will have to adapt to shortened and intensified rainy seasons coupled with dry spells, through the use of crops and cultivars that are more suitable for such conditions. Climate change will also require the use of reservoirs to capture, store and use the surface runoff for irrigating crops, as well as the conjunctive use of surface water with groundwater.

5 Agriculture, Water Resources and Climate Change

5.1 Climate Change Impacts on Crop Production and Adaptation Strategies

The latest Special Report of the Intergovernmental Panel on Climate Change indicates global warming of 1.5 °C above pre-industrial levels. Related assessment of global greenhouse gas emission pathways indicates that the earth is already experiencing the adverse impacts of 1 °C of global warming through more extreme weather, among other changes, and that warming of 1.5 °C is not 'safe' for most nations, communities, ecosystems, and sectors, posing significant risks to natural and human systems due to the likelihood of increased frequency of extreme weather events coupled with land-use change resulting in increased frequency of flood events (IPCC 2018).

Using a statistical metric of multidimensional climate change to quantify the emergence of global climate change hotspots in the CMIP5 (Fifth phase of the Coupled Model Intercomparison Project climate model ensemble), Diffenbaugh and Giorgi (2012) observed that the Sahel and tropical West Africa are among the persistent regional climate change hotspots throughout the twenty-first century of the RCP8.5 and RCP4.5 forcing pathways in the 2016–2035, 2046–2065 and 2080–2099 periods. These hotspots are characterized by widespread increases in mean temperature, the intensification of extreme hot seasons and changes in intensity and distribution of precipitation. These analyses, however, did not consider non-climatic factors that will ultimately determine the impacts of climate change.

The West African region is particularly vulnerable to climate change and extreme climate events such as floods and droughts. The region is highly reliant on rain-fed agriculture and has little economic and institutional capacity to address the consequences of climate variability and change (Sultan and Gaetani 2016; Zougmoré et al. 2016). These factors make the region particularly vulnerable to frequent food crises, especially in the drier areas, including in northern sections of the Volta Basin (Caritas 2018; van de Giesen et al. 2010). These impacts invariably result in loss of human life and livelihoods, which undermine advances toward sustainable development objectives (CCAFS 2013). Poverty is defined as unacceptable deprivation in one or several dimensions of human welfare (Ravallion 1996). The African Development Bank has noted that agricultural growth in sub-Saharan Africa has been demonstrated to be a principal driver of poverty reduction and stimulus of growth across economic sectors on the continent. In fact, agricultural growth in this region has been demonstrated to be twice as effective in reducing poverty as growth based in other sectors (AfDB 2010). Between 2007 and 2017, the share of the national labor force attributed to the agricultural sector ranged from 33.6% in Mali to 47.5% in Togo, with an average of 41.6% across the six countries of the Volta basin. In 2017 alone, the sector generated about 42.5% of the basin's economic output (AfDB 2018). Several factors, including climate shocks and loss of employment in agriculture, have adversely affected agricultural productivity in Sub-Saharan Africa, which is expected to drop by nearly 10% this decade, with a corresponding increase in the numbers of people running small household businesses (World Development Report 2018).

Data submitted by all six riparian countries to the UNFCCC over the past 15 years reveals that increasing temperatures and changes in precipitation in the Volta Basin are very likely to reduce cereal crop productivity with the potential to undermine food security (UNFCCC 2011, 2015a, b, 2017, 2018). In spite of these challenges, agriculture continues to make an important contribution to the Gross Domestic Product (GDP) of all the Volta Basin countries (Lemoalle and de Condappa 2010). Between 2007 and 2017, the share of the national labor force of the agriculture sector ranged from 33.6% in Mali to 47.5% in Togo, with an average of 41.6% across the six countries of the Volta basin (IMF 2018).

In 2017 alone, the sector generated about 42.5% of the basin's economic output (AfDB 2018). Food crops grown in the basin include maize, millet, sorghum, rice, yam, groundnut, Bambara beans, cowpea, soybean and vegetables including tomato, onion, cabbage, pepper and okra. Cocoa, cotton and coffee are the predominant cash crops (UNFCCC 2011, 2015a, b, 2017, 2018). Planting is usually done in the wet season and the vast majority of crop farms are small in size, with farmers typically using rudimentary tools such as hoes and cutlasses with limited application of agrochemicals (Chamberlin 2008; Zougmoré et al. 2016). Total chemical fertilizer use in 2014 amounted to 558,800 metric tons, equivalent to 12.2% of African fertilizer consumption and 2% of global fertilizer consumption (FAOSTAT 2019). The Fifth Assessment Report of the IPCC confirms that climate change exacerbates prevailing stress on water availability and agricultural systems, particularly in semi-arid environments of Africa, such as the northern parts of the Volta River Basin. Increasing temperatures and changes in precipitation in the Volta Basin are very likely to reduce cereal crop productivity, with strong adverse effects on food security (Niang et al. 2014). Crop growing periods in West Africa may decrease by an average of up to 20% by 2050, resulting in cereal yield losses of 40% as well as cereal biomass losses for livestock due to a decline in rainfall and frequent, long-lasting and intense droughts (Niang et al. 2014).

In its Fifth Assessment Report, the IPCC projects increases in precipitation in West Africa to increase millet and sorghum yields; however, projected temperature increases greater than 2 °C (relative to the 1961–1990 baseline) are estimated to counteract positive effects on the yields of millet and sorghum (for B1, A1B, and A2 scenarios) (Niang et al. 2014). These adverse effects will be felt more strongly in the savannah ecological zone than in the Sahel, with new millet and sorghum varieties subject to greater adverse impacts compared to their traditional counterparts (Sultan et al. 2013). Cassava (*Manihot esculenta)* is very adaptable to relatively marginal soils, high temperatures and erratic rainfall conditions compared to many cereal crops, making tuberous root crop a better substitute for cereals as an adaptation response to climate change under rain-fed agricultural conditions. For peanuts, both positive and negative impacts have been projected by researchers under A2 and B2 scenarios. Suitable agro-climatic zones for growing economically important perennial crops such as cocoa and coffee are estimated to diminish significantly, largely as a result of the effects of rising temperatures. Under an A2 scenario, by mid-century, suitable agroclimatic zones that are currently classified as very good to good for perennial crops may become marginal, and zones that are currently marginally suitable may become unsuitable. The report further concludes that crop growing periods in West Africa may decrease by an average of 20% by 2050, causing a 40% decline in cereal yields and a reduction in cereal biomass for livestock (Niang et al. 2014); these findings have been largely confirmed by Zougmoré (2016). Detailed atmospheric modeling for the West African region shows that the onset of rainy season will shift to later in the year, roughly from April to May (van de Giesen et al. 2010). Since the end of the rainy season, as well as the total amount of rainfall, remain more or less fixed, the rainy season is decreasing in duration, leading to reduced rain-fed agriculture, the dominant mode of food production in the region. The IPCC Special Report projects that warming of up to 2 °C could result in accumulated heatwave duration up to three months; warming of up to 3 °C would lead to drastic reductions in the maize crop in Africa (IPCC 2018).

Schroth et al. (2016) observed that within the Volta Basin, cocoa is now mostly grown in climates that have a maximum of four consecutive dry months, defined as less than 100 mm of rainfall per month; in Ghana and Ivory Coast, the northern limit of the cocoa belt also coincides roughly with the 4 months of dry season. The difference between rainfall and ETP during the driest quarter is projected to decrease in the savannah in the 2050s compared to the present climate, especially in Ghana (Schroth et al. 2016). Drier-than-average years could become more frequent in West Africa than they are now, possibly increasing the risk of drought (Abiodun et al. 2013). Under present climatic conditions, it is generally assumed that drought is a greater threat to cocoa than is high temperature in West Africa (Carr and Lockwood 2011). Drought has affected cocoa yields regularly in West Africa, particularly during the severe El Niño years of the 1980s (Ruf 2011). The only practical way of protecting cocoa trees from high temperatures is through overhead shade from appropriately selected, spaced and managed companion trees and selected crops (especially bananas and plantains) co-located on the cocoa farm so that shading can reduce coca leaf temperatures by up to 4 °C (Almeida and Valle 2007). Adequate ventilation is also an important complement because it can reduce fungal disease on cocoa plantations (Zhang and Motilal 2016).

Because of drought periods and erratic rainfalls without effective water management to promote crop production, the Volta Basin countries can no longer rely on rain-fed agriculture to feed their rapidly increasing populations. Mitigation practices may include irrigation and water harvesting technologies to enhance agricultural productivity. This is especially applicable to Northern Ghana and Burkina Faso, where the rainy season lasts for only four to five months. Adaptive strategies would therefore include continued efforts to produce faster-growing rain-fed crop cultivars, mainly corn and sorghum, as well as better storage of surface run-off in small reservoirs during the wet season for use during the dry season. The use of (shallow) groundwater during the dry season is a highly complementary adaptation strategy (van de Giesen et al., 2008) for building the resilience of crop production systems against climate change. Other such strategies in the Volta Basin include conservation agriculture, alternate wetting and drying approach (AWD) for rice production, agroforestry, "zaï" water conservation technology and the development of varieties of cultivars that are resilient to higher temperatures, drought, pest, weeds, salinity, flooding, etc., with a greater emphasis on new varieties that can withstand multiple stresses.

Trees in or near agricultural fields reduce the vulnerability of cropping systems to climate variations, so the introduction of trees in agriculture, such as in agroforestry and silvopastoralist systems, is considered to be an effective adaptation strategy. Tree roots reach into deep soil for water and nutrients, which benefits crops during droughts. Trees improve fertility and protect soils from erosion by increasing soil organic matter, porosity, infiltration and soil cover (Verchot et al. 2007). Nitrogen-fixing trees contribute to the resilience to droughts of crops due to improvements in soil nutrients and water infiltration – in, for example, Ghana and Burkina Faso (Garrity et al. 2010). Studies of agroforestry systems highlight the trade-offs of trees between increasing soil protection and reducing the light available to crops under the canopies. In agroforestry, tree ecosystems may contribute differently to the adaptation of crops to climate change depending on climate scenarios and production systems (Verchot et al. 2007). Within the Volta Basin, farmers mostly maintain agroforestry parkland systems, characterized by the deliberate retention on farmlands of economically valuable and multipurpose trees such as Shea (*Vitellaria paradoxa*),

Dawadawa (*Parkia biglobosa*) and Kapok (*Ceiba pentandra*) (Poudyal 2009). There is heavy dependence on the inherent fertility of the soil due to low fertilizer application, both in terms of quantity per hectare and frequency of application. In addition, pressure on agricultural lands has reduced fallow periods and therefore decreased the quality of soil fertility.

5.2 Climate Impacts on Livestock-Rearing and Adaptation Strategies

The livestock sector remains a major contributor to both rural livelihoods and the national economies of all six Volta Basin countries, particularly in the savannah ecological zones and semi-arid regions, where the predominant livestock is cattle, sheep, goats, pigs, fowls and guinea fowls (UNFCCC 2011, 2015a, b, c, 2017, 2018). It is estimated that at least 100 million poor people in West Africa rely on livestock as part of their livelihood. In the Sahelian zone, livestock production accounts for about 30% of revenue from the agriculture sector. For both crop farmers and pastoralists, livestock serves as a productive asset to generate income, and contribute to increasing food security in the basin. Repeated droughts in the Sahel has led to the adoption of agro-pastoralism—a combination of crop farming and livestock rearing on the same farm—by pastoralists who were once solely dependent on livestock. Similarly, crop farmers have diversified into rearing livestock due to repeated crop failure associated with droughts (Thornton et al. 2014).

For both crop farmers and pastoralists, livestock rearing generates income, contributing a key element of food security in the basin, because this diversification is an effective adaptation strategy to increased prevalence of droughts and warming temperatures (Thornton et al. 2014; Turner et al. 2014; Rosenstock et al. 2016).

Livestock contributes significantly to satisfying household needs, and constitutes the second-largest export commodity for countries such as Mali and Burkina Faso (FAO and ECOWAS 2017). Livestock rearing stabilizes the socioeconomic capability of households by providing reliable income in times when prices of crops are low due to a bumper harvest. The predominant livestock reared by farmers is cattle, sheep, goats, pigs and fowl. In Burkina Faso, animal husbandry, like crop production, is an important element of the farming system in all the regions of the country. In 2011, the animal population in the West African region was estimated at 283 million head, comprising 63,128,047 cattle, 7,492,551 sheep, 130,446,369 goats, 11,225,235 pigs and 2,895,861 camels (FAO and ECOWAS 2017).

Strategies to enhance livestock-based livelihoods include the implementation of management practices suited to drought and warming temperatures; introduction and support of animal species and breeds able to adapt to drought and warming temperatures and associated synergies and trade-offs, and policy and institutional mechanisms to enhance adaptation of livestock production systems (Zougmoré et al. 2016).

5.3 *Climate Change Impacts on Inland Fisheries and Adaptation Strategies*

Fishing is the primary livelihood of many residents around the main surface water bodies, both natural and artificial, such as the Lake Volta (Mul et al. 2015). The majority of fish species found in the basin are edible. Along the shoreline of Lake Volta are over 1,200 fishing villages inhabited largely by impoverished rural populations whose livelihood is primarily fishing in the lake. It is estimated that over 80,000 fishers and 20,000 fish processors and traders are involved in Lake Volta fishing (UNEP-GEF Volta Project 2013). Inland fisheries are the major livelihood of many communities along the Volta River, and fish provide about 60% of total protein requirements of Ghana; inland fisheries and aquaculture contribute about 20% of Ghana's total production of fish (UNEP-GEF Volta Project 2013).

Fisheries encompass a wide range of ecological and socioeconomic components within the Volta Basin; artisanal fishery dominates employment in the fishing industry and is active year-round. Artisanal fishermen use traditional wooden boats, sometimes motorized, with a variety of gear, including nets, lines and seines. In addition to natural fish resources in the basin, aquaculture development, particularly caged fish production, is increasing, especially in Lake Volta and in downstream reaches of the river. Traditional processing methods include smoking, drying, salting and curing (UNEP-GEF Volta Project 2013).

Climate change is projected to impact fisheries and aquaculture through water stress and competition for limited water resources, which have the potential to adversely impact aquaculture operations and inland fisheries production and increase the likelihood of conflicts among water users Poff et al. (2002) and Cochrane et al. (2009). The projected impacts of climate change on fisheries are both social and economic for fishing communities.

5.4 *Ecosystem-Based Adaptation Issues*

The freshwater ecosystems of the Volta Basin provide a number of services that contribute to local livelihoods and larger-scale basin objectives, including food, fuel and construction materials, regulating flows (reducing peak and increasing baseflows) and habitats for animals, such as migrant birds, which may attract tourism and recreational activity.

Medicinal and pharmaceutical products are derived from various forest and wetland products in the basin. The leaves, bark, seeds and roots of a range of plant species are used for a variety of pharmaceutical and medicinal products, which are also derived from the fauna that inhabits the ecosystems in the basin. But flora species are threatened in Burkina Faso, Mali and Togo. In Burkina Faso, a study in the southern Sahelian zone (axis Kaya-Tougouri-Yalgo) corresponding to the White Volta-Nakambe watershed indicates that some flora species have shown a great decline in numbers so are considered to be at high risk (UNEP-GEF Volta Project 2013).

6 Sustainable Development Issues

In 2013, the United Nations Economic Commission for Africa conducted a country-level survey on sustainable development priorities in the member states of the Economic Community of West African States (ECOWAS). These were then aligned with the Poverty Reduction Strategies (PRSs) that integrate the economic, social and environmental pillars of sustainable development. This analysis revealed that the priorities for sustainable development, in order, are: education; health; sustainable infrastructure development (energy, water, transport); inclusive economic growth, diversification and transformation; good governance and the rule of law; agriculture and food security; environment and natural resource management (forest, water and soils); social protection for the poor and vulnerable; sanitation and urban management; peace and security. Poverty is recognized as an overarching issue, while the others are considered interrelated (UNECA 2015 and World Bank 2015).

In September 2015, UN Member States adopted the 2030 Agenda for Sustainable Development, otherwise known as the Global Goals to end poverty, fight inequality and injustice, and tackle climate change by 2030, and more commonly called the Sustainable Development Goals (SDGs). These build on the Millennium Development Goals (MDGs) adopted in 2000. Countries within the basin have made good progress toward the MDGs, but the indignity of poverty is the reality for millions of people within the basin. The SDGs go much further than the MDGs, addressing the root causes of poverty and the universal need for development that works for all people. The SDGs also set higher goals, seeking to eradicate, and not just reduce, poverty, as well as embracing higher aspirations to meet obligations for future generations. The SDGs seeks for no one to be left behind and is considered an agenda for shared prosperity, peace and partnership. It conveys the urgency of climate action, underscoring the potential for dire consequences of mismanagement of natural resources in the Volta Basin. Moreover, the SDGs emphasize that poverty is not merely a lack of income but also takes the form of low educational attainment, poor health and nutrition, exposure to physical insecurity and natural hazards, and substandard living conditions. Poor people also lack access to improved water sources, better sanitation facilities and electricity, the shortage of all of which contribute to poor labor productivity, amplifying the cycle of income poverty (see Table 1).

The Sustainable Development Goals include a goal on water (SDG6) with ambitious targets for universal access to drinking water and sanitation (targets 6.1 and 6.2, respectively) by 2030. Achieving sustainable universal access under the influence of climate change will be a defining challenge for the SDGs. SDG6 also includes targets to improve water quality (6.3), improve water-use efficiency (6.4), implement integrated water resources management (IWRM) (6.5), and restore water ecosystems

Table 1 Selected water and sanitation indicators in the six riparian countries

Countries	HDI[1] 2017	Mortality rate attributed to unsafe water, sanitation and hygiene Services (per 100,000 population) SDG 3.9	Population using improved water sources (%) (2015) SDG 6.1	Population using improved sanitation facilities (%) (2015) SDG 6.2
Benin	0.515	59.7	67.0	13.9
Burkina Faso	0.423	49.6	53.9	22.5
Cote d'Ivoire	0.492	47.2	73.1	29. 9
Ghana	0.592	18.8	77.8	14.3
Mali	0.427	70.7	74.3	31.3
Togo	0.503	41.6	62.8	13.9

http://hdr.undp.org/sites/default/files/2018_human_development_statistical_update.pdf

(6.6). All of these will be impacted by climate change and in turn, influence the resilience of drinking water and sanitation services. Inadequate water supply and limited sanitation services have important consequences for human health: the WHO (2014) estimates that globally nearly 1,000 children under five years die every day as a result of diarrhea caused by poor water and sanitation. At the end of the MDG period in 2015, it was estimated that while 91% of the world's population has access to an improved water supply (UNICEF and WHO 2015), only 68% has access to improved sanitation. The corresponding figures within the Volta River Basin for access of the population to an improved water supply and improved sanitation were 68% and 16%, respectively (IMF 2018), indicating that these six countries have a long way to go toward improving water and sanitation services. Open defecation remains a major public health concern, and its elimination has been explicitly targeted in the SDGs.

Notwithstanding the ongoing challenges, there have been significant advances in human development over the past decades in all six countries. For the period 2000–2017, the human development index (HDI) increased by 18.4% for Togo and by 47.9% for Burkina Faso (IMF 2018). In 2010, the HDI score for Ghana rose to 0.554, making it the only country within the Volta River Basin to transition from low human development to medium human development, with the HDI growing at an average annual rate of 1.14% for the past 17 years (IMF 2018). On average, all six countries within the basin achieved modest positive economic growth rates. Over the period 2000–2016, the average annual real GDP growth rate for the subregion was 4.51%; the most progress was made by Ghana, followed by Burkina Faso. Ivory Coast and Togo made the least progress in the period. Despite these moderate gains, all basin countries still face the challenge of raising and sustaining economic growth over the long term (IMF 2018). Growth over this decade may be due to strong global demand for export commodities such as oil and cocoa, better macroeconomic management and new mining ventures. Increased agricultural productivity and transition of the labor force to higher earning sectors of manufacturing and services is essential to create higher earnings, employment and demand for the value chain of outputs from other sectors. This transition will lead to a reduction in the share of agriculture to national economies and the increase in the share of manufacturing and services, which has the potential to increase employment and incomes and reduce poverty (IMF 2018).

Across the basin countries, people are living longer and have greater livelihood opportunities than 20 years ago, but they still lag behind most countries in the world in human development measures, including education, health and access to drinking water and other basic infrastructure services. These persistent social challenges have hampered the efforts of these countries to accelerate growth and reduce poverty, as envisaged in the national Poverty Reduction Strategies (PRSs) of each country. According to World Bank revised estimates of global poverty from 1981 to 2015 based on 2011 Purchasing Power Parities (PPPs) released in September 2018, the proportion of the population living on less than US$1.90/day (SDG 1.1) in Ghana, Ivory Coast, Burkina Faso, Mali, Togo and Benin are 12.0%, 28.2%, 43.7%, 49.7%, 49.2% and 49.6%, respectively (IMF 2018). Another indicator of poverty is the global Multidimensional Poverty Index (MPI), an international measure of acute poverty that goes beyond income poverty and takes into account multiple deprivations at the household level, including health, education and standard of living. As expected, computed MPI figures may vary significantly from SDG 1.1 values. MPI values, taking into account broader parameters of deprivation, are 9.6%, 24.5%, 24.5%, 38.1%, 56.7% and 64.8% for Ghana, Ivory Coast, Togo, Benin, Mali, and Burkina Faso, respectively (IMF 2018). Note that the MPI assesses poverty at the individual level, and also permits comparisons both across countries and within countries by ethnic group and urban/rural location. National life expectancy figures in the basin countries are 63 years in Ghana, 61.2 years in Benin, 60.8 years in Burkina Faso, 60.5 years in Togo, 58.5 years in Mali and 54.1 years in Cote d'Ivoire.

7 Climate Change, Water, Health and Sanitation

All six riparian countries within the basin have witnessed some improvement in access to sanitation over the past few decades. Notwithstanding, about 30% of the populations of Mali and Ivory Coast, and only 15% of the populations of Benin, Togo and Ghana, had access to improved sanitation facilities as of 2015. Significant progress has been made by all the riparian countries to provide their citizens with improved water sources. As of 2015, Ghana, Mali and Ivory Coast had provided about 78%, 74% and 73% of their populations with improved water sources, respectively, with Burkina Faso providing up to 54% of the population with improved water sources, the lowest in the river basin (see Table 1) (UNECA 2015). In the Upper East and Upper West regions of Ghana, the water supply used for drinking and other domestic purposes, serving approximately 80% of rural and urban populations, comes from groundwater sources (Johnston and McCartney 2010).

African women are particularly vulnerable to the impacts of climate change because they shoulder an enormous but imprecisely recorded portion of responsibility for subsistence agriculture, the productivity of which can be expected to be adversely affected by climate change and overexploited soil (Viatte et al. 2009). Global financial crises, such as the one experienced in 2007–2008, as well as downturns in economic trends at the national level, may cause job losses in the formal sector, driving men into the informal sector to compete for jobs that were previously performed by women, increasing the vulnerability of women (AfDB et al. 2010).

The health implications of the projected rise in temperature and rainfall variability are alarming, and include an increase in the incidence rate of measles, diarrheal disease, guinea worm infestation, malaria, cholera, cerebral-spinal meningitis and other water-related diseases. These impacts of climate change and variability worsen the plight of the poor, who are mostly women and children (Niang et al. 2014). Climate change amplifies existing health vulnerabilities, including insufficient access to safe water and improved sanitation, food insecurity and limited access to health care and education. The most effective strategies to build resilient development integrate consideration of climate change risks with land and water management and disaster risk reduction. Oppressive temperatures and increasing heatwave duration will also likely impact human health, mortality and productivity directly. With the warming of up to 2 °C, there is a substantial increase in the risk of potentially deadly heatwaves.

Climate change also poses significant risks to water and sanitation services. Risks to water services include damage to infrastructure due to flooding, loss of water sources and increasing water demand as a result of growth in the population *and* economy and as well as changes in water quality. Adverse impacts on sanitation services also include damage and loss of services from floods and reduced carrying capacity of waters receiving wastewater. Measures to reduce risks may include the integration of climate resilience strategies into water safety plans, as well as improved accounting and management of water resources. Technological options for enhancing service delivery and changes in management models offer the potential to reduce risk, particularly in low-income settings.

The IPCC-projected global warming of 1.5 °C clearly states that the earth is already experiencing the adverse impacts of 1 °C of global *warming* through more extreme weather, among other impacts. It further asserts that warming of 1.5 °C is not considered "safe" for most nations, communities, ecosystems and sectors, and poses significant risks to natural and human systems. It is clear from the report that warming of 1.5 °C or higher increases the risk associated with long-lasting or irreversible changes, such as the loss of some ecological systems (IPCC 2018). With the increased frequency of extreme weather events, coupled with land-use change-related flood events and continued growth of settlements, people will be exposed to the negative consequences of climate change increasingly (Milly et al. 2002, Jiménez Cisneros et al. 2014).

Howard et al. (2016) assert that the impacts of climate change on sanitation infrastructure can be both positive and negative, depending on the nature of the impacts as well as changes in the types of technologies employed by households. In the drier parts of the basin, the impact on rudimentary onsite sanitation infrastructure may be positive due to the fact that groundwater pollution risks may be minimized as the distance between the base of pits and groundwater (and hence travel time for pathogens) increases (Mahmud et al. 2007). Seasonal groundwater flooding of pits become less frequent (Sherpa et al. 2014), but increased flooding will pose major threats to sewage and septic systems that rely on water (Howard et al. 2016).

Various policy options could address the resilience of water and sanitation services. In rural areas, where small water supplies continue to be the norm, key policy decisions revolve around which technologies are acceptably resilient. Howard et al. (2010) suggest that some, such as dug wells, are less resilient because they are vulnerable to contamination, drought and/or long-term reduction in water volume, and inability to prevent damage during flooding. Other technologies, such as raising the wellheads of tubewells, may be more resilient, though they have yet to be demonstrated to be effective.

Groundwater resources are widely expected to be less vulnerable to climate change because of their inherent buffer against unpredictable rainfall, and are therefore believed to form the basis of adaptation programs (Rockström et al. 2009 and Jaliffer-Verne et al. 2015). In fact, groundwater resources are likely to be impacted as much by increases in demand for groundwater as by changes to groundwater recharge (Kundzewicz and Döll 2009). A caveat is that detection of and attribution to climate change are difficult, given that surface and groundwater hydrology are governed by multiple, interacting drivers and factors, such as land use change, water withdrawals, and natural climate variability.

8 Conclusions

Studies have shown that the climate of the basin has been warming since the 1980s when the temperature anomalies from the mean of the reference period of 1976–2005 are examined for the entire historical period of 1951–2005, with temperatures projected to increase by as much as 1.5 to 6 °C by the end of 2100 under mid-level and high-level IPCC greenhouse gas forcing scenarios relative to 1976–2005. Precipitation has also varied, with the 1930s and 1950s being relatively wet and 1970s and 1980s being dry. There is some evidence of migration of isohyets southwards during some decades, however, a trend has not been established with available information. There have been positive and negative anomalies in rainfall records without any clear trend, even though positive anomalies appear to dominate in the recent past. The onset of the rainy season has been shifting forward without any significant change in rainfall amount, even though there is an increase in the number of extreme rainfall events. Projections using IPCC scenarios show an increase in future extreme rainfall events and an increase in periods of dry spells within rainy seasons, which may be detrimental to rain-fed agriculture.

Although agricultural productivity is a principal driver out of the vicious cycle of poverty within the Volta basin, all six riparian countries have agricultural systems that are highly vulnerable to climate change and climate variability with associated extreme climate events such as floods and droughts. This is due to a great reliance on rain-fed agriculture, and little economic and institutional capacity to effectively manage environmental and natural resources. The sub-region, therefore, requires well designed and adequately funded adaptation strategies, such as changes in crop varieties, sowing dates and density, enhanced water harvesting and irrigation techniques, fertilizer management, access to technologically appropriate agricultural mechanization, emphasis on crop and livestock breeding to address multiple biotic and abiotic stresses, livelihood diversification and crop substitution to enhance food security and livelihood. Higher costs of food production due to climate change impacts within the basin will have a major adverse impact on food security in these countries. Climate change is a threat multiplier requiring greater investment to reach the aspirations of the Sustainable Development Goals.

References

Abiodun BJ et al (2013) Potential influences of global warming on future climate and extreme events in Nigeria. Reg Environ Chang 13:477–491. https://doi.org/10.1007/s10113-012-0381-7

AfDB (2010) Agriculture Sector Strategy 2010–2014. https://www.afdb.org/fileadmin/uploads/afdb/Documents/Policy-Documents/Agriculture%20Sector%20Strategy%2010-14.pdf

AfDB (2018) African Development Bank Group. http://comstat.comesa.int/wiqcbkg/afdb-socio-economic-database-1960-2019

Almeida AAF, Valle RR (2007) Ecophysiology of the cacao tree. Braz J Plant Physiol 19:425–448. https://doi.org/10.1590/S1677-04202007000400011

Caritas (2018) West African Food Crisis. https://www.caritas.org/2018/06/west-africa-food-crisis-threatens-6-million-people/

Carr MKV, Lockwood G (2011) The water relations and irrigation requirements of cocoa (Theobroma cacao L).: a review. Exp Agric 47:653–676. https://doi.org/10.1017/S0014479711000421

CCAFS (Climate Change, Agriculture and Food Security) 2013 West Africa: our work. https://ccafs.cgiar.org/west-africa-our-work#.VwKVfPmLRD8. Accessed 16 Nov 2018

Chamberlin J (2008) It's a small world after all: defining smallholder agriculture in ghana. Discussion Paper 00823. IFPRI, Washington, DC. https://pdfs.semanticscholar.org/b8f0/775e9fe38b0f40504d389e59282466ed5ff2.pdf. Accessed 24 Nov 2018

Cochrane K et al (eds) (2009) Climate change implications for fisheries and aquaculture: overview of current scientific knowledge. FAO Fisheries and Aquaculture Technical Paper. No. 530. FAO, Rome

de Condappa D et al (2008) Decision-support tool for water allocation in the Volta Basin. Volta Basin Focal Project Report No. 10. IRD, Montpellier, France, and CPWF, Colombo, Sri Lanka, 28p

Dhonneur G (1981) Les amas nuageux mobiles principale composante de la meteorologie du Sahel. La Meteorologie 27:75–82

Diffenbaugh NS, Giorgi F (2012) Climate change hotspots in the CMIP5 global climate model ensemble. Clim Change 2012(114):813–822. https://doi.org/10.1007/s10584-012-0570-x Accessed 20 Dec 2018

FAO and ECOWAS (2017) Review of the livestock/meat and milk value chains and policy influencing them in West Africa. http://www.fao.org/3/a-i5275e.pdf

FAO/GIEWS (1998) Sahel Report No. 5/98 (10 October 1998), General Text. www.fao.org/docrep/004/x0059e/x0059e00.htm. Accessed 05 Jan 2019

FAOSTAT (2019) Fertilizers by nutrients. http://www.fao.org/faostat/en/#data/RFN

Garrity DP et al (2010) Food Sec 2:197. https://doi.org/10.1007/s12571-010-0070-7

Gbobaniyi E et al (2014) Climatology, annual cycle and interannual variability of precipitation and temperature in CORDEX simulation over West Africa. Int J Climatol 34:2241–2257. https://doi.org/10.1002/joc.3834 Accessed 01/01/2019)

Giorgi F et al (2009) Addressing climate information needs at the regional level: The CORDEX framework. WMO Bull 58(3):175–83

Gyau-Boakye P, Tumbulto JW (2000) The Volta Lake and declining rainfall and stream flows in the Volta Basin. Environ Dev Sustain 2:1–10

HAP (Hydrological Assessment Report of the Northern Regions of Ghana) (2006) A Bibliographical Review of Selected Papers. CIDA, WRC, SNC-LAVALIN International

Harris I, Jones PD, Osborn TJ, Lister DH (2014) Updated high-resolution grids of monthly climatic observations—the CRU, TS3.10 Dataset. Int J Climatol 34:623–642. https://doi.org/10.1002/joc.3711

Hayward D, Oguntoyinbo J (1987) The Climatology of West Africa. Hutchinson, London

Howard G et al (2010) Securing 2020 vision for 2030: climate change and ensuring resilience in water and sanitation services. J Water Clim 1(1):2–16. Accessed 02 Jan 2019

Howard G et al (2016) Climate Change and Water and Sanitation: Likely Impacts and Emerging Trends for Action. Annu Rev Environ Resour 41(1):253–276. https://doi.org/10.1146/annurev-environ-110615-085856

IMF (2018) The World Economic Outlook (WEO) database, October 2018 Edition. https://www.imf.org/external/pubs/ft/weo/2018/02/weodata/weorept.aspx?pr.x=57&pr.y=12&sy=2000&ey=2023&scsm=1&ssd=1&sort=country&ds=.&br=1&c=638%2C748%2C678%2C662%2C742%2C652&s=NGDP_RPCH%2CNGDPPC%2CPPPPC%2CPPPSH&grp=0&a=#download

Intergovernmental Panel on Climate Change (IPCC) (2014) Climate change 2014: impacts, adaptation and vulnerability. IPCC WG II AR5 technical summary. http://ipccwg2.gov/AR5/images/uploads/WGIIAR5-TS_FGDall.pdf. Accessed 19 Sept 2018

IPCC (2000) Special report on emissions scenarios: a special report of working Group III of the intergovernmental panel on climate change. [Nakićenović N, and Swart R (eds).] 570 pp

IPCC (2018) Summary for policymakers. In: Global warming of 1.5 °C. An IPCC Special Report on the impacts of global warming of 1.5 °C above pre-industrial levels and related global greenhouse gas emission pathways, in the context of strengthening the global response to the threat of climate change, sustainable development, and efforts to eradicate poverty [Masson-Delmotte V, Zhai P, Pörtner HO, Roberts D, Skea J, Shukla PR, Pirani A, Moufouma-Okia W, Péan C, Pidcock R, Connors S, Matthews JBR, Chen Y, Zhou X, Gomis MI, Lonnoy E, Maycock T, Tignor M, Waterfield T (eds).]. World Meteorological Organization, Geneva, Switzerland, 32 pp. http://www.ipcc.ch/pdf/special-reports/sr15/sr15_spm_final.pdf)

Jaliffer-Verne I et al (2015) Impacts of global change on the concentrations and dilution of combined sewer overflows in a drinking water source. Sci Total Environ 508:462–476

Jiménez Cisneros et al (2014) Freshwater resources. In: Climate change 2014: impacts, adaptation, and vulnerability. part a: global and sectoral aspects. Contribution of Working Group II to the fifth assessment report of the intergovernmental panel on climate change [Field CB, Barros VR, Dokken DJ, Mach KJ, Mastrandrea MD, Bilir TE, Chatterjee M, Ebi KL, Estrada YO, Genova RC, Girma B, Kissel ES, Levy AN, MacCracken S, Mastrandrea PR, White LL (eds).]. Cambridge University Press, Cambridge, United Kingdom and New York, NY, USA, pp 229–269

Johnston R, McCartney M (2010) Inventory of water storage types in the Blue Nile and Volta river basins. Colombo, Sri Lanka, International Water Management Institute. 48 p. (IWMI Working Paper 140). http://dx.doi.org/10.5337/2010.214

Kasei RA (2009) Modeling impacts of climate change on water resources in the Volta Basin, West Africa. PhD thesis, Mathmatisch-Naturwissenschaftlichen Fakultat, RheinischenFriedrick-Wilhelms-Universitat Bonn

Klutse NAB et al (2015) Daily characteristics of West African Summer monsoon precipitation in CORDEX simulations. Theoret Appl Climatol 1–18:369–386. https://doi.org/10.1007/s00704-014-1352-3

Kundzewicz ZW, Döll P (2009) Will groundwater ease freshwater stress under climate change? Hydrol Sci J 54(4):665–675

L'Hote Y, Mahe G (1996) Afrique de l'Ouest et central, precipitation moyennes annuelles (periode 1951–1989). Map scale 1:6,000,000. Paris Orstom Editions

Laux P et al (2007) Predicting the regional onset of the rainy season in West Africa. Int J Climatol 2007:329–342. https://doi.org/10.1002/joc.1542

Lemoalle J, de Condappa D (2009) Water atlas of the Volta Basin-Atlas de l'eau dans le bassin de la Volta. Challenge Program on Water and Food and Institut de Recherche pour le Developpement, Colombo, Marseille, 96 p

Lemoalle J, de Condappa D (2010) Farming systems and food production in the Volta Basin. Water Int 35(5):655–680. https://doi.org/10.1080/02508060.2010.510793 Accessed 21 November 2018)

Mahmud SG et al (2007) Development and implementation of water safety plans for small water supplies in Bangladesh: benefits and lessons learnt. J Water Health 5(4):585–597

Maranan M et al (2018) Rainfall types over southern West Africa: Objective identification, climatology and synoptic environment. Quart J Royal Met Soc 144(714). https://doi.org/10.1002/qj.3345

Milly PCD et al (2002) Increasing risk of great floods in a changing climate. Nature 415:514–517

Min SK et al (2011) Human contribution to more- intense precipitation extremes. Nature 470(7334):378–381. https://doi.org/10.1038/nature09763

Mul M et al (2015) Water resources assessment of the Volta River Basin. International Water Management Institute (IWMI), Colombo, Sri Lanka, 78 p. (IWMI Working Paper 166). http://dx.doi.org/10.5337/2015.220

Niang I et al (2014) Africa. In: climate change 2014: impacts, adaptation, and vulnerability. part b: regional aspects. contribution of working group ii to the fifth assessment report of the intergovernmental panel on climate change [Barros VR, Field CB, Dokken DJ, Mastrandrea MD, Mach KJ, Bilir TE, Chatterjee M, Ebi KL, Estrada YO, Genova RC, Girma B, Kissel ES, Levy AN, MacCracken S, Mastrandrea PR, White LL (eds)]. Cambridge University Press, Cambridge, United Kingdom and New York, NY, USA, pp 1199–1265

Oguntunde PG (2004) Evapotranspiration and complementary relations in the water balance of the Volta Basin: field measures and gis-based regional estimates, ecology and development Series 22. Cuvillier Verlag Gottingen

Oguntunde PG et al (2006) Hydroclimatology of the Volta River basin in West Africa: trends and variability from 1901 to 2002. Phys Chem Earth 31:1180–1188. https://doi.org/10.1016/j.pce.2006.02.062

Peters M, Tetzlaff G (1988) The structure of West African Squall Lines and their environmental moisture budget. Meteorl Atmos Phys 39:74. https://doi.org/10.1007/BF01041933

Poff NL et al (2002) Aquatic ecosystems and global climate change. Potential impacts on inland freshwater and coastal wetland ecosystems in the United States. https://www.c2es.org/site/assets/uploads/2002/01/aquatic.pdf

Poudyal M (2009) Tree tenure in agroforestry parklands: implications for the management, utilisation and ecology of shea and locust bean trees in Northern Ghana. PhD thesis, University of York. http://etheses.whiterose.ac.uk/861/1/M.Poudyal_PhD.Thesis_Dec2009.pdf

Ravallion M (1996) 'Comparaisons de la Pauvrete´, Concepts et Methodes', LSMS Document de travail No122. Banque Mondiale, Washington, D.C. http://siteresources.worldbank.org/INTPA/Resources/429966–1092778639630/Ravallion88French.pdf

Rockström J et al (2009) A safe operating space for humanity. Nature 461:472–475

Rosenstock TS et al (2016) The scientific basis of climate-smart agriculture: a systematic review protocol. CCAFS working paper no. 138

Ruf FO (2011) The myth of complex cocoa agroforests: the case of Ghana. Hum Ecol 39:373–388. https://doi.org/10.1007/s10745-011-9392-0

Schroth Götz et al (2016) Vulnerability to climate change of cocoa in West Africa: Patterns, opportunities and limits to adaptation. Sci Total Environ 556(2016):231–241. https://doi.org/10.1016/j.scitotenv.2016.03.024 (Accessed 05/11/2018)

Sherpa AM et al (2014) Vulnerability and adaptability of sanitation systems to climate change. J Water Clim Change 5(4):487–495

Sonabel (Societe nationale d'electricite du Burkina) (2009) http://www.sonabel.bf/Statist/sour_prod.htm

Sultan B, Gaetani M (2016) Agriculture in West Africa in the twenty-first century: climate change and impacts scenarios, and potential for adaptation. Front Plant Sci 7:1262. http://dx.doi.org/10.3389/fpls.2016.0126225

Sultan B et al (2013) Assessing climate change impacts on sorghum and millet yields in the Sudanian and Sahelian savannas of West Africa Assessing climate change impacts on sorghum and millet yields in the Sudanian and Sahelian savannas of West Africa. Environ Res Lett 8:014040. https://doi.org/10.1088/1748-9326/8/1/014040

SWAC-OECD/ECOWAS (2008). Livestock and regional market in the Sahel and West Africa-potentials and challenges. Sahel and West Africa Club/ OECD, Paris

Sylla MB et al (2016) Climate variability and change over the Volta River Basin in The Volta River Basin, water for food, economic growth and environment. In: Williams TO, Mul ML, Biney CA, Smakhtim V (eds). www.routledge.com/series/ECMRBW

Taylor KE et al (2012) An overview of CMIP5 and the experiment design. Bull Am Meteor Soc 93(4):485–498. https://doi.org/10.1175/BAMS-D-11-00094.1

Thornton PK et al (2014) Climate variability and vulnerability to climate change: a review. Glob Change Biol 20:3313–3328

Turner MD et al (2014) The role of livestock mobility in the livelihood strategies of rural peoples in semi-arid West Africa. Hum Ecol 42(2):231–47

UNECA (2015) Sustainable development goals for the West Africa subregion, Summary report. https://www.uneca.org/sites/default/files/PublicationFiles/west_africa_sdg_summary_report_english.pdf

UNEP-GEF Volta Project (2013) Volta basin transboundary diagnostic analysis. UNEP/GEF/Volta/RR 4/2013

UNFCCC (2011) Second national communication of benin to the UNFCCC. https://unfccc.int/sites/default/files/resource/SNC_%20BENIN%20novembre%202011.pdf

UNFCCC (2015a) Second national communication of burkina faso to UNFCCC. https://unfccc.int/sites/default/files/resource/bfanc2engl.pdf

UNFCCC (2015b) Third national communication of ghana to the UNFCCC. https://unfccc.int/resource/docs/natc/ghanc3.pdf

UNFCCC (2015c) Third national communication of togo to the UNFCCC. https://unfccc.int/sites/default/files/resource/tgonc3.pdf

UNFCCC (2017) Third national communication of ivory coast to the UNFCCC. https://unfccc.int/sites/default/files/resource/3069145_C%C3%B4te%20dIvoire-NC3-1-Cote%20dIvoire%20-%20Third%20National%20Communication_0.pdf

UNFCCC (2018) Third National Communication of Mali to the UNFCCC. https://unfccc.int/sites/default/files/resource/Rapport%20TCN%20%202018%20-%20copie.pdf

UNICEF and WHO (2015) Progress on Sanitation and Drinking Water: 2015 Update and MDG Assessment. Geneva, Switz, WHO, 90 pp http://files.unicef.org/publications/files/Progress_on_Sanitation_and_Drinking_Water_2015_Update_.pdf

van de Giesen N et al (2010) Adapting to climate change in the Volta Basin, West Africa. Curr Sci 98(8), (Special Section: climate change and water resources)

van de Giesen N et al (2010) Adapting to Climate Change in West Africa. https://www.zef.de/uploads/tx_zefportal/Publications/8b40_21S8-P1-vandeGiesenetalACC.pdf

van Vuuren DP et al (2014) A new scenario framework for climate change research: scenario matrix architecture. Clim Change 122(3):373–386. https://doi.org/10.1007/s10584-013-0906-1

Verchot LV et al (2007) Climate change: linking adaptation and mitigation through agroforestry mitigation adaptation strategies for global. Change 12:901. https://doi.org/10.1007/s11027-007-9105-6

Viatte G et al (2009) Responding to the food crisis: synthesis of medium-term measures proposed in inter-agency assessments. Food and Agriculture Organization of the United Nations (FAO), Rome, Italy, 100 pp. http://www.fao.org/fileadmin/user_upload/ISFP/SR_Web.pdf

VRA (Volta River Authority) (2009) http://www.vra.com/Power/akohydro.php

World Bank (2015) Improved water supply (% of population with access to a source of water that is considered protected) Data for access to improved water supply are from 2012 except for Somalia, using 2011 data. http://data.worldbank.org/indicator/SH.H2O.SAFE.ZS

World Development Report (2018). http://www.worldbank.org/en/publication/wdr2018

World Health Organization (2014) Preventing diarrhoea through better water, sanitation and hygiene: Exposures and impact in low- and middle-income countries. Geneva, Switz, WHO. http://apps.who.int/iris/bitstream/handle/10665/150112/9789241564823_eng.pdf;jsessionid=4D4BD64631612823386FB2FD89C27A56?sequence=1

Zhang D, Motilal L (2016) Origin, dispersal, and current global distribution of cacao genetic diversity. In: Bailey BA, Meinhardt LW (eds) Cacao Diseases. Springer, Switzerland, pp 3–31. http://dx.doi.org/10.1007/978–3-319-24789-2_1

Zougmoré et al (2016) Toward climate-smart agriculture in West Africa: a review of climate change impacts, adaptation strategies and policy developments for the livestock, fishery and crop production sectors. Agric Food Sec 5:26. https://doi.org/10.1186/s40066-016-0075-3

Chapter 16
Potential Transboundary Impacts of the Grand Ethiopian Renaissance Dam Under Climate Change and Variability

Khaled M. AbuZeid

Abstract The Blue Nile originates in Ethiopia and flows downstream through Sudan where it joins the White Nile to form the Main Nile which flows downstream into Egypt, the most downstream country on the Nile River. The Blue Nile represents the largest tributary to the Main Nile, providing an average annual flow of about 50 billion cubic meters (BCM), equivalent to about 60% of the natural average flow of the Main Nile at Aswan in Egypt. Egypt depends on Nile River waters as its main source of renewable water, utilizing 55.5 BCM annually as per the 1959 agreement which entitles Sudan to 18.5 BCM per year, making Sudan also highly dependent on the Nile waters. Ethiopia embarked on the construction of the Grand Ethiopian Renaissance Dam (GERD) on the Blue Nile in 2011, with the announced objective to generate hydropower. The GERD's reservoir maximum storage capacity is 74 BCM and is projected to start filling in 2020, with full operation planned for 2022. The combined effect of the GERD filling and operation, together with the effect of climate change and variability on the flows of the Blue Nile, calls for careful assessment of the transboundary implications on downstream. This chapter presents an assessment of the transboundary impacts of GERD filling and operation on downstream flows to Sudan and Egypt. Climate variability exacerbates the impacts of the GERD, especially during years of drought. The analysis was based on 105 historical years of Blue Nile flows, simulation of five different initial water levels of the High Aswan Dam (HAD) reservoir in Egypt at the start of first filling of the GERD upstream the Blue Nile, and four different scenarios for annual operation. Modeling shows that climate change effects, if they materialize with more precipitation, may naturally mitigate the negative impacts of the GERD on downstream countries so that they can meet their basic water needs during droughts. In contrast, if climate change effects materialize with less precipitation this will result in the reduction of flows of the Blue Nile, and the GERD's transboundary negative impacts will be exacerbated.

Keywords GERD · Impacts · Nile river · Climate · Transboundary · Renaissance dam · Blue Nile · Egypt · Sudan · Ethiopia · High Aswan Dam

---

K. M. AbuZeid (✉)
Regional Water Resources Director, Centre for Environment & Development for the Arab Region (CEDARE), CEDARE, 2 Elhegaz Street, Heliopolis, Cairo, Egypt
e-mail: kabuzeid@cedare.int

S. Diop et al. (eds.), *Climate Change and Water Resources in Africa*,
https://doi.org/10.1007/978-3-030-61225-2_16

1 The Nile River Basin and the Blue Nile Basin Climate Variability

The Nile Basin in Ethiopia receives about 450 BCM of rain per year out of a total of 970 BCM falling on Ethiopian lands. Ethiopia has several aquifers and river basins other than the Blue Nile, Sobat, and Atbara Basins associated with the Nile Basin in Ethiopia (AbuZeid 2016). Ethiopia, with the largest livestock production in Africa, depends on rainfed natural pasture lands for feeding. The geographical climate variability and the distribution of rainfall on the Nile Basin countries created the dependence of Egypt and Northern Sudan on the river's water and Ethiopia's reliance on direct rainfall, which contributes to vast areas of forests, pasture, and rainfed agriculture in Ethiopia, as well as recharging a vast reservoir of renewable groundwater. It is therefore expected that upstream countries such as Ethiopia rely on "rainwater" in the Nile Basin, whereas downstream countries such as Egypt and Sudan rely on "running water" from the Nile River itself. Egypt and large areas of Sudan are considered arid and hyper-arid areas where there is practically no rainfall. Egypt and Sudan use only about 4.6% of the total rainfall in the 11 countries of the Nile Basin, which is about 1660 BCM per year on average. The rainfall in the Nile countries, including other basins, is about 7000 BCM annually on average, with annual variation corresponding to temporal climate variability. Figure 1 shows the

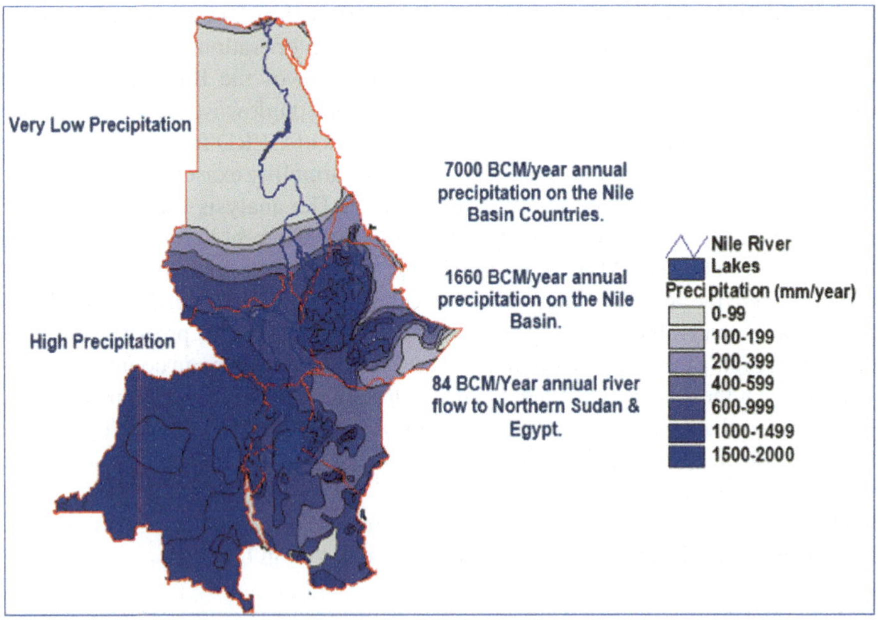

Fig. 1 Distribution of rainfall on the Nile Basin countries (AbuZeid 2012)

distribution of rainfall on the Nile Basin countries, reflecting the wide range of spatial geographical climate variability.

According to the 1959 agreement, Sudan‘s annual share from the Nile is 18.5 BCM. Egypt‘s annual share is 55.5 BCM, and is governed by the 1959 agreement and confirmed by historical uses for decades. This share is not enough to meet Egypt's increasing water needs. In 2015, Egypt imported agricultural food products that would have required about 50 BCM of (virtual) water to grow (AbuZeid 2018). Egypt is the only country among those in the Nile Basin that is forced to reuse wastewater and agricultural drainage to meet its water demand. It has also desalinated seawater for decades to fill the water gap on the coasts of the Red Sea and recently in cities on the Mediterranean Sea. Satellite images confirm Ethiopia‘s use of the Blue Nile Basin water in agriculture, industry, and urban development, but no data is published showing the exact abstractions or water uses in the Blue Nile Basin in Ethiopia. However, Ethiopia's agriculture is more dependent on rainfall. The Blue Nile is the main tributary of the Nile River, and has the largest flow volume among all major tributaries, discharging about 50 BCM/year on average. The effect of annual climate variability on the Blue Nile can be inferred from the recorded flows from 1911 to 2015, with a range from 20.69 BCM/year in 1913 to 69.85 BCM/year in 1929. The Blue Nile originates in Ethiopia, and flows downstream to Khartoum in Sudan, where it joins the main Nile, which then flows into Egypt.

2 The "GERD" and The "Agreement on the Declaration of Principles"

Ethiopia launched the construction of the Grand Ethiopian Renaissance Dam (GERD) on the Blue Nile in Ethiopia in April 2011 and the project was about 65% complete by November 2017. Figure 2 shows the progress of construction on the GERD and the auxiliary Saddle Dam as of 2017.

GERD has become controversial because of its potential impacts on downstream countries, Egypt and Sudan. In an attempt to assess the impact of the GERD on Egypt and Sudan during filling and operation, the Center for Environment and Development of the Arab Region and Europe (CEDARE) conducted a comprehensive study on the potential impacts of the GERD on Egypt and Sudan (AbuZeid 2017a). Some of the results of this study are presented here.

The "Agreement on the Declaration of Principles on the GERD," signed by the leaders of Egypt, Ethiopia, and Sudan in March 2015, stipulates that the three countries will agree on "the first filling rules, the annual operation rules and will establish a mechanism to coordinate the management of dams and reservoirs in the three countries." It is therefore important to study the potential impacts of the GERD on the downstream countries of Sudan and Egypt, so that determinations about filling and operating rules can be made based on the results. A consortium of two consulting firms was hired by the three countries to conduct formal impact assessment studies.

Fig. 2 Progress of construction in the GERD and the auxiliary Saddle Dam as of 2017

However, progress on the studies has been delayed due to disagreements among countries.

Figure 3 shows the High Aswan Dam (HAD) that lies downstream of the GERD in Egypt; HAD is the control structure that regulates the release of flows for Egypt's water uses as per the 1959 agreement with Sudan which allocates 55.5 BCM/year for Egypt and 18.5 BCM/year for Sudan.

Fig. 3 High Aswan Dam (HAD) (Egypt)

3 GERD Impact Assessment Methodology, Scenarios, and Assumptions

The study of the impacts of the first filling and operation of the GERD is based on simulating more than 100 possible scenarios (AbuZeid 2017b). The amount of the first seepage losses, annual seepage, and annual evaporation from the GERD were estimated based on the volume of storage and the water surface area of the GERD Reservoir. The study presents scenarios for the first filling volume of up to 15, 25, and 62 BCM over a period of 6 years, 10 years, and in some cases, 4 or 5 years. The study is based on simulating the historical series (1911–2015) of Blue Nile flows on which GERD is being built in Ethiopia (Abu-Zeid 2010; Said 1993). Different water levels in the High Aswan Dam (HAD) reservoir of 150 meters (m), 160, 165, 170, and 175 m (above mean sea level) were used as simulations for the HAD level at the beginning of the first filling of the GERD reservoir in the Upper Blue Nile. Different combinations of these variables created the scenarios that were studied and modeled into the future when the GERD is in place. The impact of the GERD on the flows of the Blue Nile downstream and the water levels and storage volumes of the HAD reservoir was predicted by the model, taking into account the other Nile tributaries flows and Egypt's and Sudan's shares of the Nile water. The results were compared to the baseline scenario (without GERD). The number of years of water deficit for Egypt and Sudan, and the volume of the deficits were assessed for every scenario. Since there is no agreement yet on the operating rules of the GERD, assumptions for operation scenarios were made based on lowering the reservoir levels of GERD by the end of the hydrological year back to different scenarios for the first filling volume of 15, 25, and 62 BCM.

The "first filling" is defined as filling up the dead storage volume in addition to a safety height of about 20 m above the level of the main turbines, to 590 m above mean sea level (amsl), which is equivalent to total storage of around 15 BCM. The relationship between the water level in the GERD reservoir and the water surface area, and the storage volume in the GERD reservoir is depicted in Fig. 4.

The "first seepage" and "annual seepage" from GERD were assumed as per the "average scenario" in the charts given in Figs. 5 and 6.

According to Wale (2008), the evaporation rate from Lake Tana in Ethiopia, the surface water level of which is 1786 m (amsl), is about 4.63 mm/day. According to Bashar and Mustafa (2009) , the evaporation rate from the Roseires Dam Reservoir in Sudan, the water level of which is 480 m (amsl), has an average evaporation rate of about 6.62 mm/day. As the GERD highest water level is about 640 m (amsl), by interpolation the evaporation rate of the GERD reservoir may be estimated to be 6.38 mm/day. This estimated evaporation rate is reflected by the "interpolation scenario" for GERD "annual evaporation" as shown in Fig. 7. The assessment in this study uses the "average" scenario.

The relationship among the water level in the HAD reservoir, the water surface area, and the storage volume is presented in Fig. 8.

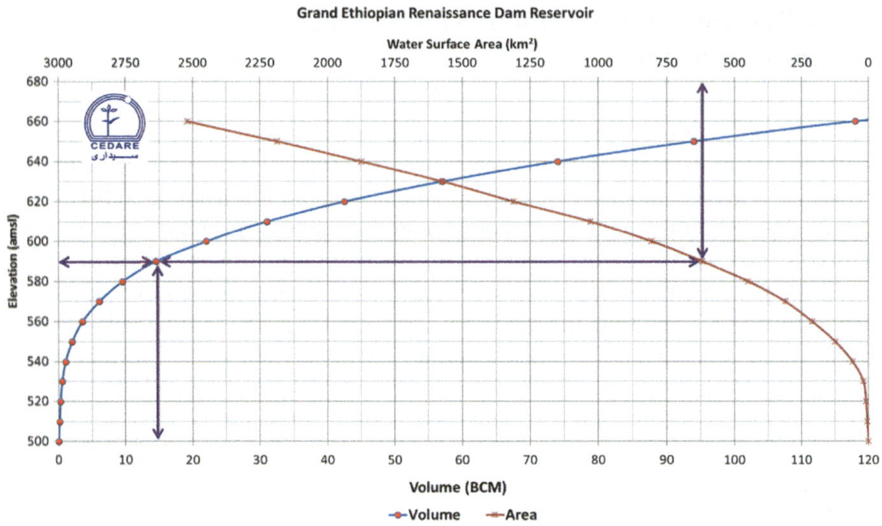

Fig. 4 The relationship between the water elevation, surface area, and volume in the GERD reservoir

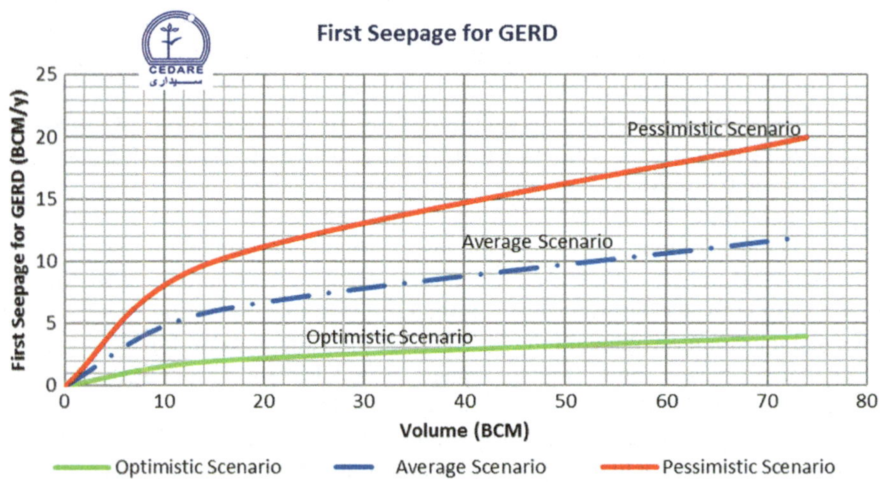

Fig. 5 The "first seepage" scenarios from GERD related to the storage volume

According to Abou El-Magd and Ali (2012), the rate of evaporation from the HAD reservoir is about 4.45 mm/day, about 11.1 BCM per year at the maximum water level and surface area of the reservoir. This is in line with Fig. 8 for calculating the evaporation and seepage losses from the HAD Reservoir, assuming seepage losses of about 2 BCM/year at the maximum water level. Assumptions for evaporation and seepage losses from the HAD reservoir are shown in Fig. 9.

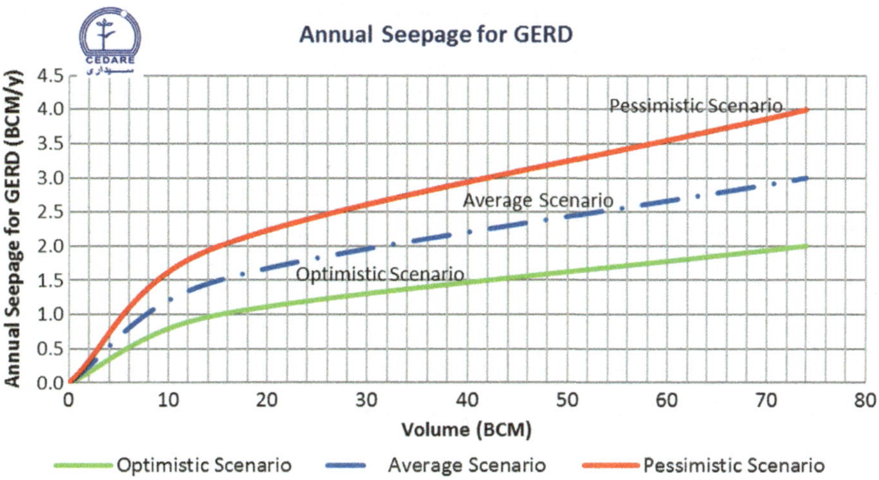

Fig. 6 The "annual seepage" scenarios from GERD related to the storage volume

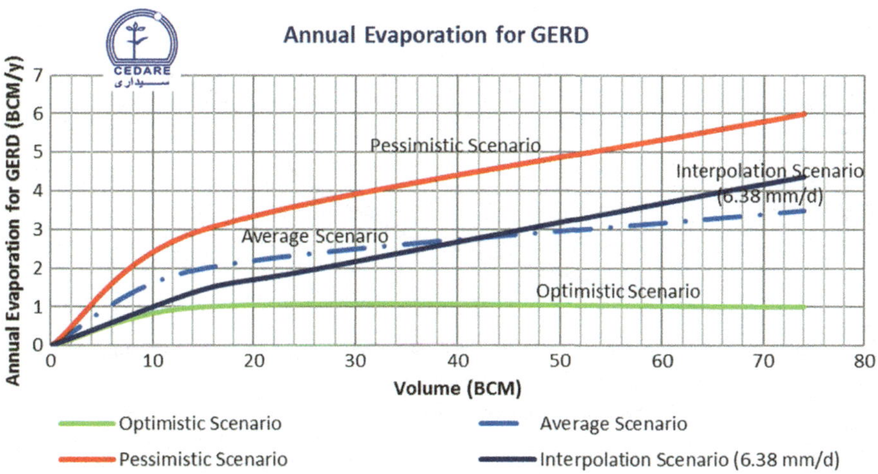

Fig. 7 GERD "annual evaporation" scenarios related to storage volume

GERD operating rules are based on Blue Nile flow storage during the high flow period (July–October), which represents about 80% of total annual discharge, to take advantage of the high levels of hydropower generation from GERD, and to be released throughout the year.

The annual average flow of the White Nile River was considered to be 28.5 BCM/year at Malakal, the annual average flow of the Atbara River to be 12 BCM/year, and that of the Rahad and Dinder Rivers at Khartoum to be 4 BCM/year. The average annual losses from the Roseires and Sennar Dam Reservoirs were estimated at 0.9 BCM/year. The average annual losses were estimated to be 1.0 BCM/year from the

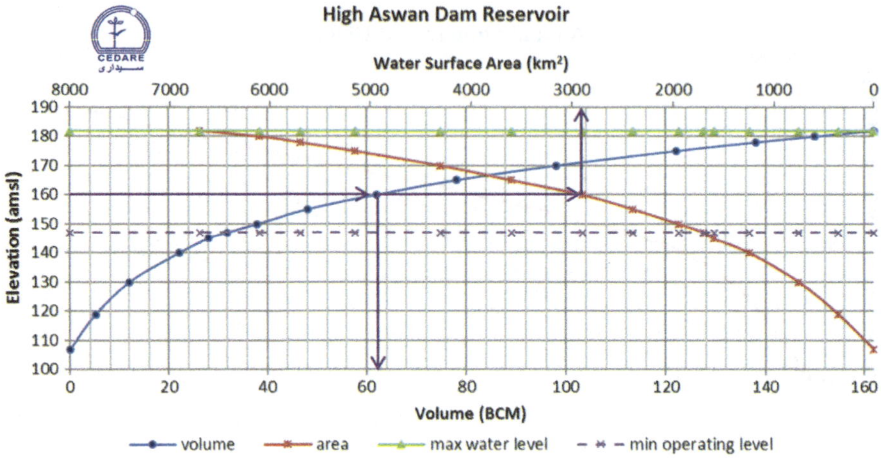

Fig. 8 Relationship between water elevation, surface area, and volume in the HAD reservoir

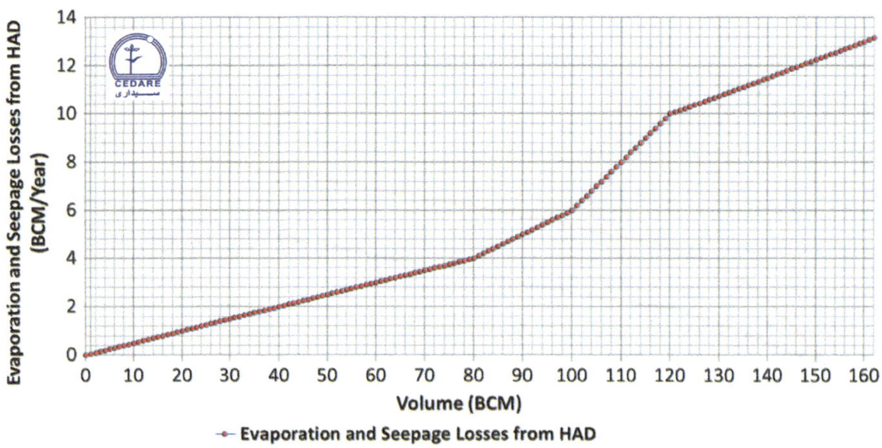

Fig. 9 Relationship between volume and evaporation and seepage losses from the HAD reservoir

Atbara, Tekizi, and Khashm El-Girba Dam Reservoirs, 2.5 BCM/year from Jebel El-Awliaa Dam Reservoir, 1.8 BCM/year from the Marwi Dam Reservoir, and 3.3 BCM/year from the upstream Nile River reaches to Aswan (MWRI 2004).

Projections for Blue Nile flows were assumed to begin with 10-year average flows of 38 BCM/year (lowest historical 10-year average), 45 BCM/year (moderate historical 10-year average), and 50 BCM/year (historical 10-year average). It was assumed that the flow volumes of the Blue Nile between 1911 and 2015 will be repeated in the same order for the next 105 years starting as soon as the GERD filling starts.

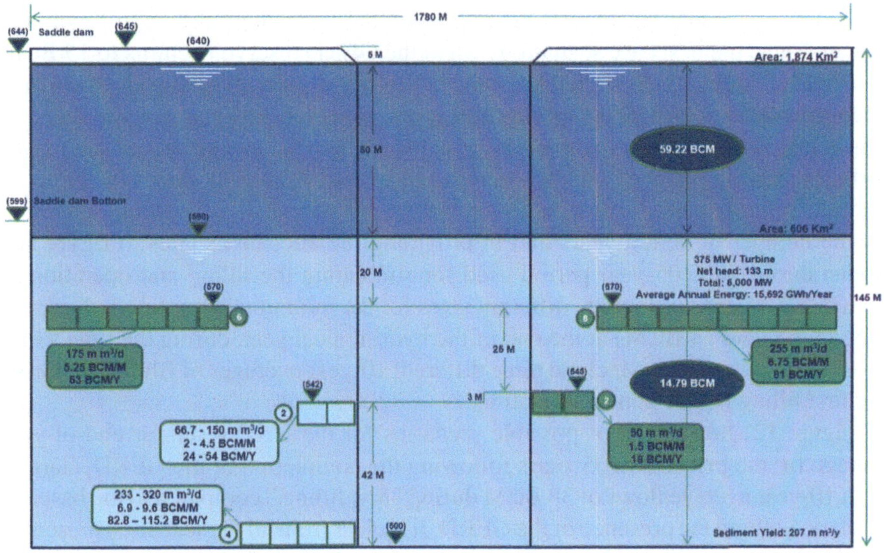

Fig. 10 The cross-sectional levels and technical specifications of the GERD main features

According to the 1959 agreement, Egypt and Sudan share any agreed deficit caused by another riparian state in the Nile Basin equally. Scenario simulations were based on the assumption that these Egypt and Sudan's shares are to be met and that number of deficit years in meeting these shares is to be determined. Deficit scenarios ranging from 0.5 to 5 BCM/year were assumed for each country during deficit years.

It was assumed that the purpose of the GERD is to generate hydropower only, which was the basis of the Declaration of Principles on the GERD signed by Egypt, Sudan, and Ethiopia in March 2015. It was understood that Ethiopia commits not to withdraw water from the GERD reservoir for any other purposes.

The engineering specifications of GERD in terms of water levels, elevations, heights, and the allowable flow according to the capacity of the turbines and gates are outlined in Fig. 10.

4 Hydrological Impact Scenarios of the GERD Under Climate Variability

The average annual storage volume of the GERD and the cumulative evaporation and seepage losses from GERD reservoir are the most influential variables in the ability of Egypt and Sudan to fulfill their shares from the Nile in the presence of GERD. The storage volume is assumed to return to the minimum storage of the "first filling" at the end of the hydrological year to accommodate the next year's flood. The smaller the storage amount, the smaller the evaporation and seepage losses from the

GERD reservoir and the smaller the impact on the flows downstream to Egypt and Sudan. Thus, the best case scenario is when the GERD reservoir returns to 15 BCM at a level of about 590 m at the end of the hydrological year. In reality, the level may need to drop to 7 or 3 BCM in very dry years, however, this was not simulated in this study. When the level is this low, it is insufficient to operate any of the top 14 turbines, leaving just the two lower turbines operable. Figure 11 shows the effect of the remaining storage volume at the end of the hydrological year during operation on the number of deficit years and in providing the allocated shares to Egypt and Sudan during the 105-year period used for simulating the filling and operation of the GERD. Scenarios include different sizes for the minimum operating volumes of (15, 25, 62, and 74 BCM) at the end of the hydrological year, during the time series modeled. Scenarios also include three different 10-year averages of Blue Nile during the first filling before annual operation.

Figure 12 shows a set of possible scenarios for the HAD reservoir end-of-year storage for the projected 105 years following the completion of the GERD, starting with 10-year average flows of 38 BCM during "first filling," compared to the baseline scenario without the presence of the GERD. It is clear from all the scenarios presented that HAD storage falls under the dead storage level of 31 BCM (i.e., below the level at which turbines can operate) due to the expected cumulative effect of evaporation losses and seepage in the GERD reservoir, which will be evident in years of low Blue Nile flows.

Figure 13 shows a set of possible scenarios for the HAD reservoir end-of-year storage in the 105 years following the completion of the GERD starting with 10-year average flows of 45 BCM during "first filling," compared to the baseline scenario without the presence of the GERD. It is clear from all the scenarios presented that HAD storage falls under the dead storage level of 31 BCM (i.e., below the level at which turbines can operate) due to the expected cumulative effect of evaporation

Operating after first filling period of different volumes and keeping this storage as the minimum during operation	Average flows for Blue Nile in the first 10 years of filling and operation (m^3 / year)	Water level in front of the HAD at the beginning of the time series 150 m			Water level in front of the HAD at the beginning of the time series 160 m			Water level in front of the HAD at the beginning of the time series 165 m			Water level in front of the HAD at the beginning of the time series 170 m			Water level in front of the HAD at the beginning of the time series 175 m			
		Filling volume scenarios, remaining minimum range during operation BCM			Filling volume scenarios, remaining minimum range during operation BCM			Filling volume scenarios, remaining minimum range during operation BCM			Filling volume scenarios, remaining minimum range during operation BCM			Filling volume scenarios, remaining minimum range during operation BCM			
		15	25	74	15	25	74	15	25	74	15	25	74	15	25	62	74
		Deficit years in providing Egypt and Sudan shares			Deficit years in providing Egypt and Sudan shares			Deficit years in providing Egypt and Sudan shares			Deficit years in providing Egypt and Sudan shares			Deficit years in providing Egypt and Sudan shares			
Filling Period in 10 Years	50	20	22	37	15	16	34	15	16	30	15	16	21	15	16		21
	45	18	20	28	16	18	24	15	17	23	14	17	23	14	16		22
	38	19	20	24	16	17	22	15	16	21	14	15	20	11	15		19
Filling Period in 6 Years	50	21	23	29	15	18	28	15	16	26	15	16	25	15	16		21
	45	19	21	29	16	23	26	15	17	25	14	16	24	14	16		23
	38	19	20	24	16	18	23	15	18	22	14	16	21	12	15	20	20
Filling Period in 5 Years	50																
	45															23	
	38																
Filling Period in 4 Years	50															21	
	45																
	38																

Fig. 11 Scenarios for the number of deficit years during GERD operation

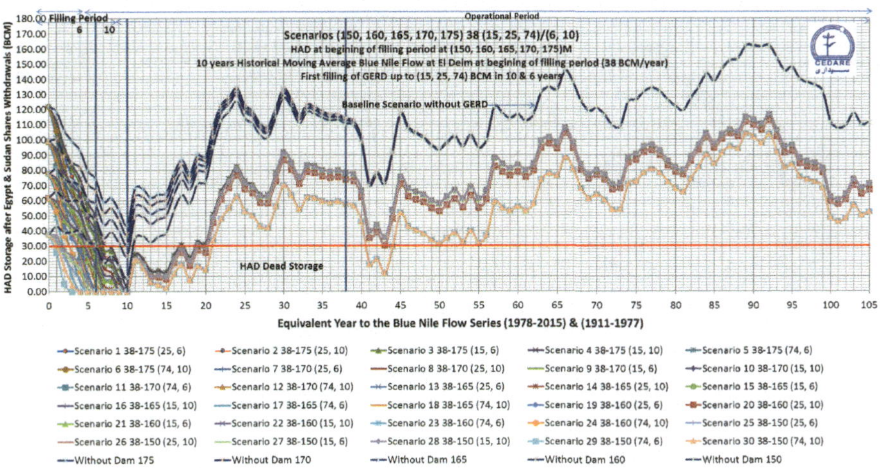

Fig. 12 Scenarios: (150, 160, 165, 170, 175) 38 (15, 25, 74)/(6,10)

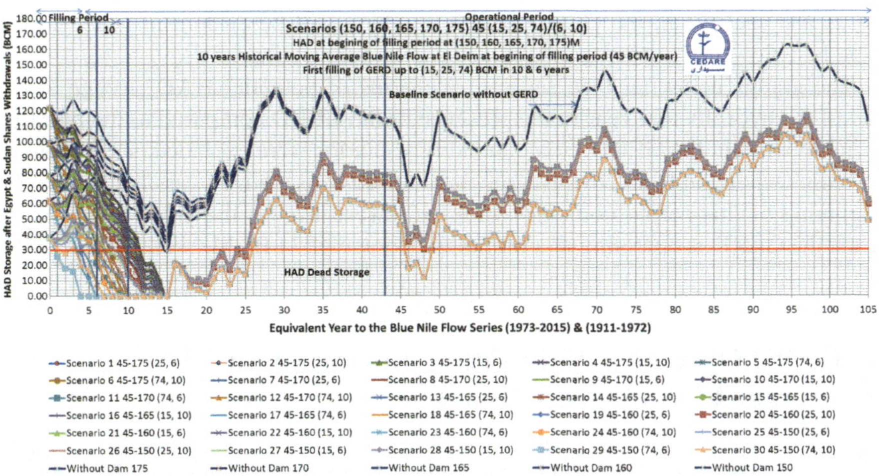

Fig. 13 Scenarios: (150, 160, 165, 170, 175) 45 (15, 25, 74)/(6,10)

losses and seepage in the GERD reservoir, which will be evident in years of low Blue Nile flows.

Figure 14 shows a set of possible scenarios for the HAD reservoir end-of-year storage in the 105 years following the completion of the GERD, starting with 10-year average flows of 50 BCM during "first filling," compared to the baseline scenario without the presence of the GERD. It is clear from all the scenarios presented that HAD storage falls under the dead storage level of 31 BCM (i.e., below the level at which turbines can operate) due to the expected cumulative effect of evaporation

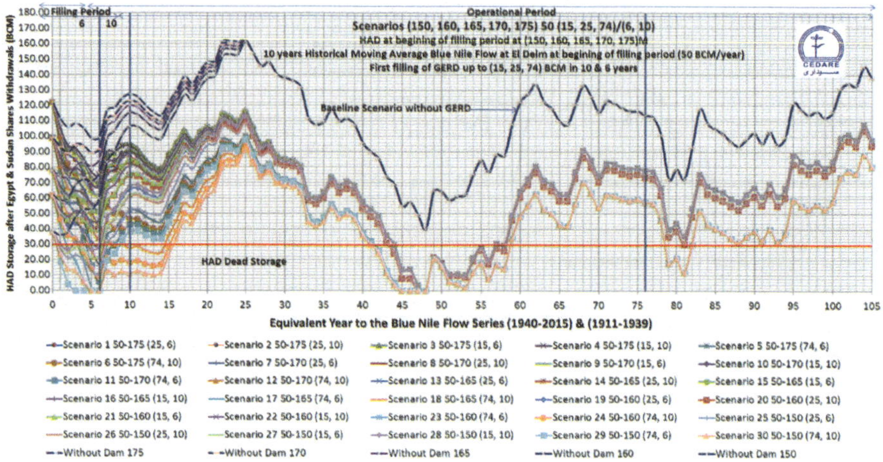

Fig. 14 Scenarios: (150, 160, 165, 170, 175) 50 (15, 25, 74)/(6,10)

losses and seepage in the GERD reservoir, which will be evident in years of low Blue Nile flows.

The graphs below analyze separate scenarios of potential HAD storage at the end of the hydrological year in the presence of the GERD compared to the baseline scenario without the GERD. Some unlikely scenarios were excluded, such as those scenarios when the Blue Nile average mean of 10 consecutive years is low and the beginning of this period coincides with the lowest level in the HAD reservoir, or when the GERD is operating at a minimum storage of 74 BCM.

5 GERD Operation After Filling During a Low 10-Year Average Flow (38 BCM/Year)

Figure 15 assumes a series of flows at the beginning of the filling period with a 10-year annual average equal to 38 BCM/year (similar to the period from 1978 to 1987), which was the lowest 10-year historic average, coinciding with a HAD reservoir level at the beginning of the series of about 165 m (equivalent to HAD storage of 78 BCM). This scenario assumes annual operation at 15 BCM minimum storage. Deficit years where the Blue Nile flows through GERD will be short of meeting the shares of Egypt and Sudan reached about 15 years during the first 20 years of the series of 105 years of the simulated historical flows used.

There will be a deficit in the shares of Egypt and Sudan to avoid the falling of the HAD reservoir levels below the dead storage level.

Figure 16 shows an annual deficit ranging from 2.5 BCM to 3 BCM for both Egypt and Sudan for a period of 10 years, with a total deficit during the first 10 years of about 56 BCM.

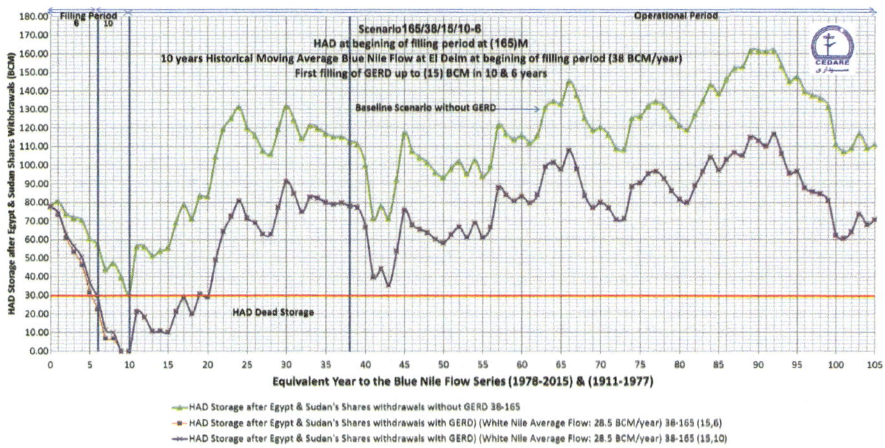

Fig. 15 Scenarios: (165) 38 (15)/(6,10)

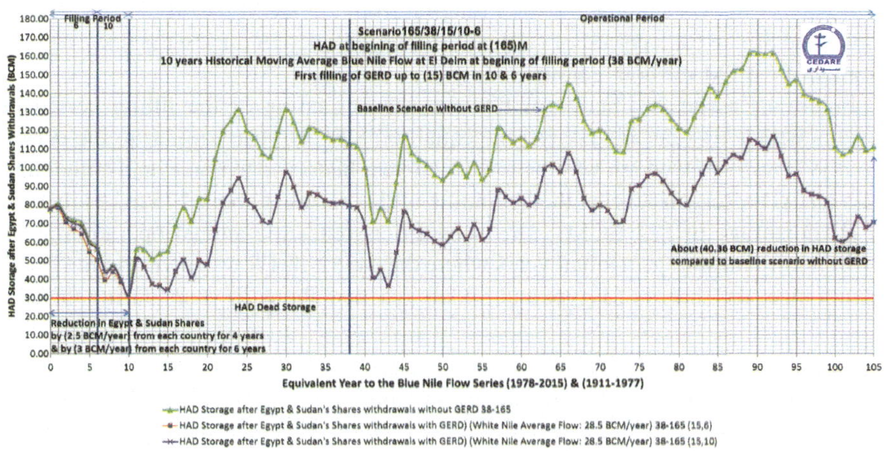

Fig. 16 Scenarios: (165) 38 (15)/(6,10) with deficit in the shares of Egypt and Sudan

Figure 17 assumes a series of flows at the beginning of the filling period of 6 years with a 10-year annual average equal to 38 BCM/year (similar to the period from 1978 to 1987), which was the lowest 10-year historical average, and coinciding with a HAD reservoir level at the beginning of this series at about 170 m (equivalent to HAD storage of about 99 BCM). This scenario assumes the annual operation at 25 BCM minimum storage. Deficit years where Nile water shares of Egypt and Sudan cannot be met reached about 16 years during the first 20 years of the series of 105 years of the simulated historical flows used.

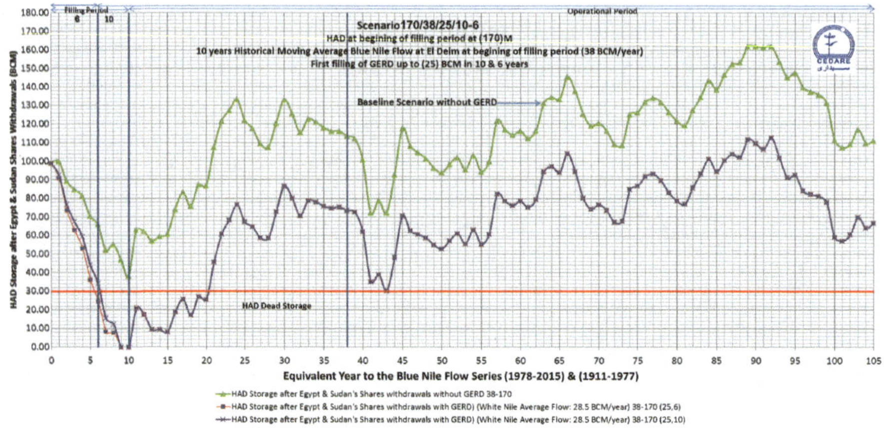

Fig. 17 Scenarios: (170) 38 (25)/(6,10)

Figure 18 shows an annual deficit ranging from 2.5 BCM to 3 BCM for each of Egypt and Sudan for a period of 10 years, with a total deficit during the first 10 years of about 57 BCM for both countries.

Figure 19 assumes a series of flows at the beginning of the filling period of 6 years with a 10-year annual average equal to 38 BCM/year (similar to the period from 1978 to 1987), which was the lowest 10-year historical average, and coinciding with a HAD reservoir level at the beginning of this series at about 175 m (equivalent to HAD storage of 122 BCM). This scenario assumes the annual operation at 62 BCM minimum storage. Deficit years where the Blue Nile Flows downstream GERD cannot meet the shares of Egypt and Sudan, reached about 20 years during the first

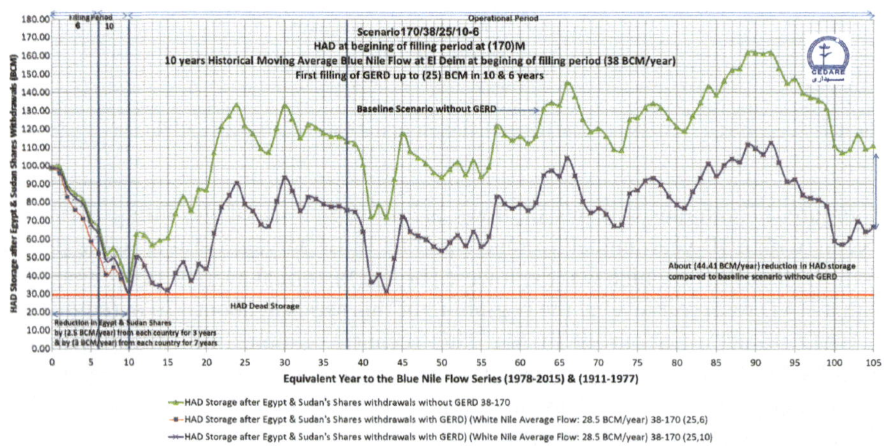

Fig. 18 Scenarios: (170) 38 (25)/(6,10) with deficit in the shares of Egypt and Sudan

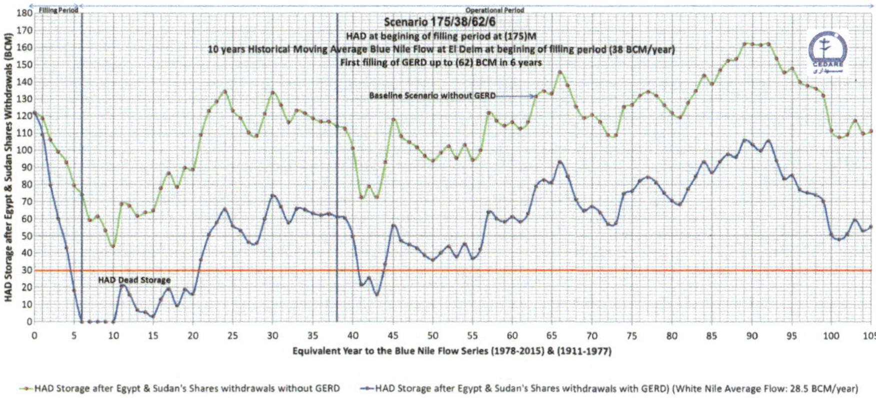

Fig. 19 Scenario: (175) 38 (62)/(6)

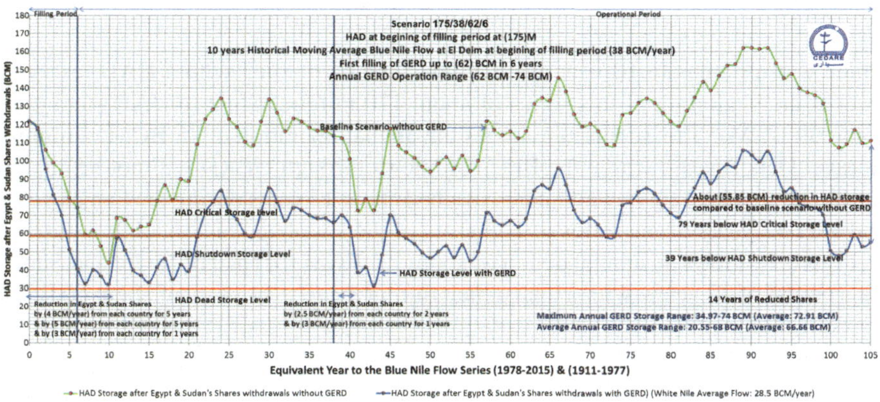

Fig. 20 Scenario: (175) 38 (62)/(6) with deficit in the shares of Egypt and Sudan

20 years and again after 40 years of the series of 105 years of the simulated historical flows used.

Figure 20 shows an annual deficit ranging from 2.5 BCM to 5 BCM for each of Egypt and Sudan for a period of 11 years, at the beginning of the series and for 3 years after 38 years with a total deficit of about 112 BCM for both countries.

6 GERD Operation After Filling During a Moderate 10-Year Average Flow (45 BCM/Year)

Figure 21 assumes a series of flows at the beginning of the filling period with a 10-year annual average equal to 45 BCM/year (similar to the period from 1973 to 1982),

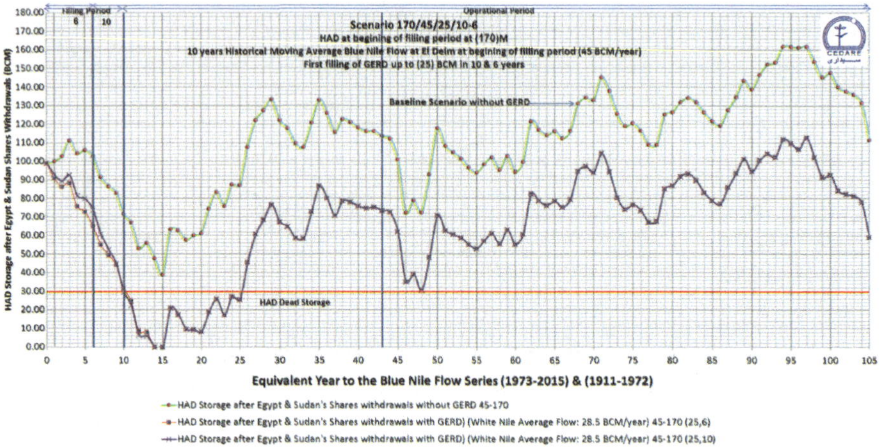

Fig. 21 Scenarios: (170) 45 (25)/(6,10)

which was a moderate 10-year average flow, and coinciding with a HAD reservoir level at the beginning of this series of about 170 m (equivalent to HAD storage of about 99 BCM). This scenario assumes the annual operation at 25 BCM minimum storage. Deficit years where the Blue Nile flows through GERD will not be able to satisfy the shares of Egypt and Sudan, reached about 17 years during the first 25 years and after 47 years of the series of 105 years of the simulated historical flows used.

Figure 22 shows an annual deficit of about 2.5 BCM for each of Egypt and Sudan for a period of 12 years, with a total deficit during the first 14 years of about 60 BCM for both countries.

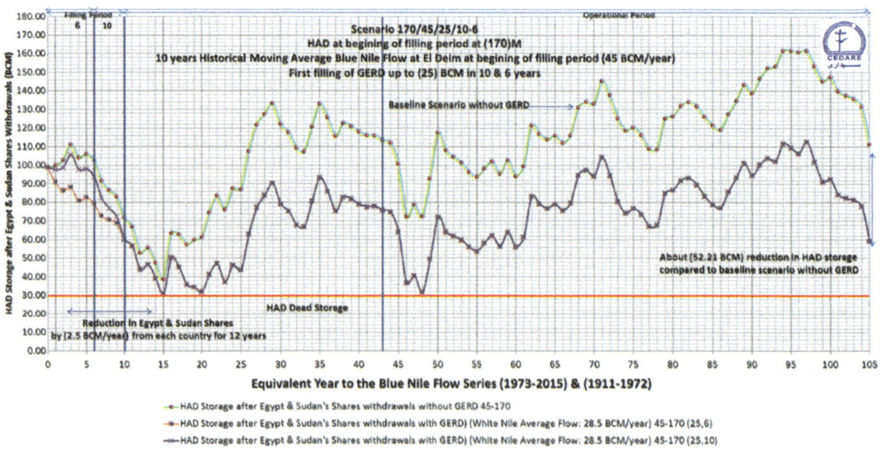

Fig. 22 Scenarios: (170) 45 (25)/(6,10) with deficit in the shares of Egypt and Sudan

Figure 23 assumes a series of flows at the beginning of the filling period with a 10-year annual average equal to 45 BCM/year (similar to that which occurred in the period from 1973 to 1982), which was a moderate 10-year historical average, and coinciding with a HAD reservoir level at the beginning of this series at about 175 m (equivalent to HAD storage of 122 BCM). This scenario assumes the annual operation at 15 BCM minimum storage. Deficit years of inability of Blue Nile flows through GERD to meet the shares of Egypt and Sudan reached about 14 years during the first 17 years of the series of 105 years of the simulated historical flows used.

Figure 23 also shows an annual deficit of about 2.5 BCM for each of Egypt and Sudan for a period of 8 years, with a total deficit during the first 17 years of about 40 BCM for both countries.

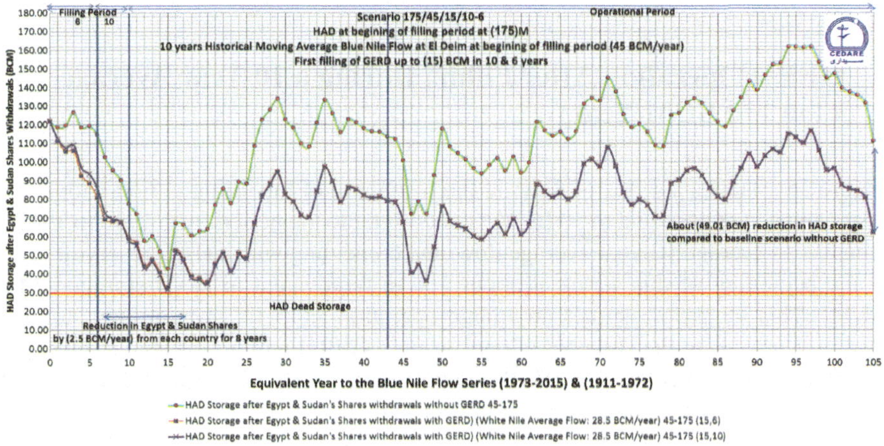

Fig. 23 Scenarios: (175) 45 (15)/(6,10) with deficit in the shares of Egypt and Sudan

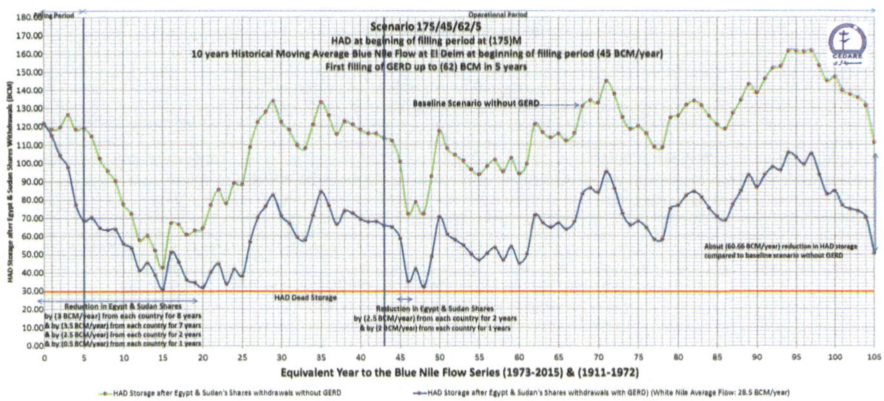

Fig. 24 Scenario: (175) 45 (62)/(5) with deficit in the shares of Egypt and Sudan

Figure 24 assumes a series of flows at the beginning of the filling period with a 10-year annual average equal to 45 BCM/year (similar to the period from 1973 to 1982), which was a moderate 10-year historical average, and coinciding with a HAD reservoir level at the beginning of this series at about 175 m (equivalent to HAD storage of 122 BCM). This scenario assumes annual operation at 62 BCM minimum storage. Deficit years where the Blue Nile flows through GERD will not be able to meet the shares of Egypt and Sudan reached about 23 years during the first 20 years and after 42 years of the series of 105 years of the simulated historical flows used.

Figure 24 also shows an annual deficit ranging from 0.5 BCM to 3 BCM for each of Egypt and Sudan for a period of 18 years at the beginning of the series and a period of 3 years after 42 years, with a total deficit of about 122 BCM for both countries.

7 GERD Operation After Filling During a 10-Year Average Flow (50 BCM/Year)

Figure 25 assumes a series of flows at the beginning of the filling period with a 10-year annual average equal to 50 BCM/year (similar to the period from 1940 to 1949), which is an average 10-year average flow, and coinciding with a HAD reservoir level at the beginning of this series at about 160 m (equivalent to HAD storage of 62 BCM). This scenario assumes an annual operation at 15 BCM minimum storage. Deficit years where the Blue Nile flows through GERD will not be able to meet the shares of Egypt and Sudan reached about 10 years after 38 years of the series of 105 years of the simulated historical flows used.

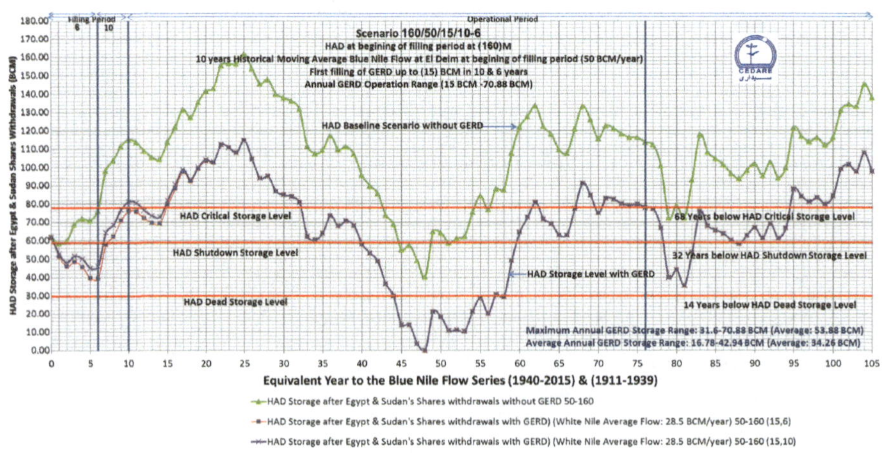

Fig. 25 Scenarios: (160) 50 (15)/(6,10)

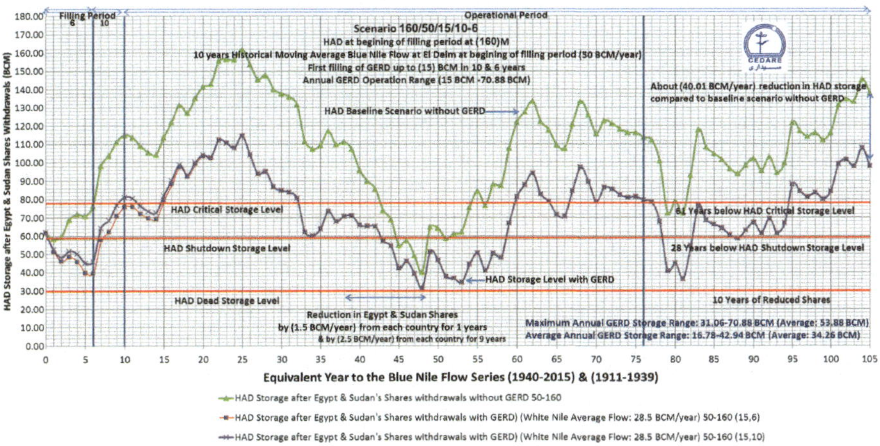

Fig. 26 Scenarios: (160) 50 (15)/(6,10) with deficit in the shares of Egypt and Sudan

Figure 26 shows an annual deficit ranging from 1.5 to 2.5 BCM for each of Egypt and Sudan for a period of 10 years, with a total deficit after 38 years of about 48 BCM for both countries.

Figure 26 assumes a series of flows at the beginning of the filling period with a 10-year annual average equal to 50 BCM/year (similar to the period from 1940 to 1949), which is an average 10-year average flow, and coinciding with a HAD reservoir level at the beginning of this series of about 175 m (equivalent to HAD storage of 122 BCM). This scenario assumes annual operation at 25 BCM minimum storage. Deficit years where the Blue Nile flows through GERD will not be able to meet the shares of Egypt and Sudan reached about 16 years after 36 years of the series of 105 years of the simulated historical flows used.

Figure 27 shows an annual deficit of about 2.5 BCM for each of Egypt and Sudan for a period of 12 years, with a total deficit after 36 years of about 60 BCM for both countries.

Figure 28 assumes a series of flows at the beginning of the filling period with a 10-year annual average equal to 50 BCM/year (similar to the period from 1940 to 1949), which is an average 10-year average flow, and coinciding with a HAD reservoir level at the beginning of this series of about 175 m (equivalent to HAD storage of 122 BCM). This scenario assumes the annual operation at 62 BCM minimum storage. Deficit years where the Blue Nile flows through GERD will not be able to meet the shares of Egypt and Sudan reached about 21 years at different periods of the series of 105 years of the simulated historical flows used.

Figure 28 shows an annual deficit ranging from 0.5 BCM to 3 BCM for each of Egypt and Sudan for a period of 29 years, with a total deficit in the different period of the series about 124 BCM for both countries.

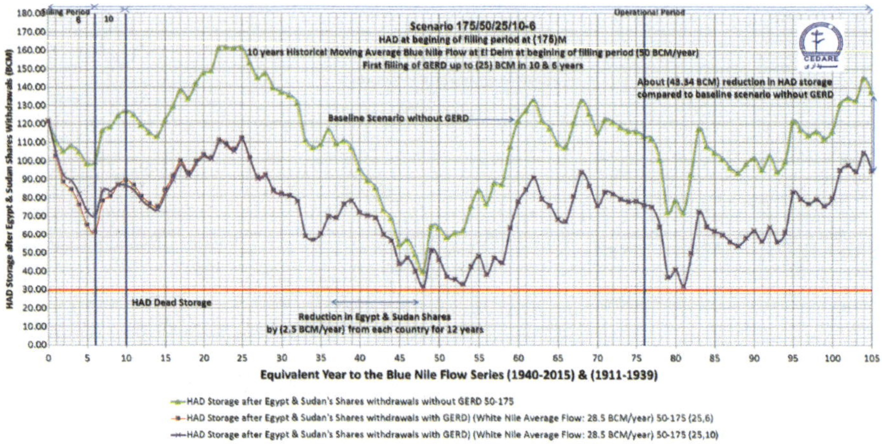

Fig. 27 Scenarios: (175) 50 (25)/(6,10) with deficit in the shares of Egypt and Sudan

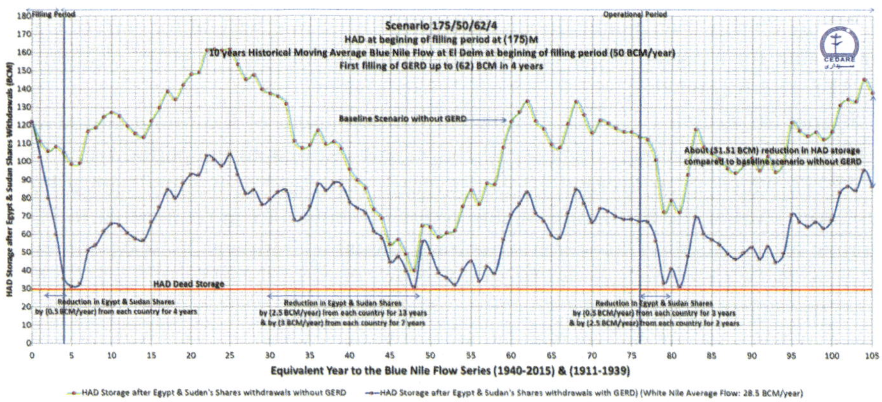

Fig. 28 Scenario: (175) 50 (62)/(4) with deficit in the shares of Egypt and Sudan

8 Deficit Years and Volumes for Selected GERD Filling and Operation Scenarios

Figure 29 shows water-deficit years and volumes of deficit for the selected scenarios for Egypt and Sudan applied to the simulated historical series of flows used during the "operating period" after "first filling" of an amount of 15 BCM and keeping this storage as a minimum operating level.

Operating after first filling period of 15 BCM and keeping this storage as the minimum during operation	Average flows for Blue Nile in the first 10 years of filling and operation (m³ / year)	Water level in front of HAD at the beginning of time series at 160 m				Total Deficit Volume per country	Water level in front of HAD at the beginning of time series at 165 m				Total Deficit Volume per country	Water level in front of HAD at the beginning of time series at 175 m				Total Deficit Volume per country
		Withdrawals by Egypt and Sudan					Withdrawals by Egypt and Sudan					Withdrawals by Egypt and Sudan				
		55.5	54	53	52.5		55.5	54	53	52.5		55.5	54	53	52.5	
		18.5	17	16	15.5		18.5	17	16	15.5		18.5	17	16	15.5	
		Deficiency Volume (BCM/year/each country)					Deficiency Volume (BCM/year/each country)					Deficiency Volume (BCM/year/each country)				
		Variable	1.5	2.5	3		variable	1.5	2.5	3		variable	1.5	2.5	3	
		Deficiency Years					Deficiency Years					Deficiency Years				
Filling Period in 10 Years	50	15	1	9		24										
	45											14		8		20
	38						15		4	6	28					
Filling Period in 6 Years	50	15	1	9		24										
	45											14		8		20
	38						15		4	6	28					

Fig. 29 Deficit years and volumes for Egypt and Sudan while GERD is operating at 15 BCM end of year

Figure 30 shows water-deficit years and volumes of deficit for the selected scenarios for Egypt and Sudan applied to the simulated historical series of flows used during the "operating period" after "first filling" of an amount of 25 BCM and keeping this storage as a minimum operating level.

Figure 31 shows water-deficit years and volumes of deficit for the selected scenarios for Egypt and Sudan applied to the simulated historical series of flows used during the "operating period" after "first filling" of an amount of 62 BCM and keeping this storage as a minimum operating level.

Operating after first filling period of 25 BCM and keeping this storage as the minimum during operation	Average flows for Blue Nile in the first 10 years of filling and operation (m³ / year)	Water level in front of HAD at the beginning of time series at 170 m				Total Deficit Volume per country	Water level in front of HAD at the beginning of time series at 175 m				Total Deficit Volume per country
		Withdrawals by Egypt and Sudan					Withdrawals by Egypt and Sudan				
		55.5	54	53	52.5		55.5	54	53	52.5	
		18.5	17	16	15.5		18.5	17	16	15.5	
		Deficiency Volume (BCM/year/each country)					Deficiency Volume (BCM/year/each country)				
		variable	1.5	2.5	3		Variable	1.5	2.5	3	
		Deficiency Years					Deficiency Years				
Filling Period in 10 Years	50						16		12		30
	45	17		12		30					
	38	15		3	7	28					
Filling Period in 6 Years	50						16		12		30
	45	16		12		30					
	38	16		3	7	28.5					

Fig. 30 Deficit years and volumes for Egypt and Sudan while GERD is operating at 25 BCM end of year

Operating after first filling period of 62 BCM and keeping this storage as the minimum during operation	Average flows for Blue Nile in the first 10 years of filling and operation (m³ / year)	Water level in front of the HAD at the beginning of time series at 175 m								Total Deficit Volume per country
		Withdrawals by Egypt and Sudan								
		55.5	55	53.5	53	52.5	52	51.5	50.5	
		18.5	18	16.5	16	15.5	15	14.5	13.5	
		Deficiency Volume (BCM/year/each country)								
		Variable	0.5	2	2.5	3	3.5	4	5	
		Deficiency Years								
Filling Period in 4 Years	50	21	7		15	7				62
Filling Period in 5 Years	45	23	1	1	4	8	7			61
Filling Period in 6 Years	38	20			2	2		5	5	56

Fig. 31 Deficit years and volumes for Egypt and Sudan while GERD is operating at 62 BCM end of year

9 GERD Impacts on HAD Hydropower

The impact on a shortage of hydropower generated from the HAD is evident, unless parties agree to discharge the entire flow of the Blue Nile in the same months it arrives, without distributing the flows throughout the year. However, uniform hydropower production throughout the year from GERD leads to a decrease in average water levels in the HAD, even when the entire flow is discharged annually after the filling period. The HAD's hydropower may decrease by 24–40% due to the GERD. Hydropower of the HAD does not represent a large proportion of the generated power at the national level in Egypt, but it does account for about one-third of the hydropower to be generated by the GERD and contributes to meeting the demands of two or more governorates in Upper Egypt.

Figure 32 shows a decrease in the average levels and contents of the HAD affecting hydropower generation due to the operation of the GERD at a minimum operating level equivalent to a storage volume of 25 BCM and ensuring that GERD levels return to that level by the end of the hydrological year. These scenarios lead to a maximum reduction in the average reservoir storage of HAD of about 50 BCM, and a maximum drop of about 11.5 m in the average water level in the HAD. Minimum reductions are also shown in Fig. 32.

Figure 33 shows a decrease in the average levels and contents of the HAD affecting hydropower generation due to the operation of the GERD at a minimum operating level equivalent to a storage volume of 15 BCM and ensuring that GERD levels return to that level by the end of the hydrological year. This operating rule at 15 BCM has the least impact on the hydropower of HAD downstream in Egypt. These scenarios lead to a maximum reduction in the average reservoir storage of HAD of about 43

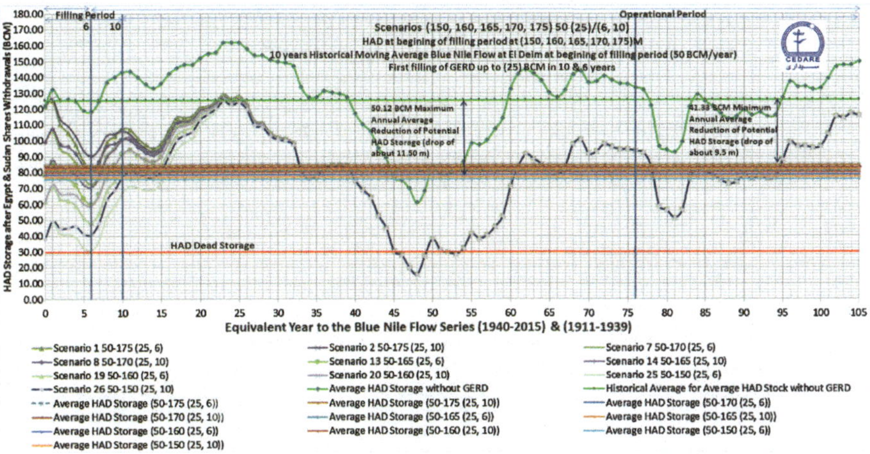

Fig. 32 HAD average content/level scenarios: (150, 160, 165, 170, 175) 50 (25)/(6,10)

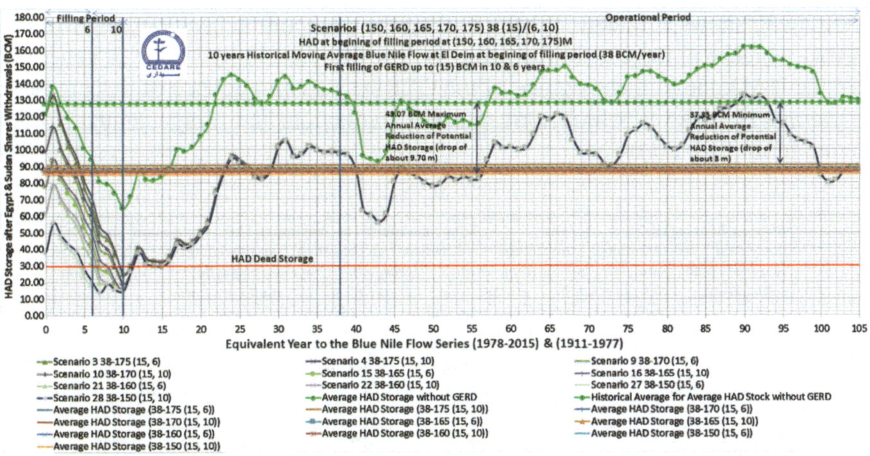

Fig. 33 HAD average content/level scenarios: (150, 160, 165, 170, 175) 38 (15)/(6, 10)

BCM, and a maximum drop of about 9.7 m in the average water level in the HAD. Minimum reductions are also shown in Fig. 33.

Figure 34 shows a decrease in the average levels and contents of the HAD affecting hydropower generation due to the operation of the GERD at a minimum operating level equivalent to a storage volume of 15 BCM and ensuring that GERD levels return to that level by the end of the hydrological year. This operating rule at 15 BCM has the least impact on the hydropower of downstream countries. These scenarios lead to a reduction in the average reservoir storage of HAD of about 30 to 37 BCM, and a drop of about 6.5 to 8 m in the average water level in the HAD.

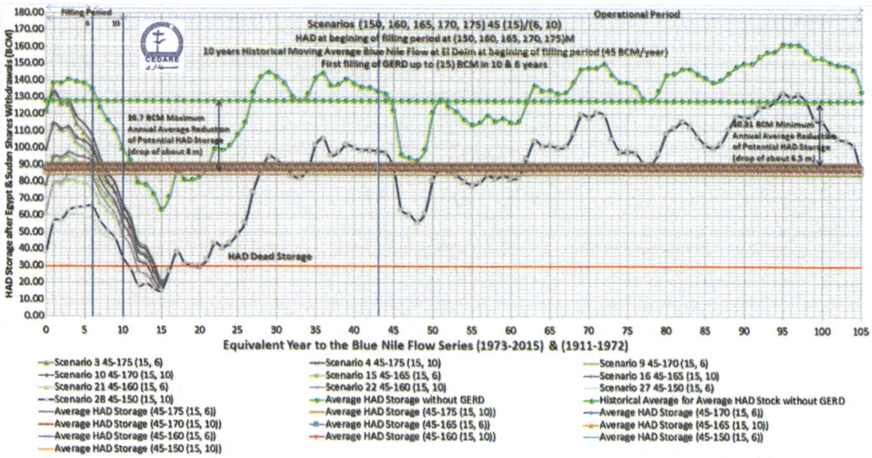

Fig. 34 HAD average content/level scenarios: (150, 160, 165, 170, 175) 45 (15)/(6,10)

Figure 35 shows a reduction in the hydropower generation head ranging from 24 to 40%, which will have a similar effect on the decrease in hydropower generated from the HAD due to the presence of the GERD as a consequence of a decrease in the average HAD level, which corresponds to reductions in average annual storage ranging from 30 to 50 BCM.

	Average flows in the first 10 years of filling and operation	The minimum reduction in average HAD storage (BCM)	The maximum reduction in the average HAD storage (BCM)	The minimum reduction in the average level in the HAD (m)	The maximum reduction in the average level in the HAD (m)	The minimum percentage reduction in hydropower generation head	The maximum percentage reduction in hydropower generation head
Operating at 15 BCM	50	37.28	45.17	8.30	10.30	29%	36%
	45	30.31	36.70	6.50	8.00	24%	29%
	38	37.35	43.07	8.00	9.70	27%	33%
Operating at 25 BCM	50	41.33	50.12	9.50	11.50	33%	40%
	45	34.30	40.89	7.00	9.00	25%	33%
	38	41.69	46.82	9.20	10.20	32%	35%

Fig. 35 Reduction estimates in HAD hydropower generation reflected by reduced volumes and levels

10 Potential Climate Change Impacts on the Blue Nile Flows

According to ESCWA (2017a, b), in the Blue Nile Basin, the change in mean temperature for Representative Concentration Pathway (RCP) 4.5 shows an increase of 1.5 °C at mid-century and 1.8 °C at end century. For RCP 8.5, temperatures increase by 2 °C for mid-century and 3.6 °C at end century. At the seasonal level, the highest increase in temperature is shown to occur in winter with an increase of as much as 3.9 °C by the end of the century for RCP 8.5. Precipitation results for RCP 4.5 projected a change of −6 and −5% at mid- and end centuries, respectively. For RCP 8.5, precipitation change is −3% for mid-century and −5% by end century, reaching the greatest reduction in winter, with a 7% decrease.

The headwaters of the Blue Nile in the Ethiopian Highlands show broad variation among individual ensemble members for runoff in both models (HYPE and VIC); no conclusive trend can be perceived. The mean values from discharge changes show a decrease over time, but, given the broad ranges, this trend cannot be considered conclusive. For instance, the change in mean discharge for end century is −8% for RCP 8.5, but values range from −68 to 86%, as shown in Fig. 36.

The direction of change of flows of the Blue Nile has inverse impacts when reconciled with GERD impacts as described above. If the direction of the change is positive (increased flow), the result can be a reduced impact of GERD on Egypt and Sudan. But if there is a decrease in Blue Nile flows (as suggested by the direction of the mean discharge of −8%), GERD impacts on Egypt and Sudan are exacerbated.

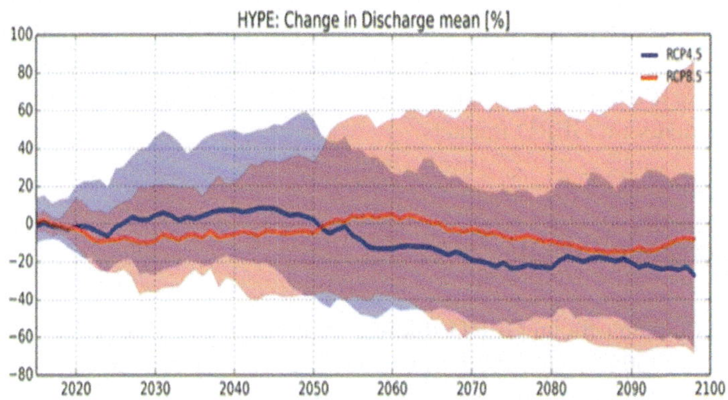

Fig. 36 HYPE: change in Blue Nile discharge mean (%), (ESCWA 2017a, b)

11 Conclusion

There is a failure to fulfill the shares of Egypt and Sudan in almost all scenarios. The largest impact on the annual shares of Egypt (55.5 BCM) and Sudan (18.5 BCM) after the completion and operation of the GERD is the volume of dead storage (storage up to the highest level of the turbines), the first filling volume, and the average annual storage volume, which affects the volume of evaporation losses and seepage. The cumulative effects of these losses have a great impact on the volume of storage in the HAD reservoir. The lower the annual average storage of the GERD reservoir, the less the evaporation and the seepage losses of the GERD and the less the effect on the downstream Blue Nile flows to Egypt and Sudan. GERD operations will lead to a decrease in hydropower generated from the HAD due to the operation of the HAD turbines at a lower average level than the baseline scenario before the construction of the GERD.

The volume (first filling/minimum operating level) of the GERD has the largest impact on the cumulative effect of evaporation losses and seepage during operation; its biggest impact is the reduction of Blue Nile flows to Egypt and Sudan. The agreement among the three countries on the annual operating rules is as important, if not more important, than the agreement on the first filling rules. The greatest risk may occur in the operating period and not in the filling period, so the real impact of the GERD may occur in the years following the filling and during the operation period. In the long run, all filling scenarios will converge, and the greatest impact of the GERD will be during the drought years of low natural flows of the Blue Nile.

Deficit modeling shows that the current shares of Egypt and Sudan, due to GERD, could be affected and will vary in time according to climate variability impacts on Blue Nile flows within the year, but may also last for periods ranging from 10 to 20 consecutive years during low-flow years due to climate variability from 1 year to another. There is a high level of uncertainty regarding the impacts of climate change when combined with GERD impacts.

12 Recommendations

It is critical that the official GERD studies, commissioned by the three countries, be completed promptly so that the parties can agree on the first filling and operating rules to minimize any negative impacts on Egypt and Sudan.

Egypt, Sudan, and Ethiopia must agree on the size of the first filling before consideration of the filling rules. Agreement on the rules of filling must take into account and coordinate the operating rules. The feasibility of reducing the overall size of GERD below the current design size of 74 BCM should be considered to enable operating at low average levels to reduce the negative impacts on Egypt and Sudan downstream, where the maximum storage capacity is only large enough to absorb high flows during the high-flood months.

It is recommended that the volume of the first filling of the GERD should be established at just above the dead storage level, i.e., the lowest storage above which turbines can operate, estimated at less than 15 BCM. It is important to reduce GERD storage to the first filling volume at the end of each hydrological year and to discharge the annual flows to reduce the cumulative impact of evaporation and seepage losses and to absorb the flows of the next year's flood.

A comprehensive climate change impact study on the Blue Nile Basin is needed to reduce the level of uncertainty about the predicted flows, and to assess GERD impacts on downstream countries in light of climate change impacts.

Egypt, Sudan, and Ethiopia must agree on the parameters for the first filling, the operating rules, and the mechanism for coordinating dam management in the three countries. Among the details that are important to include in the agreement is Ethiopia's commitment to limit the use of the GERD reservoir for hydropower generation as per the Declaration of Principles signed by the heads of states to ensure water security for the downstream countries. The three countries must agree to compensation principles that account for the possible negative impacts on Egypt's and Sudan's water shares, rights, existing uses, and hydropower generation from the HAD in Egypt, as well as other dams in Sudan pursuant to the provisions of the Declaration of Principles.

References

Abou El-Magd I, Ali E (2012) Estimation of the evaporative losses from reservoir Nasser, Egypt using optical satellite imagery. National Authority for Remote Sensing & Space Sciences, Suez University, Article in International Journal of Digital Earth (March)

AbuZeid K (2012) The watercourse/blue water & river basin/green water approach to win-win solutions in transboundary river basin management. Offi J Arab Water Council 3(1):1–15, July 2012, ISSN 1996-5699

AbuZeid K (2016) Chapter IV: Impacts of the operation of the GERD, "Economic Impacts of the GERD on Water, Agriculture and Egypt's National Security", Center for Environment & Development for Arab Region & Europe-(CEDARE) & Institute of Research and Strategic Studies on the Nile Basin Countries (Fayoum University)

AbuZeid K (2017a) Potential impacts of grand ethiopian renaissance dam (GERD) on Egypt & Sudan. In: Water Resources Program, Centre for Environment and Development Center for the Arab Region and Europe (CEDARE)

AbuZeid K (2017b) Potential impacts of grand ethiopian renaissance dam on Nile water availability for Egypt and Sudan. In: 4th Arab Water Forum AWC Shared waters in the Arab Region: potential transboundary impacts & cooperation opportunities, 27 Nov 2017, Cairo, Egypt

AbuZeid K (2018) Egypt state of the water indicators. Egyptian Water Partnership Workshop, Cairo, Egypt

Abu-Zeid M (2010) Arab and African water in the twenty first century. El-Tobgy Foundation for Trading, Printing and Publishing, Cairo, Egypt

Bashar KE, Mustafa MO (2009) Water balance assessment of the roseires reservoir. In: Intermediate results dissemination workshop, UNESCO, Ministry of Irrigation and Water Resources (Sudan) 5–6 Feb 2009, Addis Ababa, Ethiopia

Ministry of Water Resources and Irrigation (MWRI) (2004) Central Directorate for Nile Water Control Yearly Report, 2003–2004, 44 p

Said R (1993) The River Nile geology, hydrology and utilization. Pergamon Press, Oxford, 123 p

United Nations Economic and Social Commission for Wester Asia (ESCWA) (2017a) Chapter 4: Findings for selected shared water basin in the Arab region, 4.1: Nile River Basin, Arab Climate Change Assessment Report–Main Report. Beirut, E/ESCWA/SDPD/2017/RICCAR/Report, 125 p

United Nations Economic and Social Commission for Western Asia (ESCWA) (2017b) Chapter 5: Nile River: blue nile headwaters", Arab Climate Change Assessment Report, Technical Annex. Beirut, E/ESCWA/SDPD/2017/RICCAR/Report/Annex, 70 p

Wale A (2008) Hydrological balance of reservoir tana upper blue Nile Basin. Thesis submitted to the International Institute for Geo-Information Science and Earth Observation (ITC), Ethiopia, April

Chapter 17
Strengthening Flood and Drought Risk Management Tools for the Lake Chad Basin

Abou Amani, Justin Sheffield, Aleix Capdevilla, Mohammed Bila, Colby Fisher, Ming Pan, Hylke Beck, Abdou Ali, Mohamed Hamatan, Bernard Minoungou, Anil Mishra, Koen Verbist, Eric Wood, and Blanca Jimenez-Cisneros

Abstract Lake Chad is extremely sensitive to climate variability because it is a shallow inland lake, and about 97.5% of its water supply depends on the Chari-Logone River System and other tributaries. Any increase or decrease in lake volume inflow means a substantial increase or decrease in lake area. Droughts in the Sahel Region and within the basin after the 1970s had great impact on discharges of different tributaries, which led to a drastic decrease of water inflow in the lake, as well as significant seasonal and inter-annual variation of the lake area over the last 50 years. Information gaps about the water system and uncertainties about climate variability and change remain a challenge. Hydrological extremes, both floods and droughts, present a threat to agriculture and water resource management within the Lake Chad Basin. Drought and flood monitoring over the basin is difficult because of the shortage of observational data, both historic and in real time. Satellite remote sensing and hydrological modelling are techniques used to compensate for the data collection shortcomings of the region. The Africa Flood and Drought Monitoring (AFDM) provides drought and flood monitoring, and short-term and seasonal forecasting that combine climate prediction, hydrological modelling and remote sensing data in the sub-Saharan African continent. For the Lake Chad Region, the system was adapted

A. Amani · A. Mishra · K. Verbist · B. Jimenez-Cisneros
Water Sciences Division, UNESCO, Paris, France

J. Sheffield (✉)
University of Southampton, Southampton, UK
e-mail: justin.sheffield@soton.ac.uk

A. Capdevilla
World Bank, Washington, DC, USA

M. Bila
Lake Chad Basin Commission, N'Djamena, Chad

J. Sheffield · C. Fisher · M. Pan · H. Beck · E. Wood
Princeton Climate Analytics, Princeton, USA

A. Ali · M. Hamatan · B. Minoungou
AGRHYMET Regional Centre, Niamey, Niger

S. Diop et al. (eds.), *Climate Change and Water Resources in Africa*,
https://doi.org/10.1007/978-3-030-61225-2_17

with higher resolution to provide near-real-time water levels, as well as short-term forecast of flood risks, as well as medium-term forecasts of drought hazards and long-term projections of climate change impacts. Preliminary results are very encouraging; the system will continue to be updated, tested and validated to enable its operational use by decision-makers at all levels.

Keywords Flood · Drought risk management · Early warning · Medium-term forecasts · Long-term projections · Lake chad · Climate variability

1 Introduction

Lake Chad and its basin are located in the eastern Sahel Region on the southern edge of the Sahara Desert. It is a vast resource of freshwater shared by Cameroon, The Central African Republic, Chad, Niger, Nigeria and Libya. The average depth is only 1.5 m, making it a shallow lake with a relatively low volume of water and therefore extremely sensitive to any variation of its inflows. The hydrological regime of the lake and related basin, which constitutes the largest surface water input, is influenced by rainfall patterns in the wet tropics and equatorial southern regions of the basin (Leblanc et al. 2011). Ninety-five percent of the lake's water supply depends mainly on the Chari-Logone River System flowing from the Adamawa Plateau and draining from the Central African Republic, Cameroon and Chad, while the Komadugu-Yobe River and its tributaries draining from Nigeria and Niger provide less than 2.5% of the lake's water. The remaining water comes from direct precipitation falling on the lake (LCBC 1998). Therefore, any variation of the discharge of the Chari-Logone and other tributaries has a direct impact on the volume and area of the lake. The most distinctive feature of Lake Chad is its water volume variability.

The lake has experienced a long history of wet and dry periods over geological, annual and seasonal time scales. Observed droughts in the Sahel Region and within the basin after the 1970s had great impact on lake discharges of the different tributaries, particularly the Chari-Logone ensemble. These droughts led to a drastic decrease in the amount of freshwater available in the lake basin; as a consequence, the hydrology of the lake has experienced progressive and extensive change over the last 50 years (Lemoalle et al. 2012; L'Hôte et al. 2002). The lake area has varied between 2,500 km^2 and 25,000 km^2 during this period, with significant consequences for the livelihoods of the millions of people dependent on the resources of the lake and its basin. Policelli et al. (2018) observed that despite various evidence of changing lake size, an updated time series of the lake's total surface water area does not exist except for 1986–2001, which was derived from thermal remote sensing (Leblanc et al. 2011). However, the EU Joint Research Center Global Surface Water Dataset (Pekel et al. 2016) provides data for the 32 years from March 1984 to October 2015 based on the Landsat archive that could provide a new understanding of the spatio-temporal variability of inundation in Lake Chad.

Lake Chad has been a centre for development, trade and cultural exchange among established populations in the north and south of the Sahara for thousands of years. The lake is a direct source of livelihood to 2.5 million people living on its shores and islands. It is also a net exporter of food and a source of seasonal jobs, contributing to the food and job security of about 13 million people living in its hinterlands, including Ndjamena and Maiduguri. Therefore, understanding the dynamics of the lake, its natural variability and the dynamics of water resources within the basin on seasonal to multi-annual timescales is of critical importance for adaptation. While climate change models do not show a clear rainfall trend in the region, rainfall is an important component of natural hydro-climatic variability. Importantly, human intervention within the basin includes water resource development such as large-scale irrigation, which will also continue to influence the basin's water balance and dynamics. An example of this is the agricultural development of the Komadougou-Yobe Upper Basin and the decline and sometimes disappearance of the river, resulting in the extinction of Lake Chad's Northern Basin.

Due to its critical importance for the livelihood of the population, the water resources dynamics of Lake Chad must be understood and challenges facing the basin must be addressed through an inclusive and evidence-based ecosystem and Integrated Water Resource Management (IWRM) approach. The Lake Chad Basin Commission (LCBC) has been working with its member states and partners to improve transboundary water resource management. Milestones have included the adoption of a common vision in 2008 and the development of the Lake Chad Basin Strategic Action Programme 2025. The Lake Chad Water Charter, currently developed and approved, is a binding framework for the member states to promote sustainable development in the basin through the integrated, equitable and coordinated management of natural resources, in particular, water resources. Various plans and projects have been developed in response, including the Lake Chad Development and Climate Resilience Action Plan (LCDAP) for the period 2016–2025, which seeks to turn Lake Chad into a centre of regional rural development, known as the LCBC Rehabilitate and Strengthen the Resilience of Socio-ecological Systems in the Lake Chad Basin (PRESIBALT) Programme. A Global Environment Facility (GEF) project, 'Improving Lake Chad management through Building Climate Change Resilience and Reducing Ecosystem Stress through Implementation of the SAP for the Lake Chad Basin', has also been approved.

Uncertainties about the water resources of the Lake Chad Basin and its changes under climate variability and climate change remain a challenge. In particular, floods and droughts present significant challenges to agriculture and water resource management, and monitoring hydrological extremes in near-real-time and short-term-to-seasonal forecasts is therefore essential to be able to manage risks. However, data collection mechanisms are currently very limited, in part because of unreliable monitoring networks. Similarly, climate and hydrological forecasting has not been implemented and therefore its potential to provide early warning is untested.

Satellite remote sensing and hydrological modelling has been introduced into regions that have limited, if any, ground observations, and little capacity to develop other adequate monitoring and analysis systems. Satellite remote sensing can provide near-real-time estimates of nearly all aspects of the hydrological cycle (Sheffield et al. 2018), and, when coupled with hydrological modelling systems, can provide consistent and continuous estimates of the hydrological cycle, including its extremes. The effectiveness of this approach has been demonstrated in various settings, such as the African Flood and Drought Monitor (AFDM) system (Sheffield et al. 2014). The AFDM provides drought and flood monitoring, and short-term and seasonal forecasting that combines climate predictions, hydrological models and remote sensing data in the sub-Saharan African continent. This chapter describes how the AFDM is adapted to the Lake Chad Basin, a mechanism known as the Lake Chad Basin Flood and Drought Monitor (CHADFDM), to provide timely and useful information on water resources and hydrological hazards and improve management and preparedness in the region. The CHADFDM is designed specifically for the Lake Chad Basin, building on representations of the lake dynamics, updated soil and vegetation parameters, and incorporation of existing ground data.

1.1 Supporting the Lake Chad Basin Commission and Its Member States

The *Lake Chad Basin Flood and Drought Monitor* (CHADFDM) is a platform to develop monitoring and forecasting for planning and decision-making in Lake Chad. The platform supports the Lake Chad Basin Commission (LCBC), the organization that represents member states, by providing data so that it can fulfill its mandate to implement the Water Charter.

The Water Charter highlights the importance of monitoring water dynamics in the lake and recognizes that, in addition to natural climate variability, an uncontrolled increase in irrigation abstractions could impact the water level and footprint of Lake Chad. This in turn will have an impact on dependent ecosystems and the livelihoods they support. The Charter explicitly recognizes the importance of data generation and information exchange among member states in this transboundary basin, which is essential to develop a common understanding and manage water resources and aquatic ecosystems effectively. Specific obligations of the LCBC include (World Bank Group 2018) the following:

a. Monitoring and analysing the hydrologic system and its components (rainfall, ET, streamflow, soil moisture, vegetation).
b. Monitoring the hydro-ecological status of the basin in real time and in the context of short-term and seasonal forecasts.
c. Managing relevant information (e.g. hydro-meteorological data, model results, water uses and abstraction impacts) among member states through a water resources information management system.
d. Supporting evidence-based decision-making for sound management of water resources in the Lake Chad Basin.

This chapter describes the development of the CHADFDM and its main features, including the online interface, and provides an overview of the validation of the system outputs. It concludes by discussing the potential use of the system in decision-making and policy development, and future plans for enhancing and further refining the system.

2 Overview of the Lake Chad System

2.1 Rationale and Background

The CHADFDM is an outgrowth of various continental systems that have been developed since about 2010: the Africa Flood and Drought Monitor (AFDM) (Sheffield et al. 2014), the Latin America Flood and Drought Monitor (LACFDM 2019) and the US Flood and Drought Monitor (USFDM) (Sheffield et al. 2012). In general, these systems leverage the coverage of satellite remote sensing, the consistency of hydrological modelling and the power of climate forecasts, combined with available on-the-ground information, to estimate water availability (primarily flood potential and drought conditions), historically, in near-real time and as forecasts. The CHADFDM draws from the long legacy of operational and experimental systems developed for the U.S. by the Terrestrial Hydrology Group at Princeton University (Luo and Wood 2007), part of the North American Land Data Assimilation System Phase 2 (NLDAS-2; Xia et al. 2012) and the Climate Forecast System (NCEP) Climate Test Bed Program.

2.2 Main Components of the System

The CHADFDM consists of four parts (Fig. 1): 1. A historic, multi-decadal reconstruction of the terrestrial water cycle is generated by forcing a land surface hydrological model (Variable Infiltration Capacity (VIC)) (Liang et al. 1994 and Hamman et al. 2018) with a merged reanalysis/satellite/observation dataset of precipitation and other meteorological variables. This forms the climatology against which current conditions are compared. Then the real-time monitoring system (2018–present) is driven by a merged precipitation dataset and atmospheric analysis data that tracks flood and drought conditions in real time. 2. The simulated outputs are augmented by satellite remote sensing of soil moisture and vegetation indices. 3. A 7-day forecasting component provides hydrological forecasting of floods and water availability, and 4. A seasonal forecast component provides hydrological predictions and derived drought products out to 6 months, based on bias-corrected and downscaled climate model forecasts that are used to drive the hydrologic and potentially other impact.

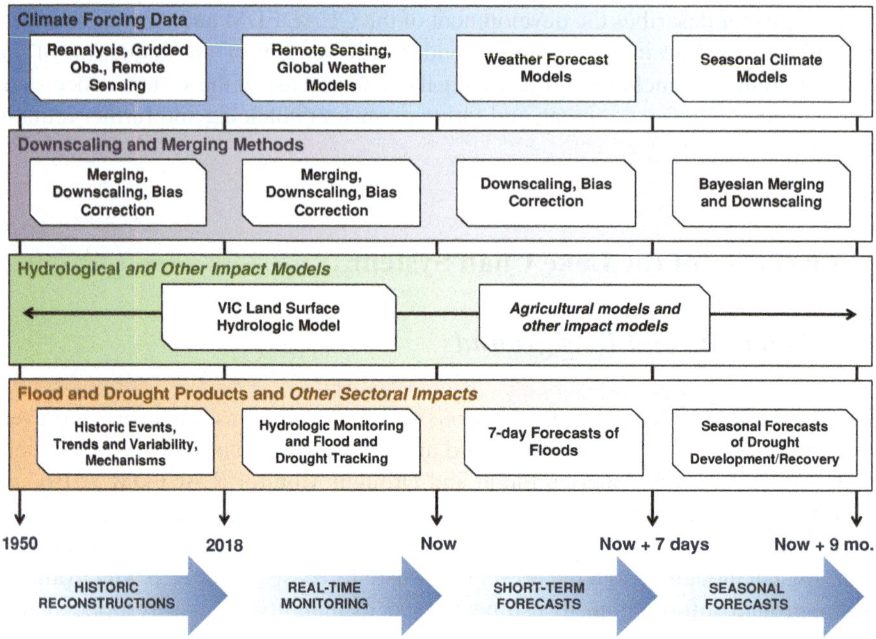

Fig. 1 Flowchart of the CHADFDM. The system comprises four parts: 1. Historic reconstructions of the terrestrial hydrological cycle that is derived from simulations of the VIC land surface model forced by a hybrid reanalysis-observational meteorological dataset. The datasets can be used for a variety of applications including analysis of historic drought events, estimation of trends and variability, and investigation of drought mechanisms. 2. A real-time monitoring component that updates the model runs to 2–3 days from real-time forced by a merged in situ, satellite and reanalysis fields of precipitation (MSWEP: Beck et al. 2017, 2019), and GFS analysis fields of temperature and wind speed. There is also potential to include other impact models such as crop models. 3. Short-term (7-day) hydrological forecasts focused on flood prediction, derived from downscaled and bias-corrected GFS precipitation and temperature forecasts that are used to drive the VIC model and provide ensemble forecasts of precipitation, soil moisture and streamflow. 4. A seasonal climate forecast component that uses bias-corrected and downscaled NMME climate model forecasts of precipitation and temperature to generate ensemble predictions of meteorological drought conditions. Existing components are shown in standard font, and potential future components in *italic* font (Sheffield et al. 2014)

The regional system uses the same framework as the continental systems. Several upgrades have been implemented, in part to improve the fidelity and accuracy of the system relative to the continental systems, and also to address user needs at regional scale. These include (i) increasing the spatial resolution of the monitor from 25 to 5 km. This was accomplished by upgrading the forcing precipitation data, the soil properties and the vegetation land cover. The new precipitation data merges improved historical data, a large suite of satellite precipitation products and weather model analysis information; (ii) using historic in situ and upgraded data to fine-tune the hydrological modelling system and its parameters, and evaluating the model output against data provided by the Lake Chad Basin Commission and from the Global Runoff Data Centre (GRDC) and (iii) inclusion of representation of Lake Chad, as the most important hydrological feature of the region, via a component of the modelling system and using satellite imagery. These are discussed in detail below.

2.3 Details of Each Component

a. Historic and real-time climate data

Historically, the system is driven by Version 2 of the Multi-Source Weighted-Ensemble Precipitation (MSWEP) dataset (Beck et al. 2019) that significantly enhances Version 1 (Beck et al. 2017). MSWEP is a precipitation product (1979–present) recently completed with a focus on use in regional monitoring systems, with the potential to be used globally. Its new (and unique) features include: (i) optimal merging of an unprecedented broad range of gauge, satellite and reanalysis precipitation products; (ii) high spatial (0.1°) and temporal (three hourly) resolution; (iii) fully global coverage; (iv) correction for distributional precipitation biases by probability matching; (v) correction for long-term terrestrial precipitation biases using discharge observations from 13,762 stations around the globe; (vi) incorporation of daily (rather than monthly) gauge observations from 66,993 gauges around the globe and (vii) accounting for regional differences in the 24-hour accumulation period of gauges. Version 2 of the software is described and validated by Beck et al. (2019). A near-real-time extension has also been developed and is used to force the CHADFDM operationally.

The VIC model is driven by MSWEP plus other meteorological variables (minimum and maximum air temperature and wind speed) that are derived in the same way as is the AFDM, that is, historically they are based on downscaled and bias-corrected data from the Princeton Global Forcings (PGF; Sheffield et al. 2006), which is a hybrid reanalysis-satellite-gauge dataset. In real time, they are based on bias-corrected analysis fields from the US National Oceanic and Atmospheric Administration (NOAA) NCEP Global Forecast System (GFS).

b. VIC model, lake model and streamflow routing

The VIC model is a land surface hydrological model that was originally developed as the land component of coupled climate models (Liang et al. 1994), but has evolved through its use in applied hydrological research to be a fully distributed hydrological model (Sheffield et al. 2014; Hamman et al. 2018). Given inputs of precipitation and other near-surface meteorology, the model solves the surface water balance and estimates the partitioning of precipitation into evapotranspiration, surface runoff, baseflow (subsurface flow) and the change in soil moisture content. The model is parameterized in terms of the soil type and land cover spatial fields and associated soil and vegetation parameters. These have been updated to align with the AFDM. The updated soil data is from SoilGrids, a global collection of updatable soil property and class maps with a spatial resolution of 250 m (Hengl et al. 2017). The soil maps were produced using machine learning algorithms based on about 150,000 global soil profiles and 158 remote sensing soil covariates. The maps consist of about 280 raster layers of global soil information, including organic carbon, bulk density, cation exchange capacity (CEC), pH, soil texture fractions and coarse fragments. Soil properties are provided at seven standard depths (0, 5, 15, 30, 60, 100 and 200 cm). Soil texture data was converted into soil hydraulic properties needed by the land surface model using standard pedotransfer function relationships and coarsened to 5 km using bilinear interpolation.

Land cover data is obtained from GlobeLand30, which consists of open-access, high-resolution maps of land cover across the globe (Chen et al. 2015). It is generated using a machine learning algorithm based on the integration of pixel- and object-based methods, by the use of more than 20,000 Landsat and Chinese HJ-1 satellite images. The accuracy assessment, using nine types and over 150,000 test samples, shows that the overall accuracy of the product is 80.33%. The spatial resolution is as high as 30 × 30 metres, but the land cover is aggregated into tiles of different land cover types at the 5 km modelling grid scale. Other model parameters that are specific to the VIC model and its parameterizations are taken from the AFDM and interpolated to 5 km.

c. Short-term (GEFS) and seasonal (NMME) forecasts

Short-term forecasts are based on the U.S. Global Ensemble Forecast System (GEFS) which produces global weather forecasts out to 16 days operationally. We use GEFS precipitation, temperature and wind speed forecasts out to 7 days to drive the VIC model to produce weather and hydrological forecasts for the LCB; evaluations for some example rainfall events are shown in Sect. 3.3. Weather forecast data are downscaled and bias-corrected to align with the statistics of the historic CHADFDM data, using a quantile mapping approach (Wood et al. 2002). Seasonal forecasts are taken from the North American Multi-Model Ensemble (NMME; Kirtman et al. 2014), which consists of hindcasts (historic forecasts) and operational forecasts from a set of six global climate models. The hindcasts are for approximately the past 30 years, depending on the climate model, with 10–20 ensemble members per model per monthly forecast, with lead times of up to 12 months. The forecasts of precipitation and temperature are downscaled and bias-corrected based on the ensemble mean forecasts for each model.

d. Web interface

The website (Fig. 2) provides an intuitive interactive interface for visualizing, analysing and accessing data. The interface is built on Google Maps, which provides the functionality to pan, zoom and interrogate maps, including animation of maps of historic, current and forecast data and indices. The user can select among daily, monthly and annual data, and derived statistics. Users can also display time series of data at any time step for a chosen location selected via a mouse click or latitude-longitude selection. The data can also be downloaded in standard formats such as Netcdf, Arc-ASCII and csv to be used with other software.

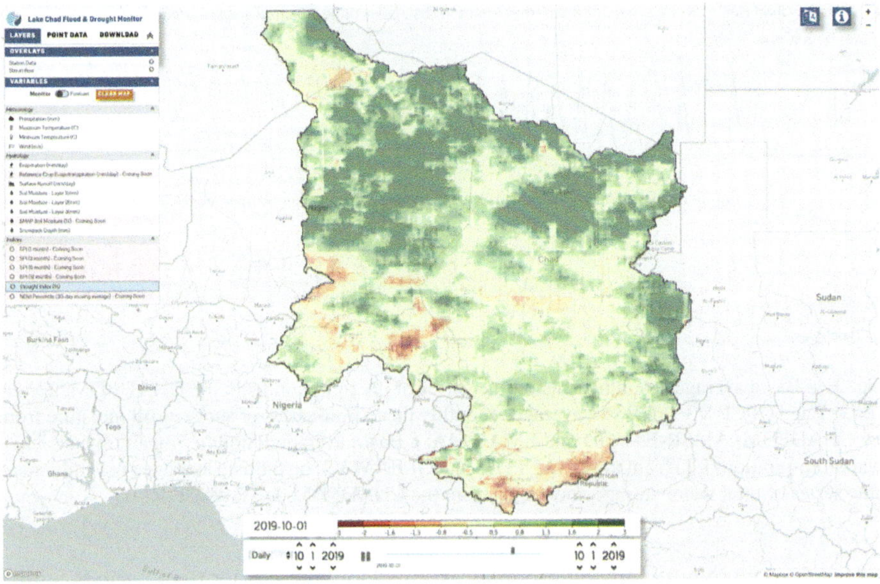

Fig. 2 Web user interface for the Lake Chad Flood and Drought Monitor showing a map of SPI6 for July 2018

3 Validation of Historic Data and Evaluation of Forecast Skill

In this section, we show preliminary validation of the system outputs on a range of time scales, including historic data and short-term forecasts. Evaluation of the seasonal forecasts is ongoing and will be reported elsewhere.

3.1 Evaluation of Historic Data

Historic data is compared to independent datasets of various components of the water budget, including precipitation, surface soil moisture, evapotranspiration and terrestrial water storage changes derived from satellite remote sensing (datasets described in the appendix). Simple comparisons between the CHADFDM and these datasets are made at the basin average and spatial map level. All datasets represent monthly means, if not already available at monthly resolution. For the basin average comparisons, all datasets were interpolated to the CHADFDM resolution (0.05-deg, ~ 5 km), masked to the basin, and then spatially averaged, to produce monthly time series. The spatial map comparisons are based on average maps calculated over all years of the comparison period. All datasets were compared over the period 2001–2016,

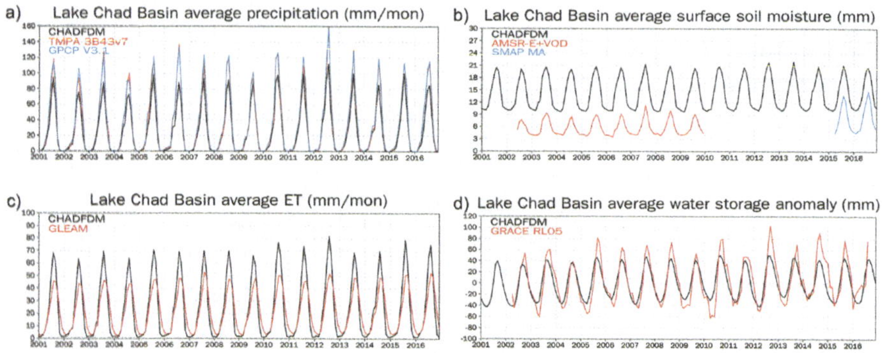

Fig. 3 **a** Basin averaged monthly mean time series of precipitation from the CHADFDM, TMPA 3B43v7 and GPCP V3.1. **b** Basin averaged monthly mean time series of surface soil moisture from the CHADFDM, AMSR-E+VOD and SMAP MA. **c** Basin averaged monthly mean time series of evapotranspiration (ET) from the CHADFDM and GLEAM V3.2a. **d** Basin averaged monthly mean time series of total water storage anomalies from the CHADFDM and GRACE RL05

although some remote sensing products are available for shorter time periods within these years.

Figure 3 shows basin average time series comparisons of the CHADFDM with the various remote sensing-based products. Figure 3a indicates that the CHADFDM precipitation (based on MSWEP) is generally lower than the coarser resolution GPCP data but comparable to the TMPA data, at least in the latter part of the time period. Figure 3b shows the surface soil moisture from the CHADFDM, which is for the top 10 cm model layer, compared to the AMSR-E + VOD and SMAP data (both valid for the top few cms of soil). This difference represents soil depth and explains the overall differences in the datasets. Despite this, the seasonality is consistent among the datasets. Comparison of evapotranspiration (ET) is shown in Fig. 3c, and indicates large differences among the datasets, which is typical for ET datasets. The CHADFDM and GLEAM datasets have a well-defined seasonal cycle with a peak in the middle of the wet season in July-September, and a minimum close to zero in the dry season. The CHADFDM ET is significantly higher than GLEAM at the peak season, which requires further investigation to understand whether it is attributable to specifications or parameters for vegetation distribution, or to forcing. Comparisons of total water storage anomalies are shown in Fig. 3d and indicate a reasonable agreement between CHADFDM and GRACE data on average, with a general underestimation by the CHADFDM of the seasonal range. This could be attributable in part to high ET. GRACE anomalies from year to year are not well captured by the CHADFDM, although there are large uncertainties in GRACE basin average data which have to be filtered to remove external influences in the underlying coarse scale GRACE data.

Figure 4 shows time-averaged spatial map comparisons of the CHADFDM using the various remote sensing-based products. Figure 4a shows the strong wet-to-dry gradient in precipitation from south to north across the basin, which is consistent

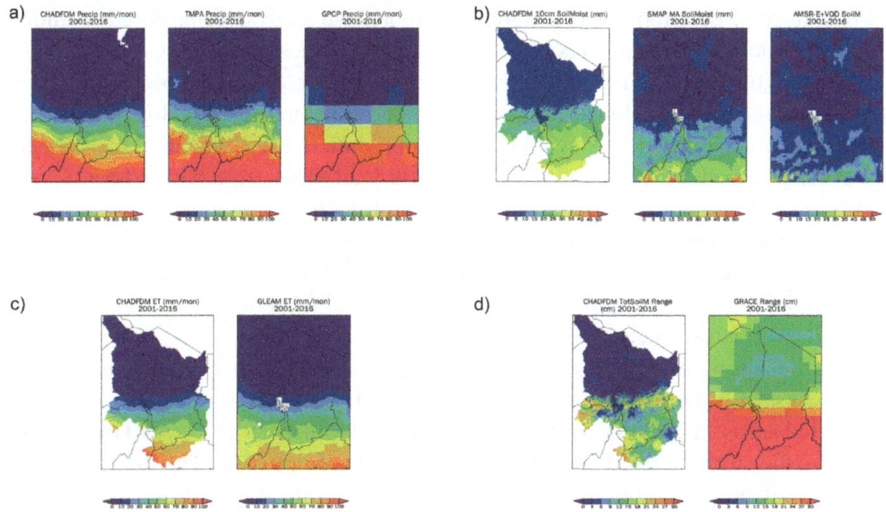

Fig. 4 **a** 2001–2016 average precipitation from the CHADFDM, TMPA 3B43v7 and GPCP V3.1. **b** 2001–2016 average surface soil moisture from the CHADFDM, AMSR-E+VOD and SMAP MA. **c** 2001–2016 average evapotranspiration (ET) from the CHADFDM, and GLEAM V3.2a. **d** 2001–2016 average total water storage anomalies from the CHADFDM and GRACE RL05

across all three products. GPCP data is relatively coarse at a 2.5° resolution and so does not capture the fine scale variability found even at the scale of the CHADFDM. Figure 4b shows the surface soil moisture maps from the CHADFDM and the two remote sensing datasets. The difference in mean levels shown in Fig. 3 is apparent spatially, with a slightly less distinct north-south gradient in the CHADFDM. The two remote sensing products have a similar wet-dry gradient to the CHADFDM data, and is generally wetter in the southern part of the basin. Figure 4c shows the spatial maps of ET and both datasets display a wet-dry boundary in the middle of the basin that is consistent with the precipitation maps in Fig. 4a. The CHADFDM is spatially comparable to the GLEAM dataset. Comparisons of maps of total water storage in terms of the maximum range of storage change over 2001–2016 are shown in Fig. 4d. The coarse resolution (1.0°) of this version of the GRACE dataset is evident, but there is reasonable alignment of the wet-dry transition between the two datasets, consistent with the precipitation maps. As with the time series comparison in Fig. 4, the CHAFDM underestimates the GRACE data across the basin.

3.2 *Evaluation of GEFS Short-Term Forecasts*

The accuracy of the forecasts is evaluated here for three selected large precipitation events across the basin during 2018 compared to observed precipitation and model simulations driven by observed precipitation. This provides only a snapshot of the performance of the model and a more comprehensive analysis is required to understand its reliability. Nonetheless, the evaluation provides an initial estimate of the potential of the system. The GEFS provides 20 ensemble forecast members which are aggregated to daily, downscaled and bias-corrected to the LCB domain at 0.05° resolution. Bias correction is against the historic data in the Lake Chad Monitor for the overlap period. Forecasts are carried out for lead times from 6 to 0 days.

Figure 5 shows an example of the 6-day lead forecasts of precipitation for 2 May 2018: the forecast was initialized on 26 April 2018. The 20 ensemble members are shown and compared to the observed data for the forecast date. Note the spread in the ensembles, which generally encompass the spatial footprint of the observed data, but tend to underestimate it overall.

Figure 6 shows a summary of the accuracy in terms of the spatial correlation of the forecasts across the three example dates and for each lead time. Correlation is calculated across the wetter part of the basin up to latitude 15°N to avoid biasing the results to the strong wet-dry meridional gradient of precipitation. The lines represent the mean of the ensemble and the error bars are ± one standard deviation of the ensemble. Precipitation and runoff correlations are modest (<0.4), which is expected given the difficulty in forecasting convective systems. Correlation tends to reduce with increasing lead time, as is also expected. The soil moisture correlations are much higher given its persistence, with much lower uncertainty across ensemble members.

Figure 7 shows an example of the basin average forecast for the three variables with a 6-day lead, highlighting the improved accuracy when averaging over larger scales, even with this somewhat long lead time. The uncertainty for the precipitation is however quite large, with some ensemble members predicting high and early precipitation. The translation of this uncertainty into total runoff and soil moisture is dampened and their relative accuracy is higher, although with a tendency to underestimate the observed data; if this is found to be systematic, it can be bias-corrected as a post-processing step.

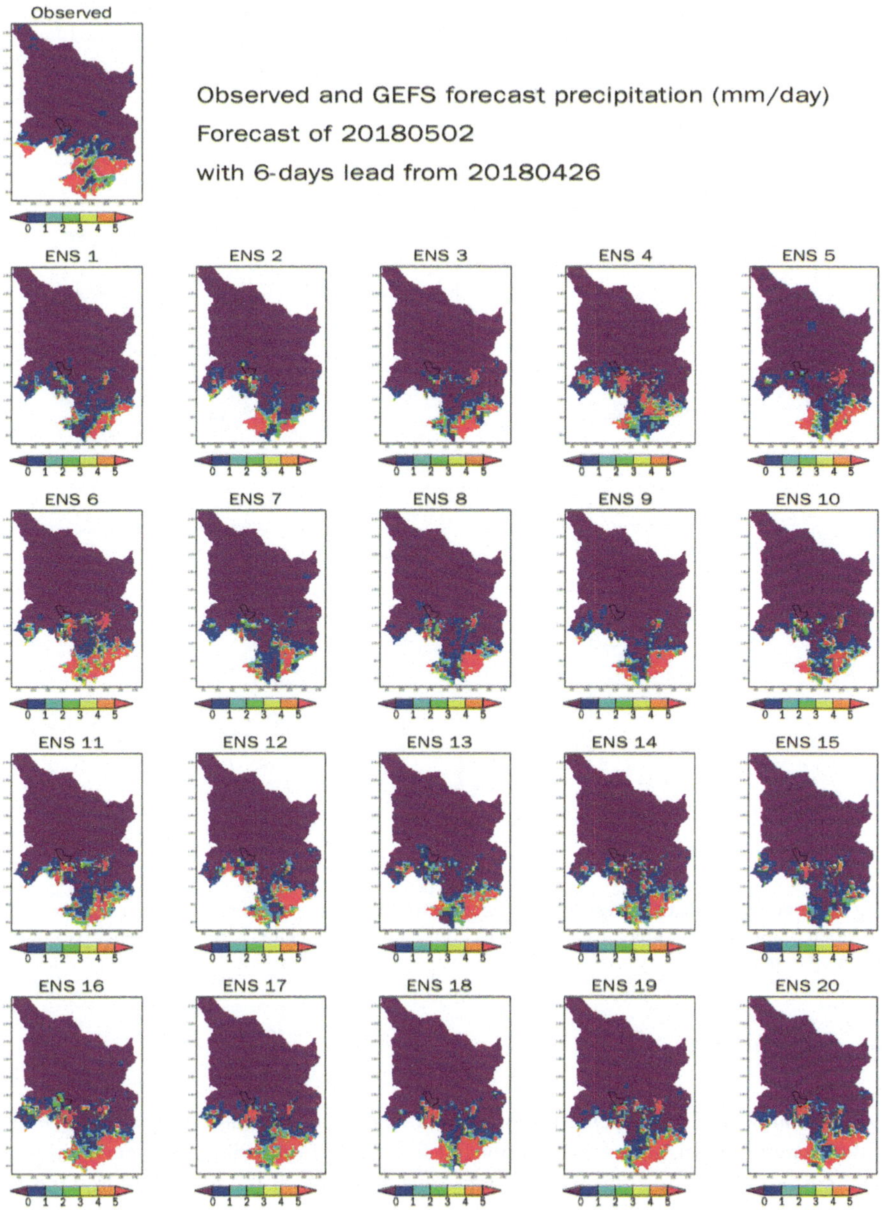

Fig. 5 Example maps of forecasts at 6 days lead of precipitation for the test forecast date of 2018-05-02. The observed data is in the top left of each panel and the 20 forecast ensemble members are below

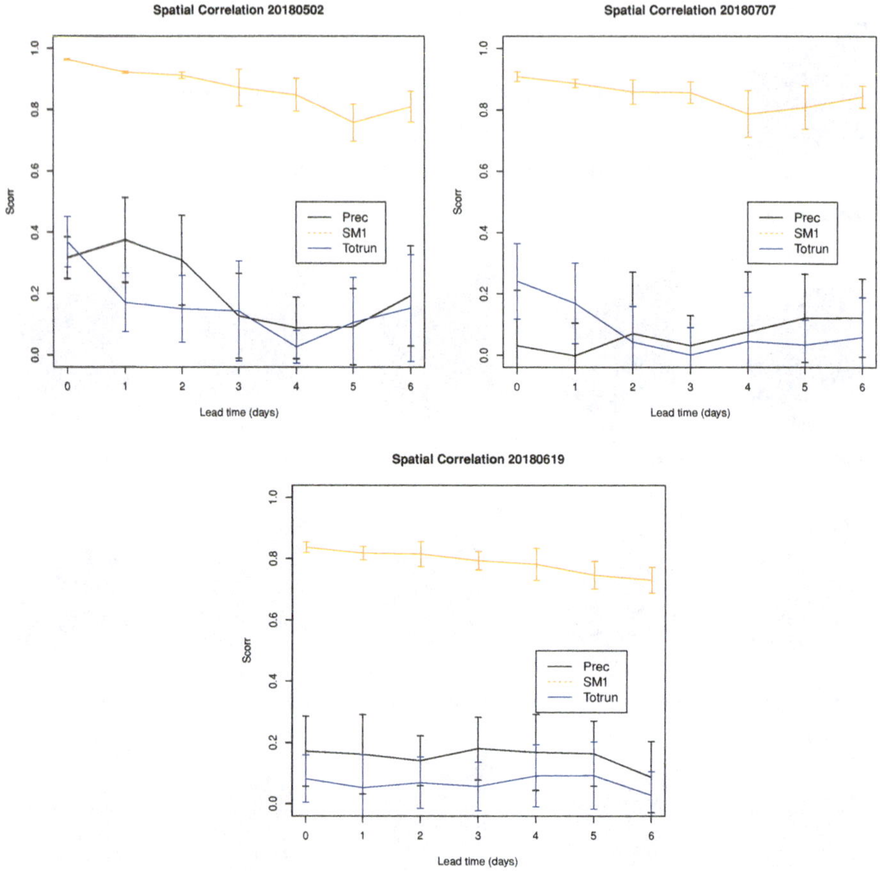

Fig. 6 Summary of spatial correlation for the three forecast test dates for precipitation, soil moisture and total runoff for 0 to 6-day lead times. The correlation is calculated across the wetter part of the basin up to latitude 15 N

4 Conclusions

In summary, the CHADFDM provides near-real-time estimates of the basin water balance and identification of droughts and floods, in monitoring and forecast mode. The system overcomes the lack of dense ground data in the basin by drawing on a wealth of remote sensing and model information that can provide consistent and continuous estimates of the surface water budget and its extremes. Initial validation of the system shows that its estimates are comparable to independent data from satellite products and are consistent with the few on-the-ground observations available. This together with further integration of in situ data (e.g. merging of precipitation gauge data) and satellite data (e.g. assimilation of SMAP soil moisture into the hydrological model) provides confidence in its use. With further calibration of the

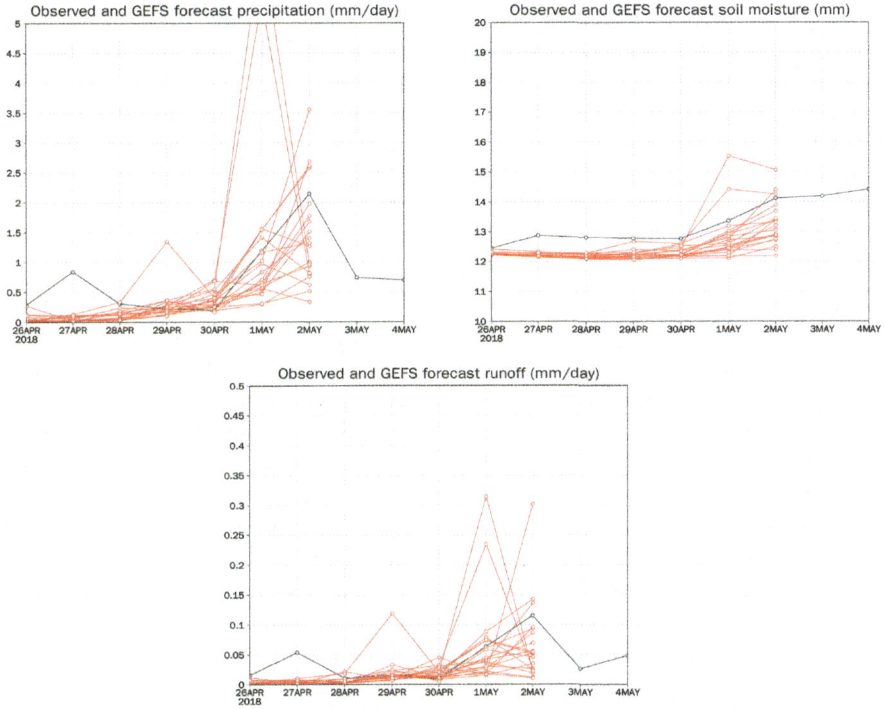

Fig. 7 Basin average precipitation, soil moisture and total runoff at 6-day lead for 20 ensembles (red lines) compared to the observed (black line) for the test forecast date of 2nd May 2018

hydrological model, estimates will be improved more. Additionally, the initial evaluation of accuracy of short-term forecasts indicates that there is potential for their use in decision-making. Systematic evaluation across a broader set of hindcasts, forecast variables (e.g. start of the rainy season) and skill metrics (e.g. probability of drought detection) is necessary to identify where and when the forecasts might be considered robust and reliable enough to inform specific planning and management decisions.

The system is being continuously improved in collaboration with LCBC and the member states in order to provide reliable and useful information for planning and decision-making. Indeed, efforts are being made to consider and mainstream all available rainfall and discharge data from LCBC member countries into the system. This is expected to provide a comprehensive view of basin water resources. Furthermore, a comprehensive validation exercise is proposed as a partnership of all meteorological and hydrological services of the LCBC member countries, with a focus on key aspects of basin hydrology.

Given the necessity for the Commission to implement the Water Charter, the system is providing useful information to improve the overall knowledge of surface water variation within the basin and on the hydrological extremes in near-real time. As the system evolves, it will improve the information it generates to enable better decision-making for management and preparedness. For example, better information on the nature of the rainy season within the basin would help policy and community leadership plan appropriately for predicted conditions.

The co-development of the CHADFDM has benefitted from continued input from the LCBC staff through periodic meetings and discussions, sometimes also attended by national agency technical staff. Operational training on the system is organized for experts from member countries and from the LCBC secretariat. The tool will be widely disseminated for use by communities and policy-makers within the basin. Given the limitation of data availability in the Lake Chad Basin, the information already provided by the CHADFDM represents a significant step towards the development of a robust transboundary information system, providing a baseline of timely information that includes historic data, near-real-time data, short-term forecasts and seasonal forecasts. This is already a great source of state-of-the-art hydrological data using remote sensing and global datasets to support the mandate of the LCBC to implement the Water Charter. The system will continue to be refined for application to specific opportunities that will arise from its use and ownership.

Priorities for next steps include the systematic integration of ground data from rainfall stations and improvement of calibration using in situ streamflow data. A continued dialogue among the LCBC and the national agencies of the member states is the key to creation of an operational and well-functioning transboundary information platform to monitor and forecast water resources.

Acknowledgements This work was supported by the World Bank's Global Water Security and Sanitation Partnership through the *Global Remote Sensing Initiative for Water Resources Management*, and the UNESCO International Hydrological Programme (IHP) through a funded project on Lake Chad by the African Development Bank.

Appendix

A.1. Validation Datasets

a. Precipitation

GPCP—Global Precipitation Climatology Project

We use the latest version (v2.3) of the GPCP monthly analysis. This improves the homogeneity of the previous version, especially since 2002, through corrections of the cross-calibration of satellite data and updates to the gauge analysis. The dataset is a merger of various satellite-based estimates globally, combined with the precipitation gauge analyses over land from the Global Precipitation Climatology Centre (GPCC). The satellite-based estimates are a combination of passive microwave estimates over ocean, passive microwave estimates over land and estimates from IR/microwave sounders, contributing at higher latitudes (above 40° latitude).

TMPA—TRMM Multisensor Precipitation Analysis

We use the latest version (V7) of the TMPA. The TMPA is a calibration-based sequential scheme for combining precipitation estimates from multiple satellites, as well as gauge analyses where feasible, at fine scales (0.25° × 0.25° and 3 hourly). TMPA is available both after and in real time, based on calibration by the TRMM combined instrument and TRMM microwave imager precipitation products, respectively. Only the after-real-time product incorporates gauge data. The dataset covers the latitude band 50°N–S for the period from 1998 to the delayed present. The data currently contains two products, 3 hourly combined microwave-IR estimates (with gauge adjustment) and monthly combined microwave-IR-gauge estimates of precipitation computed on quasi-global grids about 2 months after the end of each month starting in January 1998. We use the 3B43_7 monthly version of the dataset, which incorporates gauge adjustments, at 0.25° resolution.

b. Soil Moisture

SMAP—Soil Moisture Active Passive

The NASA Soil Moisture Active Passive (SMAP) mission launched in January 2015 was designed for global mapping of soil moisture at a 10-km spatial resolution with a 2–3-day revisit time under both clear and cloudy sky conditions. This improves on the resolution relative to AMSR-E (25 km) and SMOS (50 km) by combining an L-band radar (high resolution 1–3 km, lower accuracy) and an L-band radiometer (low resolution 40 km, higher accuracy), as well as retrievals for a wider range of vegetation conditions and for the top 5 cm of the soil. SMAP products also include level 4 (L4) root-zone estimates by merging SMAP observations with land surface model estimates via assimilation extending the utility of the data. Unfortunately, the SMAP radar failed in July 2015 thus restricting soil moisture products to the 40 km resolution of the radiometer. Here we use monthly mean values that are calculated in the AFDM based on 3-day moving average data, which are referred to as SMAP-MA.

AMSR-E + VOD

This dataset is based on a retrieval algorithm for daily global soil moisture and vegetation optical depth (VOD) derived from dual-polarized AMSR-E brightness temperatures at 10.7 GHz. It uses a simplified radiative transfer model that assumes (1) the surface to be retrieved is vegetated soil only and (2) the vegetated soil consists of a smooth bare soil surface plus a layer of vegetation. The model approximates the output from generally used, but more complex and more rigorously parameterized radiative transfer models quite well, irrespective of the vegetation density. By incorporating both soil roughness parameters and to-be-retrieved VOD implicitly, the algorithm simplifies the retrieval process since accurate prior knowledge of the global soil roughness information is unavailable at AMSR-E frequencies and footprint scales. The data are available globally, for 2002–2009, at 0.25°, daily resolution.

c. Evapotranspiration

GLEAM—Global Land Evaporation Amsterdam Model

We use the latest version of the GLEAM dataset, V3.2a, which is a global dataset spanning 1980–2017 at 0.5°, daily resolution. The dataset is based on reanalysis net radiation and air temperature, satellite and gauged-based precipitation, VOD, soil moisture and snow water equivalent. Components of evapotranspiration (transpiration, canopy interception, soil evaporation, open water evaporation and sublimation) are calculated separately. The Priestley and Taylor equation is used to calculate potential evaporation based on reanalysis surface net radiation and near-surface air temperature. Estimates of potential evaporation for the land fractions of bare soil, tall canopy and short canopy are converted into actual evaporation using a multiplicative evaporative stress factor based on observations of microwave Vegetation Optical Depth (VOD) and estimates of root-zone soil moisture. The latter is calculated using a multi-layer running-water balance. To try to correct for random forcing errors, satellite retrievals of surface soil moisture are also assimilated into the soil profile. Interception loss is calculated separately in using a Gash analytical model. Finally, estimates of actual evaporation for water bodies and regions covered by ice and/or snow are obtained using an adaptation of the Priestley and Taylor equation. We also use the surface and root-zone soil moisture calculated by GLEAM.

d. Terrestrial Water Storage

GRACE—Gravity Recovery and Climate Experiment

The GRACE mission has been monitoring changes in the Earth's gravity field since its launch in 2002 by measuring the distance between two orbiting satellites. Variations in these fields can be attributed to changes in terrestrial water storage after removal of atmospheric and ocean bottom pressure changes. We use the JPL RL05 dataset of approximately 30-day Total Water Storage (TWS) anomalies. We also use the JPL RL05-based time series for the Lake Chad Basin using a 300 km Gaussian half-width averaging kernel that is available from the University of Colorado.

References

Beck HE, van Dijk AIJM, Levizzani V, Schellekens J, Miralles DG, Martens B, de Roo BA (2017) MSWEP: 3-hourly 0.25-global gridded precipitation (1979–2015) by merging gauge, satellite, and reanalysis data. Hydrol Earth Syst Sci 21:589–615

Beck HE, Wood EF, Pan M, Fisher CK, Miralles DG, van Dijk AI, McVicar TR, Adler RF (2019) MSWEP V2 global 3-hourly 0.1° precipitation: methodology and quantitative assessment. Bull Amer Meteor Soc 100, 473–500. https://doi.org/10.1175/BAMS-D-17-0138.1

Chen J et al (2015) Global land cover mapping at 30 m resolution: A POK-based operational approach. Isprs J Photogram Remote Sens 103:7–27

Hamman JJ, Nijssen B, Bohn TJ, Gergel DR, Mao Y (2018) The variable infiltration capacity model, version 5 (VIC-5): infrastructure improvements for new applications and reproducibility. Geosci Model Dev. https://doi.org/10.5194/gmd-11-3481-2018

Hengl T et al (2017) SoilGrids 250m: Global gridded soil information based on machine learning. PLoS ONE 12(2):e0169748. https://doi.org/10.1371/journal.pone.0169748

Kirtman BP, coauthors (2014) The north american multimodel ensemble: phase-1 seasonal-to-interannual prediction; phase-2 toward developing intraseasonal prediction. Bull Amer Meteor Soc 95,585–601. https://dx.doi.org/10.1175/BAMS-D-12-00050.1

LACFDM, Latin American and Caribbean Flood and Drought Monitor (2019) Princeton Climate Analytics. http://stream.princeton.edu/lacfdm. Accessed 31 July 2019

LCBC (1998) Plan d'action stratégique. Vers un développement durable sur le bassin du Lac Tchad. Gestion intégrée et durable des eaux internationales du Bassin du Lac Tchad. N'Djamena, LCBC

Leblanc M, Lemoalle I, Bader JC, Tweed S, Mofor L (2011) Thermal remote sensing of water under flooded vegetation: new observations of inundation patterns for the 'Small' Lake Chad. J Hydrol 404:87–98

Lemoalle J, Bader JC, Leblanc M, Sedick A (2012) Recent changes in Lake Chad: observations, simulations and management options (1973–2011). Global Planet Change 80:247–254

L'Hôte Y, Mahé G, Somé B, Triboulet JP (2002) Analysis of a Sahelian annual rainfall index from 1896 to 2000; the drought continues. Hydrol Sci J Des Sci Hydrol 47:563–572

Liang X, Lettenmaier DP, Wood EF, Burges SJ (1994) A simple hydrologically based model of land surface water and energy fluxes for GCMs. J Geophys Res 99(D7):14415–14428

Luo L, Wood EF (2007) Monitoring and predicting the 2007 U.S. drought. Geophys Res Lett 34:L22702. https://doi.org/10.1029/2007gl031673

Pekel JF, Cottam A, Gorelick N, Belward AS (2016) High-resolution mapping of global surface water and its long-term changes. Nature 540:418–422. https://doi.org/10.1038/nature20584

Policelli F, Hubbard A, Jung HC, Zaitchik B, Ichoku C (2018) Lake Chad total surface water area as derived from land surface temperature and radar remote sensing data. Remote Sens 10:252

Sheffield J, Xia Y, Luo L, Wood EF, Ek M, Mitchell KE, and the NLDAS Team (2012) Drought monitoring with the North American Land Data Assimilation System (NLDAS): a framework for merging model and satellite data for improved drought monitoring. In: Wardlow B, Anderson M, Verdin J (eds) Remote sensing of drought: innovative monitoring approaches. Taylor and Francis, London, UK, 270 p

Sheffield J, Wood EF, Chaney N, Guan K, Sadri S, Yuan X, Olong L, Amani A, Ali A, Demuth S, Ogallo L (2014) A drought monitoring and forecasting system for sub-Saharan African water resources and food security. Bull Am Meteor Soc 95(6):861–882

Sheffield J, Wood EF, Pan M, Beck H, Coccia G, Serrat-Capdevila A, Verbist K (2018) Satellite remote sensing for water resources management: Potential for supporting sustainable development in data-poor regions. Water Resour Res 54. https://doi.org/10.1029/2017WR022437

World Bank Group (2018) Mainstreaming the use of remote sensing data and applications in operational contexts. World Bank, Washington, DC. © World Bank. https://openknowledge.worldbank.org/handle/10986/29333 (License: CC BY 3.0 IGO)

Wood AW, Maurer EP, Kumar A, Lettenmaier DP (2002) Long-range experimental hydrologic forecasting for the eastern United States. J Geophys Res 107:4429. https://doi.org/10.1029/2001JD000659

Xia Y, Mitchell K, Ek M, Cosgrove B, Sheffield J, Luo L, .Alonge C, Wei H, Meng J, Livneh B, Duan Q, Lohmann D (2012) Continental-scale water and energy flux analysis and validation for North American land data assimilation system project phase 2 (NLDAS-2): 2. validation of model-simulated streamflow, J Geophys Res 117:D03110. https://doi.org/10.1029/2011jd016051

Chapter 18
Developing a Framework for the Water-Energy-Food Nexus in South Africa

T. Mabhaudhi, G. Simpson, J. Badenhorst, A. Senzanje, G. P. W. Jewitt, V. G. P. Chimonyo, S. Mpandeli, and L. Nhamo

Abstract The water-energy-food (WEF) nexus is a cross-sectoral approach to resource management and sustainable development. We propose a framework for linking the WEF nexus to the Sustainable Development Goals (SDGs), emphasizing SDG 2, 6 and 7. We further propose indices and models to evaluate WEF nexus performance. A systematic analysis of existing WEF nexus frameworks in academic and gray literature resulted in the development of a South African framework that considers the three sectors as well as technological innovation, human well-being, the SDGs, and different drivers of the WEF nexus. It is proposed that this framework be utilized as a point of departure for future research related to the WEF nexus in South Africa. Future research on the WEF nexus should focus on (i) developing an integrated model and indices to assess WEF resources in South Africa and creating a WEF nexus database; (ii) translating existing knowledge to inform policies for integrated sustainable resource management; and (iii) undertaking participatory research

T. Mabhaudhi (✉) · V. G. P. Chimonyo · L. Nhamo
Centre for Transformative Agricultural and Food Systems, School of Agricultural, Earth & Environmental Sciences, University of KwaZulu-Natal, P/Bag X01, Pietermaritzburg 3209, Scottsville, South Africa
e-mail: mabhaudhi@ukzn.ac.za

T. Mabhaudhi · G. Simpson · G. P. W. Jewitt · L. Nhamo
Centre for Water Resources Research, School of Agricultural, Earth & Environmental Sciences, University of KwaZulu-Natal, P/Bag X01, Pietermaritzburg 3209, Scottsville, South Africa

G. Simpson · J. Badenhorst
Jones and Wagener Engineering and Environmental Consultants, P.O. Box 1434, Johannesburg 2128, Rivonia, South Africa

A. Senzanje
School of Engineering, University of KwaZulu-Natal, P/Bag X01, Pietermaritzburg 3209, Scottsville, South Africa

S. Mpandeli · L. Nhamo
Water Research Commission of South Africa, Daventry Street, Lynnwood Manor 0081, Pretoria, South Africa

L. Nhamo
International Water Management Institute, Southern Africa (IWMI-SA), 141 Cresswell Street, Pretoria 0184, Silverton, South Africa

S. Diop et al. (eds.), *Climate Change and Water Resources in Africa*,
https://doi.org/10.1007/978-3-030-61225-2_18

to demonstrate the applicability of the WEF nexus at the local level, focusing on the poor. The development of a WEF nexus model and indices is a pathway for unlocking the value of existing Resources data indices to integration generate sustainability new datasets.

Keywords Resources · Indices · Integration · Metrics · Models · Sustainability

1 Introduction

Sustainable resource management is a major worldwide governance concern as the demand for natural resources has increased exponentially with population and economic growth Sustainable resource management since the 1960s, when the Green Revolution was ushered in. Human resource consumption is disproportionate to the size of populations, with the wealthiest continents (North America and Europe) consuming on average ten times as much as poor continents (Africa , West Asia, etc.) (UNEP 2016), while the wealthy countries also contribute disproportionately to environmental pollution. International trade of natural resources contributes significantly to the Gross Domestic Product (GDP) of many developing countries, but is often perceived to be a "curse," because countries with an abundance of natural resources also tend to suffer from poverty, inequality, conflict, and insecurity (Ross 1999). An integrated approach to managing natural resources is required if the Sustainable Development Goals (SDGs) are to be realized by 2030.

Since 2011, the water-energy-food (WEF) nexus has been investigated from many points of view, each interest framing its analysis through the lens of its own economic, political, social, or scientific perspective. The WEF nexus is broadly defined as a framework that accounts for the interactions, synergies, and trade-offs among water, energy, and food when managing these resources. Water, energy, and food securities are inextricably linked, with usage within one sector influencing the use and availability of the others. Unlike Integrated Water Resource Management (IWRM), which is by definition water-centric, the WEF nexus approaches resource management holistically. Each resource sector within the nexus is considered equally important. The WEF nexus presents an opportunity for policymakers, researchers, and development agencies to optimize resources, maximize synergies and trade-offs, minimize ineffective redundancy of effort, and reduce conflicts among stakeholders representing each resource.

The WEF nexus is closely aligned with the SDGs, particularly SDG 2 (zero hunger), 6 (clean water and sanitation), and 7 (affordable and clean energy). Originally, the SDGs (the successor to the Millennium Development Goals (MDGs)) were established as a response to world poverty, inequality, and insecurity, but they have evolved into guiding principles for resource management. Developing countries, including South Africa , are likely to benefit greatly from the integrated resource management approach that the WEF nexus provides, particularly those countries experiencing significant trade-offs among water, energy, and food. The WEF nexus framework is particularly relevant when considering the recent proposed policy shift on land expropriation, which if adopted will significantly influence land utilization. Currently, various government agencies generally approach resource management in isolation, without accounting for the usage of water, energy, and land by other sectors. This is a major challenge in South African policymaking, especially considering the country's limited water availability, threats posed by both climate variability and change across climate-sensitive sectors, the scarcity of high-potential arable land, and the country's reliance on fossil fuel-based energy generation. Furthermore, climate change is expected to further compromise the availability of resources in South Africa, through changes to ecosystem services, rainfall frequency and distribution, and natural disasters.

South Africa is a water-scarce country with approximately 13% arable land, much of which is in regions that have a high concentration of mineral resources, e.g., coal. About 30% of South Africa's crops are grown on irrigated land, accounting for approximately 75% of total national agricultural water use (Ololade et al. 2017). From 1985 to 2008, South Africa was a net food exporter (Ololade et al. 2017). However, this has changed in recent years due to a reduction in agricultural yields and an increase in population (Ololade et al. 2017), on top of the country's heavy reliance on non-renewable energy resources (Ololade et al. 2017). Due to projected population growth and economic development, multiple countries within the Southern African Development Community (SADC) will be water-stressed by 2040, with South Africa experiencing a high ratio of water withdrawals to supply (World Resources Institute, 2015). It is therefore necessary to operationalize an inter-sectoral framework to understand dynamics among water, energy, and food systems for sustainable resource management and regional development.

This chapter seeks to develop a draft framework for implementing the WEF nexus for South Africa , linked to the Sustainable Development Goals (SDGs) and emphasizing SDGs 2, 6, and 7. We also propose indices and/or metrics as well as models that can be used to evaluate WEF nexus performance.

1.1 Current Status of the WEF Nexus in South Africa

Achieving water, energy, and food security is essential to South Africa's developmental agenda and aspirations to address existing inequalities and improving the quality of life for all. Approaches to WEF security must therefore be integrated with South Africa's primary focus on employment, poverty alleviation, equality, and the elimination of corruption. Responsibility for tackling these challenges is shared by the government at the national, provincial, and local levels. Progress toward achieving the SDGs consequently also depends on breaking down the "silo" mentality between government sectors. Despite the relative sophistication and success of South Africa in developing sound public policy, it is still too often compartmentalized by sector. This was the case in the adoption of the Millennium Development Goals (MDGs), which often lacked harmonization and thus failed to deliver as effectively as it otherwise could have.

In 2014, the Worldwide Fund/South Africa (WWF-SA) published a series of documents, funded by the British High Commission in Pretoria, entitled *Understanding the Food Energy Water Nexus* (Carter and Gulati 2014). The framework for approaching the WEF nexus was through the relationship of climate change, waste management, financial flows, and integrated planning to various disciplinary perspectives. South Africa For example, for energy, South Africa is taking advantage of its geographic location and ecosystem to develop renewable energy generation projects (see Fig. 1). The focus is on wind power generation and photovoltaic energy conversion, with relatively few concentrated solar power (CSP) and photovoltaic energy projects in the southern portion of the Northern Cape. Similarly, energy generated through biomass must be rainfed and not use irrigation (Nhamo et al. 2018).

Figure 2 shows changes in WEF nexus elements since 1999 in South Africa, revealing a definitive decrease in the food deficit over time, and an improving trend for sanitation facilities, as well as improved access to water and electricity. Simpson and Berchner (2017) indicate that South Africa currently fulfills its own demand for cereals. Less than 5% of the population is undernourished, and most of the population has access to clean and safe drinking water sources (in 2015, 93.2% of the population had access to improved water sources) and reliable electricity (in 2014, 86% of the population had access to electricity, albeit predominantly fossil fuel-based).

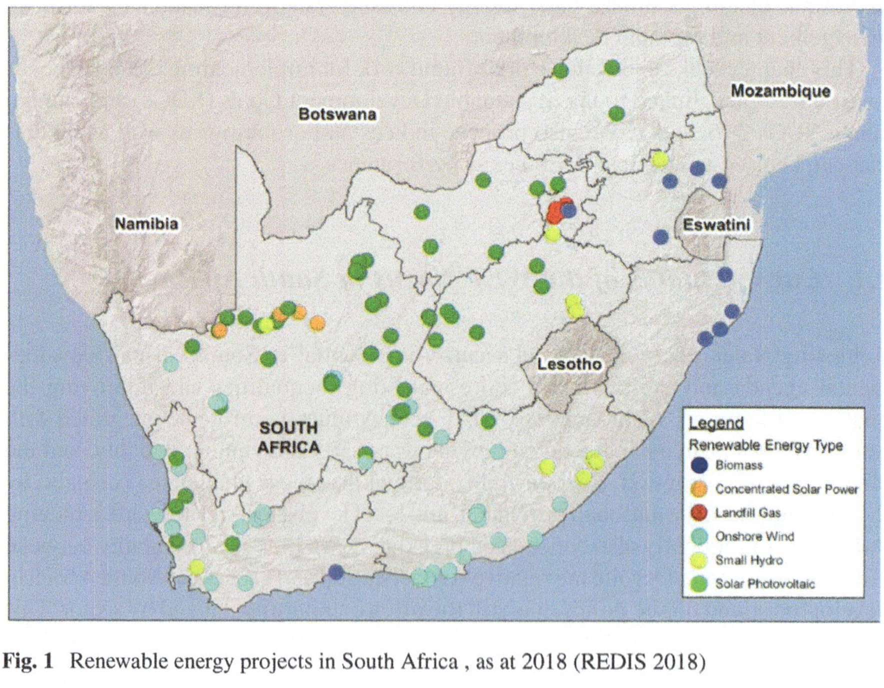

Fig. 1 Renewable energy projects in South Africa , as at 2018 (REDIS 2018)

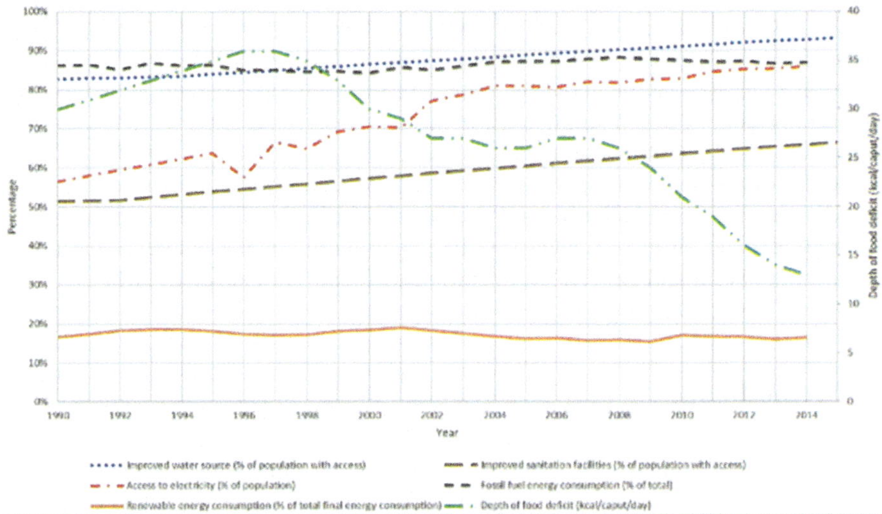

Fig. 2 Annual changes in WEF nexus indicators (improved water source, access to electricity, renewable energy consumption, improved sanitation facilities, fossil fuel energy consumption, and depth of food deficit) from 1990 to 2014 relating to SDGs 2, 6, and 7 over time in South Africa (FAO 2017; The World Bank 2018)

Obstacles to achieving WEF security differ by province. Gauteng is the physically smallest province in South Africa, but includes Johannesburg, the economic capital of the country; the province is home to more than a fifth of the nation's population. Due to isolation from significant water sources, Gauteng imports approximately 88% of its water via various inter-basin transfer schemes. The province contributes approximately 3% of agricultural production but consumes about 20%. Electricity consumption is high South Africa , accounting for 24% of South Africa's total electricity delivered in March 2018 (StatsSA 2018). Electricity for Gauteng is supplied predominantly by coal-fired power stations in Mpumalanga (Von Bormann and Gulati 2014).

The Western Cape Province generates approximately 25% of the agricultural sector's gross income, and exports more than 50% of its produce, 75% of which is destined for European markets. The provincial government has invested significantly to ensure good water quality within this region, as the income from produce exports is between R190 million and R570 million annually (Von Bormann and Gulati, 2014). The sustainability of food exports is controversial, since this practice indirectly exports water from a water-scarce area. The threat of exporting "virtual water" was demonstrated with the drought of 2017–'18 in the Western Cape, including Cape Town. These trade-offs must be studied further from a water perspective to determine the continued viability of fruit and vegetable exports.

In the Karoo, no large-scale electricity generation projects exist; the region relies on small solar farms and access to the national power grid. The energy generation of this region has been the subject of debate, focusing on unconventional energy sources such as shale gas or coalbed methane. Water is required to enable drilling and hydraulic fracturing, and it is a very scarce resource in the arid, semi-desert Karoo, leading to concern about the impact of these energy generation methodologies on the quality and quantity of groundwater. Water resource systems and the supporting infrastructure within the Karoo are already extremely strained: only 14% (16 million m^3) of the storage capacity of the Welbedacht Dam is currently available due to unmitigated siltation (Ololade et al. 2017). Smaller towns in the Karoo generally depend on groundwater supply, which amplifies the potential threat that unconventional oil and gas (UOG) operations present (Ololade et al. 2017).

2 Methodology

A framework for analysis of the WEF trade-offs was developed by stakeholders during a consultative workshop. Existing WEF nexus frameworks were examined to ascertain their applicability to South Africa . Based on the literature review[1], 20 frameworks were identified and screened, leaving 18 frameworks for assessment for their applicability to the WEF ecosystem of South Africa based onAfricamultiple factors:

- Inclusion of all WEF nexus sectors, with equal weight to water, energy, and food;
- Comprehensive consideration of the economic and demographic drivers of change (industrialization, global climate change, population growth, urbanization);
- Applicability to the economic and demographic circumstances of South Africa (based on the above drivers of change);
- Applicability to other socio-economic, environmental, and technical characteristics of South Africa (livelihoods [rural poverty], data requirements, sectoral compartmentalization [governance/policy], fossil fuels, etc.);
- Integration of the framework among the different sectors;
- Breadth of sectors considered, e.g., environment/ecosystems, land, climate change, livelihoods, waste management, recycling/reuse;
- Integration of the SDGs and MDGs; and
- Innovation and infrastructure, e.g., power stations, improved technology.

A ranking template was applied, giving each framework a score (on a ten-point scale) based on its relevance to the established criteria. The South African framework was adapted from the most relevant of the frameworks based on the above-stated criteria. Those selected are discussed in detail in Sect. 3.1.

A specific aim South Africa was to yield a framework linking the WEF nexus to the SDGs, particularly SDGs 2, 6, and 7. Land is a key consideration in South

[1] Search words such as WEF nexus model, WEF nexus frameworks, WEF nexus tools, WEF nexus analytical tools, WEF nexus governance, WEF nexus synergies and trade-offs, WEF nexus sustainability indicators, WEF nexus application, WEF nexus data, WEF nexus operationalisation, WEF nexus case studies, and WEF nexus financing were searched in Google.

Africa, as is Africathe incorporation of innovation in the WEF nexus to supplement advances achieved by just rebalancing among WEF trade-offs, which in itself would be insufficient. Innovation is necessary to ensure resource security, and access to water, energy, and food for all people.

3 Results and Discussion

A wide range of international initiatives exist to analyze the close relationships among the WEF components. The WEF nexus has gained prominence as an approach to consider solutions to global challenges in part because it can be used as an analytical tool, a conceptual framework, and as a basis of discussion and debate (Keskinen et al. 2016). As an analytical tool, the WEF nexus systematically and quantitatively reveals the interactions among WEF resources; as a conceptual framework, it is a pathway to understand WEF linkages, promote coherence in policymaking, and enhance sustainability , and as a discussion platform it is a tool for problem-framing and the promotion of cross-sectoral collaboration (Albrecht et al. 2018). For these reasons, the WEF nexus contributes substantively to understanding the complex and dynamic interlinkages among issues related to the security of water, energy, and food.

3.1 Applicability of Other Existing Frameworks to the WEF Nexus in South Africa

Based on the methodological criteria, five existing frameworks were identified as most applicable and were evaluated.

(a) Smajgl et al. 2016

The study by Smajgl et al. (2016) presents a cross-sectoral, balanced, and dynamic WEF nexus framework where sectoral objectives are given equal weight, revealing the emergences and/or changes in cross-sectoral connections as a result of single sector interventions. The dynamic WEF nexus framework describes interactions between (a) the three sectors and (b) the nexus core and the three sectors (Fig. 3). The nexus core in this framework consists of the major drivers of the water, energy, and food sectors, as well as cross-sector factors.

This framework, however, limits consideration of the interacting variables in the nexus core as it looks at only climate change and population growth as the main drivers of ecosystem services. The framework also displays how sectoral outcomes provide feedback and control core drivers, thereby creating sustained interactions. Three distinct entry points are depicted in the framework, introducing sector-specific interests. This conceptualization coordinates across sectors, to avoid unintended trade-offs and consequences. The framework requires that stakeholders account for sectoral interdependencies.

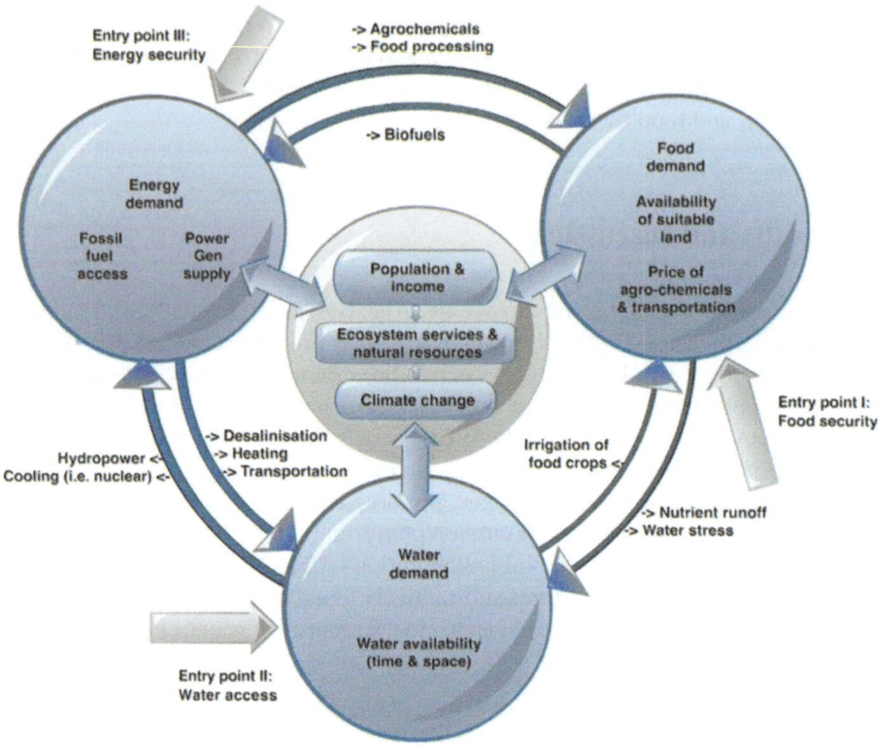

Fig. 3 The energy-water-food nexus presented by (Smajgl et al. 2016)

(b) Ringler et al. (2013)

Ringler et al. (2013) present the concept of the water-energy-land and food (WELF) nexus, emphasizing that this concept plays out differently in various parts of the world. The WELF nexus framework evaluates the linkages that exist among the water, energy, land, and food sectors (Fig. 4). The direct and indirect drivers of change, which affect these linkages, are clearly depicted in the framework. Standard WEF nexus frameworks do not include the land dimension; its inclusion in this framework is recognition of its importance not only in the production of food, but also for water (underground water storage, reservoirs) and energy supply (shale gas or biofuels), as well as shedding light on the importance of land scarcity. In contrast to many WEF nexus frameworks that tend to be water-centric, this framework puts food at the center.

(c) Karabulut et al. (2018)

In this study, a synthesis WELF matrix system describes the complex and closely related relationships among the natural resources used for food (most importantly, water and land), energy, and ecosystems . The matrix can be set for different scales (from global to local) and includes the impacts and nexus with climate change. The

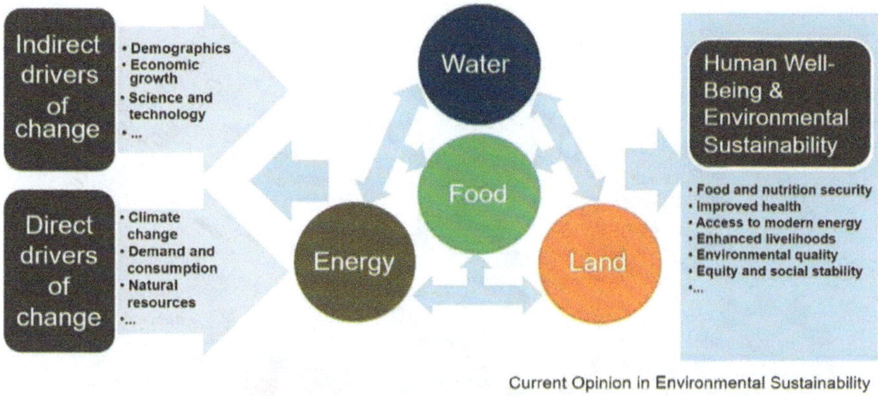

Fig. 4 The extended water, energy, food, and land nexus presented by (Ringler et al. 2013)

aim of the matrix is to integrate quantitative and qualitative aspects, which are often neglected in conventional approaches of impact assessment. Due to the complexity of interactions among the different components, quantitative and expert assessments are both required.

This framework centers on ecosystems (Fig. 5), because they represent all features of water, energy, land and food availability, and production. Land is included within the concept of ecosystems, since the term "land" embraces different land users, land covers, and soil ecosystems. The centrality of ecosystems is also recognized by their incorporation into environmental policies and initiatives internationally. Figure 5 uses three matrices for the WEFL (water-energy-food-land) nexus in the form of a double-entry table to identify relationships among sectoral uses of resources as well as the provision of ecosystem services. Potential service flows are classified according to types/sub-types of sectoral uses, which can refer to either final or intermediate services, with direct or indirect effects on human well-being. The main purpose of this framework is to support a comprehensive nexus approach, called the Ecosystem-Water-Food-Land-Energy (EWFLE) Nexus.

(d) Martinez-Harnandez et al. (2017)

This study, called "NexSym," employs simulation and analytics to depict its Nexus Simulation System. The purpose is to develop a tool for integrated resource assessment, and to account for integration within and across WEF sectors, ecosystems, and consumption components that interact with local systems. A premise of this study (Martinez-Hernandez et al. 2017) is the need for a nexus tool on a local scale, tailored to local conditions to more easily achieve synergistic techno-ecological interactions.

The NexSym software allows users to simulate how a part of a nexus is affected by a change in another part, as well as to evaluate key interactions the synergies of which may be further integrated. It offers a conceptual and modeling framework, taking into account local use of energy, water and food production, and waste treatment, as well as interactions of local components of the WEF nexus, such as ecosystems and

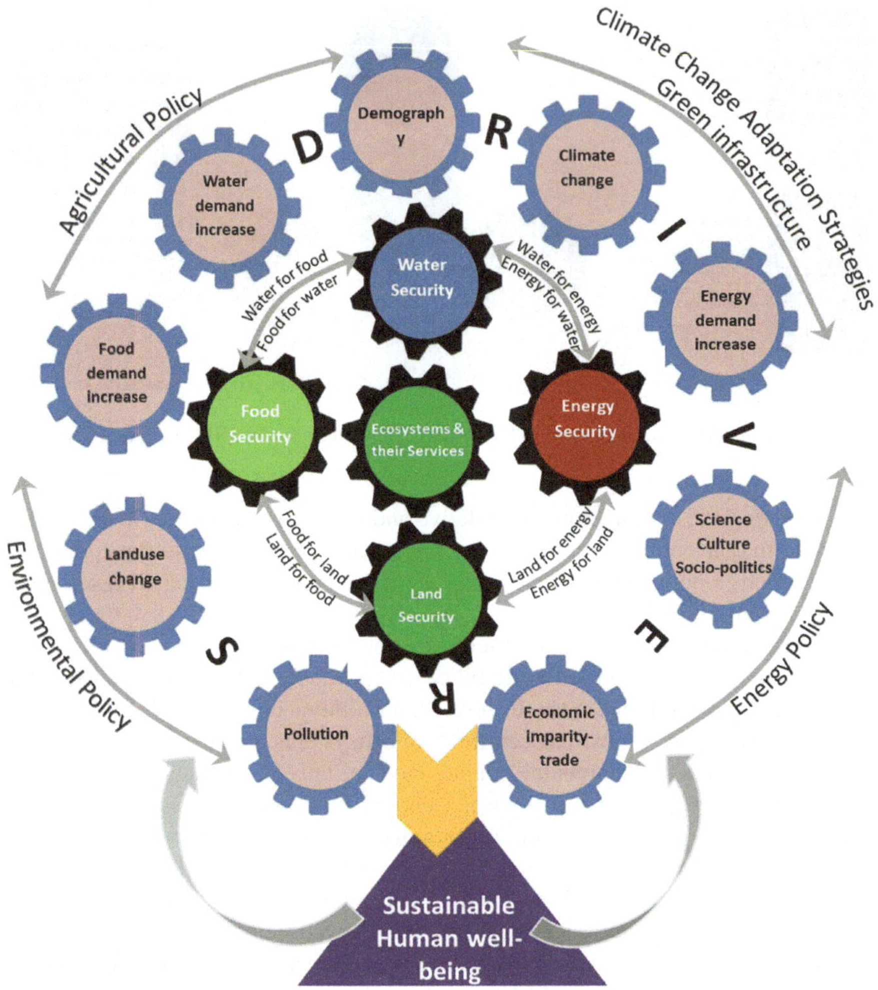

Fig. 5 Framework for the ecosystem-water-energy, land and food security nexus presented by Karabulut et al. (2018)

consumption. The approach combines data inputs, predictive models, and integrated outcome analysis (Fig. 6).

The NexSym models a local WEF system using ecological (managed or natural ecosystems such as heathlands or forests), technological (man-made industrial and municipal facilities), and consumption components ("sinks" of products and services, such as residential and commercial activities). The framework is primarily focused on local systems, and therefore requires detailed input from a subject locale to enable meaningful assessment. The study summarizes its focus by stating that, "engagement

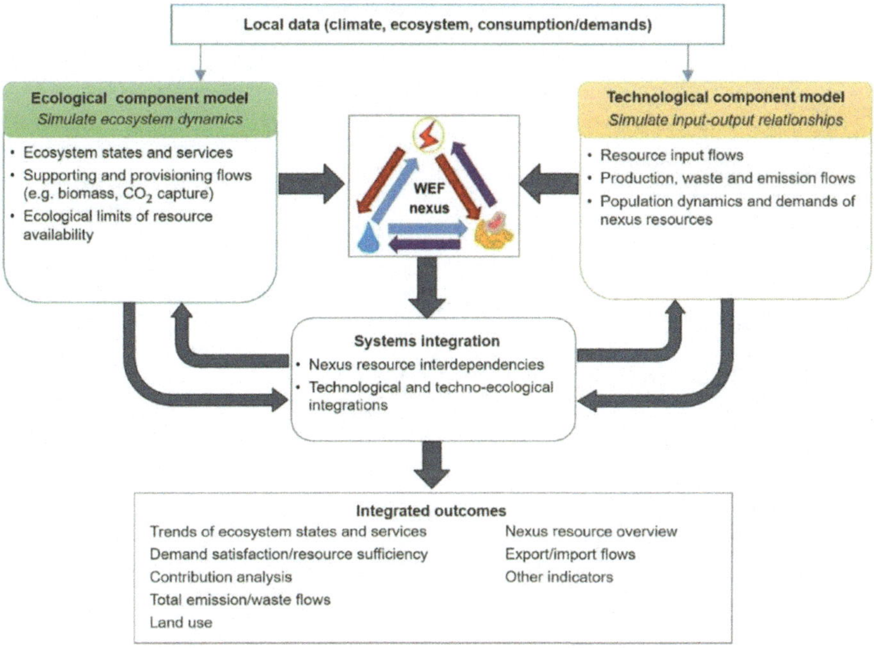

Fig. 6 The NexSym model's intended input, output, and techno-ecological view of the WEF subsystem and their interactions (Martinez-Hernandez et al. 2017)

with researchers and local communities to develop datasets specific to local contexts is crucial for the successful application of the tool."

(e) Conway et al. (2015)

Conway et al. (2015) approach southern Africa's nexus from the perspective of climate, based on a modified Hoff's nexus framework (Hoff, 2011), which integrates global trends (drivers) with actions, to highlight the role of climate as a driver. The framework considers the main elements of intra-regional interdependencies in WEF sectors within each country, while it highlights connections on the river basin scale and draws attention to case studies of specific trade-offs and synergies.

The importance of climate in determining potential agricultural production, medium-term water availability, and some components of energy production and demand are emphasized in this framework. Figure 7 depicts how climatic variability drives fluctuations in WEF elements, with secondary effects on the whole nexus.

3.1.1 Ranking of Frameworks

Table 1 indicates the rankings of the selected frameworks, features of which are used to develop a modified framework for South Africa . Of the 25 identified frameworks,

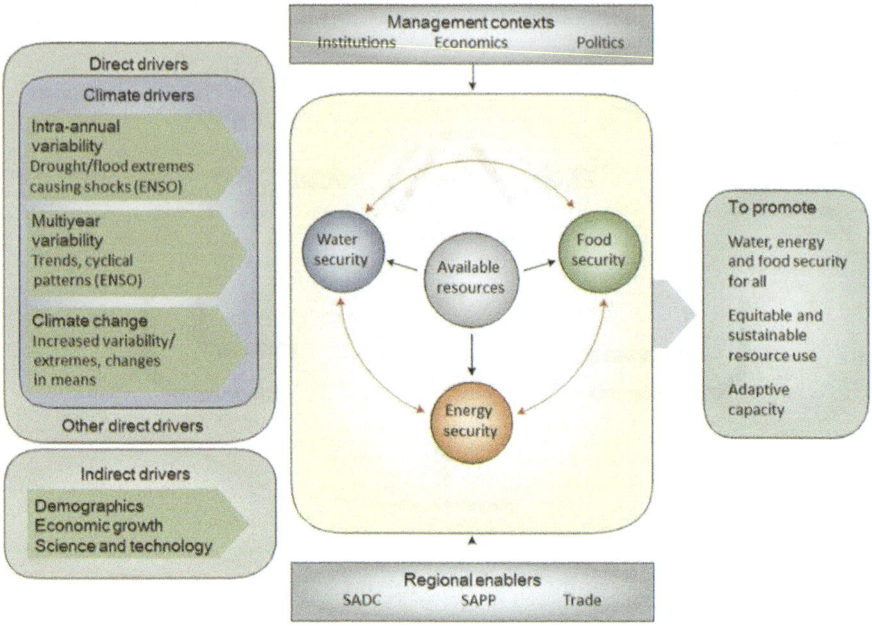

Fig. 7 A modified version of the (Hoff 2011) nexus framework presented by Conway et al. (2015)

only five were sufficiently suitable to the South Africa context using the established criteria (Table 1).

The framework developed by Smajgl et al. (2016) scored relatively well on the "applicability to South Africa" criterion, but was weak Africa in its consideration of innovation.

The next three studies, by Ringler et al. (2013), Karabulut et al. (2018), and Martinez-Hernandez et al. (2017), ranked second through fourth, respectively, in terms of their applicability to South Africa, but would have benefited from South Africa deeper consideration of potential innovation and its connection to the SDGs. The framework by Conway et al. (2015) would have been more applicable to use in South Africa by integrating sectors, acknowledging adjacent sectors (e.g., livelihoods, land, ecosystems), and including innovation. All these shortcomings were accounted for by modifying the selected frameworks for more applicability and relevance to South Africa.

3.1.2 Innovation

South Africa Consideration of innovation encompasses improved infrastructure (e.g., power stations with lower emissions and/or dry-cooled power plants), renewable energy mechanisms (biofuels, wind, tidal, and better exploitation of abundant solar

Table 1 Criteria for ranking the existing WEF nexus frameworks

Framework name	Sectors considered	Scale of application	Sector integration (Y/N)	Weaknesses	Strengths	Score 1—low 5—high
Smajgl et al. 2016	All three	National	Y	None, possibly innovations	Integration, inclusion of SDGs, drivers, and other sectors	5
Ringler et al. (2013)	All three	National	Y	To a small degree, innovations and SDGs	Inclusion of drivers and integration	4
Karabulut et al. (2018)	All three	National	Y	Innovations and inclusion of SDGs	Integration of other sectors and drivers	3
Martinez-Harnandez et al. (2017)	All three	National	Y	Innovations and inclusion of SDGs	Applicability of South Africa and integration	3
Conway et al. (2015)	All three	National	Y	Innovations, other sectors, and integration	Addressing challenges and applicability to South Africa	3

energy), and data generation improvements to support the big data needs to develop useful output to inform decision-making and facilitate data sharing and dissemination. There is also potential to further develop technologies to improve water efficiency (desalination, establishing dry-cooled power plants) and seasonal climate forecasting (climate change for farmers).

3.1.3 Sustainable Development Goals (SDGs)

The WEF nexus is a key tool for regional integration and development of the SDGs, as well as delivery of SDG goals at the national level (Mabhaudhi et al. 2016). SDGs will drive future policies related to the WEF nexus through SDGs 2, 6, 7, 8, and 9.

SDG 2 (zero hunger); SDG 6 (clean water and sanitation); SDG 7 (affordable and clean energy); SDG 8 (affordable work and economic growth); and SDG 9 (industry, innovation, and infrastructure) illustrate how the SDGs connect with the WEF nexus. For example, achieving SDG 2 requires the eradication of food insecurity and improving nutrition. SDG 6 requires basic access to water and sanitation to overcome water scarcity. SGD 7 necessitates the development of greater renewable energy sources, and access to them.

3.1.4 Challenges

The frameworks offered by Karabulut et al. (2018) and Martinez-Hernandez et al. (2017) scored low for the inclusion of key societal challenges and would be strengthened by taking into account livelihoods (rural poverty, high rates of unemployment, barriers to educating the poor, electricity shortages, land issues), nutrition, health and food insecurity (agricultural sector), as well as improving economic growth, water scarcity within the context of climate change, and data requirements and accessibility.

3.1.5 Integration

Conway et al. (2015) had the lowest score for integration among the five top-ranked frameworks; in order to improve its applicability to South Africa, it must account for integration among the three WEF sectors by illustrating how they interrelate, and offer recommendations to improve integration and its potential to enhance livelihoods, land management, environmental ecosystems, climate change resilience, and waste management.

3.2 A Proposed South African-Based WEF Nexus Framework

Figure 8 is a schematic of a proposed WEF nexus framework for South Africa based on the criteria Africa used to select the top five WEF nexus frameworks delineated in Sect. 2.

The work by Smajgl et al. (2016), Ringler et al. (2013), and Karabulut et al. (2018), in conjunction with the work by Hoff (2011), was built upon to design the WEF nexus framework for South Africa. Figure 8 illustrates the vital drivers of change and the biggest challenges facing the country Africa, strongly influencing the WEF nexus. It also illustrates that with proper policies, strategies, and the consideration of alternative clean, renewable options, vastly improved human well-being and environmental sustainability can be achieved. The framework developed for South Africa puts emphasis on SDGs 2, 6, and 7, describing drivers and feedback among these sectors. Direct and indirect drivers of change, which affect these linkages, are also illustrated in the framework (Fig. 8).

3.3 Potential Indices, Metrics, and Models for Evaluating the WEF Nexus

It is imperative to define the priorities, aims, scale of application, and data availability in order to effectively adopt the WEF nexus framework in South Africa (Endo et al. 2017). No one approach is applicable to all situations and the suitability of a given framework or methodology will vary in response to the aims, priorities, and scale of application, from global to local. As one would expect, data requirements and needs for their aggregation and analysis will increase as one move from local to global scale. Likewise, required tools and models increase in complexity from local to global scale (Zhang et al. 2018).

To accurately model and assess the WEF nexus, it is useful to generate data that quantifies flows of energy and materials, makes numerical predictions, and estimates the associated costs (Keairns et al. 2016). When developing or considering models to guide data generation, it is important to restrict scope to the relevant aspects of the WEF nexus to eliminate complexity, but to be aware of risks associated with these omissions and to develop assumptions associated with these risks. All stakeholders must contribute to the assessment process, which will inevitably require trade-offs between indicator-based assessments and quantitative approaches (Keairns et al. 2016).

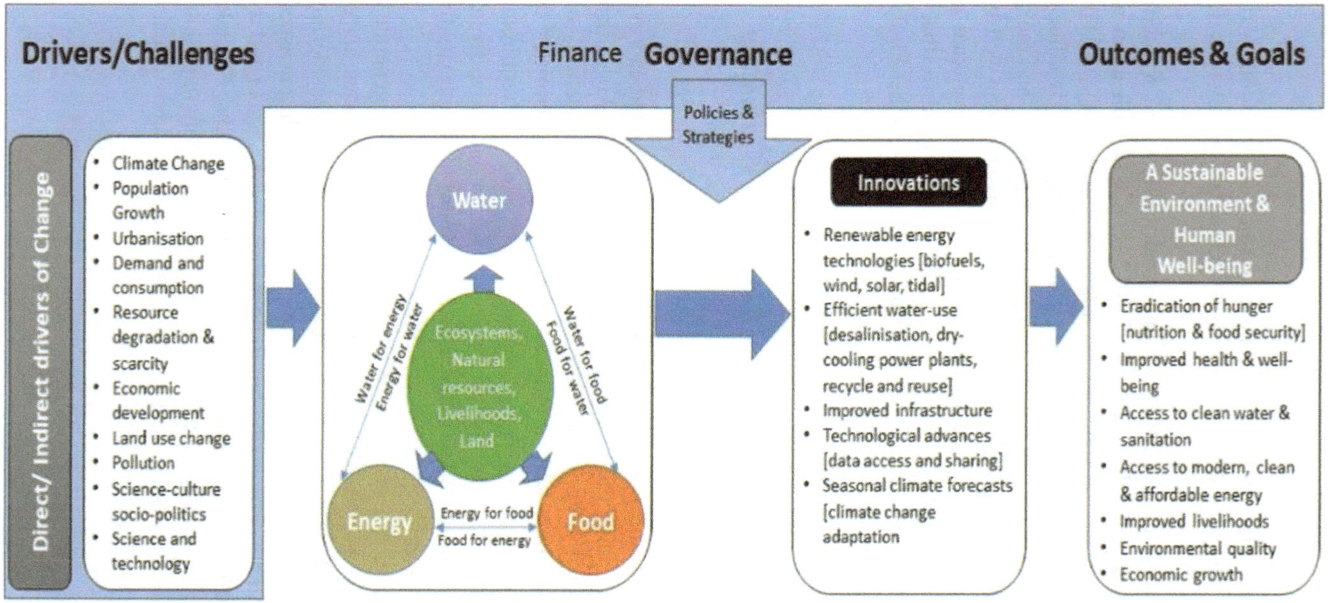

Fig. 8 A Proposed WEF nexus framework for South Africawith particular emphasis on Sustainable Development Goals (SDGs) 2, 6, and 7 (modified based on Smajgl et al. (2016, Ringler et al. (2013), Karabulut et al. (2018) and Hoff (2011))

3.3.1 Tools

A number of indices, metrics, and models can be used to evaluate the WEF nexus. The MuSIASEM (Multi-Scale Integrated Assessment of Society and Ecosystem Metabolism) tool was developed to simulate the WEF nexus by means of depicting the metabolic patterns of WEF in relation to ecological and socio-economic variables. It was originally developed for the energy sector, but can be adapted to the WEF nexus by including water and food in its methodology (FAO 2013). MuSIASEM allows the simultaneous use of demographic, ecological, and social variables even if they were defined on different levels and scales, thus enabling effective analysis of the interdependencies among water, energy, and food at a national or sub-national level. Furthermore, MuSIASEM provides feasibility, viability, and desirability checks of each proposed scenario. It was used to generate an integrated assessment of the contribution and convenience of Concentrated Solar Power (CSP) and woody biomass as alternative sources for electricity production in South Africa (LIPHE4, 2013). This case study uses quantitative data from published research to evaluate electricity consumption in South Africa , as well as production factors of CSP and woody biomass-based electricity. The maximum short-term potential of CSP and woody biomass were calculated to be 3,000 GWh and 5,900 GWh, respectively; their respective input requirements as well as the trade-offs between production factors, such as water and land requirements, are described in Table 2.

The WEF Nexus Tool 2.0 was developed by QEERI to evaluate the water, energy, land, financial, and carbon production requirements for food supply in Qatar (Wicaksono et al. 2017). It is a scenario-based tool created primarily to quantify the resources required for food supply at a national scale; it allows the user to create various scenarios by setting parameters for inputs of water, energy, and foods. In Qatar, it assessed water, energy, and land requirements, carbon footprint, financial cost, energy consumed through import, and carbon emission through import (Daher and Mohtar 2015). Multiple scenarios were generated, with the most realistic scenarios depending on scientific and policy inputs. Assumptions and limitations include the following:

- Food products assessed are only agricultural crops and exclude meat, dairy, etc.
- There is no calculation of water and soil quality.
- Relationships among system components are based on empirically based data.
- The tool assumes linear relationships among systems.
- Projections of prices, population growth, and resource demands are not included.

Table 2 Requirements of production factors for scenarios concentrated solar power (CSP) and woody biomass-based electricity (LIPHE4, 2013)

Scenario	Labor (Mhr/y)	Water (hm^3/y)	Land (ha)
CSP	2.7	9.1	5 100
Woody biomass	120	NA	9 241 000

- It does not capture the financial costs associated with the use of different water and energy sources.

The WEF Nexus Tool 2.0 represents an opportunity to develop a tool for South Africa specifically. It may be further improved by including prediction analyses of population growth, resource demand increase, and financial considerations.

The water and energy nexus is applied to a Sankey diagram (Fig. 9) that quantifies flows in each stage of the water and energy supply chains (Hu et al. 2013). The diagram depicts the water-energy nexus at a household scale in Australia (Kenway et al. 2013), a regional scale in China (Hu et al. 2013), and a national scale in the USA (USDoE, 2014). More recently, it was used to generate the relationships among water, energy, and food for the United Kingdom at a capital, government, and household levels, using multiregional input-output (MRIO) databases, as can be seen in Fig. 9. If applied to South Africa, it would provide a graphical representation of the complexity of the interlinkages among water, energy, and food. The Foreseer Tool (https://www.foreseer.group.cam.ac.uk) can be used to create Sankey diagrams.

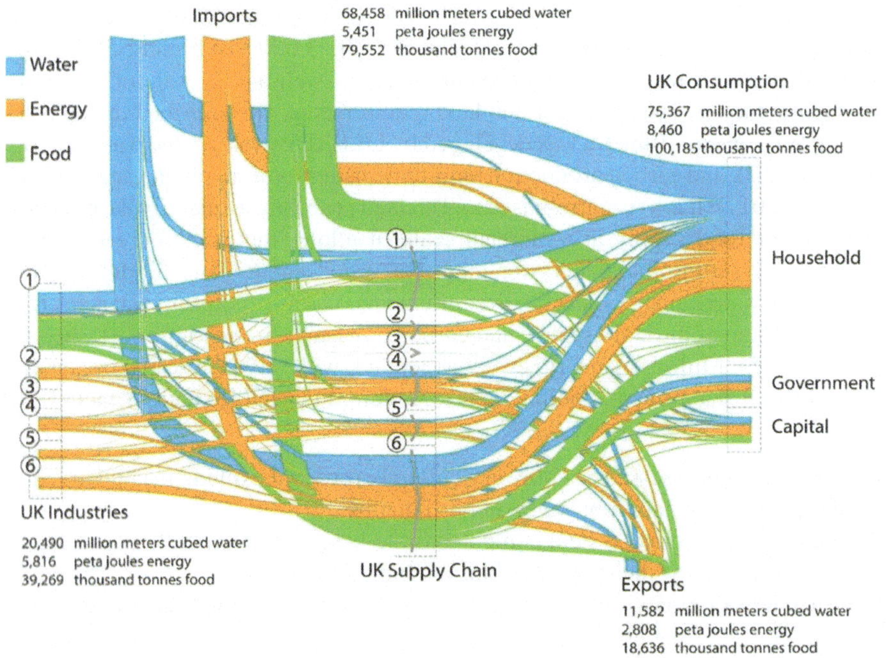

Fig. 9 Sankey diagram showing water, energy, and food flows, from industry to final consumption for the UK in 2013, where 1 = agriculture and food processing, 2 = power generation and distribution, 3 = primary material industries, 4 = manufactured goods and recycling, 5 = transport, 6 = other services (Owen et al. 2018)

3.3.2 Indices

The Nexus City Index (NXI), developed by the United Nations, considers food, energy, and water resources and includes an equity index. The UN-Habitat approach forms the basis for NXI and offers indices to monitor the development of productivity, infrastructure development, quality of life, equity, and environmental sustainability. The approach is based on urban resilience, which is targeted in Goal 11 of the SDGs (Schlör et al. 2018). Along with the NXI_{region}, the World City Prosperity Index, the Regional City Prosperity Index, and the Regional City Index (NXI_{city}) were developed to assess the resilience of various regions and cities in the world (Schlör et al. 2018). They provide data to support decision-making to identify, monitor, plan, and manage urbanization in cities and regions, with special attention to developments within the WEF sectors (Schlör et al. 2018). These indices are useful for South Africa, but do not consider policy implications, nor do they include scenario-based predictions of population growth, climate change, or economic growth.

3.3.3 Models

The South Africa TIMES model (SATIM-W) tool considers trade-offs between water and energy systems for cost-effective sustainable planning (Ahjum et al. 2018). As the name suggests, it is specific to South Africa and incorporates large amounts of quantitative data relating to the country's water supply, usage, and costs (including water quality and treatment). Scenarios include climate change impacts, economic growth, local environmental best practice, policy compliance, and low carbon technologies (Ahjum et al. 2018). To address the hydrological gaps of the model, the World Bank, with the SADC secretariat, has launched a project to build sustainable groundwater management in the region (The World Bank, 2016). This model may be altered to include the food sector of the WEF nexus as well as social aspects; it has great potential to effectively evaluate the WEF nexus in South Africa.

A new model for integrated assessment of global change (ANEMI) was established as an integrated assessment model to simulate all relevant variables, including climate, the carbon cycle, the economy, population, land use, the hydrological cycle, and water demand and quality (Davies and Simonovic, 2011). It emphasizes the interconnections and feedback of each element, and is a significant improvement over previous models because it includes food production and considers factors that optimize the energy-economy factor (Akhtar et al. 2013).

Ozturk (2017) formulated simple non-linear regression equations using a set of explanatory variables of agricultural sustainability to create an understanding of the water-energy-food nexus within six sub-Saharan African countries. The study utilized three panel regressions, including the least squares regression ("common constant method"); fixed effects ("least squares dummy variables"); and the random effects model ("Dynamic Model").

The Climate Land-use Energy and Water Strategies (CLEWS) modeling framework integrates existing models and systems, such as Water Evaluation and Planning (WEAP), Long-range Energy Alternatives Planning System (LEAP), and Agro-ecological Zoning (AEZ), by simulating and comparing data among them in multiple iterations to find a convergent solution (Keairns et al. 2016). It analyzes interlinkages among resource sectors to determine the effect that one sector might have on the others, and identifies counterintuitive findings in these integrated systems. It is a free online tool that creates scenarios based on (UN DESA 2013):

- Global estimates of CO_2 emissions, water use, and investment in energy and material production.
- Estimates of CO_2 emissions and water use by energy source.
- Estimates of mix of energy supply.

This model has been applied to a case study in Mauritius, focusing on two policy goals: renewable energy production and renewable fuel standards that mandate the blending of ethanol into gasoline. Similarly, case studies for Kenya and Bolivia investigated SDG 7 (energy access to all). This model may be adaptable to South Africa ; however, it is mainly limited to the energy sector.

Simpson and Berchner (2017) propose to develop a composite indicator to report on the WEF nexus. Their study highlights that indices should be based on quantitative data and represented by a single numeric value, for consistency among evaluations of different cities and countries. Mitigation scenarios could be tested to ensure the establishment of achievable and measurable goals to improve the WEF nexus index over time.

For all these and any tools, models, and indices, data storage and accessibility will play a significant role in understanding and analyzing the WEF nexus. It is also important to consider temporal and spatial scale differences of different elements, suggesting the need to integrate multiple available models and tools, as well as input from stakeholders and policymakers. Table 3, adapted from Martinez-Hernandez et al. (2017), summarizes key models and indices that may be used for evaluation.

4 Conclusions

A WEF nexus framework for South Africa is urgently required so that it can be applied as a tool for resource management to optimize benefits and minimize conflicts. Water, land, and energy resources in South Africa are highly emotional issues and therefore must be approached and managed judiciously; the WEF nexus offers such an opportunity. An understanding of the WEF nexus in South Africa is of great value when developing a framework that is specific for the country.

The WEF nexus framework was developed with emphasis on Sustainable Development Goals 2 (zero hunger), 6 (clean water and sanitation), and 7 (affordable and clean energy). Its framework considers the importance of livelihoods and human well-being and eminent threats to sustainable development, especially in South

Table 3 Potential models and indices that could be used to evaluate the water-energy-food nexus in South Africa, adapted from Martinez-Harnandez et al. (2017)

Tool	Modeling framework	Scale	System breadth	Analytical capability	Flexibility	Applicability to WEF nexus in South Africa
GLOBIOM	Dynamic multiregional partial equilibrium model	Global	WEF nexus and other interacting systems such as ecosystems	Geographically explicit and long-term management of global land uses	Focused on land uses	No, only applicable at a global scale
WEF Nexus Tool 2.0	Input-output	National	WEF nexus components	Scenario-based for given food self-sufficiency level calculates nexus resource flows and interactions, and greenhouse gas (GHG) emissions	Focused on food as entry point and Qatar country	Yes
MuSIASEM	Input-output, nested hierarchical view of the economy	Aggregated to national or sub-national level	WEF nexus components, land, economy, human capital, and ecosystems	Accounting of flows and funds and their ratios as indicators. GHG emissions and land use	Adaptable to various contexts	Yes, it has already been applied to South Africa
CLEWS	Integrates detailed models from different tools (including WEAP, LEAP, and AEZ)	National	Climate, land, energy, and water	Depend on the tools used for the CLEW assessment	Depend on the tools used for the CLEW assessment	Yes, if the model can be changed to evaluate the inter-sectoral influences of the WEF nexus components

(continued)

Table 3 (continued)

Tool	Modeling framework	Scale	System breadth	Analytical capability	Flexibility	Applicability to WEF nexus in South Africa
Quantitative assessment framework	Input-output based on Lontief matrices	National	WEF nexus components	Scenario-based, accounting of nexus resource consumption, and interdependency indicators	Fixed defined technologies and interactions	Yes, could be extended to analyze the influence of socio-economic factors
DEA	Data envelopment analysis model	Local (city level)	WEF nexus components	Input-output efficiency		No, cannot be used for national evaluation of the WEF nexus
PRIMA	Integrates regional climate, hydrology, agriculture and land use, socioeconomics, and energy systems sector models	Regional	WEF components, economy, land use	Climate change related analyses and costs, land use, greenhouse gas emissions	Flexible, portable, and modular	No, only relevant for regional decision-making
ANEMI	Integrated assessment model	All scales	Climate, carbon cycle economy, population, land use, hydrological cycle, and water demand and quality	Reveals the interconnections and feedback of each element	System dynamic simulation	Yes
Sankey diagram	Graphically represents the complex conversion pathways, flows, and interdependencies between variables	All scales	WEF nexus components	Based on the data input	Adaptable to various contexts	Yes

(continued)

Table 3 (continued)

Tool	Modeling framework	Scale	System breadth	Analytical capability	Flexibility	Applicability to WEF nexus in South Africa
Nexus City Index	Measures the prosperity and sustainability of the FEW nexus for 69 cities	All scales	WEF nexus component, prosperity	A top-down urban WEF nexus approach which aggregates the WEF sectors to a single indicator	Flexible, and includes likewise indices World City Prosperity Index, the Regional City Prosperity Index, and a regional city index	Yes
Message	Modeling potential future energy scenarios	Global and Regional	Energy and greenhouse gas emissions	Dynamic linear programming model and can be linked with MAGICC (a separate program for predicting GHG-induced climate change) and GLOBIOM		No, does not consider all WEF nexus components

Africa . Current literature shows that policies, strategies, and plans have not fully embraced the applicability of the WEF nexus to sustainable resource management. More research is needed and must include participation of policymakers, researchers, and other stakeholders to provide a comprehensive perspective on the desirability of and approach to implementing the WEF nexus in South Africa.

The most relevant and applicable indices , models, and metrics relating to the WEF nexus in South Africawere reviewed and summarized. Most would require modifications to be applicable to South Africa. Data availability and quality will be a factor in the reliability of any model, emphasizing the necessity of a central database where data can be compared and validated. Temporal and spatial scale differences among data inputs also require further consideration and may be best resolved by integrating multiple models and tools.

References

Ahjum F, Merven B, Cullis J, Goldstein G, DeLaquil P, Stone A (2018) Development of a national water-energy system model with emphasis on the power sector for south africa. Environ Progr Sustain Energy 37:132–147

Akhtar MK, Wibe J, Simonovic SP, MacGee J (2013) Integrated assessment model of society-biosphere-climateeconomy-energy system. J Environ Model Softw 49:1–21

Albrecht TR, Crootof A, Scott CA (2018) The water-energy-food Nexus: A systematic review of methods for nexus assessment. Environ Res Lett 13:043002

Carter S, Gulati M (2014) Climate change, the food energy water nexus and food security in South Africa. Understanding the Food Energy Water Nexus. WWF-SA, South Africa. http://awsassets.wwf.org.za/downloads/1__a16231_wwf_climate_change_few_and_food_security_in_sa__online.pdf

Conway D, Van Garderen EA, Deryng D, Dorling S, Krueger T, Landman W, Lankford B, Lebek K, Osborn T, Ringler C (2015) Climate and southern Africa's water–energy–food nexus. Nat Clim Change 5:837

Daher BT, Mohtar RH (2015) Water–energy–food (WEF) Nexus Tool 2.0: guiding integrative resource planning and decision-making. Water Int 40:748–771

Davies EGR, Simonovic SP (2011) Global water resources modeling with an integrated model of the social–economic–environmental system. Adv Water Resour 34:684–700

Endo A, Tsurita I, Burnett K, Orencio PM (2017) A review of current state of research on the water, energy and food nexus. J Hydrol Region Stud 11:20–30

FAO (2013) An innovative accounting framework for the food-energy-water nexus. Application of the MuSIASEM approach to three case studies

FAO (2017) Food security indicators. Vol. 2018

Hoff H (2011) Understanding the nexus: Background paper for the Bonn Nexus Conference. SEI

Hu G, Ou X, Zhang Q, Karplus VJ (2013) Analysis on energy–water nexus by Sankey diagram: the case of Beijing. Desalin Water Treat 51:19–21

Karabulut AA, Crenna E, Sala S, Udias A (2018) A proposal for integration of the ecosystem-water-food-land-energy (EWFLE) nexus concept into life cycle assessment: a synthesis matrix system for food security. J Clean Prod 172:3874–3889

Keairns DL, Darton RC, Irabien A (2016) The Energy-Water-Food Nexus. Annu Rev Chem Biomol Eng 7:239–262

Kenway S, Scheidegger R, Larsen TA, Lant P, Bader HP (2013) Water-related energy in households: a model design to understand the current state and stimulate possible measures. J Energy Build 58:378–389

Keskinen M, Guillaume JH, Kattelus M, Porkka M, Räsänen TA, Varis O (2016) The water-energy-food nexus and the transboundary context: insights from large Asian rivers. Water 8:193

LIPHE4 (2013) The Nexus between energy, food, land use, and water: application of a multi-scale integrated approach South Africa case study, vol 2018

Mabhaudhi T, Mpandeli S, Madhlopa A, Modi AT, Backeberg G, Nhamo L (2016). Southern Africa's water-energy Nexus towards regional integration and development. Water 8

Martinez-Hernandez E, Leach M, Yang A (2017) Understanding water-energy-food and ecosystem interactions using the nexus simulation tool NexSym. Appl Energy 206:1009–1021

Nhamo L, Bekithemba B, Nhemachena C, Mabhaudhi T, Mpandeli S, Matchaya G (2018) The water-energy-food nexus: climate risks and opportunities in Southern Africa. Water 10

Ololade OO, Esterhuyse S, Levine AD (2017). The water-energy-food Nexus from a South African perspective. In: Water-energy-food Nexus. John Wiley & Sons, Inc, pp 127–140

Owen A, Scott K, Barrett J (2018) Identifying critical supply chains and final products: An input-outputapproach to exploring the energy-water-food nexus. Appl Energy 210:632–642

Ozturk I (2017) The dynamic relationship between agricultural sustainability and food-energy-water poverty in a panel of selected Sub-Saharan African Countries. Energy Policy 107:289–299

REDIS (2018) Location and contracted capacities of renewable energy projects in South Africa, vol 2018. Deparment of Energy, South Africa

Ringler C, Bhaduri A, Lawford R (2013) The nexus across water, energy, land and food (WELF): potential for improved resource use efficiency? Curr Opin Environ Sustain 5:617–624

Ross ML (1999) The political economy of the resource curse. World Polit 51:297–322

Schlör H, Venghaus S, Hake J-F (2018) The FEW-Nexus city index measuring urban resilience. Appl Energy 210:382–392

Simpson G, Berchner M (2017) Measuring integration towards a water-energy-food nexus index. Water Wheel 16:22–23

Smajgl A, Ward J, Pluschke L (2016) The water–food–energy Nexus-Realising a new paradigm. J Hydrol 533:533–540

StatsSA (2018) "Electricity generated and available for distribution (Preliminary). South Africa, Pretoria

The World Bank (2016) Southern Africa energy-water Nexus

The World Bank (2018) Indicators, vol. 2018

UN DESA (2013). Global climate, land, energy & water strategies, vol. 2018

UNEP (2016) Global material flows and resource productivity. An Assessment Study of the UNEP International Resource Panel, Paris

USDoE (2014) The water-energy Nexus: challenges and opportunities. Washington

Von Bormann T, Gulati M (2014) The food energy water Nexus: understanding South Africa's most urgent sustainability challenge. South Africa

Wicaksono A, Jeong G, Kang D (2017) Water, energy, and food nexus: review of global implementation and simulation model development. Water Policy 19:440–462

Zhang C, Che X, Li Y, Ding W, Fu G (2018) Water-energy-food nexus: Concepts, questions and methodologies. J Clean Prod 195:625–639

Chapter 19
Mainstreaming Climate Change into Transboundary River Basins: A SADC Regional Case Study

Motlole Christopher Moseki

Abstract Member States that are signatory to the Revised Protocol on Shared Water Courses in the Southern African Development Community (SADC) have to collaborate on development, management, use, and conservation of water resources of the shared river system and to enhance a common and better understanding of the system. Additionally, international water and sanitation agreements must be robust and resilient with respect to climate change. This chapter seeks to investigate climate change considerations in regional water policy formulation and implementation, and to provide an account of the extent to which climate change has been considered in transboundary water agreements, SADC regional protocols and frameworks. Recommendations for assessment and subsequent mainstreaming of climate change factors into regional water policy frameworks, plans, and practice are offered. Case studies of the Orange-Senqu, Limpopo, and Incomati (Inkomati) River Basins are used to test the extent to which climate change is considered in transboundary water resources. The Orange-Senqu River is a water-scarce basin with low rainfall-to-runoff conversion, high aridity, and high climatic variability. Climate models project an increase in mean annual temperature of over 3 °C in the intermediate future (2046–2065). The Limpopo Basin is semi-arid, highly variable in climate, and prone to drought, which is projected to increase in frequency and severity, while rainfall projections are uncertain. The Incomati, a semi-arid basin with an average annual rainfall of 767 mm and temperature of 17 °C, is projected to become drier in the future. This chapter is based on climate change research at the Water Research Commission as well as literature search. The objective of this work is to encourage mainstreaming of climate change forecasting into water management and planning at the transboundary level. There is a great need to strengthen institutional capacity for sustainable management of water under a changing climate and to improve coverage of climate change scenario projections in southern Africa.

Keywords Transboundary · Mainstreaming · Climate change · Water

M. C. Moseki (✉)
Department of Water and Sanitation, Pretoria, South Africa
e-mail: mosekic@dws.gov.za

S. Diop et al. (eds.), *Climate Change and Water Resources in Africa*,
https://doi.org/10.1007/978-3-030-61225-2_19

1 Introduction

There are 16 member states of the Southern African Development Community (SADC) that are signatory to the Revised Protocol on Shared Water Courses (Fig. 1). The recently admitted country is the Union of the Comoros, which joined at the 37th SADC Summit of Heads of State and Government held in Pretoria, South Africa in August 2017. At the Summit, each member state has to ensure that she is meeting her obligations on transboundary relations in the interest of regional economic integration, peace, and security. The Water Master Plans for each country encourage all riparian states to integrate management of transboundary water resources. Member states are tasked with collaboration on development, use, and conservation of water resources of the shared river systems and to enhance common and better understanding of the Revised Protocol on Shared Water Course Systems, which is essential given the ubiquitous nature of water and the fact that it does not recognize political boundaries. For instance, about two-thirds of South Africa's streamflows in rivers are shared with neighboring countries, underscoring the essential nature of transboundary cooperation in adaptation to climate change impacts. International water and sanitation agreements must be robust and resilient with respect to climate change. Moreover, cooperation among riparian states often strengthens their adaptive capacities (Keller 2012). This chapter investigates the degree to which climate change

Fig. 1 SADC Member States (modified to include the Comoros)

strategies have been integrated into water and sanitation management strategies at the transboundary level. Based on this examination, and using case studies from the three transboundary basins, recommendations are made to address vulnerability in regional water policy frameworks.

The Orange-Senqu River Commission (ORASECOM) agreement of 2000 was the first river basin institution to be established in the SADC region among the four countries (South Africa, Botswana, Lesotho, and Namibia). The Limpopo Watercourse Commission (LIMCOM) evolved from the SADC structure and mandate, but the four economic powers (South Africa, Botswana, Zimbabwe, and Mozambique) have yet to agree on a management plan for the Limpopo River Basin. The Incomati-Maputo Agreement, ensuring cooperation on the protection and sustainable utilization of water resources, was reached among South Africa, Mozambique, and Swaziland in 2002. That was a milestone because it resolved a previously intractable conflict over water allocation of the Incomati Basin through inclusion of Maputo in the negotiations. However, neither the agreement (adopted in May 2002, signed in August 2009 and ratified in February 2010) nor a technical report by the Tripartite Permanent Technical Committee (TPTC) on systems operating rules (drafted in 2010)[1] even mentions climate change.

2 Climate Outlook for Southern Africa

Southern Africa is characterized by a highly variable climate. Francois Engelbrecht (2018, personal communication) of the CSIR contends that South Africa's present-day climate is often under the influence of high-pressure systems of the subtropical high-pressure belt, which causes air to sink over southern Africa, thereby suppressing thundercloud formations and therefore rainfall. Hence, a large part of South Africa's climate is semi-arid. Figure 2 shows key features of the climate system of southern Africa.

Chapman (2011) observes that climate change will affect different parts of southern Africa differently. Indeed, climate change model scenarios project that some areas will become dry, others wet, and others remain uncertain, particularly in the more distant future. Davis-Reddy and Vincent (2017) assert that the global mean annual temperatures have increased by 0.85 °C since 1880 and are projected to increase by 0.3 to 2.5 °C by 2050, relative to the 1985–2005 climatological average. As it is generally the case, globally, changes in rainfall are harder to detect than changes in temperature; there is no clear trend in rainfall projections, but higher variability in rainfall (spatially) is expected. Notwithstanding uncertainty in rainfall projections, a decrease in rainfall is projected over central southern Africa (e.g.,

[1] Additional information in this regard is accessible from the Agreement and the TPTC report by Hallowes et al. (2010).

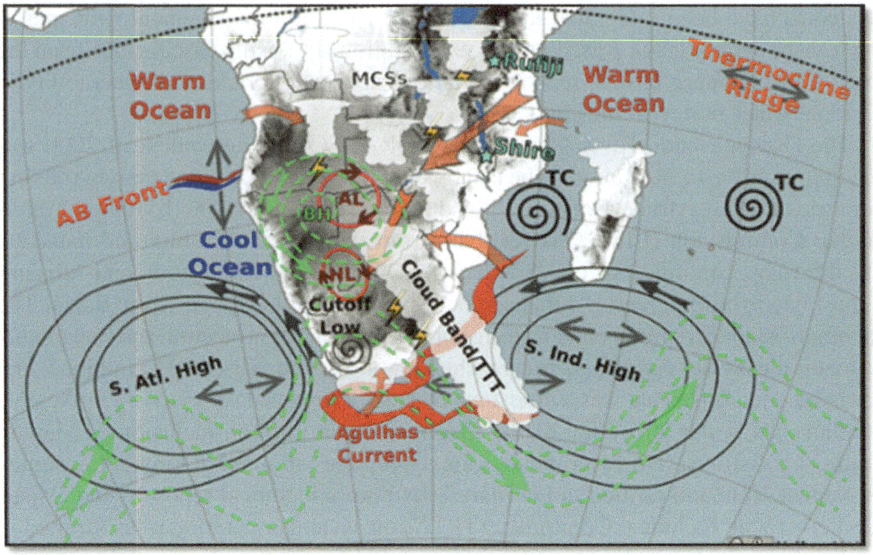

Fig. 2 Key features of southern Africa's climate system (Jack et al. 2016)

northern Botswana, Namibia, southern Zambia, and Zimbabwe) and over the southwestern Cape of South Africa, while an increase in rainfall is projected over northern Mozambique and Tanzania (Davis-Reddy and Vincent 2017).

3 The Orange-Senqu River Basin

The Orange-Senqu River Basin (Fig. 3) covers central to southwestern Namibia, southern Botswana, central South Africa, and Lesotho. It is notable that part of the Orange River Basin in Lesotho is known as the Senqu River.

The basin originates in the Lesotho Highlands at an altitude of more than 2,750 m above sea level from where it flows 2,500 km westward to its mouth at Alexander Bay on the Atlantic West Coast (ORASECOM 2011). It is the third largest river in southern Africa, after the Zambezi and the Congo. The region as a whole is committed to the overall objectives of the "Revised SADC Protocol on Shared Watercourses" to foster close cooperation among members for judicious, sustainable and coordinated management, protection and utilization of shared watercourses, and to advance the SADC agenda of regional integration and poverty eradication.

In terms of future climate outlook, the anticipated increase in temperature and resulting increase in evaporation rates in the basin is likely to impact the magnitude, frequency, and intensity of rainfall events as well as seasonal and geographical distributions of rainfall and its internal variability (ORASECOM 2008). These impacts are likely to influence the magnitude and variability of streamflow in the

Fig. 3 The map of the Orange-Senqu River Basin (Lange et al. 2007)

river basin, food and water supply security, and sustainability of the ecosystem. In response, the SADC Climate Change Adaptation Strategy aims to promote application of the Integrated Water Resource Management (IWRM) plan to reduce climate vulnerability and promote adaptation to climate variability and change. The IWRM uses the "SADC Water Adaptation Cube" mechanism which calls for adaptation measures at various levels (i.e., local, river basin, and regional), stages (i.e., preparatory, response, and recovery), and levels of authority (such as governance, development, and management). The cube was designed to raise awareness, build resilience, reduce climate risks, and facilitate coordination among stakeholders.

To ensure sustainability of the IWRM, adaptation to climate change is essential. Approaches include adjusting economic activities, national and regional reforms (IWR 2014) design and development of infrastructure, and strategic intervention to address resilience in agricultural production and enterprise development which are important for rural as well as national and basin-wide development.

3.1 Background and General Climatic Conditions in Southern Africa with Implications for the Orange-Senqu Basin

Even prior to consideration of potential global climate change impact, the natural hydrological system of the southern Africa region is highly variable, with a high-risk hydro-climatic environment. The rainfall-to-runoff conversion of less than 5% is very low, due to high aridity and highly seasonal rainfall (ORASECOM 2008). Complexities and uncertainties in water resource management are a consequence of inter-annual rainfall variability (the coefficient of variation in rainfall for southern Africa is more than 40%). According to Schulze (2011), the intra-annual variability of hydrological response is even higher. The Orange-Senqu River Basin is the largest basin south of the Zambezi, spanning an area of 896,368 km^2 and covering 43% of South Africa, the whole of Lesotho and large portions of Namibia and Botswana. Figure 4 depicts major dams (including the Vanderkloof, Gariep, Vaal, Hardop, and Katse) in the basin.

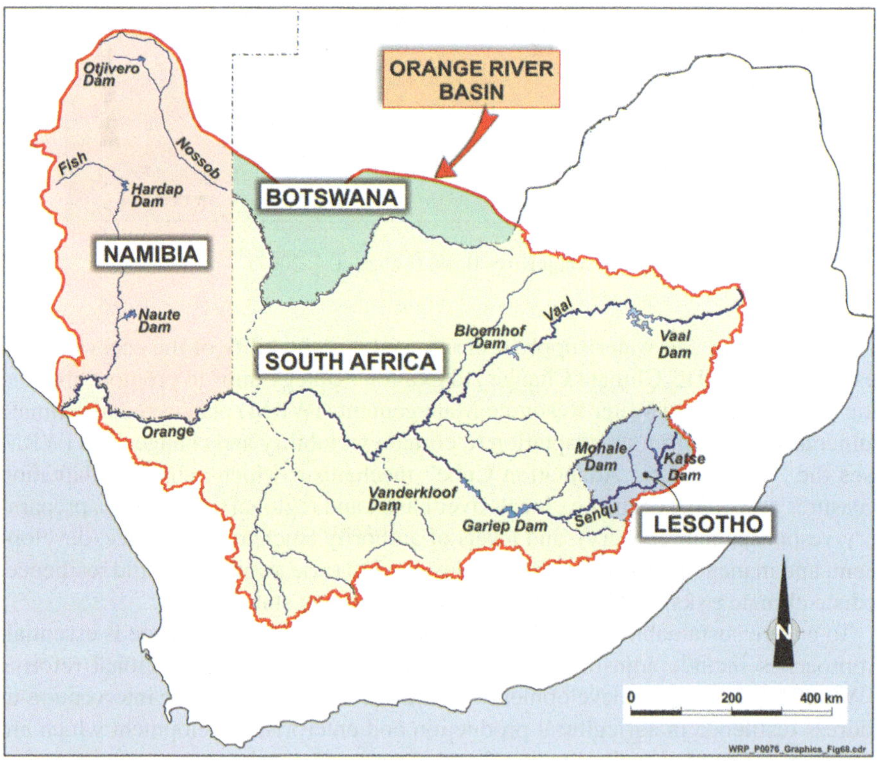

Fig. 4 The Orange-Senqu River Basin in detail showing major dams (WRP Consulting Engineers (Pty) Ltd, 2020)

The Orange River Catchment within the Orange-Senqu River Basin is generally arid, with very low annual precipitation and high evaporation rate. Seasonal rainfall is reported (Knoesen 2012) to be concentrated over only a few months, with approximately 85% of the rainfall in Lesotho occurring during the period from October to April; as much as 15% of Mean Annual Precipitation (MAP) may occur in a 24-hour period in some areas.

The complex geology of the basin consists mostly of sedimentary rock of the Karoo Supergroup with some dolerite intrusions, dolomite, and Kalahari sands. Groundwater is the main source of supply, particularly in the western parts of the basin. A high erosivity index and poor land management practices (such as overgrazing, cultivation of steep slopes, uncontrolled burning, and alluvial diamond mining) have led to high erosion rates and sediment loading (Knoesen 2012). Agriculture is typically a major water consumer at more than 60% of total use. Water is often transferred from plentiful to water-deficient areas. High variability in climate, low rainfall, a high evaporation rate, and non-climatic stress factors such as deterioration in water quality, increasing competition for water use, and challenges associated with access all render the basin vulnerable.

3.2 *Modeling Climate Change Impacts*

Most of the projected climate change-related work in the Orange-Senqu River Basin is based on research by Knoesen et al. 2009, Knoesen 2012) and Schulze (2011).

The agro-hydrological ACRU model (a daily time step physical conceptual model) was used (Fig. 5) at the University of Kwa-Zulu Natal to predict climate change impacts on hydrology and water resources.

Earlier, (i.e., prior to delineation of catchments into the current finer scale) the South African Department of Water and Sanitation had delineated South Africa, including both Lesotho and Swaziland into 1946 interlinked and hydrologically cascading Quaternary Catchments (Schulze 2011), that was deemed too coarse to represent climatologically homogeneous units and thus that could not appropriately depict hydrological responses to climate impacts (Fig. 6).

The southern Africa area modeled using ACRU also covers much of the Orange-Senqu River Basin. The entire area is now sub-delineated into quinaries (i.e., at higher resolution) that resulted in 5 838 hydrologically interlinked and cascading quinaries covering the RSA, Lesotho, and Swaziland.

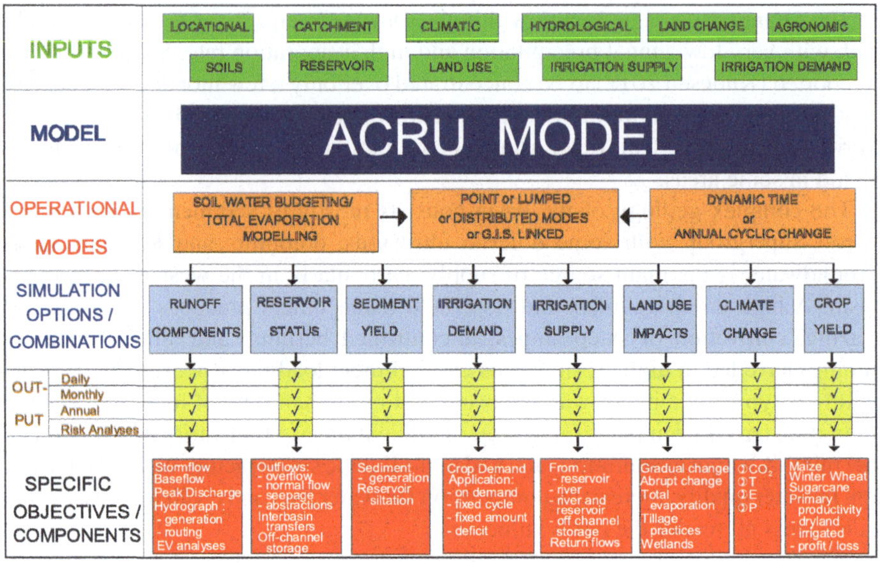

Fig. 5 The ACRU agro-hydrological model (concepts and linkages) (Schulze 2011; Koesen 2012)

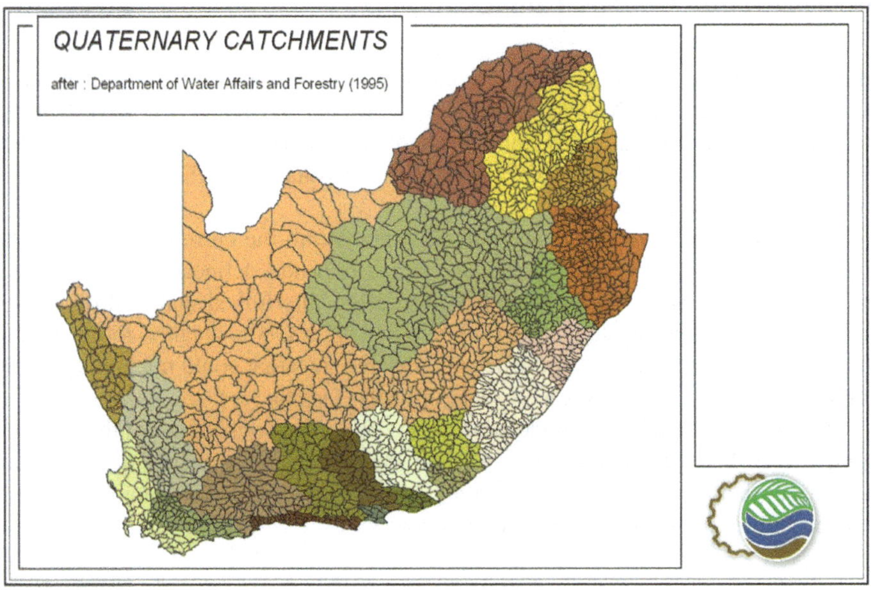

Fig. 6 Primary and quaternary catchments covering South Africa, Lesotho, and Swaziland (Schulze 2011)

3.3 *Assessing Climate Change Impacts on Key Hydrological Drivers and Responses in the Orange-Senqu River Basin*

The Orange-Senqu is a water-scarce high-risk hydro-climatic basin as a consequence of low rainfall-to-runoff conversion, high aridity, and high climatic variability. It is thus at higher risk to climate change impacts than many other regions in the world (Knoesen 2012). Each of the 1,443 hydrologically interlinked quinary catchments that make up the Orange River Catchment was simulated using data from the Quinary Catchment Database.

Streamflows for the intermediate (2046–2065) and distant (2081–2100) future climate were also simulated. However, that was only for catchments that exist in South Africa and Lesotho, while Botswana and Namibia were not included due to the lack of readily available historical data and downscaled values for these areas of the basin.

3.4 *Projected Changes Due to Climate Change in the Orange-Senqu River Basin*

3.4.1 Rainfall

Rainfall is categorized as short or long duration. Short duration rainfall occurs over 24 h or less, and depending on the intensity, it may contribute to local flooding. Long duration rainfall in contrast occurs over 1 to 7 days and, if extreme, may result in larger scale regional damage. Mean annual rainfall is projected to generally increase across the basin. However, Schulze (2011) projects precipitation in southern Africa, including the Orange-Senqu Basin area, to be highly variable, particularly in the distant future (Fig. 7).

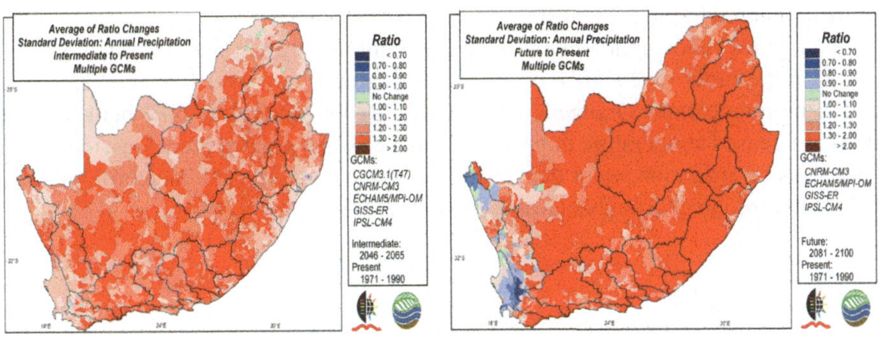

Fig. 7 Increase in variability of annual precipitation under projected future climates, derived from multiple GCMs, with variability increasing more into distant future (Schulze 2011)

In contrast, Fig. 8 shows that overall across South Africa an increase of up to 10% in short duration rainfalls may be expected in the intermediate future, with the exception of patches south of 32 °S and north of 27 °S where models show no discernible change from the present.

Spatially within the basin, relative increases in mean annual rainfall are greatest in the eastern runoff-producing regions of the basin and tend to decrease westwards. Studies also show (Knoesen et al. 2009) that rainfall across the basin is likely to become more variable in both the intermediate and more distant futures, which is of particular concern to water managers because such variability may result in drought and flood events. The number of days of rainfall above a certain amount is expressed as the ratio of the number of rain days for both intermediate and more distant futures to those experienced presently; the higher this ratio, the more rain days are likely to occur. In the Orange-Senqu River Basin, projections show that the increase in the number of rain days will be greatest in the eastern regions.

Koetze et al. (2009) also assert that in the intermediate future about 40% of the basin is likely to experience an increase in the amount of rainfall per short duration event, while 38% of the basin is projected to experience a decrease. In the more distant future, however, about 72% of the basin is projected to show an increase in the amount of rainfall per short duration event. Regarding long duration rainfall, extreme events are likely to increase, with about 83% of the Orange-Senqu River Basin projected to experience larger extreme rainfall events with a recurrence interval of 2 years than at present. These events have implications for the safety of water infrastructures.

3.4.2 Temperature

Indications are that mean annual temperatures will increase over the entire Orange-Senqu River Basin. The projected increase in mean annual temperatures in the Orange River Catchment is between 3 and 3.5 °C in the intermediate term; this is expected to grow to more than 7 °C in the more distant future (Knoesen 2012). It is worth noting that for much of southern Africa, high increase in temperature is projected for the future compared to the present (Fig. 9).

The highest ratios indicating an increase from the present mean annual temperatures on the order of 20–40% occur in the medium term over Lesotho (Knoesen 2012), the source of the Orange River and the Caledon, and by 40–60% in the distant future. Increased temperature impacts a number of water-related processes, including soil moisture, irrigation water demands, heat wave episodes, and meteorological as well as hydrological droughts (Knoesen et al. 2009).

3.4.3 Evaporation

The projected increases in temperature will have a profound impact on evaporation, affecting runoff, recharge, and water storage. Figure 10 shows that potential evaporation in future (2016–2045) will increase compared to present (1975–2006).

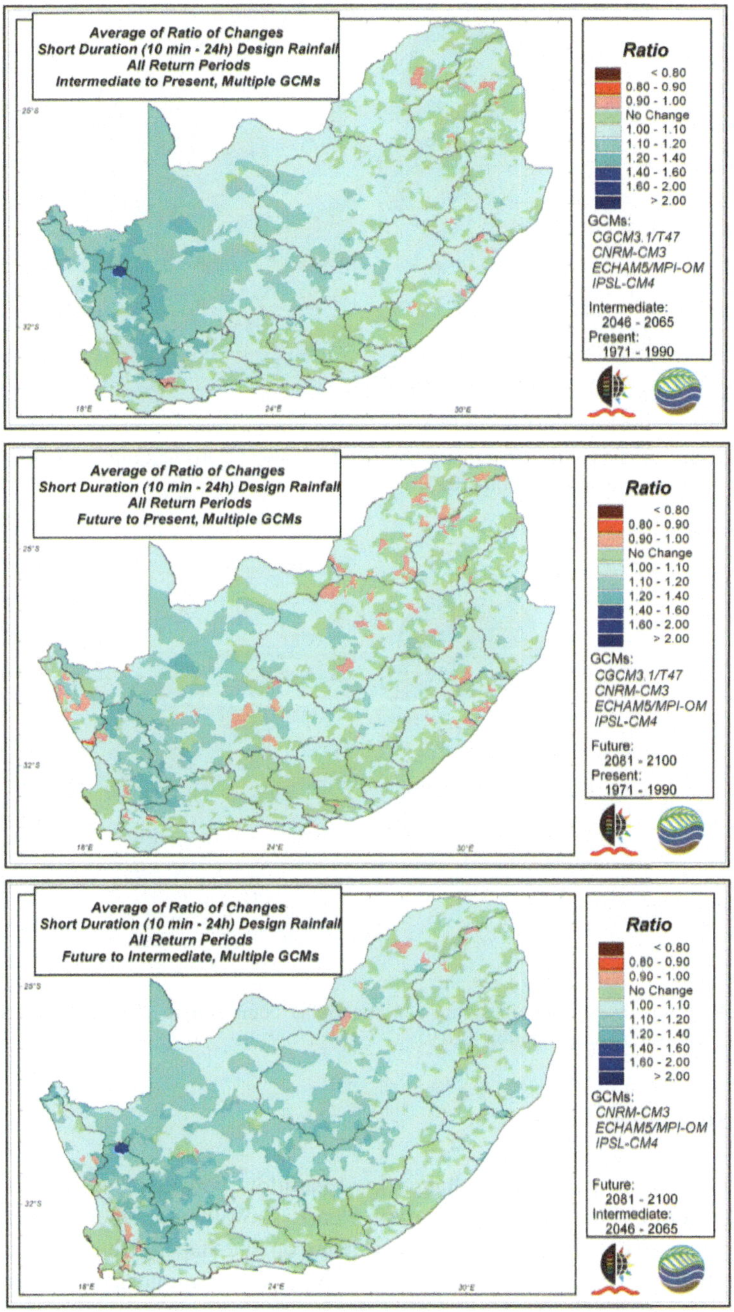

Fig. 8 Averages of ratio changes in short duration design rainfall between intermediate future and present (top), more distant future and present (middle) and distant and intermediate future (bottom) climate scenarios, using outputs from multiple GCMs (Schulze 2011)

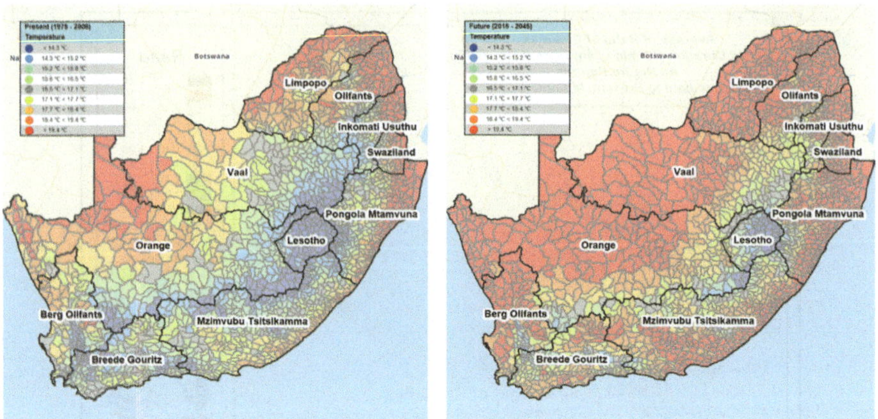

Fig. 9 Average of change in temperature between the present (left) and future (right). Source is Department of Water and Sanitation (2019)

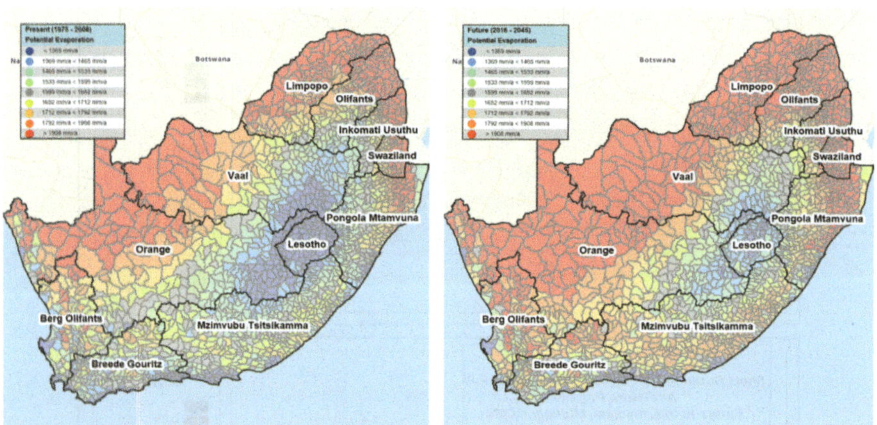

Fig. 10 Potential evaporation in future (on the right) compared to present (left). Source is Department of Water and Sanitation (2019)

3.4.4 Streamflow

Figure 11 shows a percentage change in streamflow between the present (1975–2006) and future (2016–2045). The graph clearly shows that the percentage change in streamflow will generally decrease. It is clear from the map that the largest decrease in streamflow is expected in the Vaal, eastern part of the Orange and the Mzimvubu-Tsitsikamma water management areas of South Africa. Slight increase in streamflow is projected for the southern, western, and eastern parts of the country.

Schulze (2011) projects an increase in design streamflows over large parts of South Africa. For example, for the 1-day, 2-year return period, the increase is 10–50% (i.e.,

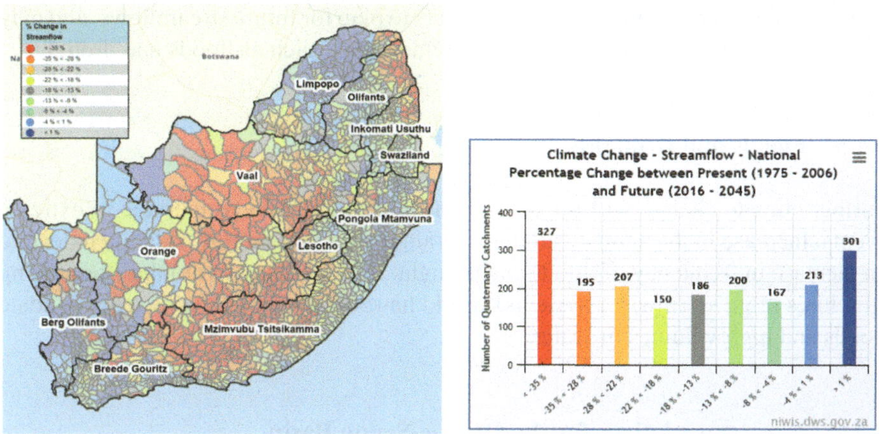

Fig. 11 Percentage change in streamflow between the present (1975–2006) and future (2016–2045). The map (left) depicts spatial percentage change while the graph (left) depicts the same changes in streamflow between present and future by also showing the number of quaternary catchments and corresponding percentage change in streamflow (Department of Water and Sanitation 2019)

ratio changes of 1.1 to 1.5) over the eastern two-thirds of the country, and 50–100% in the western one-third. The exception is the extreme southwest, where a decrease in design streamflows is projected when the daily output from multiple GCMs is used in the ACRU model with techniques to analyze extreme values (Schulze 2011). The following issues were highlighted in the study:

- As the return period increases from two through five and 10 to 20 years, the rate of increase in design streamflows decreases and becomes less pronounced, irrespective of duration (i.e., irrespective of duration of 1 or 7 days).
- The regions within South Africa where the greatest increases in design stormflows are projected are the transitional areas between summer and winter rainfall regions where inconsistent weather patterns are likely to be experienced in future.
- On a local scale, as the return period increases from 2 to 20 years, the area in the southwest expands where decreases in design streamflows are projected.
- For a given return period, the spatial patterns of change into the intermediate future do not vary among the 1-, 2-, 3-, and 7-day durations, i.e., the changes in design streamflow appear to be sensitive to return period, but not to duration.

The mean annual streamflow is likely to increase for both the intermediate future and more distant future since projections indicate higher rainfall events, particularly in the high altitude eastern parts of the basin. Projected increase in mean annual streamflow is likely to be significantly larger than those projected for mean annual precipitation because both increases and decreases in rainfall are amplified. According to Knoesen et al (2009), studies show that variability in streamflow is projected to increase more in the distant future than the intermediate future. Such increases make it difficult for water managers to plan for future streamflows, and may result in an increase in the number of extreme events such as floods and droughts.

3.4.5 Floods and Droughts

Further climate change will impact the basin that is already vulnerable to extreme events. Increase in the annual variation of rainfall and stream flow is likely to lead to an increase in frequency of floods and drought. In such cases, hydraulic engineering structures (e.g., stormwater systems) would have to be redesigned to accommodate floods and high volumes of water.

3.4.6 Recommendations for the Orange-Senqu Basin

Climate change research has been studied relatively more extensive in the Orange-Senqu Basin than in other basins in the region. The next step is to mainstream the climate change and research outcomes into basin planning and management. The Commissions for affected member states and other stakeholders in the basin should take climate change scenario projections into account for sustainable management and planning of water resources in the region.

4 Lesotho Highlands Water Project (LHWP)

The Lesotho Highlands Water Project (LHWP) is one of the world's largest intra-basin water transfer schemes within the Orange-Senqu River Basin; it supports the largest economic hub in Africa (Gauteng). Currently, the transfer of water from the LHWP into Gauteng amounts to about 27% of the total allocation for this area. It was not the original intention to include the LHWP in these case studies, but its uniqueness within the Orange-Senqu Basin justifies discussion. This water scheme is both important and at risk of depletion due to overutilization and climate change impacts. Two well-established tools that inform good and sustainable water management in the basin and within the region are the Revised SADC Protocol on Shared Watercourses and the Climate Change Adaptation Strategy in SADC—A Strategy for the Water Sector.

Because the Lesotho Highlands Water Project (LHWP) is an intra-basin water transfer scheme within the Orange-Senqu River Basin, it falls under the authority of the Orange-Senqu River Basin. However, there are issues that are specific to the LHWP, which is located upstream in the mountain catchments in Lesotho, and not necessarily important at a basin scale (Fig. 12).

The Lesotho Highlands represent an exception to the generally semi-arid region within the Orange-Senqu River Basin (ORASECOM 2008). Its climate is temperate

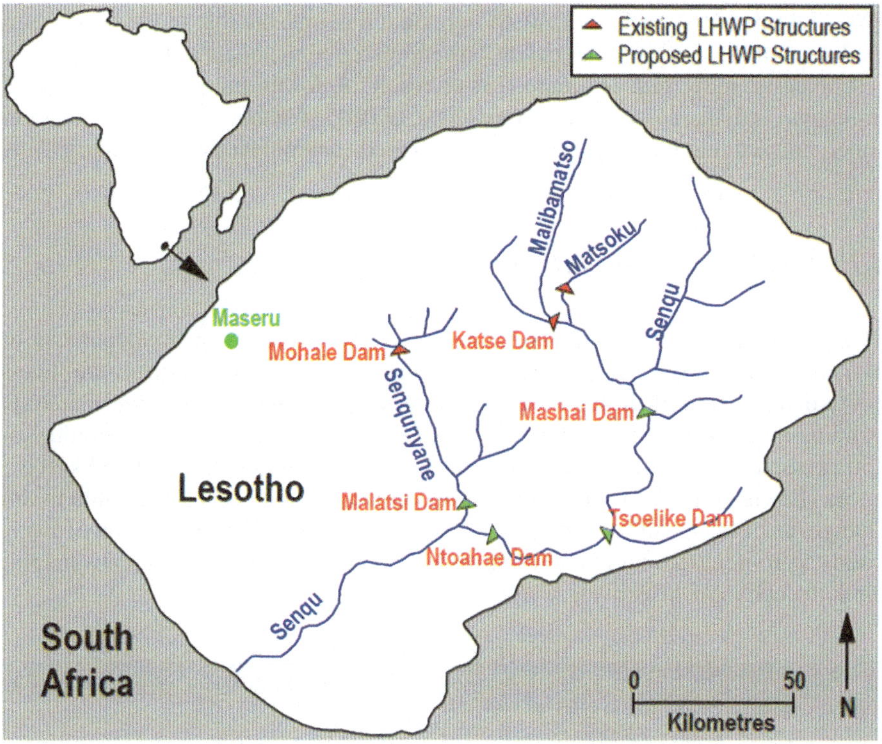

Fig. 12 The LHWP showing the water infrastructure. *Source* Dube et al. (2014)

and the annual rainfall exceeds evaporation, whereas evaporative losses in other parts of the region, such as the Lower Orange, are arid to hyper-arid.

Water quality is generally impaired by seepage, runoff, and point source discharges of municipal, industrial, and agricultural effluents. There are also high sediment loads from land degradation in the catchment.

Most dams were designed without consideration of potential future climate change projections. For example, if a dam was designed on the basis of a 30-year period that experienced significantly greater rainfall than today, the operating rules may need to be reviewed to take into account climate conditions and projections.

Climate change- related issues are not adequately addressed in any of the LHWP projects, but the SADC Climate Change Adaptation Strategy has implications for the transfer scheme because the LHWP is within the SADC region. Recent drought conditions that resulted in water supply augmentation for the South African town of Aliwal North from the LHWP, while both the Katse and Mohale Dams are at their lowest water levels ever recorded, underscore the importance of climate-related considerations in that water scheme. Climate change models project yet warmer conditions for the period from 2030 through 2050, with a projected increase in air temperature ranging from about 0.8 to 2.9 °C above the historical average of 12.7 °C

(World Bank 2016), while precipitation predictions indicate that in the future there could be either wetter or drier conditions compared to the historical record.

Forecasts show reduced runoff as a result of lower precipitation, recurring droughts, and increased temperature. The LHWP is expected to become more water-stressed into the future due to both demands and unfavorable climatic effects.

4.1 Livelihoods and Vulnerabilities

Most communities in Lesotho, particularly in rural setting markedly dependent on agriculture, with maize as the staple crop and livestock production to sustain livelihoods of households in the highlands. Resources from the wild (such as fuel materials, building materials, plants, and medicines), which are found on primary production and resource harvesting, are also critical for meeting daily nutritional and survival needs (Lewis et al. 2011a, b) (Fig. 13). Communities are hence vulnerable to any changes in climate and to extreme events such as drought. Diversification of livelihoods is promoted to reduce the resource dependency of households, thereby helping them to cope more effectively with socio-economic challenges.

Fig. 13 Grass harvested for roofing (Lewis et al. 2011)

Fig. 14 Typical thatched roof huts in Lesotho (Carraro and Marzi 2014)

4.2 Climate Change Impacts on the LHWP

An analysis by Lewis et al. (2011a, b) indicates that projected changes in climate are likely to be similar in all pilot catchment areas within the Lesotho Highlands. Mean annual temperature is projected to increase particularly in the intermediate to more distant futures with a resulting increase in evaporation of 10–15% in the medium term and 20–25% in the long term. The mean annual precipitation is projected to increase by 100–200 mm in the intermediate future and by up to 300 mm in the more distant future. Rainfall seasonality is projected to shift from mid-summer (January) to late-summer (February) and to be distributed across the months with a reduction in the summer rainfall, as well as greater frequency of high-intensity rainfall events. These increases in precipitation may impact soil moisture with a consequent increase in water clogging.

Under a changing climate, the ecosystem is likely to be impacted due to, among other things, increased temperature and evaporation. This causes concern about the potential for decrease in some of the goods and services that depend directly on the ecosystem, such as grazing and thatching. Goods such as thatching are important for many communities in Lesotho, as exemplified by the thatch-roofed houses in most villages (Fig. 14).

4.3 Key Issues to Address to Ensure Water Security Under a Changing Climate

Lesotho is currently supplying water to South Africa through the Lesotho Highlands Scheme, while Botswana also needs an allocation. Consequently, the governments of Botswana, Lesotho, and South Africa initiated a high-level planning study to evaluate the possible development and transfer of water resources from the highlands of Lesotho to the southern part of Botswana (World Bank 2016).

It is essential to account for climate change by using climate change scenario projections and potential system responses as detailed water resource balancing is considered. Adaptation strategies must also address factors that may not directly be related to climate, such as deterioration of water quality, and assessment and protection of major groundwater aquifers. Compliance with water use monitoring and enforcement should be improved in all four countries. Institutional capacity to effectively manage water quality is also a major constraint in the member states (ORASECOM 2008). Other issues requiring attention include enhanced investment in renewable energy sources while implementing technologies that increase efficiency of coal-powered stations, infrastructure optimization, water use efficiency in agriculture, and improved monitoring and use of forecasting (ORASECOM 2014).

4.4 Recommendations for the LHWP

Lewis et al. (2011a, b) dealt with climate change in the LWHP comprehensively by focusing on its impacts on ecosystem goods and services, based on the research work by Schulze et al. (2010) for the southern Africa region, including Lesotho. It is moreover recommended that a dynamic downscaling approach be applied to develop higher resolution climate change scenario projections for the LHWP to inform adaptation. Additionally, water managers and planners must develop adaptation plans and implementation strategies to address water-related challenges driven by both climatic factors and non-climatic factors (such as unsustainable land use and pollution).

5 The Limpopo River Basin

The Limpopo Watercourse Commission (LIMCOM) evolved from the SADC structure and mandate. However, it has yet to establish a shared vision among the four riparian states (South Africa, Botswana, Zimbabwe, and Mozambique) for managing the basin and developing appropriate institutional arrangements to coordinate management. The Limpopo River Basin, one of the most populated river basins in Africa, cuts across southern Botswana, most of the northern provinces of South Africa, southern Zimbabwe, and central Mozambique (Fig. 15). Economic drivers are mining, industry, agriculture, and large urban centers.

As it progressively becomes a scarce resource in the basin, water poses the biggest threat to livelihoods, economies, and ecosystems of the river basin (Petrie et al. 2014). Water supply is insufficient to meet demand, food insecurity results from climate change-driven highly variable rainfall, and livelihoods rely significantly on climate-sensitive natural resources. The basin is home to resource-poor rural communities

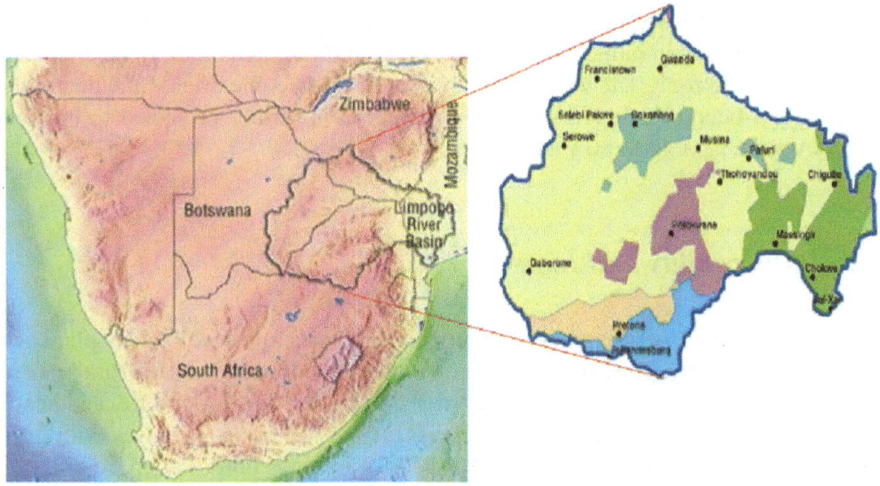

Fig. 15 The Limpopo River Basin (Gwata and Shimelis 2013)

and suffers from insufficient public and private resources, thus making the basin highly vulnerable to resource shortages and climate-related risks.

The western areas of the basin are projected to become drier than the northeastern areas. Rainfall is most uncertain due to lack of clarity around future dynamics of the intertropical convergence zone (i.e., the equatorial region where winds originating from the northern and southern hemispheres converge and generate most of the rainfall over the river basin), and cyclonic activity on the coastal plain (Chapman 2011). Additionally, rising temperatures, highly variable climate, deteriorating water quality, and over-allocated water are some of the other challenges facing the region.

By 2025, rapid growth in urban population, mining, and energy projects are anticipated (Petrie et al. 2014) to place enormous pressure on the basin's water resources. Currently, there is poor capacity to adapt to these pressures across the system, heightening the need for building resilience. Analyses indicate that no single regional or national institution is tackling the three key threats facing the Limpopo River Basin: water scarcity, declining ecosystem services, and climate change and variability.

System resilience and enhanced adaptive capacity need to be addressed, including but not limited to the use of tools such as early warning systems and forecasts and strengthening basin-wide governance and institutional arrangements for water management.

5.1 Hydrology, Climate, and Climate Change

The Limpopo Basin is semi-arid and characterized by a highly variable climate, and also prone to drought and floods. Drought is projected to increase in frequency and severity while intensity in rainfall and associated dry spells also increase. Models indicate uncertainty in rainfall due to uncertainty in future dynamics in the intertropical convergence zone. The western parts are likely to become drier while the north-eastern parts become relatively wetter. The region is further projected to become hotter over the next 40 to 80 years.

5.2 *Vulnerability and Impacts*

Vulnerability varies by country depending on its infrastructure development plans and institutional capacity. The Limpopo Region is densely populated and the rural poor in the basin rely on climate-sensitive resources such as rain-fed agriculture, and hence adaptive capacity is generally low. Vulnerability in the basin increases with poor implementation of policies. The area near the Mozambique border often experiences extreme rainfall and tropical cyclones such as the recent Dineo tropical cyclone of February 13, 2017 with an average speed of 160 km/hour (Fig. 16) that caused destruction in the region, including in much of South Africa. Rising temperatures also deteriorate water quality.

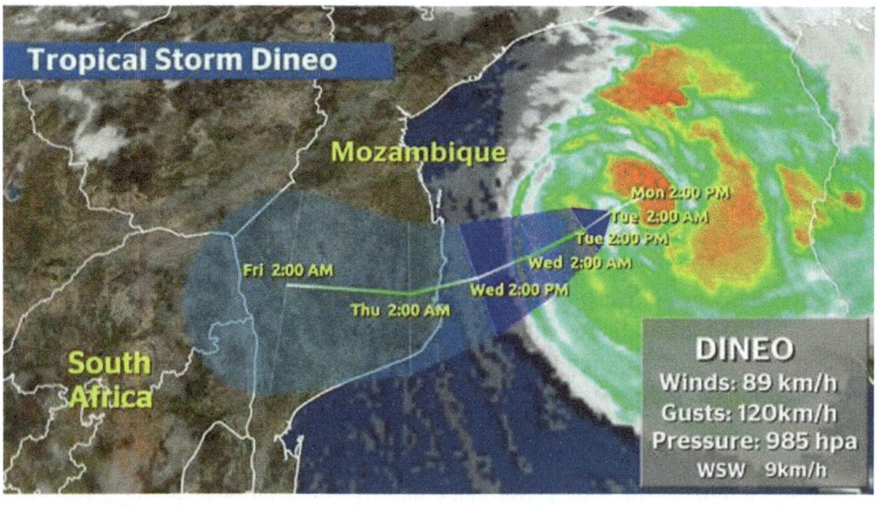

Fig. 16 Tropical cyclone Dineo (http://news.searchsa.co.za/weather-alerts/tropical-storm-dineo-to-reach-land-24h.php)

5.3 *Water Governance and Management-Related Issues*

Over-allocation and scarcity of water and poor infrastructure are some of the water-related challenges experienced in this region. Storage is lacking in certain areas and not fully functional in others, and most vulnerable and resource-poor communities do not have access to potable water.

5.4 *Adaptation Measures and Recommendations*

Strengthening of institutional arrangements for better coordination of water management processes at the transboundary level, enhanced institutional capacity and utilization of early warning systems, and preparedness are essential for improved resilience. Additionally, higher resolution climate change projections and more comprehensive studies on the changing and variable climate are recommended for the Limpopo River Basin.

6 The Incomati River Basin

The Incomati-Maputo Agreement was reached among South Africa, Mozambique and Swaziland in 2002 to resolve a conflict over allocation of Inkomati catchment water by including Maputo in the negotiations. The agreement seeks to ensure cooperation to protect and sustain water resources among the three states. The states also agreed to coordinate water security initiatives, monitor and mitigate flood events, and exchange data and information among them.

However, neither the agreement (adopted in May 2002, signed in August 2009 and ratified in February 2010), nor a technical report by the Tripartite Permanent Technical Committee on systems operating rules (drafted in 2010) even mention climate change.

The Incomati is a semi-arid transboundary river basin in southern Africa, which is water-stressed because of the competing demands of irrigated agriculture, forestry, energy, environmental flow, and basic human needs (Okello et al. 2015). The total basin area is approximately 46,750 km^2, of which 2,560 km^2 (5.5%) is in Swaziland, 15,510 km^2 (33.2%) is in Mozambique, and 28,681 km^2 (61.3%) is in South Africa. Figure 17 depicts both the Incomati Basin and adjacent Maputo Basin.

The three basins—Incomati, Maputo, and Umbeluzi—are shown in detail in Fig. 18.

All three rivers flow from east to west, from the inland highland mountains in South Africa and Swaziland to the Indian Ocean (Hake 2016).

Fig. 17 Incomati Basin (Hake 2016)

The Inco-Maputo is arid to semi-arid area with an average rainfall of about 767 mm per year and average temperature of 17 °C, while the evaporation rate varies between 1,600 and 2,000 mm per year in the southwest to eastern parts, respectively.

Communities in the region are mostly poor and vulnerable to climate change impacts with very low adaptive capacity.

6.1 Water Governance, Management, and Use

The agreement empowers the parties to individually and where appropriate jointly develop technical or administrative measures to manage, control, and protect water resources. Member states agreed to coordinate plans for water security and also to exchange data and information. All these technical issues must also take socio-political aspects of each country into consideration.

Water use in the Maputo River Basin is estimated at 1,198 mm^3/year, with agriculture accounting for about 86% of total water use.

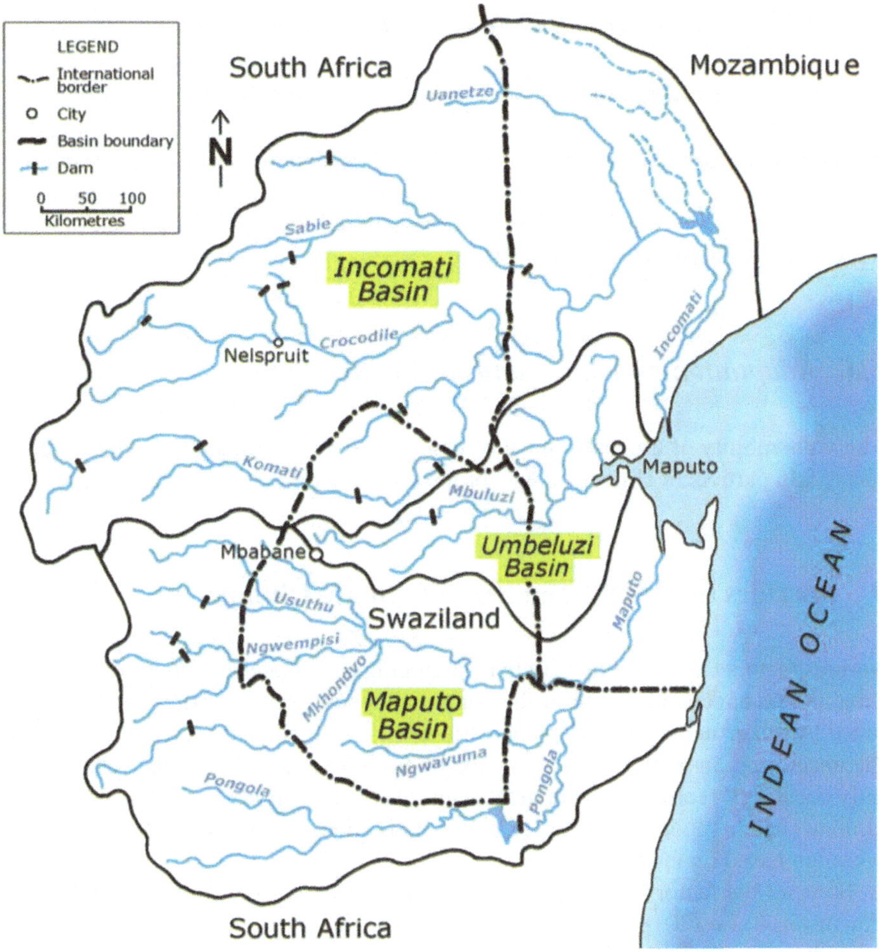

Fig. 18 Detailed map of the Incomati, Maputo, and Umbeluzi Basins (Hake 2016)

6.2 Hydrology and Current Climate

The basin is arid to semi-arid with an average rainfall (which occurs mostly in the summer) of 767 mm per year, mean annual precipitation of 400–1,000 mm, and mean annual runoff (MAR) of 3,539 mm^3/year. Evaporation varies from 1,600 mm in the southwest to 2,000 mm in the east. The mean annual temperature is 17 °C.

6.3 Climate Change, Vulnerability, and Impacts

Weak legal and institutional frameworks lead to weak adaptive capacity. In this region, poor rural communities are vulnerable to climate change as exemplified by reduced rainfall and, as a result, low crop yields (i.e., reduction in agricultural output).

This basin is generally projected to become drier in the future. Climate in the eastern parts, around Mozambique, is driven by conditions in the Indian Ocean; more study of these dynamics is required.

6.4 Adaptation Measures and Recommendations

The vulnerability of this basin must be assessed to ensure understanding of climate change. In particular, climate change projections at a more local scale are necessary.

7 Barriers to Mainstreaming of Climate Change

Various factors can limit effective mainstreaming of climate change into water planning and management. Some of the commonly encountered barriers include reluctance by water managers and planners to use climate change projections decrying the uncertainty in model predictions, the ill-advised notion that climate change will only occur or threaten people in the distant future, the conflict in time steps wherein climate change models are often based on daily metrics while hydrological models are often based on monthly metrics.

Some of the factors may be driven by narrow interests that seek to resist change. For instance, Le (2018) asserted that policy is difficult to change even in the face of such serious existential threats as climate change when the political motivation to maintain the status quo is often stronger than the drive to address challenges. Limited resources and information are also often a barrier to mainstreaming climate change. According to the Australian Government (2011), limits to the availability of, or access to, information as well as funds, expertise, and other capacity necessary to make appropriate decisions that lead to proportional action are some of the barriers to mainstreaming of climate change.

8 Conclusions

The Orange-Senqu Basin has been studied in greater depth than any of the other transboundary regions. The next logical steps are for the Commissions of the member states as well as other stakeholders to incorporate this information and data into basin planning and management. Higher resolution climate change scenario projections must be developed for the Lesotho Highlands Water Project and subsequently taken into account in the development and implementation of adaptation strategies to address water-related challenges driven by both climatic and non-climatic factors. A number of countries in the SADC region, particularly South Africa and Botswana, look to Lesotho to supplement their water supply and this calls for development and application of climate change scenario projections to support and inform decisions. Further work on climate change projections and more comprehensive studies on the changing climate should be undertaken in the Limpopo River Basin. Additionally, institutional capacity must be enhanced for better coordination of water management as well as effective use of early warning systems and preparedness to improve resilience. The Incomati River Basin needs climate change vulnerability assessment study to inform adaptation and to better understand climate change-related challenges. In particular, climate change projections at higher resolution and at local level are essential for sustainable water management and use.

References

Australian Government (2011) Barriers to effective climate change adaptation: a submission to the productivity commission. Department of Climate Change and Energy Efficiency

Carraro L, Marzi M (2014) Social protection in Lesotho: towards a coordinated and effective national system. Oxford Policy Management

Chapman, A. (2011). Transboundary water resources in southern Africa, Research brief 4. The Regional Climate Change Programme

Davis-Reddy CL, Vincent K (2017) Climate risk and vulnerability: a handbook for Southern Africa, 2nd edn. CSIR, Pretoria, South Africa

Department of Water and Sanitation (2019) The National Integrated Water Information System (NIWIS). http://niwis.dws.gov.za/niwis2/

Dube R, Maphosa B, Scott-Goldman J (2014) The role of local community institutions in the adaptation of rural and urban communities to the impacts of climate change on water access and use. WRC Report No. 1963/1/14

Earle A, Malzhender D, Turton A, Manzungu E (2005) A preliminary basin profile of the orange-senqu river. Published by AWIRU, Unicversity of Pretoria, South Africa. ISBN 1-86854-618-7

Gwata ET, Shimelis H (2013) Evaluation of pigeonpea germplasm for important agronomic traits in Southern Africa. In: Goyal A, Asif M (eds.) Crop Production, IntechOpen. https://doi.org/10.5772/56094

Hake S (2016) The mission of suppling maputo with freshwater - a political and environmental water history. Unpublished thesis; Uppsala, Department of Archaeology and Ancient History; A Master's thesis in Global Environmental History Department of Archaeology and Ancient History, Uppsala University, Uppsala, Sweden

Hallowes J, Frezghi M, Jonker V, Manley R (2010) Consulting services for system operating rules for the incomati and maputo watercourses interim technical document first draft

Jack C, Wolski P, Pinto I, Inda V (2016) Southern Africa regional overview, tools for observing and modeling climate, in future climate for Africa | Africa's climate: Helping decision-makers make sense of climate information, pp 23–30 @ future_climate www.futureclimateafrica.org

Keller EJK (2012) Critiquing cooperation: transboundary water governance and adaptive capacity in the orange-senqu basin. J Contemp Water Res Edu (149):41–55

Knoesen DM (2012). Integrating hydro-climatic hazards and climate change as a tool for adaptive water resources management in the orange river catchment. Unpublished PhD Thesis. University of Kwa-Zulu Natal. Pietermaritzburg

Knoesen D, Schulze R, Pringle C, Summerton M, Dickens C, Kunz R (2009) Water for the future: impacts of climate change on water resources in the orange-senqu river Basin. Report to NeWater. A project funded under the Sixth research framework of the European Union. Institute of Natural resources, Pietermaritzburg, South Africa

Lange G, Mungatana E, Hassan R (2007) Water accounting for the Orange River Basin: An economic perspective on managing a transboundary resource. Ecol Econ 61(2007):660–670

Lewis F, McCosh J, Pringle C, Bredin I, Nxele Z (2011a) Lesotho climate change adaptation project—ecosystems, agriculture and livelihoods in the lesotho highlands: likely futures and implications of climate change. Institute of Natural Resources, GROW and Serumula Development Association. Funded by USAID

Lewis F, McCosh J, Pringle C, Bredin I, Nxele Z (2011b) Lesotho climate change adaptation project: ecosystems, agriculture and livelihoods in the lesotho highlands—likely futures and the implications of climate change. Institute of Natural Resources NPC Report, 163 p

Okello AML, Masih I, Uhlenbrook S, Jewitt GPW, van der Zaag P, Riddell E (2015) Drivers of spatial and temporal variability of streamflow in the Incomati River basin. Hydrol Earth Syst Sci 19:657–673

Orange-Senqu River Commission (ORASECOM) (2008) Preliminary transboundary diagnostic analysis. Main Report. Tethys Environmental Consultants and the SAIEA, 197 p

Orange-Senqu River Commission (ORASECOM) (2011) Joint baseline survey-1: baseline water resources quality state of the Orange-Senqu river system in 2010. In: Quibell G, Rossouw JN (eds) for the ORASECOM. Pretoria, South Africa

Orange-Senqu River Commission (ORASECOM) (2014) Draft concept note in contribution to development of a proposal on actions towards adaptation to climate change in the Orange-Senqu River basin. By the Institute of Natural Resources in collaboration with DRFN, Kalahari Conservation Society and Native Seed Consulting

Petrie B, Chapman A, Midgley A, Parker R (2014) Risk, vulnerability and resilience in the Limpopo River Basin system: climate change, water and biodiversity—a synthesis; USAID Southern Africa. Resilience in the Limpopo River Basin (RESILIM) Program; OneWorld Sustainable Investments, Cape Town, South Africa

Seago C (2020) The Orange-Senqu Rover basin map. WRP Consulting Engineers (Personal Communication)

Schulze RE (2011) A 2011 perspective on climate change and the South African water sector. Water Research Commission, Pretoria, RSA, WRC Report 1843/2/11, 366 p

World Bank (2016) Lesotho water security and climate change assessment. World Bank, Washington, DC, License, Creative Commons Attribution CC BY 3.0 IGO

Chapter 20
Does the Use of Local Knowledge in Complex Systems Reduce Vulnerability to Climate Change? Insights from Nexus Water Management in the Niger Basin

Luca Ferrini, Euan Mackway-Jones, and Nora Van Cauwenbergh

Abstract In response to climate change and in an attempt to address complexity and uncertainty, management of water resources is increasingly framed as adaptive management. By definition, adaptive management includes experiential learning to improve understanding of systems and their responses to change. It is increasingly recognized that the knowledge of local communities contributes to system understanding. This is particularly important in Africa, where data is limited and vulnerability to climate change is high. This chapter focuses on the contribution of local knowledge in adaptive water resource management in Africa in a context of nexus governance. The hypothesis tested in this study is that a holistic approach to water management, which includes local knowledge and ensures local participation, contributes to a decrease in climate change vulnerability and can lead to greater capacity for adaptation, particularly for natural resource-based societies. The analysis focuses on water governance in the Niger Basin through three case studies. The first looks at the creation of National User Coordination structures for basin management. The second and third look at local participation in two major projects in the basin; the Kandadji Dam in Niger and the Fomi Dam in Guinea. These cases were selected to understand the roles of system boundaries and system understanding, stakeholders and levels of involvement, adaptation capacity and vulnerability in complex adaptive systems, and their governance. Results show that nexus governance structures generally facilitate inclusion of local knowledge in theory, but that this

The ideas and opinions expressed in this article are those of the authors and do not necessarily represent the view of GIZ, UNESCO, or IHE.

L. Ferrini (✉)
GIZ, Niamey, Niger
e-mail: luca.ferrini@giz.de

E. Mackway-Jones
UNESCO, Paris, France

N. Van Cauwenbergh
IHE Delft, Delft, The Netherlands

S. Diop et al. (eds.), *Climate Change and Water Resources in Africa*,
https://doi.org/10.1007/978-3-030-61225-2_20

inclusion is limited in practice. Moreover, we identify a negative impact on management outcomes when involvement is limited, with indications that such limitations increase the vulnerability of communities.

Keywords Local knowledge · Participation · Complex adaptive systems · Nexus

1 Introduction

Management of water resources under climate change is highly complex. Complexity is a result of multiple uses across multiple spatial and time scales, as well as of uncertainties. Water management is a crucial element of systems management under climate change and has the potential to either mitigate or exacerbate the effects of climate change. Given the complexity and changing nature of systems under climate change, water resources' management is increasingly framed as Adaptive Management (AM). The foundation of AM is a thorough understanding of the system that is managed, through experiential learning and knowledge exchange (Fabricius and Cundill 2014). This encompasses understanding of various uses, sectors, sources, stakeholders, and their interrelations. The AM paradigm emerged in the 1970s and is predicated on managing uncertainty through management action (Johnson 1999) centered on ongoing learning about the "processes governing the system" cited by Medema et al. 2008).

AM is one of various strategies that has been developed to holistically address complexity and interdependencies in water systems. The ecosystem services approach (e.g. de Groot et al. 2002) recognizes the highly complex interactions among people and the ecosystems on which their livelihoods depend. Integrated Water Resources Management (IWRM) focuses on integration and coordination through governance and stakeholder involvement (Medema et al. 2008).

One approach that has gained traction is the Water-Energy-Food-Environment (WEFE) nexus framework, which moves away from a water-centric approach to an issue-based approach, in which natural resources are seen as instruments for achieving human water, energy, and food security and environmental sustainability (Hoff 2011). Particularly under climate change, holistic planning of natural resources should thus take account of the trade-offs and synergies of different use (Brouwer et al. 2018). However, few analyses examine practical attempts to apply the approach, and those that do focus largely on the Global North (Kurian and Ardakanian 2015; Cairns and Krzywoszynska 2016).

This chapter focuses on the role of local knowledge in adaptive water resources management in Africa to mitigate vulnerability through multisectoral nexus governance.

There is a trend toward methods and approaches in which multiple stakeholders are involved not only in the implementation of adaptation measures, but also more fundamentally in the understanding of the system that needs to adapt, as well as the design of adaptation measures. These approaches seek to improve the outcome through better system understanding (Berkes et al. 2000; Vedwan 2006; Byg and Salick 2009; Vignola et al. 2009; Raymond et al. 2010; Alexander et al. 2011), tailoring measures, and interventions to the needs of the communities involved (Agrawal 2010; Lebel 2013), increasing ownership of implementation (Adger 2003, Chapagain et al. 2009) and strengthening sustainability, e.g., through reinforcement of local institutions (Agrawal 2010). The improvement of system understanding through integration of local knowledge might be particularly important in Africa, where data is limited and vulnerability is great, notably in natural resource-based societies (Thomas et al. 2005; Armitage 2005; Lebel 2013) where the stake of local communities is high.

This chapter describes how this premise has been tested in the Niger Basin to determine if the inclusion of local knowledge in adaptive water management promotes community empowerment and reduced vulnerability. Included is an overview of the latest research and thinking in adaptive management of vulnerable complex systems, emphasizing adaptation to climate change. The study seeks to determine how water management and governance in the Niger Basin (1) uses local knowledge and local participation; (2) uses local knowledge in a nexus approach to reduce vulnerability, and, if so, for whom, and (3) how institutional (and other) challenges are identified and addressed in inclusive nexus governance. Based on analysis of Niger Basin governance at different scales, the link between vulnerability and local knowledge integration in adaptive water management will be examined, and findings will provide elements to improve water management as a climate adaptation tool where there is great vulnerability to climate change.

2 Complex Adaptive Systems and Climate Change

A system, at its most basic, is "any group of interacting, interrelated or interdependent [elements] that form a complex and unified whole that has a specific purpose" (Kim 1999:4). All elements of a system must be present in order for its purpose to be optimized. Indeed, if elements can be removed without it affecting system performance, it can be concluded that the elements did not comprise a "system," but were rather a collection of different parts (Kim 1999; Meadows 2008).

Similarly, the arrangement and ordering of different elements within a system also affect system performance; if the elements can be combined or ordered randomly with no effect on performance, then it also cannot be considered a "system" (Kim 1999). A key feature of system behavior is the tendency to maintain stability through feedback mechanisms, which creates a virtuous cycle of information within the system that improves performance for its systemic purpose, and enables action to correct suboptimal performance.

While systems can be delineated by boundaries, they are permeable and intersecting or overlapping with the boundaries of other systems, increasingly so in the contemporary globalized world (Midgley 2004, 2008; Brincat 2017).

Because the multitude of stakeholders interacting within and across systems (and/or issue areas) influences the system as a whole, systems, which are characterized by connectedness, complexity, uncertainty, conflict, multiple stakeholders, and, thus, multiple perspectives, grow increasingly unpredictable (de Savigny and Adam 2009 as cited in Swanson et al. 2012). Systems are characterized by boundaries, perception of system complexity, and of the stakeholders and elements composing the system, as well as of the system's vulnerability and capacity to adapt, which all contribute to system understanding. Ecosystems are also "perceived as bounded by the conceptualizations and judgments of humans" (Ison et al. 2007, p.5).

It can therefore be useful to apply analysis that accounts for the complexity of the system. The conceptualizing of systems as Complex Adaptive Systems (CAS) addresses this (Arnold and Wade 2015). Unlike "regular" systems, CAS behave unpredictably owing to external influence, to the extent that cause and effect patterns within such systems are difficult if not impossible to detect (Brincat 2017). General (linear) systems and complex adaptive systems are distinguished by the constantly evolving nature of CAS that interact with and are shaped by interactions with other systems (de Savigny and Adam 2009; Brincat 2017). CAS have emergent properties which give rise to unexpected behaviors (Brincat 2017).

Climate change can be considered the ultimate destabilizing contributor to a river basin system that excludes it from a linear realm and places it in a complex *adaptive* system. The unpredictability of climate change and its localized impact is the external dimension that requires adoption of this additional layer of complexity.

3 Vulnerability, Resilience, and Adaptation

Conceptual and analytical frameworks with roots in disciplines from economics to anthropology have integrated climate change in the analysis of complex systems of natural resource management. The vulnerability of biophysical and social systems to climate change is a specific subject of analysis (e.g., Cuevas 2010; Podschun 2017).

Vulnerability to climate change can be defined as "the degree to which a system is susceptible to, or unable to cope with, adverse effects of climate change, including climate variability and extremes. Vulnerability is a function of the character, magnitude, and rate of climate variation to which a system is exposed, its sensitivity, and its adaptive capacity" (IPCC 2001, quoted in Maxim and Spangenberg 2006:4). Maxim and Spangenberg (2006) provide a good overview of vulnerability, illustrating its analysis through the use of the DPSIR framework (driving forces, pressures, state, impacts, and responses) which integrates social, political, economic, and environmental factors.

For societies reliant on direct use of natural resources, which is the case in the Niger Basin, the adaptation definition by Thomas et al. (2005:7) applies: "Adaptation is seen as the adjustment of a system to moderate the impacts of climate change, to take advantage of new opportunities or to cope with the consequences. In many cases, climate does not affect people directly but indirectly, by affecting the physical and biological systems in which they live. For societies that are reliant to a significant degree on the use of natural resources, these changes take on both direct and indirect aspects, since for such groups livelihoods do not provide a buffer against climate, but are highly reliant upon it. Adaptation can be best seen as a process that involves changes in a system to increase its coping range, rather than temporary adaptation of historically familiar measures to cope with a transient threat."

Adaptive capacity is "the ability of countries, communities, households and individuals to adjust in order to reduce vulnerability to climate variation, moderate potential damage, cope with, and recover from the consequences, including ecosystem responses to climate forcing" (Thomas et al. 2005:7). The greater a system's resilience, the more it adapts to maintain its critical functions (adaptation capacities): "resilience is the capacity of a system to absorb disturbance and reorganize while undergoing change so as to still retain essentially the same function, structure, identity, and feedbacks" (Walker et al. 2004:2). Points of no-return, at which the system is unable to maintain its critical functions even after adaptation, are described as tipping points.

4 Role of Governance, Participation, and Local Knowledge

From anthropology to social policy, an array of terminology and concepts has been adopted in the study of climate change adaptation and governance, as well as management of natural resources in vulnerable societies. Examples are "social learning" (Cundill 2010), "collective action" (Ratner et al. 2017), "social capital" (Adger 2003, 2013), "role of knowledge" (Lebel 2013), and "traditional ecological knowledge (TEK)" (Berkes et al. 2000).

"Adaptive governance" as it is used by Pahl-Wostl et al. (2012) describes water governance for "mastering complexity" in basins worldwide, including in sub-Saharan Africa.

Most analyses of the role of knowledge in adaptive system governance and management find that different types of knowledge improve system outcomes. Local knowledge (Vedwan 2006), indigenous knowledge and participation (Nyong et al. 2007), local perceptions and understanding (Byg and Salick 2009; Chapagain et al. 2009; Lebel 2013), and the combination of local and scientific knowledge (e.g., Van Cauwenbergh 2008; Vignola et al. 2009; Raymond et al. 2010; Alexander et al. 2011) all seem to enhance system outcomes, albeit with nuances, mostly related to the institutions and the individuals or groups managing the knowledge, as well as the role of knowledge brokers and moderators (Berkes et al. 2000; Thomas et al. 2005; Agrawal 2010).

Participatory management and governance is widely promoted not only in relationship to water, natural resources, and climate change, but also as a general approach to management of goods and services in a society. However, there is limited understanding of how to achieve the benefits of participation effectively (Van Cauwenbergh et al. 2018). Whereas methods for stakeholder analysis (e.g., Reed 2008) are routinely

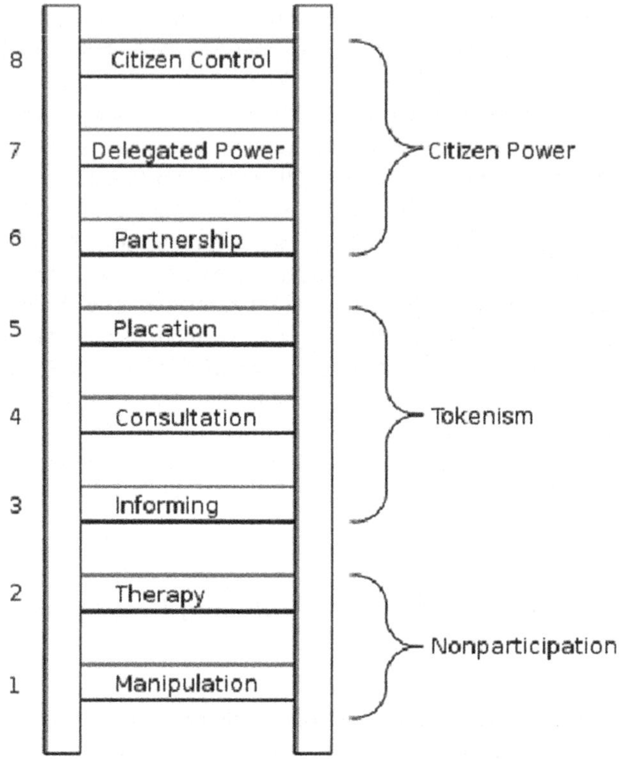

Fig. 1 The Arnstein Ladder of citizen participation (Arnstein 1969)

used, it remains a matter of debate who to involve, when, and at what level (e.g., Godinez-Madrigal et al. 2019). As early as 1969, Sherry R. Arnstein developed a ladder of citizen participation to describe the degree of citizen involvement in governance (Fig. 1). Comprised of eight rungs, the Arnstein Ladder describes increasing degrees of citizen participation, from the lowest level of *manipulation* to the highest level of *citizen control.*

The eight rungs are grouped in three types of relationships between authorities and citizens. At the lowest level is *non-participation*, where citizens are manipulated, "educated," or "cured" by power holders (Arnstein 1969:2); the middle level is *tokenism*, where a gesture toward participation is offered to citizens who are discouraged or precluded from requesting more participation or protesting; the highest level is *citizen power*, in which citizens have an influential and equal say to that of authorities. The Arnstein Ladder is the framework used in this chapter to evaluate the degree of local participation and use of local knowledge in adaptive water management.

5 Materials and Methods

The Niger Basin is the largest in West Africa and third largest in Africa. It covers an area of 2.13 million km^2 and is home to over 130 million people in nine countries in West and Central Africa: Benin, Burkina Faso, Cameroon, Chad, Ivory Coast, Guinea, Mali, Niger, and Nigeria (Fig. 2).

Six of the nine basin countries are Least Developed Countries (LDCs) and seven are among the 20 poorest countries in the world, with large income disparities in the richer basin countries. The population of the basin is projected to double by 2050 while also getting richer, meaning the region could be required to increase its food production by a factor of five (Akumaga and Tarhule 2018). Over 70% of the population lives in areas where food security depends on unreliable rainfall and highly variable inter-annual and intra-annual river flows.

The basin covers six agro-climatic zones and presents a cross-section of the complex development challenges of West African societies (Ogilvie et al. 2010). Climatic zones vary from hyper-arid to sub-equatorial; annual rainfall fluctuates from over 4,000 mm in southern Nigeria/Cameroon to less than 400 mm (with no rain in some years) on the edges of the Sahara Desert in northern Mali and Niger (Ogilvie et al. 2010). Figure 3 shows the heterogenic distribution of annual precipitation and mean temperature in the basin, posing specific and diversified challenges

Fig. 2 Niger Basin and member countries of the Niger Basin Authority (NBA) (*source* Niger Basin Authority)

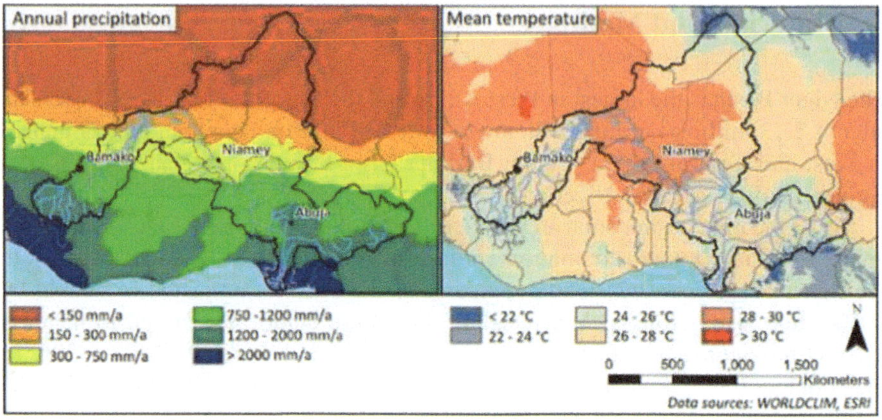

Fig. 3 Average annual precipitation and mean temperature in the Niger Basin (*source* Aich 2015)

to the sustainable management of water resources. The IPCC has declared the Niger Basin to be one of the most vulnerable regions to climate change in the world. Climate change impacts on the water resources of riparian countries are expected by some to exacerbate existing poverty and worsen inequality (Oyerinde et al. 2017).

Management and governance of water in the basin are overseen by the Niger Basin Authority (NBA), with representation of member states and civil society at regional and local levels, through National User Coordinations (NUCs). The Niger Basin is an example of a complex adaptive system. Its vulnerability to climate change ranks among the most in the world, with weak human and financial capacity and direct dependence on water resources for livelihoods. Table 1 presents an analysis of the Niger Basin as a Complex Adaptive System (CAS) (Fig. 4).

The first case study focuses on water governance at the Niger Basin scale to understand the process of inclusion of civil society, including representatives of resource users, in basin governance. It examines intended and actual levels of local participation and use of local knowledge as perceived by user organizations themselves. The next two case studies illustrate local participation in two local dam projects, the Kandadji Dam in Niger and the Fomi Dam in Guinea, both of which with regional impact. These studies reveal some of the effects of the terms of water governance as a result of varying degrees and forms of local participation (or lack thereof).

Analysis focuses on the structure of water governance in the Niger Basin, in particular, on the extent of local participation and local knowledge. Outcomes are then related to changes in vulnerability.

Analysis is based on[1] the following:

1. The history of the governance structure of the Niger Basin Authority (NBA), with a focus on the creation and functioning of the National User Coordinations (NUCs) of basin resources.

[1]More details in the Annex.

Table 1 The Niger River Basin as a Complex Adaptive System (CAS) marred by vulnerability and climate change *Source* adapted from de Savigny and Adams (2009). Niger Basin inputs from the authors of the chapter

Characteristics of a CAS	Relevance in the Niger Basin
Self-organization: system dynamics arise from the internal structure of the system	The Niger Basin's multiple climatic zones, significant percentage of arid and semi-arid zones, and large ecosystems which directly provide livelihoods for millions of people (Inner Niger Delta), as well as its size and its steeply growing population make it particularly vulnerable to changes and impacts of climate change
Constant changes: systems are constantly adapting to internal and external stimuli, often in a way that is unpredictable	Increasing water withdrawals, large dam construction, intensifying agricultural practices, and population pressures on land and forests are all internal stimuli which are only partly predictable and quantifiable, particularly across national borders. Climate change, migration, armed conflict, and geopolitical dynamics (e.g., oil and food prices) are external stimuli
Governed by feedback: systems are controlled by feedback loops, moderating behavior as elements "react and back react" on one another	The Niger Basin is experiencing changes in rainfall patterns, increasing soil erosion, desertification and sedimentation, and changes in ecosystems. These affect the choice of human settlement, create conflict on hot spots of resource availability and cross-border migration to make up for changes in pressure on resources
Non-linearity: relationships within a system cannot be arranged along a simple input-output model	A choice to increase water withdrawals for irrigation as well as fertilizer use may well increase agricultural productivity, but from there to food security there is still a long way, with trade, food prices, rainfall, and energy for storage and transformation still playing an important role. Climate conditions still play an unpredictable role to determine the output of the system
Tightly linked: high levels of connectivity between and across systems: changes in one sub-system affect others (negatively or positively)	One significant example is the worsening security and instability situation in the Sahel. Itself a dimension of vulnerability, this situation affects almost all aspects of human and natural life

(continued)

Table 1 (continued)

Characteristics of a CAS	Relevance in the Niger Basin
History dependence: effects may not be evident immediately; short-term effects may be insufficient to generate continued support for an intervention	A reduction in the rate of deforestation can have a positive long-term impact on limiting soil erosion and sedimentation. But immediate pressure for biomass and firewood most often prevail, particularly for the poorest and most vulnerable
Counter-intuitive: proven or effective interventions in one setting fail to deliver when transposed into a different context	Measures addressing conflict and instability and strengthening climate-resilient development sometimes have unintended negative effects, deteriorating the security situation even further. For example, in cases such as communities in northern Mali, Niger, and Chad at the edge of the Sahara Desert, where desertification and radicalization are perceived as going hand in hand. Efforts to channel water to some of these communities, to restore land and improve its productive use, have in various cases resulted in conflict with surrounding communities and other resource users, such as herders and cattle breeders, often belonging to different ethnic groups or communities, who have attempted to benefit from the improved infrastructure for their own economic activities
Resistant to change: interventions may not generate desired outcomes if the bottlenecks and barriers are too entrenched	Attempts to change cooking practices for moving away from traditional three-stone fireplace cooking to efficient stoves for reducing firewood consumption have often failed due to resistance to change social practices

2. Reports from workshops, consultations, and negotiations among governments and mayors, village leaders, basin user representatives, elders, women's groups, and civil society organizations. The focus is on two processes of dam construction and displacement of populations in order to analyze the modalities of involvement of local populations, concerns about the use of local knowledge, and the information needs of communities. The first wave of displacement occurred in the Kandadji Dam and the concerns of local communities in the design and preparation are examined in the context of the Fomi dam.
3. Workshop debates, recommendations, and interviews with representatives of NUCs, the NBA, and representatives of national governments of riparian countries.

The question this chapter aims to answer is whether local knowledge can improve the outcome of decisions for the complex system of water resource management

Fig. 4 The Niger River Basin and the position of the Fomi Dam at the basin's head and of the Kandadji Dam in the mid-Niger. (*source* adapted from L. Corsini 2011)

under climate change, and thus reduce vulnerability? To answer this question, analysis of vulnerability and levels of involvement of stakeholders is used. The WEFE approach mobilizes theories of complex adaptive systems and their governance; referencing system boundaries and understanding; and stakeholders and their levels of involvement, adaptation capacity, and vulnerability.

We analyze the current and past degree of local participation as called for in statutes of the NBA as well as in the case of two specific major dam projects. A timeline of important events for water management in the basin since the founding of the Niger Basin Authority (NBA) provides intended levels of involvement and how these levels materialized in practice. Using the Arnstein Ladder of citizen participation (Arnstein 1969), degree of involvement is plotted against intentions.

For purposes of this study, vulnerability is defined as changes in perceived adaptation capacity by water user groups and individuals themselves as described in written assessments and reported in interviews. Finally, we identify institutional (and other) challenges for adaptive governance and management for all stakeholders. This approach enables a broad perspective of differences in system understanding, and application of system boundaries at different levels of governance in relation to participation and vulnerability.

6 Results

6.1 *Inclusion of Civil Society and User Groups in Basin Governance of the Niger Basin Authority (Case Study 1)*

Civil society involvement in water management in the Niger basin dates back to the 1980s, when the construction of large dams was increasingly met by resistance at the local level. Local communities blocked work during the construction of dams such as the Talo Dam on the Bani (a major tributary of the Niger). It took more than two decades, however, for the role of civil society to be institutionalized. In 2005, the NBA formally recognized the need to involve civil society in the governance of the basin, and in 2006 civil society was officially recognized as a stakeholder and given a formal platform in the National User Coordinations (NUCs) and, in 2007, its participation mechanisms. The functioning of NUCs was reviewed in 2009, leading to a number of adjustments and signing of new NUC Memoranda of Understanding in 2014. This section focuses on the 2005–2014 period to compare the formal provisions for stakeholder participation to its manifestation in reality in relation to capacity building and vulnerability. The section relies heavily on the analysis conducted jointly by GIZ and ABN (Edl 2017) on state of cooperation among the NBA Executive Secretariat, civil society, and national focal structures.

The NBA (2010:3) defines the objectives of civil society in the NUCs as, "the users of the natural resources of the basin [to] participate in various decision-making stages: planning, program implementation, evaluation, etc.; and contribute, through their actions as structured organizations, to the sustainable development of the basin and in particular the actions of the Development Plan and Investment Plan which may concern them." On paper, this corresponds to a high level of participation (level 6 according to Arnstein 1969–*Partnership*). The scale of participation is defined by the geographical extent of the NBA mandate, which is the entire river basin; the type of stakeholders concentrates on main users. Users are considered to be a varied but "structured" body of stakeholders, able to make decisions and implement actions throughout the chain of water management activities. Peripheral reference is made of investments and activities at a more local level, presumably touching specific groups of users. The capacity required to participate beyond civil society participation objectives in NUCs is high, including mobilization, participation, coordination, defining interests, brokerage, planning, implementation, and evaluation. It also formalizes the definition of users.

In 2009, following a series of false starts, the NBA launched a study on capacity building and strengthening the involvement of civil society in the process of sustainable development of the Niger Basin. This study notes that the NUCs "operate timidly and are not aware of their roles and their anchoring in the process" for sustainable development of the Niger Basin. In addition, NUCs are reported to lack "an initial financial allocation to support the installation and operation" and suffer from "the absence of a formal commitment and effective involvement of Member States to provide support for the establishment and operation of the NUCs" (Tchouplaou et al.

2009:7). This reported lack of organizational, financial, and institutional capacity (Tchouplaou et al. 2009; NBA 2010; Edl 2017) indicates potential vulnerability of NUCs to decisions that reflect the interests of other, more powerful stakeholders (CRU-BN 2017; Edl 2017). This was confirmed at the NBA regional workshop on "integration of the water-energy-food Nexus in the NBA Operational Plan 2016–2024" in June 2018, where NUC representatives evaluated alongside representatives of the water, energy, agriculture, and environment ministries their respective and combined capacities to influence and implement decisions on uses of the basin's resources in each country's portion and the basin as a whole. The NBA study even notes ongoing debate about the identity of users, some of whom felt misrepresented by the term "civil society." The definition ultimately agreed upon is, "rural and urban populations whose livelihoods depend on the exploitation of the natural resources of the Niger basin" (Tchouplaou et al. 2009:4).

To improve the situation, the following interventions were proposed: "The setting up of a technical assistance mission that provides technical and organizational support/advice to the NUCs; the capacity building of the members of the NUCs on accompaniment through adequate logistical means; organization of an information and awareness campaign for users and civil society organizations on NBA, Shared Vision and investment plans; support for capacity building; implementation of income-generating activities" (Tchouplaou et al. 2009:19). These interventions reflect the NBA's drive and interest for users to support implementation of certain projects, and also serves as motivation, at least indirectly, for some state entities of the NBA member states. These interventions are however reported to have the unintended and undesirable effects of making participation more passive (KPMG 2016).

Representatives of the NUCs, particularly those who participated in the creation of the agreement following protests over the construction of dams, expressed in interviews their perception that, once users are seen as recipients of capacity building and financial support, the nature of their relationship to public authority changes. Involvement would be characterized by Arnstein's levels 1–3, *manipulation*, to *therapy* to *informing*, which are all highly passive. In addition, the definition of "user" is limited to the identified representatives, who are allocated offices and a budget, distancing them from their stakeholder base (interviews with members of NUCs networks in Mali and Niger). Finally, the system focus remains on high-level, basin-wide processes, mainly representing the view of NBA, failing to connect effectively with lower level processes and system understanding (interviews with NUC representatives in Chad, Ivory Coast, and Nigeria).

In 2014, new protocol agreements were signed among NUCs, the ministries responsible for the NBA in each country and the NBA itself. The revised protocols refine expectations for the involvement of users in the management of the Niger Basin, to promote greater local involvement as advocated by scholars (e.g., Thomas et al. 2005; Armitage 2005; Lebel 2013); formal institutions tend to welcome voices of local interests (Agrawal 2010). But NUCs are effectively under the control of the national focal structure of the relevant ministry, and made dependent on NBA supervision and oversight. The objective of the new agreements of the NUCs are to "Promote and organize consultation between users of the national portion of the basin; Contribute to the sustainable management of the natural resources of the Niger Basin; Mobilize users at the national and regional levels for greater participation in decision-making about the future of the basin; Promote the capacity building of users of natural resources at local, national and regional levels."

This language suggests a top-down approach, which threatens to significantly reduce the agency of users compared to the objectives initially formulated in 2006 (interviews with NUC members in Niger and Guinea). System boundaries are limited to the national portions of the basin, and participation modalities focus more on mobilization than on planning, decision-making, and implementation (KPMG 2016; CRU-BN 2017). The role of stakeholders—users themselves—remains a mostly passive one, emphasizing the need to co-opt them to buy into schemes developed at higher levels (interviews with CNU members in Guinea and Niger). The contributions of users to actual decision-making are channeled through modalities developed and controlled by stakeholders and structures other than the local communities themselves (Eau Vive 2006, 2007; Barry 2010). Local knowledge is thus marginalized (interviews with NUCs representatives) and its impact on decision-making is diluted. On the Arnstein ladder of participation, interviewees characterize their involvement as 3 (*informing*), 4 (*consultation*), and 5 (*placation*), which make up the second of three levels of "tokenism." There was no consensus among interviewees about the cause(s); some perceived the intention to marginalize by their NUCs (e.g., NUC Nigeria) and others blame the limited capacities of user organizations (NBA representatives).

In 2014, the NBA planned the "revitalization project of the partnership NBA-Civil Society of the Niger basin" for the period 2014–2016. The project aimed to strengthen the NBA-civil society partnership through the implementation of the NBA Investment Program. But the project was never launched due to lack of funding, with NUCs reporting "enormous difficulties due to the lack of funds as well as difficulties in the implementation of the memorandum of understanding" (CRU-BN 2017). The lack of financial resources is explained by non-payment of membership fees and dues of the NUC members, in addition to the failure to regularly finance the effort by the NBA and the member states. In addition, the NUCs report a consistent lack of regular communication with the NBA national structures that formally supervise them, as well as deficiencies in regular reporting and reaching consensus on annual work plans (KPMG 2016; Edl 2017).

At present, NUCs are not actively involved in the preparation and implementation of projects and programs. This makes them vulnerable to resource allocation decisions made at other levels and without sufficient input from users. This is in spite of policies to encourage greater stakeholder involvement and the sensitivity of technical partners to these issues, e.g., ECOWAS Directive announcement of the construction of large dams in West Africa, 2017; the Water Charter of the Niger Basin Authority, 2008); and coalitions such as the Network of Mayors on water infrastructures in West Africa. Further examples of these challenges are provided in the following case studies.

6.2 Increased Vulnerability of Displaced Communities at the Kandadji Dam (Case Study 2)

The Kandadji Program for Regenerating Ecosystems and Enhancing the Valley of Niger (P/KRESMIN) is a priority sectoral program of Rural Development in Niger. It is also one of the three dams of common regional interest identified in the NBA's Programme for Sustainable Development. The project is supported by multiple donors and will lead to the displacement of at least 38,000 people. This section documents the nexus impacts of displacement at the local level and uses the case to reflect on the relation between nexus governance at the project level and vulnerability of the communities involved.

A first displacement and resettlement operation to allow construction of the dam affected 5,500 people in villages in the river bed. The NUC deployed several information-gathering missions in Niger between 2013 and 2018, producing a wealth of information and lessons learned on the participation of local communities in decision-making around displacement and the project itself. These serve not only as key learning points for subsequent and future waves of displacement and planning in the Kandadji Dam Project, but also for the other major dam projects of common regional interest in the Niger Basin, Fomi in Guinea, and Taoussa in Mali.

In terms of system understanding and system boundaries, the sensitivity of the displacement operations means that the focus stays relatively local for all stakeholders. However, the project is generally perceived as a national utility initiative by the Nigerian authorities. The transboundary dimension is weak, an assessment confirmed by the little participation of the NBA in the process, despite the clear impacts beyond Niger.

The legal (inventory, compensation, land rights), cultural, and social (disruption of social networks by the displacement) boundaries of the system are subjects of misunderstanding and miscommunication and are identified by stakeholders as the main obstacles to deeper local participation in decision-making around the project. Populations underscored insufficiencies in three areas: preparation of the targeted sites before displacement, low participation of populations to be displaced, and unsatisfactory communication between project owners and impacted populations.

Whereas the project owner appears to have a very good understanding of the stakeholders involved and the different interlocutors and representatives of the local communities, serious shortcomings are noted in the level of involvement. Local community representatives think of themselves as recipients of information, and complain of the inconsistency and incompleteness of the information provided by project authorities. This self-perception as passive participants in management and communication places them, at best, at level 3 of the Arnstein Ladder (*informing*). Overall, the community structures (committees) created for the projects are perceived as dysfunctional (see reports in Annex). The capacity of local populations to participate and influence the process proactively is limited, particularly in the domains of legal rights and dealing with the unexpected. Both residential and agricultural land rights of displaced populations are identified by interviewees as weak. Clarity among village leaders, elders, and women's groups is necessary for equitable compensation.

These shortcomings have negative impacts for the communities on the ground. Examples include the struggle by women for even minimal access to water following displacement, and a reported increase in vulnerability for new home owners and farmers. Villagers were given the choice between receiving new houses already built, and receiving the value of their existing houses in cash to enable them to build their own houses. Many villagers choose the latter, with the unintended consequence in many cases of under-investment in new homes and therefore exposure of these villagers to new risks. The capacity to plan and invest corresponds strongly to perceptions of risk, social pressures, risk aversion, and cultural influences. Village leaders and mayors are most often highly aware of the tendencies of their constituents and, in the opinion of NUC representatives in Niger, could have seen this coming. In their opinion, discussing the relocation plan with village leaders in more detail would have allowed village leaders to better explain the pros and cons of compensation options for the villagers, leading to wiser and more sustainable choices, i.e., more villagers choosing quality housing over financial compensation. Unexpected flooding of the irrigated perimeter caused by abundant rainwater during the displacement period compromised agricultural productivity for farmers.

These outcomes increase the vulnerability of local populations through damage to their livelihoods as well as a lack of sufficient involvement of local communities in the project. The reports of the NUCs also identify that the populations of certain villages "live in fear; are very perplexed and pessimistic. Even the collaborators closest to the project [management] are worried, having witnessed the experiences of the first wave of displacement" (NUC Niger 2015—translated from French by the authors of this chapter).

In the words of NUC members, "in view of the lessons of the first wave and the number of people affected by the second wave, it should be emphasized that the enormous second wave calls for a holistic and cross-sectoral participatory approach, with close communication and strong involvement of the populations to be displaced and institutional actors. The effective inventorying of goods is one of the essential keys to the success of these operations" (NUC Niger 2015—translated from French by the authors of this chapter).

6.3 Discrepancies in System Understanding Around the Fomi Dam (Case Study 3)

Located in the Guinean Highlands, the Fomi Dam on the Niandan River is 39 km from its confluence with the Niger River, and was already identified in 1940. The current presidential administration of Guinea has made the project a political priority, with knock-on political effects in Mali and the other basin countries, due to the magnitude of the project and the downstream impacts. This section considers the objectives and potential impacts of the project at the local and regional levels and examines how its (nexus) governance leads to changes in vulnerability.

At the national level, the first priority of the Fomi Dam Project is to provide local households, markets, and the national mining industry with electricity. The significant hydroelectric potential of 100 MW was the primary motivation for Guinean authorities to conceive the project. Furthermore, the water reservoir has the potential to irrigate around 100,000 ha of agricultural land in Guinea, as well as to provide fishing and fish-farming opportunities. The initiative would require, however, the displacement of an estimated 45,000 people, which unsurprisingly is the main concern of local rural stakeholders who have long been aware of the project and associated with various activities around its preparation. Recently, a new site seems to have gained preference among the project's managers and contractors: 15 km upstream, where the population to be displaced is estimated to be only about 5,000.

The potential impact of the project is not limited to Guinea. The project regained some regional interest in 2014, notably with the call to re-evaluate the possible environmental and social impacts and discuss its shared costs and benefits among NBA countries.

The Fomi Dam Project is noted among basin-scale planning projects by the NBA for its potential to regulate the Niger's discharge beyond significant seasonal variations, along with two other dam projects (the Taoussa Dam in Mali and the Kandadji Dam in Niger). In a region that has a rainy season typically concentrated in a period of only 3 months, the need to store water to optimize its use at the economic, social, and environmental levels is significant. Regulating the discharge of Niger's main watercourse would potentially allow expansion of irrigation and agricultural production, thus improving the security and independence of food production in the region. The NBA's Sustainable Development Action Plan (in French Plan d'Action de Développement Durable) (Niger Basin Authority, 2008) estimates, for example, that the Fomi Dam could provide for the development of about 10,000 ha that could be developed as far away as the country of Niger. Discharge control throughout the year would also allow for improved navigation and regional commerce to be developed on the river.

The dam project has been validated by all stakeholders, including basin users. But the magnitude of the basin and of the number of communities affected mean that any proper consultation is ambitious and extremely costly. On a political and regional integration level, the project falls under the mandate of the NBA, but raises important questions for all impacted riparian countries—for example, who would determine policy for water retention and release, and the general regime functioning of the dam. There are many stakeholders associated with the project and they span a variety of geographies, including local, national, and basin/regional. In the recent consideration of social and economic impacts of the dam, certain stakeholders have felt consistently alienated, particularly when focus is on local impacts to the exclusion of regional stakeholders (interviews with stakeholders in Mali and Niger). Mali and Guinea have established a joint inter-ministerial consultation group; nonetheless, Mali has complained about not having access to the most up-to-date information about the planning of the dam (36th Ordinary Session of the NBA Council of Ministers 2018).

The position of the dam "en tête de bassin" (at the head of the basin) is particularly delicate, and raises questions about the nature of its impacts and the capacities required to adapt locally and regionally. Economically and socially, changes in the river's regime and discharge would demand big changes in agricultural and economic practices. Artificial control of the river's natural regime would entail a change in the relationship between the river and the ecosystems it sustains through seasonal fluctuations (Zwarts et al. 2005). It would also directly impact human activity, depending on the natural resources of each ecosystem, for example, fishing and rice farming, which depend on the seasonal flooding of certain areas of the basin. Furthermore, redirecting the main purpose of the dam from hydropower to river discharge regulation would have a significant influence on financial calculations of the economic viability of the project (NBA internal document 2018). This means, for example, that farmers, fishermen, and herders in the Inner Delta would realize impacts. While some organizations (e.g., Wetlands International and the International Union for the Conservation of Nature) regularly attempt to voice concerns, the inhabitants of the Inner Delta are only vaguely aware of the projected impacts of the dam and its implications for their economic activity. A comprehensive understanding of the project's impacts is complex for inhabitants close to the project location as well (interviews with representatives of Wetlands International and NUC Guinea). Participants in workshops with local community stakeholder representatives in Guinea, particularly those who are likely to be affected by displacement plans, raise concerns and questions (CRU-BN 2017) that indicate insufficient understanding of the dynamics of such a project, including individual and group rights, the expected benefits of the dam as well as the risks and changes it imposes on different groups.

The NUC in Guinea is considered an honest information broker and a trustworthy moderator of relationships between the local communities and national authorities (interview with member of NUC Guinea). Mayors and village leaders participate regularly in meetings and workshops, and their point of view is integrated in discussions and group exercises. But highly varying capacities to access and use technology, communicate effectively, and advocate for strategies mean that these populations have achieved *consultation* and, at best, *placation* (levels 4 and 5 according to Arnstein 1969). An illustration of this comes from the workshop on the kick-off of the dam's social and environmental impact updated study in 2017, at which contracted consultants had access to the full list of participants and a ranking of the issues that were most sensitive for their communities. The workshop was intended to demonstrate the complexity of the project to stakeholders, and only minimally affected the consultant's plans for the study (AECOM workshop report, September 2017; authors' observation at the workshop). Moreover (interview with NUC Guinea; report of the 27th Ordinary Session of the NBA Council of Ministers 2018), issues of displacement had not been communicated in detail to villagers, and many very basic questions around management and governance of the displacement processes were not addressed.

Open questions around the Fomi Dam make it difficult to predict its impact on vulnerability to climate change in the region. In view of the potential impact and the great number of stakeholders at varying levels of governance, involvement of local populations to prepare adaptation strategies for socio-economic change as a result of the dam is important.

7 Discussion

Results show the factors influencing levels of involvement of local knowledge and its relation to nexus governance. Issues of scale, system understanding, and barriers to inclusive nexus governance are identified across the case studies.

The reported lack of organizational, financial, and institutional capacity of National User Coordinations (NUCs) signals the potential vulnerability of the NUCs to the interests of other, more powerful, stakeholders. The three case studies confirm that involving local knowledge has the potential to improve multisectoral nexus governance and reduce vulnerability. This is shown in the case of the NUC in Guinea, which established itself as a trusted knowledge broker and led to a potential new location of the Fomi Dam that would significantly reduce forced displacement. Involvement of local knowledge also accommodates sensitivities that may not be accounted for in scientific analysis: cultural considerations, relationships between and within groups, social capital, and preferred modes of communication and of social learning. Involvement of local stakeholders increases ownership of the project, as was the case, with some significant positive impacts on project implementation, in Kandadji (capacity building, involving women groups, etc.). Conversely, vulnerability can increase when local knowledge structures are not involved, as was the case in giving villagers individual choice on housing compensation.

On the Arnstein Ladder of participation, relevant rungs identified by interviewees range consistently from 3 (*informing*) to 4 (*consultation*) to 5 (*placation*), representing the second of three levels of the ladder—tokenism. This generally corresponds to a feeling of increased vulnerability, mostly caused by the lack of control of local stakeholders. This has negative implications for decision-making and trust among governing authorities.

It could be argued that the increases in vulnerability in the short term are necessary evils in order to realize the long-term benefits of the completed dam projects. Dams designed at basin level are intended to result in return at a large scale, something that is difficult if not impossible for local communities to achieve. But it is also the case that displacement occurs at a very advanced phase of dam project implementation, and co-design of resource management could start from a much earlier stage of planning to include populations in the analyses, such as alternatives to the dam. This would mean initiating a partnership with local communities early and at different scales, because resource use and allocation apply at minimum to a national scale, and often a multinational scale.

Our analysis confirms that formalization helps to mainstream local knowledge into multisectoral water management processes. Once NUCs are officially established, they participate regularly in statutory meetings and have an official mandate to influence processes. They learn the "rules of the game" and build historical knowledge of the institutions with which they negotiate. On the downside, formalization of NUCs may result in introducing some rigidity to the system, in turn leading to a loss of depth of the stakeholder base. This is because representatives of the users in some cases lose a direct link to their base (farmers, fishermen, energy users and producers, cattle breeders), preferring to turn to other entities, particularly for financial support, which naturally influences their priorities. The establishment of official relationships between the users and public institutions, while legitimizing the NUCs and facilitating access to information and contact with project designers, owners, and implementers, also makes users more dependent on the state and other official bodies, such as the NBA.

With regard to involvement of local knowledge in water management at different levels, we note significant shortcomings in involving local communities meaningfully across the project spectrum. There are also barriers to an effective partnership among government/public institutions and users: inequality in financial capacities and leverage, different professional networks and legitimacy, and issues of perception and semantics (e.g., "civil society" and its connotations). The lack of extensive information exchange to reach a shared understanding of the boundaries of the system excludes or undervalues some stakeholders, and by extension the management outcomes for those stakeholder groups.

We also find some evidence of the value of increasing the stakeholder base in governance processes. While complexity can be daunting, our analysis confirms that polycentrism in governance reduces vulnerability and improves outcomes. The obstacles to maximizing the potential of polycentrism lie in the lack of established mechanisms and *fora* for an effective information exchange, in the overlapping and conflicting interests of different information brokers which can contradict each other, and in a loss of clarity and trust, particularly by those stakeholders which are less equipped in terms of capacities and analytical tools (often the users themselves). Effective information exchange is crucial to overcome different understandings of the system and ensure a more holistic view of project impacts.

Observations from the three case studies confirm that institutions with different mandates tend to frame the system differently, leading to a divergence of understandings (Boelens et al. 2016). The NBA predominantly looks at the basin in its entirety, while users (such as women, young people, pastoralists, fishermen, farmers) have the tendency to understand the system around boundaries and cleavages that bear on their interests. In general, stakeholder groups present a better understanding of aspects of the system that are closest to them (Van Cauwenbergh et al. 2018). This is illustrated by the deep awareness of the displaced populations of Kandadji of the rights they have lost, and the introduction of irrigated perimeters. Similarly, inhabitants of the Inner Niger Delta may well not yet grasp the potential impact of the distant upstream Fomi Dam on their livelihoods. Governance and system dynamics are directly influenced by who is interpreting the system, even within the same stakeholder group (Reed 2008). This has an impact on individual capacity to participate; we also observed that the demographics of the spokespersons and representatives (e.g., age, gender, social capital, socio-cultural identity, and interests) also influences the type of knowledge being communicated.

At the group level, management outcomes are determined by the capacity to organize, to make decisions internally, to build partnerships, and to negotiate. This is particularly true for local groups which find themselves facing national and international institutions which are often the official owners and/or promoters of projects such as multi-purpose dams (Boelens et al. 2016). Also important is the arbiter of knowledge exchange between groups, who understands the system differently and uses different vocabularies to describe it (Berkes 2000; Thomas et al. 2005; Agrawal 2010).

Vulnerability is perceived in concrete terms by local stakeholders. Displaced populations reported difficulty in accessing drinking water after displacement, which mainly affected women. In addition, due to concurrent climatic conditions, unexpected floods and dry spells in Kandadji directly impacted food availability for the populations displaced from their farmland. In an area that increasingly experiences food insecurity, in part due to climate change, it is all the more important to put safeguards in place to ensure that food needs are met; this effects the trust of the population in the authorities managing the dam project.

The result of the overall under-inclusion of local knowledge is a potential and, in the case of Kandadji, established increase in vulnerability. This vulnerability is directly related to the underdeveloped potential of capacity across stakeholders at different levels. This dynamic can be most effectively addressed, we believe, through more autonomous organization and agency of civil society, instead of using a top-down approach driven by central governments. There is potentially a significant role for donors and technical cooperation partners in accompanying local stakeholders, while keeping in mind the risks of over-formalization and of changing the nature of the coordination-type institutions themselves. There may be a research gap here to be filled—how best to support civil society while allowing it to maintain its status as independent representatives of under-represented, and often vulnerable, groups.

The cases presented confirm the complexity involved in the multisectoral, nexus management of water resources in contexts of vulnerability, limited capacities, and highly different scales and stakeholders. Capacities remain the main obstacle to deeper proactive participation of users to higher level decision-making. Indeed, mobilizing constituencies, obtaining financing, managing transport, formulating interests, pooling resources, and organizing collective action are capacity-intensive activities. But these are exactly the activities that translate into adaptation capacities and reduce vulnerability to climate change and unsustainable management of natural resources (Walker et al. 2004; Agrawal 2010; Lebel 2013).

8 Conclusions

These case studies set out to test the claim that more holistic approaches to nexus water management, including local knowledge and ensuring local participation, lead to reduced climate change vulnerability and greater capacity for adaptation, particularly for natural resource-based societies.

Based on theories of complex adaptive systems, interventions in the West African Niger River Basin were analyzed for implications on the nexus structure across different levels, depth of participation, use of local knowledge, and the existing governance system. The recent (2006–2018) inclusion of civil society and user associations in the governance of the Niger Basin Authority was the first such intervention. The second study, in the Kandadji Dam in Niger, where several villages were displaced in the period 2013–2018, lessons learned from this experience are used to analyze the effect of local participation on (intermediate) project outcomes in the short to medium term. The third is the case of the Fomi Dam in Guinea, scheduled to start construction in the coming months; data from stakeholder meetings examines the level of local involvement and how it bears on decisions and outcomes.

Generally, these case studies indicate the importance of involving local stakeholders. This often translates into ambitious strategies and aspirational goals for a *partnership*-type involvement of local knowledge. The reality, however, is that involvement falls significantly short of goals; *consultation* and *informing* are the norm, often with complaints from local stakeholders that information is partial, inconsistent, and difficult to digest. In these cases, a lower level of involvement does appear to impact management outcomes negatively.

Polycentrism of decision-making and governance is seen as an approach that facilitates management outcomes. Formalization of polycentric governance structures appears to improve outcomes, albeit with a risk of increasing the rigidity of the system to the point of undermining continuous learning and adaptation. Given the largely unpredictable nature of climate change, designing management systems that anticipate the need to adapt and learn seems key and should be given more attention.

Institutions with different mandates tend to frame and understand the system differently, and the weakness of information exchange to reach a shared understanding of system boundaries excludes or undervalues some stakeholders, and thus damages the outcomes for those stakeholder groups. The cases analyzed indicate that inclusion of local knowledge in the management of natural resources can help reduce vulnerability to climate change. In particular, for projects such as dams, which profoundly alter the lives of entire villages in a short period of time, greater integration of local knowledge can soften the severity of vulnerability. However, assessment (both before and after governance initiatives) needs to be improved to understand the relationship among stakeholder involvement, management outcomes, and resulting vulnerability.

Future research must examine which types of local knowledge are most crucial to the management of water and land resources in vulnerable complex adaptive systems, and look into the capacities necessary and the stakeholders best placed to collect and deliver such knowledge in governance and decision-making. This would also allow stakeholders to identify obstacles to greater involvement of local knowledge and design interventions and capacity development to support greater integration. Research must also contribute to maximizing efficiency through local knowledge, which would provide clear guidance to project developers and national governments and validate the allocation of resources required to create deeper partnerships with owners and brokers of local knowledge.

Acknowledgements Most of the merit for the analysis of the history of involvement of NUCs in the governance of the Niger Basin goes to Madeleine Edl of GIZ in her role of advisor to the NBA in the period 2016–2018.

Annex—Data for Analysis

1. Rapport De Synthese De La Visite D'echange D'experiences Des Delegations De La Societe Civile De Fomi (Guinee), De Kandadji (Niger) Et De Taoussa (Mali) A Selingue Au Mali Du 25 Au 27 Mars 2010
2. Capitalisation Des Expériences Du Processus De Déplacement Et Réinstallation Des Populations De La Zone Des Barrages De Kaleta Et Souapiti, Préfecture De Dubréka
3. Troisième Rencontre Du Cadre De Concertation Des Maires Autour Des Grands Barrages Dans Les Bassins Hydrographiques En Afrique De L'ouest Et Du Centre
4. Rapport De L'atelier De Restitution Suite A La Visite D'echange D'experiences Tenue A Selingue Entre Les Populations Riveraines Des Barrages De Fomi (Guinee) Et Selingue (Mali)
5. Atelier De Restitution Du Voyage D'etude Des Acteurs Du Projet De Barrage De Fomi (Kouroussa) Sur Le Site Du Barrage De Garafiri (Kindia)
6. Rapport De La Visite D'echange D'experiences Entre Les Futures Populations Affectees Par La Realisation Du Projet De Barrage De Fomi En Guinee Et Les Populations Du Barrage De Selingue Au Mali.
7. Concertation Avec Les Populations Affectees Pour Le Suivi De La Mise En Oeuvre Des Aspects Sociaux Des Pges/Pr/Pdl Du P/Kresmin
8. Voyage D'etude Des Acteurs Du Projet De Barrage De Fomi (Kouroussa) Sur Le Site Du Barrage De Garafiri (Kindia)
9. Débat télévisé: https://www.gwiwestafrica.org/fr/kandadji-reinstallation-des-populations-quelle-lecons-tirees
10. Documentaire Kandadji: https://www.gwiwestafrica.org/fr/kandadji-compensation-des-terres-agricoles-quelles-lecons-capitaliser-debat-canal3-niger
11. Focus Fomi à l'Assemblée Nationale de la Guinée: http://www.crubn.org/videos/la-cru-bn-a-l-assemblee-nationale-du-mali.html
12. Rapport de l'atelier d'information sur le Barrage de Kandadji, CNU Niger, Novembre 2018

13. Synthese Du Premier Atelier Regional De Dialogue Nexus Eau-Energie-Securite Alimentaire-Environnement Dans Le Bassin Du Niger
14. Dialogue Nexus dans le bassin du fleuve Niger : intégration du Nexus eau-alimentation-énergie dans le Plan Opérationnel de l'ABN, Rapport sommaire de l'Atelier régionale de l'ABN de Juin 19–20 2018
15. Edl, M., 2017. Processus d'établissement et état de lieux de la coopération entre le Secrétariat Exécutif, les Structures Focales Nationales et les Coordinations des Usagers/Usagères. GIZ/ABN, Niamey

(ITSELF BASED ON THE FOLLOWING PRIMARY SOURCES)

I. Autorité du Bassin du Niger, février 2008 : Processus de la Vision Partagée. Programme d'Investissement du plan d'action de développement durable du bassin du Niger. Synthèse.
II. Autorité du Bassin du Niger, mars 2010 : Rapport de synthèse de l'atelier de validation de l'étude sur le renforcement des capacités et la consolidation de l'implication de la société civile dans le processus de développement durable du Bassin du Niger pour la période 2009–2010.
III. Autorité du Bassin du Niger, juillet 2011 : TDRs relatifs à l'organisation des ateliers nationaux pour le renforcement des capacités des Structures Focales Nationales sur la connaissance des objectifs et résultats poursuivis par l'ABN dans les états membres (2001 à 2013).
IV. Autorité du Bassin du Niger, mars 2012 : Proposition des TDR relatifs à la mission de la Commission de réflexion interne de l'ABN par rapport aux Structures Focales Nationales et à la Société Civile.
V. Autorité du Bassin du Niger (a), 2013: Rapport final. Atelier national de renforcement des capacités de la structure focale nationale (SFN)-ABN Bénin (Cotonou du 25 au 27 mars 2013).
VI. Autorité du Bassin du Niger (b), 2013: Rapport de synthèse de l'atelier régional de concertation des structures focales nationales de l'ABN. Abidjan (Ivory Coast), du 23 au 24 septembre 2013.
VII. Autorité du Bassin du Niger, novembre 2014: Rapport du Comité ABN de réflexion sur la dynamisation des Structures Focales Nationales dans les pays membres.
VIII. Autorité du Bassin du Niger, septembre 2015: Compte rendu de la rencontre du Secrétaire Exécutif (SE) de l'Autorité du Bassin du Niger (ABN) avec les Coordinateurs des Structures Focales Nationales (SFN) des Etats membres.
IX. Autorité du Bassin du Niger, Eau Vive, novembre 2007: Offre technique et financier pour mission d'appui à la participation efficace des usagers de l'eau au processus de la vision partagée. Organisation et réalisation d'ateliers nationaux dans 5 pays de l'ABN (Bénin, Burkina Faso, Cameroun, Niger et Nigeria).
X. Barry, B. (a), Autorité du Bassin du Niger, 2008: Rapport de mission au Bénin.

XI. Barry, B. (b), Autorité du Bassin du Niger, 2008: Rapport de mission au Guinée.
XII. Barry, B. (c), Autorité du Bassin du Niger, 2008: Rapport de mission au Niger.
XIII. Barry, B. (d), Autorité du Bassin du Niger, 2008: Rapport de mission au Nigeria.
XIV. Barry, B., Autorité du Bassin du Niger, 2010: Rapport d'activités 2010 du Chargé des SFN et la Société Civile.
XV. Barry, B. Autorité du Bassin du Niger, 2012: Présentation de la mission, la vision, l'état du partenariat avec la société civile du Bassin du Niger.
XVI. Barry, B., Autorité du Bassin du Niger, 2013: Résultat de la réflexion interne du Secrétariat Exécutif sur les SFN-ABN.
XVII. Barry, B., Autorité du Bassin du Niger, janvier 2015: Rapport de passation de service du chargé du suivi des structures focales nationales (SFN) et d'implication de la Société Civile dans la GIRE au 31/12/2014.
XVIII. CRU-BN, avril 2016: Dispositif pouvant améliorer la communication et l'information au sein des coordinations nationales des usagers des ressources naturelles du bassin du Niger.
XIX. CRU-BN, janvier 2017: Bilan technique et financier du fonctionnement de la CRU-BN.
XX. Diarra, A. T., Autorité du Bassin du Niger, ACDI, mars 2006: Processus de vision partagée. Participation de la société civile. Rapport de synthèse des études nationales d'identification et de caractérisation des acteurs usagers de l'eau en vue de leur implication et participation au processus de la vision partagée. Thème: Barrages.
XXI. Eau Vive et Autorité du Bassin du Niger, juin 2006: 1ère Forum Régional des Acteurs Usagers de l'eau du bassin du fleuve Niger (FOREAU). Les 23, 24 et 25 février 2006. Fada-Ngourma. Burkina Faso.
XXII. Fraval, P., Autorité du Bassin du Niger, octobre 2008: Rapport de l'atelier de concertation entre le Secrétariat Exécutif et les Structures Focales Nationales de l'Autorité du Bassin du Niger. Conakry, les 30 et 31 octobre 2008.
XXIII. KPMG, Autorité du Bassin du Niger, Mai 2016: Audit organisationnel et institutionnel de l'ABN. Rapport modifié suite à l'atelier clinique.
XXIV. République du Niger, Ministère de l'Hydraulique et de l'environnement secrétariat exécutif, Direction Générale de l'Hydraulique, Direction des Ressources en Eau. Structure Focale Nationale, Autorité du Bassin du Niger, décembre 2011: Rapport de l'atelier national de renforcement des capacités de la structure focale nationale de l'Autorité du Bassin du Niger. Tillabéri du 01 au 03 décembre 2011.
XXV. Senes, Autorité du Bassin du Niger, AFD, ACDI, juin 2006: Etude de mise en place des Structures Focales Nationales. Rapport Final Phase –I. Référentiel des fonctions identiques pour les neuf états membres à remplir par les structures focales nationales.

XXVI. Senes, Autorité du Bassin du Niger, AFD, ACDI, juin 2007: Etude de mise en place des Structures Focales Nationales (SFN) de l'Autorité du Bassin du Niger (ABN). Rapport Finale Phase-II.

XXVII. Tchouplaou (a), Autorité du Bassin du Niger, AFD, GTZ, Union Européenne, Septembre 2009: Etude sur le renforcement des capacités et la consolidation de l'implication de la société civile dans processus de développement durable du Bassin du Niger pour la période 2009–2012. Rapport diagnostic. Compte rendu des rencontres nationales participatives (période : juillet-août 2009).

XXVIII. Tchouplaou (b), Autorité du Bassin du Niger, AFD, GTZ, Union Européenne, novembre 2009 : Etude sur des capacités et la consolidation de l'implication de la société civile dans processus de développement durable du Bassin du Niger pour la période 2009–2012. Document de programme. Version provisoire.

XXIX. Tchouplaou, Autorité du Bassin du Niger, AFD, GTZ, Union Européenne, mai 2010 : Etude sur les capacités et la consolidation de l'implication de la société civile dans le processus de développement durable du Bassin du Niger

References

Swanson RC, Cattaneo A, Bradley E, Chunharas S, Atun R, Abbas, KM, Best A (2012). Rethinking health systems strengthening: key systems thinking tools and strategies for transformational change. Health Policy Plan 27(4), iv54–iv61

Adger NW (2003) Social capital, collective action, and adaptation to climate change, economic geography; Oct 79(4). Adger NW et al (2013) Cultural dimensions of climate change impacts and adaptation. Nat Clim Change 3, February

Agrawal A (2010) Local institutions and adaptation to climate change. In: Mearns R, Norton A (eds) Social dimensions of climate change, equity and vulnerability in a warming world, new frontiers of social policy. World Bank, pp 173–197

Aich V (2015) Floods in the Niger River Basin in the face of global change—analysis, attribution and projections. Doctoral dissertation, Math-ematisch-Naturwissenschaftlichen Fakultät, Universität Potsdam

Akumaga U, Tarhule A (2018) Projected changes in intra-season rainfall characteristics in the Niger River Basin, West Africa. Atmosphere 9:497

Alexander C, Bynum N, Johnson E, King U, Mustonen T, Neofotis P, Oettlé N, Rosenzweig C, Sakakibara C, Shadrin V, Vicarelli M, Waterhouse J, Weeks B (2011) Linking indigenous and scientific knowledge of climate change. BioScience 61(6):477–484. https://doi.org/10.1525/bio.2011.61.6.10

Armitage D (2005) Adaptive capacity and community-based natural resource management. Environ Manage 35(6):703–715

Arnold RD, Wade JP (2015) A definition of systems thinking: a systems approach. Proc Comput Sci 44:669–678

Arnstein SR (1969) A ladder of citizen participation. J Amer Plan Assoc 35(4):216–224

Barry B (2010) Rapport d'activités 2010 du Chargé des SFN et la Société Civile. Autorité du Bassin du Niger, Niamey

Berkes F et al (2000) Rediscovery of traditional ecological knowledge as adaptive management. Ecol Appl 10(5):1251–1260

Boelens R, Hoogesteger J, Swyngedouw E, Vos J, Wester P (2016) Hydrosocial territories: a political ecology perspective. Water Int 41(1):1–14. https://doi.org/10.1080/02508060.2016.1134898

Brincat S (ed) (2017) Dialectics in world politics. Routledge

Brouwer F et al (2018) Commentary To SEI Report 'Where is the added value? a review of the water-energy-food nexus literature', SIM4NEXUS Project. www.sim4nexus.eu. Accessed 1 Nov 2018

Byg A, Salick J (2009) Local perspectives on a global phenomenon—Climate change in Eastern Tibetan villages. Global Environ Change 19:156–166, Elsevier

Cairns R, Krzywoszynska A (2016) Environ Sci Policy 64:164–170

Chapagain BK, Subedi R, Paudel NS (2009) Exploring local knowledge of climate change: some reflections. J Forest Livelihood 8(1), February

CNU-Niger (2015) Concertation Avec Les Populations Affectees Pour Le Suivi De La Mise En Oeuvre Des Aspects Sociaux Des Pges/Pr/Pdl Du P/Kresmin, Niamey, Niger

Corsini L (2011) L'équilibre du fleuve du Niger perturbe par les « pêcheurs de sable » , Carte du Bassin du Niger presented in IRD Fiches d'actualité scientifique 376. https://www.ird.fr/la-mediatheque/fiches-d-actualite-scientifique/376-l-equilibre-du-fleuve-niger-perturbe-par-les-pecheurs-de-sable, Accessed 8 Decemeber 2018

Cuevas S (2011) Climate change, vulnerability, and risk linkages. Int J Climate Change Strat Manag 3:29–60. https://doi.org/10.1108/17568691111107934

Cundill G (2010) Monitoring social learning processes in adaptive comanagement: three case studies from South Africa. Ecol Soc 15(3):28

De Groot RS, Wilson MA, Boumans RRJ (2002) A typology for the classification, description and valuation of ecosystem functions, goods and services. Ecol Econ 31(3):393–408

De Savigny D, Adam T (eds) (2009) Systems thinking for health systems strengthening. World Health Organization

Eau Vive (2006) 1ère Forum Régional des Acteurs Usagers de l'eau du bassin du fleuve Niger (FOREAU). Les 23, 24 et 25 février 2006. Fada-Ngourma. Burkina Faso

Eau Vive (2007) Offre technique et financier pour mission d'appui à la participation efficace des usagers de l'eau au processus de la vision partagée. Organisation et réalisation d'ateliers nationaux dans 5 pays de l'ABN (Bénin, Burkina Faso, Cameroun, Niger et Nigeria). Niamey, Niger

Edl M (2017) Processus d'établissement et état de lieux de la coopération entre le Secrétariat Exécutif, les Structures Focales Nationales et les Coordinations des Usagers/Usagères. GIZ/NBA, Niamey

Fabricius C, Cundill G (2014) Learning in adaptive management: insights from published practice. Ecol Soc 19(1):29. https://doi.org/10.5751/ES-06263-190129

Godinez-Madrigal J, Van der Zaag P, Van Cauwenbergh N (2019) Production of competing water knowledge in the face of water crises. Revisiting the IWRM success story of the Lerma-Chapala basin, Mexico. Geoforum 103:3–15. https://doi.org/10.1016/j.geoforum.2019.02.002

Hoff H (2011) Background paper for the Bonn 2011 Nexus Conference: The water, energy and food security nexus. Stockholm Environment Institute, Stockholm

Ison R, Roling N, Watson D (2007) Challenges to science and society in the sustainable management and use of water: investigating the role of social learning. Environ Sci Policy 10(6):499–511

CRU-BN, janvier 2017 : Bilan technique et financier du fonctionnement de la CRU-BN. Bamako

Johnson BL (1999) The role of adaptive management as an operational approach for resource management agencies. Conser Ecol **3**(2):8. http://www.consecol.org/vol3/iss2/art8/

Kim DH (1999) Introduction to systems thinking, vol 16. Pegasus Communications, Waltham, MA

KPMG (2016) Audit organisationnel et institutionnel de l'ABN. Rapport modifié suite à l'atelier clinique

Kurian M, Ardakanian R (eds) (2015) Governing the Nexus: water, soil and waste resources considering global change. Springer International Publishing, Switzerland

Lebel L (2013) Mitig Adapt Strat Glob Change 18:1057. https://doi.org/10.1007/s11027-012-9407-1

Maxim L, Spangenberg JH (2006) Bridging the gap between two analytical frameworks. In: Paper presented at the Ninth Biennial conference of the international society for ecological economics "ecological sustainability and human well-being", 15–18 Dec 2006, New Delhi, India

Meadows DH (2008) Thinking in systems. Chelsea Green Publishing, White River Junction, VT, pp 77–78

Medema W, McIntosh BS, Jeffrey PJ (2008) From premise to practice: a critical assessment of integrated water resources management and adaptive management approaches in the water sector. Ecol Soc 13(2):29

Midgley G (2004) Systems thinking for the 21st century. Int J Knowl Syst Sci 1(1):63–69

Midgley G (2008) Systems thinking, complexity and the philosophy of science. Emerg Complex Organ 10(4):55–73

Niger Basin Authority (2008) Plan d'Action de Développement Durable du bassin du Niger

Niger Basin Authority (2008b) Charte de l'Eau du bassin du Niger. Cogent Food Agricul 3(1):1363340

Niger Basin Authority (2010) Rapport de synthèse de l'atelier de validation de l'étude sur le renforcement des capacités et la consolidation de l'implication de la société civile dans le processus de développement durable du Bassin du Niger pour la période 2009–2010. Niamey

Nyong A, Adesina F, Elasha BO (2007) The value of indigenous knowledge in climate change mitigation and adaptation strategies in the African Sahel. Mitig Adapt Strat Glob Change 12:787–797

Ogilvie A, Mahé G, Ward S, Lemoalle J, Morand P, Barbier B, Diop T, Caron A, Namarra K, Lukasiewicz A, Paturel JE, Liénou G, Clanet JC (2010) Water, agriculture and poverty in the Niger River basin. Water Int 35(5):594–622

Oyerinde TO, Fademi IO, Denton OA (2017) Modeling runoff with satellite-based rainfall estimates in the Niger basin

Pahl-Wostl C, Lebel L, Knieper C, Nikitina E (2012) From applying panaceas to mastering complexity: Toward adaptive water governance in river basins. Environ Sci Policy 23:24–34

Podschun, S (2017) Understanding social-ecological systems under climate change—Exploring the ecosystem service concept towards an integrated vulnerability assessment

Ratner BD, Meinzen-Dick R, Hellin J, Mapedza E, Unruh J, Veening W, Haglund E, May C, Bruch C (2017). Addressing conflict through collective action in natural resource management. Int J Commons 11(2):87–906. http://dx.doi.org/10.18352/ijc.768

Raymond CM, Fazey I, Reed MS, Stringer LC, Robinson GM, Evely AC (2010) Integrating local and scientific knowledge for environmental management. J Environ Manage 91:1766–1777

Reed MS (2008) Stakeholder participation for environmental management: a literature review. Biol Cons 141:2417–2431. https://doi.org/10.1016/j.biocon.2008.07.014

Tchouplaou, Autorité du Bassin du Niger, AFD, GTZ, Union Européenne (2009) Etude sur le renforcement des capacités et la consolidation de l'implication de la société civile dans processus de développement durable du Bassin du Niger pour la période 2009–2012. Rapport diagnostic. Compte rendu des rencontres nationales participatives (période : juillet-août 2009)

Thomas D, Osbahr H, Twyman C, Adger N, Hewitson B (2005) ADAPTIVE: adaptations to climate change amongst natural resource-dependant societies in the developing world: across the Southern African climate gradient. Tyndall Centre for Climate Change Research, University of East Anglia, East Anglia

Van Cauwenbergh N (2008) Expert and local knowledge in decision support for natural resources systems: analysis of capture and use. PhD thesis, Faculty of Agronomy and Bio-engineering, Universite catholique de Louvain (UCL), 207 p

Van Cauwenbergh N, Ballester A, Ahlers R (2018) Participatory processes and supportive tools for planning in complex dynamic environments: a case study on web-GIS based participatory water resources planning in Almeria (Spain). Ecol Soc 23(2):2. https://doi.org/10.5751/ES-09987-230202

Vedwan N (2006) Culture, climate and the environment: local knowledge and perception of climate change among apple growers in Northwestern India. J Ecol Anthropol 10(1): 4–18
Vignola R, Locatelli B, Martinez C, Imbach P (2009) Ecosystem-based adaptation to climate change: What role for policy-makers, society and scientists? Mitig Adapt Strat Glob Change, Springer Verlag, 14 (8):691–696
Walker B, Holling CS, Carpenter SR, Kinzig A (2004) Resilience, adaptability and transformability in social-ecological systems. Ecol Soc 9(2):5
Zwarts L, van Beukering P, Kone B, Wymenga E (2005) The Niger, a lifeline. Effective water management of the Upper Niger Basin. Wetlands International.

Chapter 21
Proposed Research, Science, Technology, and Innovation to Address Current and Future Challenges of Climate Change and Water Resource Management in Africa

Andrea Vushe

Abstract Africa's adaptation to climate change requires significant shifts from existing management systems for water and land resources for sustainability, achieving food security, and socioeconomic development. Research and development programs must focus on practical conjunctive water management systems for salty and freshwater resources, strengthen capacities of urban and rural communities to circumvent existing institutional constraints, and socioeconomic challenges and political conflict. Research on flood- and drought-prone climates must examine water resource management at river basin scales. Soil and water conservation efforts should focus on commercializing small-scale farming, reforestation, pasture revitalization, enhancement of soil fertility, management of erosion and sediment, and development of cost-effective technologies for floodwater harvesting, treatment, recharging of aquifers, and efficient irrigation. Research on cost-effective urban wastewater treatment systems that exceeds minimal effluent quality standards and wastewater recycling may produce practical solutions that reinforce freshwater and human nutrient cycles, reduce pollution, and mitigate water stresses. Food security programs must include cost-effective mechanization; post-harvest technology; intensified agronomy; breeding; distribution; extension; commercializing production systems; and marketing of drought, pest, and disease-tolerant exotic and indigenous crops (small grains and fruit trees), livestock, wildlife, and edible insects. Developing cleaner renewable energy technologies for domestic power supply (substituting wood fuel), powering water pumps, and small-scale industries may improve the productivity and standard of life of rural and peri-urban communities. Collaboration of local and global researchers and manufacturers, designers, and testing institutions for appropriate technologies must lead to innovation and growth of small-scale industries, agriculture, creation of jobs and wealth, and improved water services.

Keywords Climate change adaptation · Conjunctive water management · Drought-tolerant crops · Rural food security · Appropriate technology · Research

A. Vushe (✉)
Namibia University of Science and Technology, Windhoek, Namibia
e-mail: avushe@nust.na

S. Diop et al. (eds.), *Climate Change and Water Resources in Africa*,
https://doi.org/10.1007/978-3-030-61225-2_21

1 Introduction

Africa is warming faster than average warming in the rest of the world, with a projected increase of 3 to 4 °C in the twenty-first century, which may cause considerable economic challenges for the continent (UNEP 2016). In some areas, warming is likely to reduce crop yields and livestock productivity, cause water scarcity, increase food insecurity, and increase reliance on food imports. Extreme weather and climate events such as droughts and floods are expected to be more frequent, with negative impacts on human life and health (UNEP 2013). Climate change may also lead to change in wildlife migration patterns due to habitat loss (UNECA 2014), thus increasing human-wildlife conflict. Demand for freshwater is increasing with the rising population and economic development, and at the same time freshwater is under threat from pollution and climate change. While some progress has been made to improve access to safe drinking water and sanitation, Africa remains plagued with waterborne diseases, such as cholera.

Competition for water resources is driving prices and local-level conflicts between communities with freshwater-dependent economic activities. Cooperation at the national and regional levels and between sectors and use of appropriate technologies can be an important intervention for conflict reduction and sustainable management of water resources in the continent. The adoption of polices and strategies to integrate river basin management approaches and consider management of surface water and groundwater resources as single systems would help African countries to prudently manage land and water resources (World Bank 2006). Several transboundary rivers in Africa are now managed by river basin organizations, which advise governments on utilization of water resources in their respective river basins, and hence modeling and coordinated management using globally successful approaches may increase the effectiveness of water resource management for economic development in Africa. For example, in China, river basin organizations hold legal status and development projects are required to be consistent with their plans, and in Italy, a 1989 law conferred river basin management as a formal government unit to regulate the programs of various sectoral and regional institutions (World Bank 2006). River basin management committees and institutions in riparian states that drive research and synchronize policy formulation must align objectives to build resilience to impacts of climate change. Understanding climate change and variability, the water, energy, food (WEF) nexus, and the urgency of providing robust and resilient solutions requires immediate, industrious, and intensive efforts to mobilize African political will, financial and human resources to effectively implement optimal solutions.

Africa has been slowly industrializing, has to diversify the region's economies, build resilience to shocks, and develop productive capacity for sustained economic growth. This commitment to industrial development led to the adoption of the New Partnership for Africa's Development (NEPAD) in 2001. Economic transformation through industrialization is a guiding principle of NEPAD as a critical vehicle for economic growth and poverty reduction. Also, in 2008, the Plan of Action for the

Accelerated Industrial Development of Africa (AIDA) was adopted (UNEP 2016), which builds on the momentum created to industrialize as well as focus on the need for African researchers to perform cutting-edge research to produce strategies for solving challenges of climate change and water resource management. These research themes will contribute to enabling African communities to mitigate their existing problems, improve their position to leverage opportunities, and adapt to and increase resilience against the threat of climate change. This would be expected to position Africa to be an active player in global markets.

In order to be competitive in the long term in the global economy, African communities facing adverse effects of climate change must urgently embrace policies and development programs that promote effective and sustainable management of water and land resources. Vulnerability to the adverse impacts of climate change may be mitigated through meticulous management of water demands, eradicating waterborne diseases and controlling water stresses caused by droughts and floods. Experiments on cost-effective coordinated management of freshwater and saltwater, treatment and reuse of polluted water for irrigation, and artificial recharge of aquifers at catchment scales are essential. This can include cost-effective urban wastewater and sludge treatment systems for nutrient and water recycling for agricultural use and urban water supply. Strategies to build national food security can examine the commercialization of indigenous African small-scale farming and enhance reforestation, pasture revitalization and diversified animal production and breeding techniques. Further efficiencies can be pursued through cost-effective mechanization of smallholder farms with low-cost technology for supplementary irrigated agriculture, tillage, environmentally friendly crop protection, and improvements to harvesting and post-harvest processes.

Forging sustainable and effective technology for rural African areas by adopting renewable energy technologies for these communities may save forests and economically empower people living in rural environments. African researchers and development partners should be at the forefront of producing and implementing effective socioeconomic models that diversify agricultural and industrial production systems, strengthen capacities of urban and rural African communities, eliminate unnecessary institutional constraints, and control economic and socio-political conflict. African communities should be producers of new, innovative products that compete and/or complement existing products in international markets. New products require new production systems that must be developed, tested, and commercialized, and hence the input of international research collaborators and institutions is indispensable.

2 Water and Land Resource Demand Management, vulnerability to climate change Impacts, Sustainable Food Security, and Socioeconomic Development

Climate change threatens to negatively impact the availability of water, which directly threatens the economic development of vulnerable developing countries in Africa. Vulnerability is exacerbated by current demographic trends, and related land-use changes. Other drivers of change are increased population pressure, economic development, and urbanization trends. Rapid urbanization and economic development have been the main drivers of water demand and pollution (spatially and temporally), so they impose cross-cutting pressure on a scarce but indispensable environmental resource, worsening imbalances between supply and demand, especially in semi-arid and arid areas. In the recent past, these drivers stressed water resources far beyond the stress caused directly by natural global climatic changes (Al-Kalbani et al. 2014). Increasing demand for vulnerable freshwater resources combined with prevalent management systems for water and land resources has resulted in the too-slow improvement of livelihoods and well-being of many inhabitants of developing countries (Gain et al. 2012).

About 500,000 km^2 of land in Africa is estimated to be degraded due to soil erosion, salinization, pollution, and desertification. Deforestation, forest fires, over-cultivation, inefficient irrigation practices, overgrazing, overexploitation of resources, and uncontrolled mining activities, as well as climate change and variability, are blamed for the degradation (UNEP 2016). Arresting the degradation and rehabilitation of the degraded lands requires shifts from current management practices to scientifically proven solutions for the various causes of degradation. Cutting-edge technology and practical and cost-effective strategies that are best developed through research are required to deal with the root causes and effects. The consequences of the degradations include reduced agricultural productivity with concomitant effects on food availability, nutrition, and human health. This research may lead to understanding cross-cutting issues like social, economic, and political tensions; migration; and the spread of diseases (UNEP 2016).

Inefficient irrigation practices and poor management of fertilizers led to loss of topsoil and salinization, resulting in loss of soil fertility and poor plant growth. The areas most affected by salinization are the arid and semi-arid regions of North and Southern Africa, where about 30% of irrigated land has been lost to salinization. Desertification, desert encroachment, and an alteration of hydrological regimes have been observed in several African ecosystems and regions (Odjugo 2010; Descroix et al. 2009; IPCC 2007). Research into a combination of agricultural management best practices and use of appropriate irrigation technology may solve salinization problems on irrigated lands. Practical solutions to desertification and salinization challenges may require reforestation, pasture revitalization, and best practices in animal production and agronomy in large-scale agricultural production systems, and management of overgrazed lands and regulation of nomadic pastoralism. The huge challenge is finding appropriate methods for utilization and/or conversion of deserts and semi-arid lands into productive industrial or agricultural lands cost-effectively and sustainably at large scales, and simultaneously fight land degradation as well as alleviate poverty and food insecurity.

According to UNEP (2016), approximately 85% of people in sub-Saharan countries live in rural areas, with farming making up their main source of income. Rainfed agriculture produces much of the food consumed and accounts for more than 95% of farmed land in sub-Saharan Africa (IWMI 2018). Therefore, the quality of life in many parts of Africa depends on climatic factors, especially frequency of rainfall events and severity of agricultural droughts. Sustainable food production is under threat from droughts and floods, especially for countries that rely on low-input and low-output subsistence production systems. Small-scale agricultural production systems are often subsistence, labor-intensive, rainfed, or informally mechanized, while many irrigation schemes have water-inefficient irrigation systems that rely on costly non-renewable energy. The cost of energy, financing and maintenance, insufficient economies of scale, and consequently modest profits remain barriers to the sustainability of many small-scale rainfed and irrigated farms.

Africa is committed to industrialization as a way of diversifying the region's economies, build resilience to shocks, and develop productive capacity for sustained economic growth. Primary industries like agriculture and the extractive sector dominate industrial activity in the region, which are likely to cause localized environmental damage such as water and land degradation (UNEP 2016). Africa's industrialization must take advantage of the region's abundant and diverse resources, including agricultural and mineral assets, but it must also embrace green economic approaches to improve resource efficiency and cleaner production, including the reduction of greenhouse gas emissions. Can improvements in technology and productivity as well as more environmentally protective approaches broaden and strengthen economic fundamentals, build wealth, create jobs, and raise the standard of life of communities with little capacity to adapt to climate change, as well as protect African river basins against pollution?

Africa has the world's fastest growing population, and hence the ratio of land and water availability to person continues to dwindle. Demand for freshwater is increasing with economic development and the rising population, and this resource is under threat from pollution and climate change. Competition for water resources is driving prices and local-level conflicts, with impacts on freshwater-dependent economic activities such as agriculture and electricity generation (Jacobsen et al. 2012; UNEP 2016). Cooperation at national and regional levels and among sectors, as well as technology development, is essential for the sustainable management of water resources in the region and in turn to improve food security and socioeconomic development. How best to cooperate in management of transboundary river basins and among communities with competing and diverging interests is a challenge for many African countries. Is equitable benefit sharing possible, and would it enhance cooperation and economic development of transboundary water resources? Case studies of the construction and management of dams on the Kunene River for Ruacana hydropower generation, inter-basin water transfer for urban and rural water supply, and irrigation development in the Cuvelai-Etosha River Basin in southwestern

Angola and northern Namibia are models that may be adoptable across the continent. Management cooperation and benefit sharing on the Kariba Dam between Zimbabwe and Zambia are not unique, but lessons learned can be applied to other projects to optimize peaceful utilization of available shared water and land resources. Identifying opportunities for intra-African virtual water trade between water-rich parts of Africa (e.g., the Congo River Basin) and water-scarce regions like the Sahel and the Sahara, or semi-arid parts of southern Africa, may be other options worth exploring.

2.1 Conjunctive Use of Surface and Groundwater as a Tool for Integrated River Basin Management

Cooperation of stakeholders improves water management, leading to well-planned recharge and recovery phases and consequently to a lower probability of harm to surface and groundwater resources. Today, conjunctive water management is less common than conjunctive use, because of weak institutions and coordination mechanisms (World Bank 2006). To improve African resilience to projected water stresses, stakeholders in the water sector must invest in and reform water management institutions to prioritize implementation of conjunctive water management before conjunctive use. New methods are required to streamline conjunctive management reforms and investments so that they optimize the balance between improving infrastructure and building conjunctive management capacities. This can be achieved through improved monitoring systems and management practices, institutional reform and better alignment of incentive to priorities (Shah et al. 2005). In many African river basins, research is still needed to quantify and understand the role of watershed management and particularly role of soil and water conservation practices on aquifer recharge, metamorphosis of river runoff and surface and aquifer storage, natural recharge, potential for induced natural and artificial recharge of the aquifer(s), and comparative economic and environmental benefits derived from the various possible recharge options (CSE 2002a). The development of GIS techniques could certainly help to assess overall water resources within a river basin and the effect of various changing human interventions such as soil and water conservation practices, land clearance for agriculture, veld fires over large areas, and construction and operation of large or small dams.

2.2 Practical Conjunctive Water Management Systems for Surface and Groundwater, Saltwater and Freshwater Resources to Adapt to Climate Change

Industrialization and the growing urban population of Africa will increase freshwater demand (Jacobsen et al. 2012). Therefore, new coping mechanisms to deal with freshwater scarcity must be explored, e.g., conjunctive water management of saltwater and freshwater resources. Abstraction from salt- and freshwater-stratified aquifers, like the Ohangwena Aquifer that straddles the northern Namibia and southern Angola border (Sorensen 2013) and commercial use of saltwater resources like the great saltwater Turkana aquifer in northern Kenya (Ruvaga 2015), is a huge challenge for the communities that may benefit from these water resources. This necessitates research into cost-effective ways to abstract and produce freshwater from saltwater aquifers. Studies must include abstraction systems for freshwater- and salty water-stratified aquifers, dilution of salty water with good quality surface water for human and animal consumption, and recharge of saltwater aquifers with treated surface runoff and commercial uses of salty water. Required may be studies on appropriate adoption and implementation of solutions from some developed economies who use brackish groundwater as cooling water for power generation, aquaculture, and a variety of applications to the oil and gas industry (such as drilling), enhancing recovery and hydraulic fracturing (USGS 2018). Other possible solutions that require studies include pumping out salty groundwater for deposition in the sea, then recharging the salty aquifer with freshwater, desalinating salty groundwater as well as extraction, and processing of the salts for industrial uses.

Many transboundary rivers in Africa pass through semi-arid and/or arid areas, and aquifers recharge zones before discharging into seas, but the recharge and extent of aquifer-river interactions are not well documented or known. Most rivers discharge their nutrient-rich sediment into sea deltas (which is important for sea and delta fauna and flora) so research is necessary to improve methods for flood water diversion, clarification for aquifer recharge, and to return sediment to the river and sea during flood seasons. Recharging depleted and salty aquifers with freshwater to respectively increase water yield and reduce salt concentration during heavy rainy seasons and flooding periods is worth exploring, especially in semi-arid and arid regions under water stress. According to CSE (2002b), the village of Balisana in India was successful when it harvested rain to reduce fluoride levels in well water by using wells that recharge artificially. Can this technique be applied to saline and polluted aquifers, by pumping out saltwater during dry seasons and using it as a raw material in the production of salts and industrial chemicals and/or dilution of the saltwater with freshwater during rainy seasons? Another option could be the deposition of water in the sea using flooded rivers during the rainy season, while diverting some flood water for treatment before recharging the aquifers for dilution of the salty groundwater. These approaches can be analyzed to determine the most cost-effective methods for diversion, treatment, and artificial aquifer recharge, potentially using renewable solar and wind energy in abstraction and saltwater treatment.

According to CSE (2002a), watershed management is an effective method to intercept dispersed runoff. Many techniques for water conservation have been developed along hill slopes with the intention to prevent soil erosion and reduce surface runoff, and then increase infiltration in the ground to recharge aquifers. This could increase recharge of aquifers as well as water retention, which, on a large scale, can change hydrological conditions within a river basin for a considerable period of time. Before such an approach could be attempted, thorough analysis of the impacts of catchment modification using soil and water conservation measures on the hydrology of each catchment would be required. In extremely arid catchments, several techniques have been tested to improve control and regulation of flash floods (spate) and reduce sediment transport. Unfortunately, current design and feasibility studies are often hampered by lack of adequate data on runoff from flash floods (Hatibu and Mahoo 2000). Therefore, data collection and more researches are required to devise better methods to harvest, treat water, and recharge aquifers for entire river catchments in a cost-effective way when there are flash floods. Other techniques that may also be tested are use of check dams, underground dams, and sand dams, as well as use of various types of artificial recharge wells.

Peak runoff is a significant proportion of the total discharge of rivers, and it usually occurs during periods with the smallest water demand (FAO 1995). Transferring water from the high-supply season to the high-demand season is a challenge usually solved by storing surface water behind dams. Unlike storage of water in aquifers, surface reservoirs have evaporation and sedimentation problems, for which practical and sustainable solutions must be developed. Groundwater storage is seldom harmful to the environment, but groundwater aquifers rarely have large storage capacity, so they generally cannot absorb great volumes from floods in a short period of time, and they also have low water yields per well (CSE 2002a). Therefore, water experts must find a balanced, practical conjunctive water storage systems for surface and groundwater, especially for water-stressed dry areas where optimizing catchment water retention is a priority. A practical example is the Omaruru Delta Reservoir and aquifer artificial recharge system, in the Namib Desert, that supplies water to Namibian coastal towns. A surface storage reservoir is used to harvest flood water and clarify the water by natural settlement of suspension for a few days. The clarified water is used to recharge an aquifer, which is a cost-effective solution for retention of water in the catchment.

Aquifers can offer natural water distribution systems to water users within and between neighboring catchments. This requires extensive knowledge of the hydrology of the catchments and of the nearest aquifers including the energy and cost implications, inter-basin transfers, and linkages. Aquifers may leak into other catchments, and hence they are natural conduits of water for inter-basin water transfer. Understanding the hydrogeology of all transboundary catchments and aquifers requires careful hydrogeological investigation to avoid conflict if equitable water allocation and abstraction are to be implemented under these conditions. One typical example is the Nubian Sandstone Aquifer in the Sahara Desert that lies beneath Egypt, Libya, Sudan, and Chad, believed to be the largest fossil aquifer ever discovered. Total and useful volumes vary greatly. Some studies suggest that the groundwater there is not entirely fossil because little recharge exists from seepage from the Nile River (Dolezal 2013) and possibly from the Tibesti Mountains and/or Ennedi Mountains in Chad. This is one of many African aquifers where factors that govern aquifer yield, recharge, and storage are not well documented or understood; thus, comprehensive studies, information sharing among stakeholders, and inclusive planning are required for strategic management of the water resource. Most likely, enhanced artificial recharge of the Sandstone Aquifer using Nile floodwater can improve water security in the Sahara Desert. Studies maybe required in order to compare the benefits of conjunctive management (use of the aquifer and Nile River surface water resources) to current independent use of surface storage reservoirs in the Nile Basin with low utilization of the Nubian Sandstone Aquifer. Artificial recharge may enable conjunctive use of the groundwater and surface water of the Nile River. This may in turn enable the building of huge settlements and the development of hinter Sahara Desert areas which may help to decongest settlements along the Nile, hence reducing the pollution risk of the Nile River. Comprehensive studies are required to formulate conjunctive management strategies in order to avoid unintended consequences such as conflicts among riparian states on the Nile River.

Storing water in disused and decommissioned mines through artificial recharge may reduce evaporation losses. However, it would require finding solutions to seepage losses, reducing pumping costs, and decontaminating water of heavy metals and mining chemicals and related contaminants. As part of an environmental management strategy effort, implementation of cleaner production systems or mine end-of-life (mining venture) clean ups that leave no water contaminants must be developed. This would enable decommissioned mines to be used as safe groundwater water storage facilities. Finding effective clean-up methods and identifying the most appropriate uses for such water (such as irrigation) are new challenges for researchers.

In South Africa, disposal of coal and gold mine wastewater is a problem because of the high concentration of chemical pollutants that make the wastewater unsuitable for direct discharge into rivers. Irrigation with mine water in suitable soils can solve the twin problems of wastewater disposal and shortage of irrigation water. A field trial in South Africa used center pivots for irrigation with coal mine wastewater on virgin (unmined) soil and mine-rehabilitated land. Several crops including wheat and beans were successfully irrigated with gypsiferous mine water on a commercial scale for 3 years, which resulted in plans to scale up the project (Shah et al. 2005). Therefore, further research into cost-effective methods for achieving practical conjunctive water management that make use of water storage in decommissioned mines and mine drainage water should be encouraged. Shah et al. (2005) further recommended that research should explore practical strategies for proper institutional and organizational development, including incentives for conjunctive management among different stakeholder groups, water pricing structures for both fresh and wastewater, and institutional reforms and investment in the capacities of local governments to lead participatory groundwater and integrated water resource management initiatives.

Energy for pumping and water treatment is required to recharge and abstract water from aquifers. To enable commercial activity, it is necessary to reduce capital and other costs by providing renewable energy from solar and wind farms, especially in arid and semi-arid areas which usually depend on groundwater for water supply. An example of the exploration of such methods is a pilot project using solar-powered abstraction and desalination of saltwater that took place in the Cuvelai-Etosha Basin in Namibia. Evaluation of sustainability of this approach is especially useful because beneficiaries are low-income communities with few skills and little capacity to manage novel technology. Additional research is necessary to build capacity, and to invigorate Africa's political framework to prioritize research in developing cutting-edge technologies and large-scale implementation of such technologies for solving major African problems.

Exploitation and pollution in urban areas can exacerbate groundwater supply challenges. According to CSE (2002b), making rainwater harvesting mandatory in Delhi helped to mitigate depleting groundwater tables. Rainwater harvesting and purification, gray water recycling, and ecological sanitation are potential solutions for mitigating water scarcity in urban areas. Demand-side management, financial and institutional support, groundwater contamination, and an integrated approach to water literacy, economics, etc. need to be addressed before harvesting rainwater for aquifer recharge in urban areas. Other relevant study themes are urban surface runoff and sewage harvesting, catchment area treatment, watershed development, artificial recharge, and water harvesting under special conditions such as groundwater harvesting and snow harvesting. Formulation of methods for sourcing of funds for subsidies and financial assistance as well as distribution and administration of funds is also of serious concern for those who implement urban water harvesting projects.

2.3 Developing Cost-Effective Technologies for Floodwater Harvesting, Treatment and Artificial Recharging of Aquifers, Sediment Harvesting, and Irrigation

Flash floods that occur in semi-arid areas can be effectively harvested and utilized, for example, use of the majaluba system devised in Tanzania (Hatibu and Mahoo 2000), but there are serious constraints. River basin-scale challenges include the optimization of water harvesting, river channel storage, runoff diversion, and aquifer recharge in a practical and economically sustainable way. During floods and high flow regimes, rivers naturally desilt certain reaches and deposit the sediment in other reaches, creating and filling up water pools simultaneously. This is why it is imperative to identify the best methods and technologies to control hydraulic deposition and desilting of pools and dams, for low-cost river channel water storage and sediment management in semi-arid areas. Technology uptake and river basin-scale implementation by communities are additional challenges for researchers, NGOs, and governments. Studies are required to determine best practices for flood water harvesting, sand trapping, enhancing aquifer recharges by maximizing retention time of floodwater, and trapping selected sediments (e.g., gravel, sand, and silt) while releasing the suspended fine particulates downstream and protecting the integrity of estuary ecosystems. One of the biggest challenges is optimizing water harvesting for entire river basins and simultaneously protect the riverine ecosystems. Therefore, researchers must identify a balanced system for reducing soil erosion and river sedimentation with sustainable river sand harvesting techniques (especially for construction) that cause negligible disruptions to riverine ecosystems, optimize subsurface water storage, and yield of subsurface water abstraction from silted rivers and sand dams.

During rainy seasons, flooded streams often overflow, and water tables are high, and hence aquifers and even some ephemeral streams can yield adequate water for supplementary irrigation seasonally. This can alleviate impacts of mid-season agricultural droughts, the main cause of crop failure for rainfed crop production in semi-arid climates. Devising commercially viable low-cost technologies and cheap energy for flood water and groundwater abstraction for supplementary irrigation is a challenge that researchers and farmers must solve in order to adapt to droughts. Small- and large-scale irrigation projects may be used for recharge of aquifers with surface freshwater as well as increase food production but environmental pollution with nutrients are major concerns. How best to derive benefits from all forms of irrigation and artificial aquifer recharge while minimizing adverse environmental effects and upscaling the benefits to catchment levels across political and administrative boundaries remains a challenge for researchers and projects managers in African countries who operate with limited resources and in stifled information sharing environments.

3 Water and Soil Conservation at River Basin Scale Through Eroded Soil and Water Harvesting, Afforestation, Soil Fertility, Erosion, and Sediment Management

According to UNEP (2016), one of the main drivers of aridity in Africa is deforestation. Changes in land cover and use in Africa are largely driven by population growth, urbanization, and investments in large-scale commercial agriculture. Overgrazed, deforested, and arable lands are susceptible to wind and water erosion. River channels and natural pools are filling up with silt, which limits surface water storage and groundwater recharge. Restoring pools in river channels (vital for survival of riverine aquatic life during dry seasons) requires practical and environmentally friendly methods. Biological and mechanical soil and water conservation measures have been proven to be effective in protecting land against degradation and desertification, but only for small areas. These measures can include veld pasture management, afforestation, and mechanical soil conservation measures, e.g., zero-grade contour construction and dugouts for water retention and low-cost check dam construction for silt trapping and gully erosion control. Large-scale conservation efforts in degraded African communal lands must account for socioeconomic and political complexities. Surface runoff in semi-arid areas often degrades fertile lands and carries fertile sediment that is deposited in river channels, behind dams, and estuaries. Conservationists must find cost-effective ways of harvesting sediment, e.g., by using sand traps, desilting surface water reservoirs, and assessing the feasibility of desilting and reusing the fertile silt/mud to rehabilitate degraded arable lands, and using infertile sand for construction in urban settlements. Probably, regulating trade in reclaimed river sediments may solve the problem of illegal sand mining (for use in construction) that has caused degradation of surroundings of many African towns.

Efforts to improve soil fertility, reduce erosion, and manage sediment to protect surface water resources must incorporate diversified production, enhanced reforestation, and pasture revitalization. Regulating livestock populations, planting grasses, reforestation, and afforestation of grazing lands with carefully selected indigenous plants that can fix nitrogen, and tolerate salty soils and droughts may succeed in reversing the deterioration of pastures. A good case study was conducted in four villages in the Mount Moorosi area in Lesotho, where communities successfully rehabilitated rangelands by constructing physical barriers on the mountain slopes to slow runoff, trap sediment, and promote water infiltration. Significant obstacles to rehabilitation of communal pastures in large catchments demand well-informed livestock management practices, policies, and community participation facilitated by broad participation in self-regulating systems which include guidance on stocking levels and community education. Research is required for refinement of regulations and their enforcement, and analyze use of property and land-use taxes to generate funding for conservation. Scholars have suggested the introduction of soil conservation incentives (such as loans), land reclamation bonds, soil loss, and legal liability for water pollution and liability insurance (UNEP 2007). These approaches have been used successfully elsewhere, but their introduction to Africa (e.g., communal areas) may require modifications.

Many streams and rivers are drying up quickly at the end of rainy seasons due to erratic and low rainfall, but siltation and deforestation of catchments (especially in riparian zones) could also be contributing factors. A long-term observed effect (indigenous knowledge) is that some wetlands, springs, and streams in semi-arid areas of Africa dry up whenever riparian vegetation like water berry (Syzygium cordatum) has been cleared. One hypothesis is that water-use-efficient riparian trees provide cool shades over the wetlands, springs, and shallow water channels which reduce evaporation rates. This means that evapotranspiration could be less than evaporation in a forested riparian zone with indigenous trees that are water use efficient. If scientifically proven, this hypothesis could lead to one method to protect riverbanks from erosion and shade streams and springs to reduce water evaporation through reforestation of the riparian zones with water use-efficient vegetation. This could be upscaled in riverbank protection and water conservation efforts. Research would be required for selecting a wide range of desirable traits in trees and grasses that are appropriate for the targeted environment, seedling production, and afforestation of degraded and deforested areas with the plant species that are economically lucrative and easy to propagate in each river basin. These traits and uses could include floods, drought, and salinity tolerance; rooting systems that aid conservation of soil and water; food security; wood fuel and/or timber production; medicinal uses; etc.

Africa's industrialization takes advantage of its abundant and diverse resources, and hence primary industrial activity in agriculture and the extractive sectors dominate economic activities in the region. These activities are likely to cause localized environmental damage such as pollution of water bodies and land degradation (UNEP 2016). Besides its effect on land, extraction activities, like mining, also result in considerable waste management challenges and water pollution, for example, drainage from acid mining challenges in South Africa (CSIR 2009). Reclaiming mine wasted lands for productive and sustainable activity and acid mine drainage in a cost-effective manner presents challenges that require serious attention in Africa's mining areas. African governments must devise management strategies that ensure that agreements to extract natural resources adequately fund restoration and compensation for environmental damage. This ensures that the short-term benefits from extractive industries are not outweighed by long-term penalties on future generations. African think tanks must devise resource management strategies to guarantee that the wealth generated from the extractive sector is used to protect the environment, reclaim already polluted lands, tackle water scarcity challenges, improve adaptation to climate change, integrate the extractive sector with improved environmental conservation and agriculture production, raise the standard of living in rural areas, and increase the number of people with productive livelihoods in affected communities.

3.1 Conjunctive Management of Fresh, Salty and Wastewater for Irrigation and Artificial Recharge of Aquifers

Conventional desalination plants may seem to be an attractive option to supply water to urban areas and large settlements in arid coastal zones, but Akinaga et al. (2018) suggested that more comprehensive desalination, which is coupled with agricultural and salt production may be more sustainable by improving food security, increasing economic activity, and reducing pollution of land for inland desalination plants. Chauhan et al. (2016) stated that this approach mitigates the potentially negative impact on the marine environment from brine discharges. Seawater greenhouses that are used to produce high-value food crops at commercial scales in hot and dry coastal regions exist, but disposal of brines from the greenhouses is considered one of the major challenges associated with all desalination plants today (Akinaga et al. 2018). Also, seawater greenhouse technology may require significant modification to be suitable for year-round use in cold and dry coastal regions of northern and southern Africa, which have Mediterranean climates. Akinaga et al. (2018) indicated that designs have been optimized for smallholder farmers, even suitable for a small family-run farm. Scaling up the technology will require state backing, political will, and affordable funding systems. Use of locally available or manufactured materials may reduce capital costs and could ensure sustainability and lower production costs, and hence they may be applicable to low-value crops. To advance this potential, collaborative research must be nurtured for owners of the technology and interested African institutions.

Coastal seawater greenhouses could be a cost-effective mechanism for turning vast desert coasts of Africa into productive green belts. Further studies are required for adopting this technology for larger-scale production of low-cost staple crops and/or complement rainfed production systems in order to reduce the cost of production and risk of drought-induced failure. Finding cost-effective ways to scale up technology under existing patent laws for mutual benefit of patent holders and poorly resourced African farmers is a challenge that must be tackled. Widespread use of such seawater greenhouse projects could complement environmental conservation efforts to halt desertification.

Generally, desalination is an energy-intensive process, and hence irrigation with desalinated water is often too costly for production of low-value crops. Proposed sustainable solutions for resilience against droughts and flooding lie in provision of renewable energy like solar pumps which can drive simple but efficient seawater green houses and irrigation systems, and installation of small-scale local grids that can power cold storage and post-harvesting processing mechanisms. These can enable smallholder farmers in remote areas to add value to their crops and store and market their produce more profitably, thus creating jobs; increasing income; socially stabilizing communities; and potentially decreasing poverty, hunger, crime, and extremism while increasing resilience to socioeconomic turbulence (UN 2016; Thomson Reuters Foundation 2018; The Economist 2014). Major challenges facing the developing countries of Africa are sustainably funding implementation of these food security measures, and seamlessly developing and adopting technologies productively, commercially, and competitively.

4 Strengthening Capacities of Urban and Rural Communities to Circumvent Institutional Constraints and Economic and Socio-Political Turbulence

FAO (2016) stated that eliminating poverty in rural and peri-urban areas is essential to eradicating hunger and poverty globally. Peri-urban poverty and rapid growth of informal settlements in African urban areas are closely linked to rural-to-urban migration as the rural population seeks jobs and improved standards of living in the cities. Vibrant land markets are slowly developing in peri-urban communities, but it is unclear if this is resulting in increased land productivity. The lack of political will to prioritize agriculture, particularly to transform peri-urban and rural smallholder agriculture, may be among the most serious post-independence misjudgements by African nations. Smallholder agricultural development, job and wealth creation, and food security in African rural areas should be taken seriously by researchers and extension services. These developments cannot be divorced from issues of democracy, politics, and governance (Rukuni 1997) as well as other cross-cutting issues like enabling socioeconomic environments that ensure adequate risk management and market linkages. Also, relevant research must focus on mobilization of new capital, funding models, specialization and commercialization of diversified production systems, and entrepreneurial training and extension.

Some studies have shown that smallholder farm investment is financially viable, but performance is better where rainfed systems are integrated with irrigation, and the irrigation system is owned and operated by farmers. More successful still is where farmers either develop their own individual water source and/or manage their irrigation independent of other farmers (Rukuni 1997). These findings suggest that smallholder farms need more flexible machinery and equipment design to optimally service 1–3 ha smallholder farming systems. Technology must enable the farmer to diversify into high-value crops and to integrate full scale, supplementary irrigation and dryland farming systems.

Security of land rights is also important, and hence smallholder farming communities must adopt governance systems that assure rights to land resources (Namubiru-Mwaura 2014). Long-term tenure will encourage direct investment in new technology and increase agricultural productivity, as well as increase the capacity for fair compensation and ensure that restoration rights and benefits are shared by affected communities. Most African governments, after political independence from colonial masters, have inadvertently maintained the colonial legacy of undermining indigenous tenure systems (Rukuni 1997), which has discouraged investment and productivity in rural areas. Most African governments have been slow to reform these systems and rights relating to natural resources, particularly water. Reforms that strengthen tenure security and allow greater control by African communal farmers over the land and water that they use will likely yield greater investment and more efficient utilization of land, water, and irrigation systems.

Research to strengthen community-based legal and institutional mechanisms for tenure and property rights is required to demonstrate the interests of African governments in decentralizing and strengthening traditional institutions. Such interests include conflict resolution and the coordination of the government and farmers in the planning and governance of land and water resources. Best practices such as Zimbabwe's Communal Areas Management Programme for Indigenous Resources (CAMPFIRE) could be adopted elsewhere to enable local communities to manage and utilize their natural resources to their exclusive economic benefit. The approach is proven for the management of wildlife held on communal land, but its broader effects, such as its robustness on sustainable natural resources management under a changing climate, socio-political, and other cross-cutting challenges linked to Zimbabwe's turbulent macroeconomics, must be studied, in order to fully understand if the CAMPFIRE concept is adaptable to macroeconomic challenges, is applicable to management of water resources, and can be copied and applied in other countries. What is the best mechanism for governments to decentralize and strengthen traditional institutions (including the ability to resolve major conflicts) for conserving natural resources? Research is required to determine appropriate targeted solutions that include institutional support and enforcement of regulations for effective management of communally shared resources like pastures and water, especially in cases where administrative and river basin boundaries overlap.

Rukuni (1997) stated that the chronic inability of smallholder farmers to represent their economic interests in the political process is cause for serious concern, particularly in societies with both commercial and independent agriculture sectors. Factors that support the production and product markets of commercial farmers are better represented in the national economy. Similarly, institutions that serve agriculture are also historically better suited for the needs of large-scale agriculture. But in sub-Saharan Africa, the next-generation agricultural population will be younger with smaller areas of land to farm (FAO 2016). Hence, research is required to improve and modernize transport, storage, and processing, and build marketing capability cost-effectively for smallholder farmers. How can cost-effective financing methods, like micro-credit schemes, be tailored to support smallholder farming? Micro-finance schemes could be the decisive factor for sustained adoption and sustainable low-cost mechanization in Africa for supporting production, marketing, and processing of produce from small-scale farms. This includes devising practical methods that enable African smallholder farmers to consistently produce high-quality products that can generate high market value (Rukuni 1997), and hence complement and/or compete in global markets, currently dominated by large-scale commercial farmers and multinational corporations. In West Africa, cocoa is produced predominantly by smallholder farmers on six million hectares, and this provides about 70% of the global cocoa production (Mfegue 2018). Probably, with the aid of research, extension, and well-coordinated marketing, African smallholder farmers may supply more products into the global markets by commercializing and consistently supplying new indigenous organic crop and animal products.

5 Commercializing and Mechanizing Indigenous African Small-Scale Farming

Small-scale farming is expected to dominate agriculture in Africa. With the world's fastest growing population, per person land availability in Africa continues to dwindle and hence efficient food production technologies are needed to meet the region's nutritional needs while also preserving the integrity and health of land resources (UNEP 2016). This requires injection of innovation to increase productivity and adaptation to climatic, socioeconomic, and political turbulence. The agriculture sector is the major employer in many African countries; unfortunately, the majority are small-scale peasant farmers with land ranging 1–3 ha (FAO 2016). Smallholder family farms are often more productive than larger units in terms of yield per hectare and profit per hectare and therefore have greater potential for alleviating poverty, hunger, and unemployment (Hazell et al. 2006). Hence, appropriate agricultural technologies that improve productivity and reduce the drudgery of farm work, development of finance capital, robust credit schemes, and building wealth and industrial development in smallholder farming areas are challenges that need to be explored to address the demographic shifts of migration of young labor from African rural areas.

According to Fisher and IAEA (2018), some Kenyan farmers are increasing crop yields and enhancing the fertility and quality of the soil in an environmentally friendly and cost-effective way through an integrated cropping-livestock production system that recycles the nutrients present in both animal manure and crop residues. This reduces the need for synthetic fertilizers that release large quantities of greenhouse gases and thereby contribute to climate change. Can the Kenyan model be upscaled to all African smallholder farms and even pastoral environments? A critical piece of this challenge is the integration of nomadic pastoralists into a cropping-livestock production system.

Introduction of exotic crops to Africa has displaced many indigenous African crops (Odeigah and Osanyinpeju 1998). Agricultural research, particularly by scientists in developed countries, has traditionally focused on staple foods, while little attention has been given to underutilized and neglected crops, such as cereals, legumes, and fruit tree species indigenous to African arid and semi-arid agroclimatic conditions (Heller et al. 1997). Africa has a number of forgotten plants (trees and grasses) that can be used for food and stock feed production. Commercial agriculture, particularly of exotic crops that require large capital investment of irrigation, agrochemicals, and technical training, is largely responsible for directing attention and resources away from forgotten plants and their past and potential contribution to generating value and reducing food insecurity. Attention must be given to the commercialization of organic production, agro-processing, and marketing of forgotten crops. Successful uptake in Africa may lead to profitable introduction to global markets.

Forgotten cereals include Oryza glaberrima Steud (African rice), Sorghum bicolor (sorghum), Eleusine coracana (finger millet), and Pennisetum glaucum (pearl millet). Sorghum and finger millet are drought-tolerant but are currently produced and marketed in few African countries at low profit margins and with few commercial applications (like making beverages), if compared to maize and wheat. African rice has unique traits that make it suitable for low-input agriculture, such as a tolerance to salt, drought, flooding, pest resistance, weed competitiveness, and the ability to grow on infertile, acidic soils. African rice also matures faster than Asian varieties and its wide leaves shut out weeds (Harlan 1995; Linares 2002; Sarla and Swamy 2005). Namibia has successfully commercialized the production of an indigenous African long grain rice variety, cultivated mainly in the Zambezi valley.

Indigenous legume crops that may be candidates for commercial production are cowpeas and Vigna subterranean (bambara groundnuts). Biochemical analysis of the composition of carbohydrate, fat, and protein reveals that bambara groundnut produces an almost complete diet, owing, in addition to its nutritional composition, to its functional properties and antioxidant potential. Bambara groundnut is composed of 15–25% protein, and it has been used in high-protein foods like porridge for children and household breads, biscuits, and soups in sub-Saharan Africa. Its organic vegetable milk has been rated higher in acceptability than milk from other legumes like soybean and cowpea (Murevanhema and Jideani 2013). The crop is underutilized because of its "hard-to-cook" (HTC) quality in combination with inadequate processing techniques (Mubaiwa et al. 2017). Further research may provide cost-effective methods to optimize farming and food-processing techniques at industrial scale. As a drought-tolerant crop that is relatively resistant to disease and pests, the bambara groundnut can be a new product for the food industry that can contribute to fighting malnutrition in disadvantaged communities (Murevanhema and Jideani 2013) or even stock feed manufacturing. Pilot experiments are also being conducted on commercializing field production, processing, and marketing of other indigenous plants crops, such as Colocasia esculenta (dumbbell yams), Cucumis metuliferus (horned melon), and Cucumis Africanus (African spiked cucumber) that may improve national food security and unlock new revenue streams that may economically empower small-scale farmers in Africa.

African indigenous fruit trees with similar potential are Strychnos spinosa (monkey oranges), Sclerocarya birrea (marula), Lannea microcarpa (tree grapes), and many others. These trees can be found in sub-Saharan Africa around villages, and their potential for commercial production on large- and small-scale farms may have to be evaluated. Moreover, the leaves, bark, roots, and fruits of these trees have various medicinal properties, and they can also be used for timber. Above all, they are hardy and drought-tolerant, and are often undamaged by natural savannah fires (Styslinger 2011). Research and development are needed to identify and select desirable and more nutritious varieties of these trees. There is potential to plant these trees on deforested rural and peri-urban areas to conserve soil and water resources, enhance food security, and provide input for industrial production of wine and beverages, which may provide new flavors and new products untainted by chemicals and with low environmental costs. Research must optimize agronomic practices, breeding, and artificial selection to attain maximum yield and financial sustainability. Innovative, low capital-intensive, and commercially viable agricultural enterprises, adapted to African climatic conditions, are required for low-income rural communities, so that they can enter a niche in the global organic agriculture market.

Farmers in some parts of the Sahara Desert and in the Namib Desert in Namibia and South Africa produce dates at a commercial scale, while relatively wetter regions of sub-Saharan Africa use date trees as ornamental plants or as wild vegetation. South Africans have demonstrated that commercialized production of juices and wine from marula fruits is possible. The marula, monkey orange, tree grape, dates, and other indigenous trees may be the appropriate tree types for planting in degraded lands of Africa as part of a reforestation program to counter the increasing aridity and to build and diversify income, and hence increase resilience to impacts of climate change. Successful cultivation may require adoption of scientifically proven intercropping techniques that optimize yield, diversity, and environmentally friendly crop protection techniques for exotic and indigenous fruit trees.

6 Diversified Products, Enhanced Reforestation, and Pasture Revitalization

African forests and natural ecosystems are expected to decline (UNEP 2007), for example, a forest clearance rate of 18 million hectares per year was stated by Swanborough (2016). Arresting the decline may require scientific selection and planting, and reforestation of degraded woodlands and grasslands with appropriate trees and grasses preferably indigenous non-invasive plants. Grasslands may require preferably indigenous, nutritious, and palatable pasture plants that are fast-growing, drought- and salinity-tolerant, and with rooting systems ideal for soil and water conservation. Top priority could be given to plants that provide multiple benefits, which may include uses for fodder, timber, foodstuff, processing of fruit juices and other industrial products, water clarification, and medicinal products (e.g., moringa tree has many of these uses). Concerted studies would be required on agronomy, post-harvest processes, produce marketing, water use efficiency, and drought tolerance of the trees, as well as their ability to restore natural pastures. The research may include formulation of methods to scale up reforestation programs, implement targeted reforestation incentives, possibly using reforestation bonds, forest management bonds (UNEP 2007), and cost-and-benefit sharing schemes. An assessment of the success of conservation and reclamation of degraded lands should determine the total value of benefits from the ecosystem and net benefits of an intervention that alters ecosystem conditions, and it can identify winners and losers (for devising appropriate compensatory measures) and potential financing sources for conservation and produce strategies that make ecosystem conservation financially self-sustaining (UNEP 2007).

For reforestation and revitalization of pastures, research must reveal best agronomic practices for selecting trees that have multiple desirable traits for reclamation and reforestation. For example, Makhado et al. (2016) showed that Colophospermum mopane leaves, twigs, pods, and seeds are nutritious, and hence the mopane may be a suitable tree for reforestation and revitalization of pastures in semi-arid areas. Some challenges are that fresh mopane leaves are less favored by livestock during the wet season due to the high levels of secondary metabolites such as phenols and tannins (Hooimeijer et al. 2005; Wessels et al. 2007), which decline during the dry season as the leaves mature and age. The mopane leaves are rich in crude protein (8–16%) which is useful for livestock and wild animals, and for mopane worm feeding. Research on how best to restore deforested lands and pastures can intensify and optimize commercial production of stock feeds from indigenous tree products (like twigs, leaves, pods, etc.) with high crude protein content. Research may also experiment on partial domestication of some wild animals (like antelopes) that are capable of feeding on crude protein-rich fresh mopane leaves (Hooimeijer et al. 2005; Wessels et al. 2007). Antelopes such as oryx, kudu, and eland are adapted to drier environments, and hence pilot studies are required on the compatibility and viability of substituting traditional livestock (e.g., cattle) with harder animals like antelopes and/or mixed production of antelopes and harder traditional livestock like goats.

These researches may encompass large-scale reforestation of degraded lands with trees that have high-protein leaves and fruits/pods, like the African ebony (Diospyros mespiliformis), also called jackalberry (National Research Council 2008), as well as the mopane tree and other hardy legume trees, e.g., some accacia species. These studies could be complemented by studies on management of plant and animal diseases and pests. Reforestation pilot projects can also be used to examine soil and water conservation and the production of edible insects like mopane worms (Wessels et al. 2006). These edible, protein-rich insects can increase the availability of low-cost protein diets. Incorporating antelope domestication and breeding can increase global trade in African agricultural products, ecotourism, and food and financial security for arid regions of Africa. These efforts are likely to lead to the creation of new industrial products that do not compete directly with but rather complement existing production systems, diversify revenue streams and create new, environmentally friendly products, economically empower communities, and reduce migration away from African rural areas.

7 Cost-Effective Urban Wastewater and Sludge Treatment Systems and Reuse in Agriculture

Practical solutions that support the freshwater and human nutrient cycles while reducing pollution, especially in water-stressed regions, are urgently required. Many urban and peri-urban settlements in Africa use pit latrines which fill up after a few years and are sometimes reused after manual desludging. African cities should investigate the feasibility of adopting integrated urban water management methods which have worked well in Indian slums (Jacobsen et al. 2012), especially where there are pit latrines. Many African settlements that use waterborne sanitation and conventional wastewater treatment facilities often release effluents that do not meet prescribed environmental standards, especially on nutrient load. Wastewater and sludge treatment and reuse in agricultural systems to enhance food production and promote a cleaner urban environment remains an acceptable concept but has low global industrial applications. Upscaling this ecological sanitation concept requires serious researches in pilot projects in cities like Accra which produces 90% of its vegetables from wastewater irrigated plots (Jacobsen et al. 2012). Urban centers must develop integrated urban water management systems, with a prime objective of reaching nearly 100% recycling of macronutrients (NPK) using low-cost sanitation technologies that disinfect pathogenic organisms and remove odor from fecal matter and wastewater.

Theoretically, nutrients in domestic wastewater and organic waste are nearly sufficient to fertilize the crops needed to feed the world population (Refsgaard et al. 2005). Direct recycling of nutrients from human fecal matter as fertilizer has been found to be cost ineffective because of the availability of affordable and user-friendly synthetic fertilizers. The benefits of using fecal matter as an agricultural fertilizer were observed in Medieval cities in Asia, some of which were more hygienic than most European cities of that time that experienced higher recurrent water pollution due to the urban fecal matter (Sedlak 2014). Today, research challenges are in devising cost-effective methods for treatment and reuse of sewage in organic agriculture in a more environmentally friendly, socially acceptable, and technically and economically sustainable at a large commercial scale, especially in developing countries of Africa that have limited technology for collection, treatment, and handling of the sewage (in conventional waterborne systems) and pit latrine sludge. Innovators must develop reuse processes that are simultaneously economically viable, and user and environmentally friendly to compete with and/or complement synthetic fertilizers that dominate the market and are the main source of nutrient pollution of water resources. The costs and benefits of implementing new waste management systems versus substitution of old systems is another challenge (Refsgaard et al. 2005), and hence the new methods may have to compliment or improve existing systems in order to avoid huge capital outlays and enhance uptake by urban authorities. Commercial production and marketing of new ecological sanitation technologies due to low social acceptability is another area for examination. New policies and regulations, as well as new processing methods, may be required to reduce nutrient and water waste in current wastewater treatment.

Quality control of ecological sanitation systems could be based on contract farming or long-term agreements to use and manage urban fertilizers. This would be a vital element especially for poor countries where urban agriculture plays an important role in sustaining food security (Refsgaard et al. 2005). Formulation of appropriate new quality control and certification processes is required, starting with the commercial scale manufacture of organic fertilizers from urban waste, particularly fecal matter, to be used in agriculture. The industry would start with production of low-risk crops such as non-food plants and fodder crops as pilot projects to innovate and test for socially acceptable and economically optimized systems. Hence, researchers must formulate methods that would partially or fully replace government subsidies on urban waste collection, treatment, and reuse so that urban waste management systems are absolutely sustainable. Government subsidies are not always available especially during periods of macroeconomic hardships, which adversely affect performance of water utilities that rely on government subsidies. It would be perfect if the novel systems are transparently monitored and evaluated, as well as ensure that the farmers and/or food consumers are well educated and informed, and actively participate in monitoring and evaluation. This may enhance the economic and environmental sustainability of water services by ensuring total wastewater recycling in freshwater and wastewater services, improve effluent quality standards, and increase water supply security. Some such programs have already been partially tested in the southern African cities of Harare and Windhoek that have improved urban water supply security by recycling wastewater indirectly and directly, respectively. Therefore, pilot research on cost-effective urban wastewater treatment systems that surpass minimum effluent quality standards, combined with wastewater recycling, may lead to practical solutions that help to close the freshwater and human nutrient cycles, reduce pollution, mitigate water stresses, and improve resilience to drought.

According to Sedlak (2014), current pricing methods and price control mechanisms for water services are not sustainable. For example, U.S. pricing models fail to incentivize water utilities to invest in existing water reticulation infrastructure or expand services, and hence the utilities rely heavily on federal subsidies. African countries with growing urban populations require new, well-thought-out water service arrangements, water pricing, and revenue collection strategies. Conventional waterborne and pit latrine systems are failing, especially in water-stressed semi-arid urban areas. For this reason, African water engineers and scientists must develop new cost-effective designs and management systems suitable for drought-prone areas.

8 Mitigate Water Stresses and Improve Resilience Against Droughts

Proactive and reactive measures are required to mitigate the effects of drought-induced water shortages and hunger. Wada et al. (2014) propose strategies that were distinguished as "soft path" and "hard path" measures. Soft strategies that may be applicable to Africa include increasing agricultural water productivity of irrigated and indigenous crops and breeding new cultivars that adapt to the changing climate and are highly nutrient-efficient. There are serious concerns about the impact of eutrophication, so nutrient pollution from agriculture must be monitored and controlled. Genetically Modified Organisms (GMOs), for food production, can be a fitting soft path solution, but there are consumer health and environmental safety concerns (Lödige 2018). The biotechnology industry must ensure that there is full understanding, which can only be established through participative long-term transparent scientific analyses and uncontested safety assurance for GMO foodstuff to be freely accepted by the African majority.

New energy- and water-efficient irrigation technologies for large-scale and smallholder farms (as small as 0.5 ha), as well as development of wastewater recycling and saltwater desalinization, are approaches worth exploring. Reduction of domestic and industrial water could be achieved in water-stressed areas through, for example, reducing leakage in the water infrastructure and improving water recycling facilities. Limiting the rate of urban population growth, especially in water-stressed areas, may be achieved through family planning and tax incentives. Other solutions that may be tested are decentralized economic development that discourage urban expansion in water-stressed areas in favor of industrial expansion in water-rich regions and/or strengthening the virtual water trade.

Hard strategies include increasing water storage in surface and groundwater reservoirs, which could, in principle, be applicable to any water-stressed basin with a reservoir. Research will be required to make existing reservoirs larger and/or build new reservoirs and reduce sedimentation with minimum socioeconomic and environmental costs. Low-cost desalination of seawater could be developed in coastal water-stressed basins through modification of existing technologies like seawater greenhouses and membrane technologies, while eliminating wastewater that would cause disposal challenges (Wada et al. 2014). The right mix of hard and soft path solutions is required for cost-effective mitigation of water stresses and improve Africa's resilience against agricultural and hydrological droughts.

9 Food Security Anchored in Cost-Effective Supplementary Irrigated Agriculture

A well-managed supplementary irrigation is a cost-effective system to mitigate the effects of droughts. Unlike full scale irrigation, the timing and amount of supplementary irrigation cannot be determined in advance, owing to rainfall stochasticity. Researchers have to find solutions to these key challenges for implementing successful supplementary irrigation in each agroclimatic region. Major research tasks include (i) determination of optimal scheduling for farm conditions, (ii) selection of crops and cropping patterns to maximize profitability to determine the socioeconomic feasibility of occasionally supplying extra water to rainfed crops, (iii) promotion of water user associations to manage water use in a sustainable manner, (iv) setting fixed and efficient water delivery schedules in rainfed dry areas, (v) maximization of yield from a unit of water as well as yield from a unit of land, (vi) exploring the cost-benefit balance of supplementary irrigation, and (vii) optimizing other inputs and cultural practices (World Bank 2005).

Other challenges include the identification of on-demand, efficient water delivery systems that are best suited to supplementary irrigation, especially for small areas such as 1–3 ha smallholder fields. Commercially available efficient irrigation systems like drip, center pivots, and linear sprinklers are too costly for small-scale farmers who mainly produce low-value staple crops, because these systems are designed for large-scale commercial agricultural systems. To be effective and sustainable, supplementary irrigation must be properly integrated with improvement of crops and soil management and improvement of other production inputs like germplasm and fertilizers in order to achieve the desired output. Also, farmers must understand the technology and how to use it, and hence researchers and extension service providers should develop simple and practical tools for supplementary irrigation scheduling, since scheduling is the key to improving water use efficiency. Extension services and human capacity building should play a major role in this respect. Long-term training and advisory programs that build the capacity of extension workers and farmers to install, operate, and maintain their irrigation systems should be designed and implemented. Use of incentives for farmer participation, technology transfer, and water cost recovery may prompt adoption of improved management options. Innovative researchers have to develop low-cost, low-energy irrigation systems such as drip or sprinkler technologies, including pumping sets. Development of low-cost and simple but effective systems for monitoring groundwater and surface water quality and quantity to avoid pollution and overexploitation (World Bank 2005) is essential.

10 Cost-Effective Mechanization of Small-Scale Farms with Low-Cost Technology

For a sustainable and profitable farming in semi-arid climates; new mechanization efforts must improve on old approaches to prepare land, irrigation systems (as insurance against drought); new techniques to harvest crops; crop management in the post-harvest phases as well as livestock production in order to convert the prevailing high risk (of failure); and labor-intensive and subsistence farming systems into modern, productive, and efficient comercialized agricultural units. The conversion may be possible if smallholder farmers acquire know-how as well as the following: robust, cost-effective, energy-efficient, ergonomic, water-efficient irrigation systems, machinery for land preparation and crop protection and post-harvest technologies, and best practises in livestock production. The low-cost technologies should be suitable for 1–3 ha dryland plots and should enable small-scale farmers to operate independently. Hence, researches for production of appropriate technologies and teaching best practices have to incorporate field-based prototype testing and farmer participation. Monitoring and evaluation, redesigns and modifications should have farmer inputs as part of the project cycle, for easy uptake of the technologies and best practices by the small-scale farmers. Research institutions, government departments that provide extension services and donors should support such protracted testing and enter into agreements with the private sector for mass production and distribution of successful prototypes (Rukuni 1997).

Utilization of robust, ergonomic, and low-cost technologies may enhance innovation and productivity of African agriculture as well as retain labor in rural areas. Land and water productivity can be improved by building capacity in government and private agencies to provide technical support and advisory services to farmers. Remaining challenges include building the African small-scale mechanized farming environment; capacity to establish national standards and control parameters for mechanization processes; growing capacity in the private sector to manufacture, import, and market machinery and equipment to provide reliable services for users and to develop business environments that support African small-scale farmer investment in new technologies through the careful use of subsidies and soft loans (World Bank 2005). Hence, political will at national and local community levels has to be cultivated so that African governments successfully provide incentives, guarantees for loans, and enabling environments.

The new technologies should be cost-effective for dryland smallholder farming even in semi-arid climates. The ecology of the rainfed dry areas is fragile and the farming business in these climates is highly susceptible to drought-induced failures.

Hence, the main challenge in this environment is to enhance and stabilize productivity, income, and expenditures even during dry seasons (World Bank 2005). The following topics may be ripe for further research in rainfed agriculture: (i) intensified agronomic practices that include breeding exotic and indigenous livestock and crops like small grains and fruit trees for drought, pest, and disease tolerance; (ii) commercialization of smallholder production systems, distribution, extension, and marketing; and (iii) enhancement of commercial production of edible, nutrition-rich insects such as locusts and mopani worms.

11 Adopting Renewable Energy Technologies to Save Forests and Economically Empower Communities

Sub-Saharan Africa continues to rely heavily on low-quality traditional sources of energy, such as wood fuel. Seventy six percent (76%) of the population depends on wood fuel as a source of energy (ENDA 2005). Adopting renewable energy technologies such as biomass, solar, wind, and seawater wave power for domestic electric power supply (to replace wood fuel), powering water pumps, small-scale water desalination plants, small tractors, and small-scale industries may improve productivity of rural and peri-urban communities (Watson et al. 2010). Methods must be devised to improve poor renewable energy adoption. These may include inadequate policymaking and/or oversight, lack of awareness among rural households about the benefits of renewable energy, the high cost of the technologies, and the undeveloped nature of renewable energy markets (Mfune and Boon 2008).

Well-thought-out programs to deploy renewable energy technologies in rural areas must be integrated into wider rural development programs to ensure suitability and harmonization. African governments must create enabling government policies to stimulate uptake of renewable energy technologies in rural areas, both on the supply and demand sides. This may include provision of subsidies, support for research, public awareness campaigns, pilot programs, and regulations and/or waivers of import duties. African countries must develop local knowledge for effective and sustainable use of renewable energy technologies, so it is essential to build local capacity among both technology suppliers and users. If local input is incorporated into program design, it will enhance local support and understanding, which can lead to job creation and cost reduction (Watson et al. 2010) as well as sustainable large-scale uptake of the technologies.

12 Forging Sustainable and Effective Technology for Rural African Areas

Forging sustainable and effective technology for rural African areas will require collaboration of local and global researchers, as well as designers, manufacturers, quality assurance testing experts, and commercial industrial producers. Local manufacture of machinery will depend on commercial agreements, complexity, skills, and competitive advantages. It could be mutually beneficial for equipment and components to be designed and manufactured both in Africa and overseas. Producing the appropriate technologies locally may enhance innovation and productivity of African small-scale industries and agriculture, which can create jobs and wealth, raise standard of rural life, and retain labor in small-scale farming areas and small towns. African governments must optimize policies, regulatory frameworks, and prioritize adoption of modern energy and technology systems (World Bank 2005) in order to build sustainable agriculture at various scales and in various African environments.

All stakeholders have roles, for example, local communities must provide input into designs, evaluation of prototypes, and manufacturing. The local and international private sectors must participate in developing hardware as well as standards and accreditation schemes to both raise quality and promote competition. Governments, in collaboration with local stakeholders, must ensure that the economic framework for technology development can survive after any subsidies have been phased out. The technology market structure must ensure a reduction in overall production cost, development of robust equipment and machinery, with competition among manufacturers not just on price but also on longer product guarantees and warranties. New programs and funding mechanisms must be developed to introduce financial incentives for users and developers of the technologies. Access to bank loans can make the technology affordable to rural customers provided there are adequate guarantees to reduce associated costs and risks (Watson et al. 2010). Researchers, practitioners, and other role players (various stakeholders) have to work together in multidisciplinary teams for planning and implementation of strategies for technology development and dissemination, so as to build resilience of Africa to adverse climatic conditions and other disasters.

13 Conclusion

Research, science, technology, and innovation to address current and future challenges of climate change and water resource management in Africa must focus on establishing sustainable resilience, food security, and socioeconomic development that strengthens the capacity of urban and rural communities and minimizes institutional constraints and economic and socio-political turbulence. African governments must establish practical conjunctive water management systems for saltwater and freshwater resources and focus on river basin-scale water and soil conservation.

Policies for commercializing small-scale farming, floodwater harvesting, and treatment and artificial recharging of aquifers and irrigation will improve water and food security. Integrated urban water management and cost-effective urban wastewater treatment systems that surpass minimum effluent quality standards, combined with wastewater recycling, may result in cleaner cities and reduce the incidence of waterborne disease in many urban settlements. Adopting renewable energy technologies for domestic power, running water pumps, agricultural machinery, and small-scale industry will reduce deforestation and pollution, and may enhance cleaner production and productivity in remote rural areas. Collaboration of local and global researchers, designers, manufacturers, and evaluators may produce commercially appropriate and ergonomic technologies that create wealth and jobs in rural areas and small towns, curbing the rate of migration from rural to urban areas of Africa.

References

Akinaga T, Generalis SC, Paton C, Igobo ON, Davies PA (2018) Brine utilisation for cooling and salt production in wind-driven seawater greenhouses: design and modeling. Desalination 426:135–154. http://dx.doi.org/10.1016/j.desal.2017.10.025. Accessed 30 July 2018

Al-Kalbani MS, Price MF, Abahussain A, Ahmed M, O'Higgins T (2014) Vulnerability assessment of environmental and climate change impacts on water resources in Al Jabal Al Akhdar, Sultanate of Oman. Water 6:3118–3135. https://doi.org/10.3390/w6103118. www.mdpi.com/journal/water. Accessed 21 June 2018

Chauhan MV, Sabla YA, Patrick L, Ahmed AW (2016) Synthesis of integrated membrane desalination and salt production networks. Desalination 400:25–37. https://doi.org/10.1016/j.desal.2016.09.010

CSE (Centre for Science and Environment) (2002a) The CSE campaign for people's water management. Catch Water 5 (1) (CSE). http://www.rainwaterharvesting.org/catchwater/feb2002/technology1.htm. Accessed 30 July 2018

CSE (Centre for Science and Environment) (2002b) The CSE campaign for people's water management. Catch water. http://www.rainwaterharvesting.org/catchwater/feb2002/technology1.htm. Accessed 30 July 2018

CSIR (2009) Acid mine drainage in South Africa. Briefing Note 2009/02. https://www.environment.co.za/documents/acid-mine-drainage-amd/AMD-Acid-Mine-Drainage-South-Africa-CSIR-draft.pdf. Accessed 12 January 2020

Descroix L, Mahé G, Lebel TG, Favreau G, Galle S, Gautier E, Olivry J-C, Albergel J, Amogu O, Cappelaere B, Dessouassi R, Diedhiou A, Le Breton E, Mamadou I, Sighomnou D (2009) Spatio-temporal variability of hydrological regimes. Around the boundaries between sahelian and sudanian areas of West Africa: a synthesis. J Hydrol, AMMA Special 375:90–102. https://doi.org/10.1016/20j.jhydrol.2008.12.012. Accessed 10 Mar 2018.

Dolezal RC (2013) The Nubian sandstone aquifer dispute. Pangaea J. https://sites.stedwards.edu/pangaea/the-nubian-sandstone-aquifer-dispute. Accessed 28 Oct 2018

ENDA (2005) Energy issues in Africa. http://www.enda.sn/energie/plaq-gb.htm#ourwork. Accessed 02 Nov 2017

FAO Food and Agricultural Organisation (1995) Land and water integration and river basin management. In: Proceedings of an FAO informal workshop Rome, Italy 31 Jan–2 Feb 1993. http://www.fao.org/docrep/V5400E/v5400e00.htm#Contents Accessed 12 Mar 2018

FAO Food and Agricultural Organisation (2016) The state of food and agriculture 2016 climate change, Agriculture and Food Safety

Fisher M, IAEA Office of Public Information and Communication (2018) Using nuclear techniques to combat climate change and improve crop yields. https://www.iaea.org/newscenter/news/using-nuclear-techniques-to-combat-climate-change-and-improve-crop-yields. Accessed 13 July 2018

Gain AK, Giupponi C, Renaud FG (2012) Climate change adaptation and vulnerability assessment of water resources systems in developing countries: a generalized framework and a feasibility study in Bangladesh. Water 4:345–366. https://doi.org/10.3390/w4020345. Accessed 18 July 2018

Harlan J (1995) The living fields: our agricultural heritage. Cambridge University Press, Cambridge, United Kingdom

Hatibu N, Mahoo HF (2000) Rainwater harvesting for natural resources management: a planning guide for Tanzania. Relma, Kenya

Hazell P, Poulton C, Wiggins S, Dorward A (2006) The future of small farms : synthesis paper. World Development Report 2008 "Agriculture for Development". https://www.odi.org/sites/odi.org.uk/files/odi-assets/publications-opinion-files/1848.pdf. Accessed 12 Jan 2020

Heller J, Begemann F, Mushonga J (eds) (1997) Bambara groundnut. Vigna subterranea (L). Verdc. Promoting the conservation and use of underutilised and neglected crops. 9. In: Proceedings of the workshop on conservation and improvement of bambara groundnut (Vigna subterranea (L). Verdc)., 14–16 Nov 1995, Harare, Zimbabwe. Institute of Plant Genetics and Crop Plant Research, Gatersleben/Department of Research & Specialist Services, Harare/International Plant Genetic Resources Institute, Rome, Italy, 166 p

Hooimeijer JP, Jansen FA, De Boer WF, Wessels DCJ, Van Der Waal C, De Jong CB, Otto ND, Knoop L (2005) The diet of kudus in a mopane dominated area, South Africa. Koedoe 48:93–102

Intergovernmental Panel on Climate (IPCC) (2007) Climate change 2007. The fourth assessment report (AR4). Synthesis report for policymakers. "http://www.ipcc.ch/pdf/assessment-report/ar4/syr/ar4_" www.ipcc.ch/pdf/assessment-report/ar4/syr/ar4_syr_spm.pdf. Accessed 20 Mar 2018.

IWMI (2018) Summary rainfed agriculture: IWMI. http://www.iwmi.cgiar.org/issues/rainfed-agriculture/summary/. Accessed 28 Nov 2018

Jacobsen M, Webster M, Vairavamoorthy K (2012) The future of water in African Cities: why waste water? directions in development. World Bank, Washington, DC. https://doi.org/10.1596/978-0-8213-9721-3. Accessed 20 Nov 2018

Linares (2002) African Rice (Oryza glaberrima): history and future potential. Proc Nat Acad Sci 99:16360–16365. https://doi.org/10.1073/pnas.252604599

Lödige M (2018) The great debate: yes GMO or no GMO. https://www.foodunfolded.com/things-you-did-not-know/the-great-debate-yes-gmo-or-no-gmo. Accessed 12 Jan 2020

Makhado RA, Potgieter MJ, Luus-Powel WJ (2016) Nutritional value of colophospermum mopane as source of browse and its chemical defences against browsers: a review. J Anim Plant Sci 26(3):569–576. ISSN: 1018-7081

Mfegue CV (2018) How plant science is helping cocoa farmers in West Africa. In: 2018 Annual report plant science post helping farmers grow #FoodHeroes public private partnerships public health & vector control blog ICCM4. https://croplife.org/cocoahero-how-plant-science-is-helping-cocoa-farmers-in-west-africa/. Accessed 12 Jan 2020

Mfune O, Boon EK. (2008) Promoting renewable energy technologies for rural development in Africa: experiences of Zambia. J Hum Ecol 24(3):175–189. http://environmentportal.in/files/PromotingRenewableEnergyTechnologies.pd. Accessed 22 July 2018

Mubaiwa J, Fogliano V, Chidewe C, Linnemann RA (2017) Hard-to-cook phenomenon in bambara groundnut (Vigna subterranea (L). Verdc). processing: Options to improve its role in providing food security. Food Rev Int 33(2):167–194. https://doi.org/10.1080/87559129.2016.1149864. Accessed 28 July 2018

Murevanhema YY, Jideani VA (2013) Potential of Bambara groundnut (Vigna subterranea (L). Verdc) milk as a probiotic beverage—a review, Crit Rev Food Sci Nutri 53(9):954–967. https://doi.org/10.1080/10408398.2011.574803

National Research Council (2008) Ebony. In: Lost crops of Africa, vol. III, fruits. Lost crops of Africa, 3. National Academies Press. ISBN 978-0-309-10596-5. Accessed 25 July 2008

Namubiru-Mwaura E (2014) Land tenure and gender: approaches and challenges for strengthening rural women's land rights. Women's voice, agency, & participation research series 2014 No.6. http://hdl.handle.net/10986/21033. Accessed 12 Mar 2018

Odeigah PGC, Osanyinpeju AO (1998) Evaluating the genetic biodiversity of bambara groundnut accessions from Nigeria using SDS–polyacrylamide gel electrophoresis. Genet Resour Crop Evol 45:451–458

Odjugo PAO (2010) General overview of climate change impacts in Nigeria. J Hum Ecol 29:47–55. Accessed 20 Mar 2018

Refsgaard K, Jenssen DP, Magid J (2005) Possibilities for closing the urban-rural nutrient cycles. In: CAB International 2005 global development of organic agriculture: challenges and promises.. Accessed 21 July 2018

Rukuni M (1997) Irrigation technology transfer in support of food security (Water Reports -14). creating an enabling environment for the uptake of low-cost irrigation equipment by small-scale farmers. In: Proceedings of a sub-regional FAO (1997). Harare, Zimbabwe. www.fao.org/docrep/w7314e/w7314e09.htm Accessed 16 Mar 2018

Ruvaga L (2015). Tests show kenya's turkana water unfit for consumption. Voice of America. https://www.voanews.com/a/kenya-turkana-water-unfit-for-human-consumption/2674350.html. Accessed 6 Nov 2018

Sarla N, Swamy MBP (2005) Oryza glaberrima: a source for the improvement of Oryza sativa. Curr Sci 89:955–963. https://www.jstor.org/stable/24110748 Accessed 26 June 2018

Sedlak D (2014) Water 4.0: the past, present, and future of the world's most vital resourse. Yale university Press, New Haven & London

Shaanika H (2012) Ombike a potent traditional brew. New Era. Archived from the original on 28 Oct 2012

Shah T, Darghouth S, Dinar A (2005). Investment note 4.3, shaping the future of water for agriculture: a sourcebook for investment in agricultural water management. or email. Accessed 11 July 2018

Sorensen P (2013) The massive Ohangwena II aquifer in northern Namibia. Int J Environ Stud 70(2):173–174. https://doi.org/10.1080/00207233.2013.779149. Accessed 9 Nov 2018

Styslinger M (2011) Tree grape: form, function, and flavor of a grape on an African tree. State of the world 2011: innovations that nourish the planet http://blogs.worldwatch.org/nourishingtheplanet/tree-grape-form-function-and-flavor-of-a-grape-on-an-african-tree-tree-grape-tree-raisin-african-grape-west-african-indigenous-crop-burkina-faso-lannea-microcarpa-dye-fruit-medicinal-plant/. Accessed 16 Aug 2018

Swanborough J (2016) We need to save Africa's forests. here's how. lead, environment initiatives, environment initiatives, world economic forum. https://www.weforum.org/agenda/2016/11/deforestation-africa-palm-oil/ Accessed 12 Jan 2020

The Economist (2014) Poverty seen as major driver of extremism. http://country.eiu.com/article.aspx?articleid=712567855&Country=Morocco&topic=Politics&subtopic=Forecast&subsubtopic=Political+stability&u=1&pid=1401838124&oid=1401838124

Thomson Reuters Foundation (2018) In the Sahel, Solar power can help ward off extremism

UN (2016) Links between extreme poverty, violent extremism can be broken by creating jobs, reducing inequalities, general assembly hears as debate concludes. General Assembly Plenary. 86th Meeting (PM). GA/11761 J Unit Nat. https://www.un.org/press/en/2015/ga11761.doc.htm

UNEP (2016) GEO-6 Regional Assessment for Africa. United Nations Environment Programme, Nairobi, Kenya

UNEP (United Nations Environment Programme) (2007). GEO-4, Global Environment Outlook: environment for development assessment report

USGS (2018) Water availability and use science program. National Brackish groundwater assessment. https://water.usgs.gov/ogw/gwrp/brackishgw/use.html. Accessed 28 Nov 2018

Wada Y, Gleeson T, Esnault L (2014) Wedge approach to water stress. Nat Geosci 7:615–617. https://doi.org/10.1038/ngeo2241. Accessed 25 July 2018

Watson J, Johnson O, Wu D, Anne Miroux A, Hamdi M (2010) Renewable energy technologies for rural development. In: United nations conference on trade and development. Unctad

Current Studies on Science, Technology And Innovation. http://www.unctad.org/en/Docs/dtlstict20094_en.pdf. Accessed 22 July 2018

Wessels D, Mushongohande M, Potgeiter M, Gardiner A, Ghazoul J, Pearce J, Stack J, Frost P, Kozanayi W, Gondo T, Kurebgaseka N, Dorward A, Poole N, Taylor F, Gardiner A, Hope R, Kozanayi W, Gondo T, Robert K (2006) Mopane woodlands and the mopane worm: enhancing rural livelihoods and resource sustainability. Final Technical Report

Wessels DCJ, Van Der Waal C, De Boer WF (2007) Induced chemical defences in Colophospermum mopane trees. Afr. J. Range Forage Sci. 24:141–147

World Bank (2005) Shaping the future of water for agriculture: a sourcebook for investment in agricultural water management. https://openknowledge.worldbank.org/handle/10986/7298. Accessed 11 July 2018

World Bank (2006) Conjunctive use of groundwater and surface water. In: Agricultural & rural development notes. http://siteresources.worldbank.org/INTARD/Resources/Notes_Issue6_web.pdf. Accessed 12 July 2018

Index

S. Diop et al. (eds.), *Climate Change and Water Resources in Africa*,
https://doi.org/10.1007/978-3-030-61225-2

P

Q

R

S

T

V

W

The manufacturer's authorised representative in the EU is Springer Nature Customer Service Centre GmbH, Europaplatz 3, 69115 Heidelberg, Germany. If you have any concerns regarding our products, please contact ProductSafety@springernature.com

Printed and bound by CPI Group (UK) Ltd, Croydon, CR0 4YY
30/09/2026
02225297-0019